Lecture Notes in Computer Science 2075

Edited by G. Goos, J. Hartmanis and J.

T0230313

Springer

Berlin
Heidelberg
New York
Barcelona
Hong Kong
London
Milan
Paris
Singapore
Tokyo

José-Manuel Colom   Maciej Koutny (Eds.)

Applications and Theory of Petri Nets 2001

22nd International Conference, ICATPN 2001
Newcastle upon Tyne, UK, June 25-29, 2001
Proceedings

 Springer

Series Editors

Gerhard Goos, Karlsruhe University, Germany
Juris Hartmanis, Cornell University, NY, USA
Jan van Leeuwen, Utrecht University, The Netherlands

Volume Editors

José-Manuel Colom
Departamento de Informática e Ingeniería de Sistemas
Centro Politécnico Superior, Universidad de Zaragoza
María de Luna, 3, 50015 Zaragoza, Spain
E-mail: jm@posta.unizar.es
Maciej Koutny
Department of Computing Science
University of Newcastle upon Tyne
Newcastle upon Tyne, NE1 7RU, UK
E-mail: Maciej.Koutny@ncl.ac.uk

Cataloging-in-Publication Data applied for

Die Deutsche Bibliothek - CIP-Einheitsaufnahme

Application and theory of Petri nets 2000 : 22st international conference ;
proceedings / ICATPN 2001, Newcastle upon Tyne, UK, June 5 - 29, 2001. José
Manuel Colom ; Maciej Kountny (ed.). - Berlin ; Heidelberg ; New York ;
Barcelona ; Hong Kong ; London ; Milan ; Paris ; Singapore ; Tokyo :
Springer, 2001
    (Lecture notes in computer science ; Vol. 2075)
    ISBN 3-540-42252-8

CR Subject Classification (1998): F.1-3, C.1-2, G.2.2, D.2, D.4, J.4

ISSN 0302-9743
ISBN 3-540-42252-8 Springer-Verlag Berlin Heidelberg New York

Springer-Verlag Berlin Heidelberg New York
a member of BertelsmannSpringer Science+Business Media GmbH

http://www.springer.de

© Springer-Verlag Berlin Heidelberg 2001
Printed in Germany

Typesetting: Camera-ready by author, data conversion by PTP Berlin, Stefan Sossna
Printed on acid-free paper   SPIN 10781797   06/3142   5 4 3 2 1 0

Preface

This volume contains the proceedings of the 22nd International Conference on Application and Theory of Petri Nets. The aim of the Petri net conferences is to create a forum for discussing progress in the application and theory of Petri nets. Typically, the conferences have 100–150 participants – one third of these coming from industry while the rest are from universities and research institutions. The conferences always take place in the last week of June.

This year the conference was organized jointly with the 2nd International Conference on Application of Concurrency to System Design (ICACSD 2001). The two conferences shared the invited lectures and the social program.

The conference and a number of other activities are co-ordinated by a steering committee with the following members: G. Balbo (Italy), J. Billington (Australia), G. De Michelis (Italy), C. Girault (France), K. Jensen (Denmark), S. Kumagai (Japan), T. Murata (USA), C.A. Petri (Germany; honorary member), W. Reisig (Germany), G. Rozenberg (The Netherlands; chairman), and M. Silva (Spain).

Other activities before and during the 2001 conference included tool demonstrations, a meeting on "XML Based Interchange Formats for Petri Nets", extensive introductory tutorials, two advanced tutorials on "Probabilistic Methods in Concurrency" and "Model Checking", and two workshops on "Synthesis of Concurrent Systems" and "Concurrency in Dependable Computing". The tutorial notes and workshop proceedings are not published in these proceedings, but copies are available from the organizers.

The 2001 conference was organized by the Department of Computing Science at the University of Newcastle upon Tyne, United Kingdom. We would like to thank the members of the organizing committee (see next page) and their teams.

We would like to thank very much all those who submitted papers to the Petri net conference. We received a total of 48 submissions from 21 different countries. This volume comprises the papers that were accepted for presentation. Invited lectures were given by S. Donatelli, R. Milner, and M. Nielsen (whose papers are included in this volume), and G. Holzmann, J. Kramer, and A. Sangiovanni-Vincentelli (whose papers are included in the proceedings of ICACSD 2001).

The submitted papers were evaluated by a program committee. The program committee meeting took place in Newcastle upon Tyne, United Kingdom. We would like to express our gratitude to the members of the program committee, and to all the referees who assisted them. The names of these are listed on the following pages.

We would like to acknowledge the local support of J. Dersley, D. Faulder, P. Jacques, L. Marshall, and R. Poat. Finally, we would like to mention the excellent co-operation with Springer-Verlag during the preparation of this volume.

April 2001

Maciej Koutny
José-Manuel Colom

Organizing Committee

Anke Jackson
Maciej Koutny (co-chair)
Alber Koelmans

Dan Simpson
Jason Steggles
Alexandre Yakovlev (co-chair)

Tools Demonstration

Albert Koelmans (chair)

Program Committee

Didier Buchs (Switzerland)
Gianfranco Ciardo (USA)
José-Manuel Colom (Spain;
          co-chair; applications)
Jordi Cortadella (Spain)
Michel Diaz (France)
Hartmut Ehrig (Germany)
Serge Haddad (France)
Jane Hillston (UK)
Jetty Kleijn (The Netherlands)

Maciej Koutny (UK; co-chair; theory)
Lars M. Kristensen (Denmark)
Sadatoshi Kumagai (Japan)
Charles Lakos (Australia)
Vladimiro Sassone (Italy)
Sol Shatz (USA)
Carla Simone (Italy)
Rüdiger Valk (Germany)
Antti Valmari (Finland)
Murray Woodside (Canada)

Referees

Felice Balarin
Paolo Baldan
Kamel Barkaoui
Twan Basten
Luca Bernardinello
Marco Bernardo
Bernard Berthomieu
Andrea Bobbio
Marc Boyer
Julian Bradfield
Roberto Bruni
Stanislav Chachkov
Giovanni Chiola
Søren Christensen
Piotr Chrząstowski-Wachtel
Andrea Corradini
Sandro Costa
Juliane Dehnert
Jordi Delgado

Giorgio De Michelis
Jörg Desel
Catalin Dima
Susanna Donatelli
Claude Dutheillet
Emmanuelle Encrenaz
Joost Engelfriet
Claudia Ermel
Javier Esparza
Rob Esser
Joaquín Ezpeleta
Jean Fanchon
Berndt Farwer
Carlo Ferigato
Jörn Freiheit
Rossano Gaeta
Maike Gajewsky
Qi-Wei Ge
Holger Giese

Stephen Gilmore

Steven Gordon

Pierre Gradit

Marco Gribaudo

Luuk Groenewegen

Xudong He

Keijo Heljanko

Jesper Gulmann Henriksen

Kunihiko Hiraishi

Kathrin Hoffmann

Hendrik Jan Hoogeboom

Andras Horvath

Jarle Hulaas

Henrik Hulgård

David Hurzeler

Nisse Husberg

Jean-Michel Ilié

Matthias Jantzen

Guy Juanole

Gabriel Juhas

Tommi Junttila

Joost-Pieter Katoen

Victor Khomenko

Mike Kishinevsky

Olaf Kluge

Michael Köhler

Alex Kondratyev

Fabrice Kordon

Olaf Kummer

Juliana Küster-Filipe

Michael Lemmon

Glenn Lewis

Johan Lilius

Bo Lindstrøm

Louise Lorentsen

Moez Mahfoudh

Thomas Mailund

Axel Martens

Giovanna Di Marzo Serugendo

Giuseppe Milicia

Andrew Miner

Toshiyuki Miyamoto

Daniel Moldt

Patrice Moreaux

Kjeld Mortensen

Isabelle Mounier

Morikazu Nakamura

Atsushi Ohta

Peter Ölveczky

Katsuaki Onogi

Julia Padberg

Emmanuel Paviot-Adet

Wojciech Penczek

Michele Pinna

Laure Petrucci

Denis Poitrenaud

Agata Pólrola

Lucia Pomello

Jean-François Pradat-Peyre

Antti Puhakka

Laura Recalde

Marina Ribaudo

Heiko Rölke

Isabel Rojas

Stefan Römer

Heinz Schmidt

Karsten Schmidt

Amal El Fallah Seghrouchni

Shane Sendall

Matteo Sereno

Radu Siminiceanu

Jiří Srba

Ramavarapu Sreenivas

Mark-Oliver Stehr

Perdita Stevens

Ichiro Suzuki

Shigemasa Takai

Enrique Teruel

Kohkichi Tsuji

Emilio Tuosto

Naoshi Uchihira

Milan Urbasek

Toshimitsu Ushio

Robert Valette

Gabriel Valiente

Kimmo Varpaaniemi

François Vernadat

Walter Vogler

Toshimasa Watanabe

Yosinori Watanabe

Lisa Wells   Xiande Xie
Herbert Wicklicky   Haiping Xu
Aiguo Xie   Tomohiro Yoneda

Sponsoring Institutions

City Council of Newcastle upon Tyne
Comisión Interministerial de Ciencia y Tecnología (TIC98-0671), España
Department of Computing Science, University of Newcastle
One NorthEast – The Development Agency for the North East of England

Table of Contents

Kronecker Algebra and (Stochastic) Petri Nets: Is It Worth the Effort?

Susanna Donatelli

Dip. di Informatica, Università di Torino, Corso Svizzera 185, 10149 Torino, Italy
susi@di.unito.it

Abstract. The paper discusses the impact that Kronecker algebra had and it is having on the solution of SPN, how this has influenced not only the solution of the stochastic process associated to an SPN, but also the algorithms and the data structures for reachability of untimed Petri nets. Special emphasis is put in trying to clarify the advantages and disadvantages of Kronecker based approach, in terms of computational complexity, memory savings and applicability to the solution of real systems models.

1 Introduction

It is well known that one of the major factors that has limited and it is limiting the impact of Petri nets on practical applications is the so-called "state-space explosion problem", stemming from the fact that even very "innocent" nets, with a small number of places and transitions, can lead to very large state spaces, and that the construction of the state space is very often a mandatory step in proving model properties and reachability for untimed nets or computing performance indices for stochastic Petri nets. A number of counter-measures have been devised by different researchers, and in this paper we present one of this solution, known as "Kronecker-based" or "structured" method, discussing the maturity of the technique with respect to the range of applicability to the whole class of Petri nets, to the complexity of the solution and to the availability of solution tools.

The *Kronecker algebra* approach was introduced in the Petri net world in the context of exact solution of stochastic models, to express the infinitesimal generator Q of an SPN in terms of Q_i matrices coming from some *components* (subnets of smaller state space) combined by Kronecker operators and to implement the solution of the characteristic steady-state solution equation $\pi \cdot Q = 0$ without computing and storing Q [5,16,17,7,23]: this technique allows to move the storage bottleneck from the infinitesimal generator matrix to the probability vector. All the works for SPN cited above were inspired by the work of Plateau [34, 33] on Stochastic Automata Networks. The ideas presented in her work have been applied first to a simple subset of SPN called Superposed Stochastic Automata [16], and later on to the larger Superposed Generalized stochastic Petri nets (SGSPN) class [17,23,24].

The approach is based on a construction of the state space in which a certain Cartesian product of reachable states of components is performed, leading

J.-M. Colom and M. Koutny (Eds.): ICATPN 2001, LNCS 2075, pp. 1–18, 2001.

eventually to a *product space* $\widehat{S}$ that includes the actual *reachability set* S. According to this structured view of the state space the $\mathbf{Q}_i$ matrices can be derived, that form the building blocks of the Kronecker algebra formula of the complete infinitesimal generator. Interestingly enough, it was shown [25] that the same approach can be used to describe the reachability matrix of a net starting from the reachability matrices of the component nets, so that Kronecker algebra constitute another example of cross-fertilization between the field of performance evaluation, in which the technique was initially devised, and the analysis of untimed behaviour for Petri nets.

In summary, two beneficial effects are produced by using Kronecker algebra: the space complexity of the solution is lowered, moving the bottleneck from the infinitesimal generator to the probability vector for performance evaluation and from reachability graph to reachability set in untimed nets, with possibly a gain also in time complexity (depending on the structure of the model [6]), moreover the existence of an expression for the infinitesimal generator and for the reachability matrix provides insight on the relationships between the structure of a GSPN and that of its underlying CTMC and reachability graph.

The paper is organized as follows: Section 2 introduces the simple notions of Kronecker algebra used later in the paper, with special attention to the complexity of multiplying a vector by a matrix given as a Kronecker expression, Section 3 describes the use of Kronecker algebra in the field of Petri nets, while Section 4 summarizes problems and extensions of the method. Section 5 concludes the paper discussing whether the use of Kronecker algebra in the Petri net field was, and is, "worth the effort".

2   Preliminaries

Kronecker algebra is an algebra defined on matrices, with a product operator denoted by $\otimes$, and a sum operator denoted by $\oplus$. In the following definitions we consider matrices on real values.

Definition 1. *Let $\mathbf{A}$ be a $n \times m$ matrix $(\mathbf{A} \in \mathbb{R}^{n \times m})$, and $\mathbf{B}$ be a $p \times q$ one $(\mathbf{B} \in \mathbb{R}^{p \times q})$; $\mathbf{C}$ is the Kronecker product of $\mathbf{A}$ and $\mathbf{B}$ and we write $\mathbf{C} = \mathbf{A} \otimes \mathbf{B}$ iff $\mathbf{C}$ is a $n \cdot p \times m \cdot q$ matrix $(\mathbf{C} \in \mathbb{R}^{n \cdot p \times m \cdot q})$ defined by:*

$$\mathbf{C}_{i,j} = \mathbf{C}_{\bar{\imath}\bar{\jmath}} = a_{i_1 j_1} b_{i_2 j_2}$$

with $\bar{\imath} = (i_1, i_2), \bar{\jmath} = (j_1, j_2)$, and $i = (i_1) \cdot p + i_2$ (similarly, $j = (j_1) \cdot q + j_2$).

As a simple example consider the Kronecker product of a 2×2 matrix, with a 2×3. We have

$$\mathbf{A} = \begin{pmatrix} a_{00} & a_{01} \\ a_{10} & a_{11} \end{pmatrix} \qquad \mathbf{B} = \begin{pmatrix} b_{00} & b_{01} & b_{02} \\ b_{10} & b_{11} & b_{12} \end{pmatrix}$$

$$\mathbf{C} = \mathbf{A} \otimes \mathbf{B} = \begin{pmatrix} a_{00}b_{00} & a_{00}b_{01} & a_{00}b_{02} & a_{01}b_{00} & a_{01}b_{01} & a_{01}b_{02} \\ a_{00}b_{10} & a_{00}b_{11} & a_{00}b_{12} & a_{01}b_{10} & a_{01}b_{11} & a_{01}b_{12} \\ a_{10}b_{00} & a_{10}b_{01} & a_{10}b_{02} & a_{11}b_{00} & a_{11}b_{01} & a_{11}b_{02} \\ a_{10}b_{10} & a_{10}b_{11} & a_{10}b_{12} & a_{11}b_{10} & a_{11}b_{11} & a_{11}b_{12} \end{pmatrix}$$

An important matter is that there are two different ways to address a line (or a column) in a product matrix: either with a natural number i, with $i \in [1 \ldots n \cdot p]$ (j, with $j \in [1 \ldots m \cdot q]$) or with a vector $[i_1, i_2]$, with $i_1 \in [1 \ldots n]$ and $i_2 \in [1 \ldots p]$ (and analogously for columns). The transformation from i to $[i_1, i_2]$ is well defined since $[i_1, i_2]$ is nothing but the representation of i in the mixed base (n, p).

The generalization to K matrices is immediate: Let $\mathbf{A} = \bigotimes_{k=1}^{K} \mathbf{A}^k$ denote the Kronecker product of K matrices $\mathbf{A}^k \in \mathbb{R}^{n_k \times m_k}$. Let $n_l^u = \prod_{k=l}^{u} n_k$, $n = n_1^K$. If we assume a $(n_1, \ldots, n_K)$ mixed-base numbering scheme, the tuple $\bar{i} = (i_1, \ldots, i_K)$ corresponds to the number $(\ldots ((i_1)n_2 + i_2)n_3 \cdots)n_K + i_K$. A $(m_1, \ldots, m_K)$ mixed-base numbering scheme can be similarly be associated to $\bar{j}$. If we assume that $\bar{i}$ ($\bar{j}$) is the mixed-based representation of i and j, respectively, the generic element of $\mathbf{A} \in \mathbb{R}^{n \times n}$ is

$$a_{i,j} = a_{\bar{i},\bar{j}} = a_{i_1,j_1}^1 \cdot a_{i_2,j_2}^2 \cdots a_{i_K,j_K}^K \tag{1}$$

The Kronecker sum is defined instead for square matrices only, in terms of the Kronecker product operator:

Definition 2. *Let $\mathbf{A}$ be a $n \times n$ matrix, ($\mathbf{A} \in \mathbb{R}^{n \times n}$) and $\mathbf{B}$ be a $p \times p$ one ($\mathbf{B} \in \mathbb{R}^{p \times p}$); $\mathbf{D}$ is the Kronecker sum of $\mathbf{A}$ and $\mathbf{B}$ and we write $\mathbf{D} = \mathbf{A} \oplus \mathbf{B}$ iff $\mathbf{D}$ is a $n \cdot p \times n \cdot p$ matrix ($\mathbf{D} \in \mathbb{R}^{n \cdot p \times n \cdot p}$) defined by:*

$$\mathbf{D} = \mathbf{A} \oplus \mathbf{B} = \mathbf{A} \otimes \mathbf{I}_p + \mathbf{I}_n \otimes \mathbf{B}$$

where $\mathbf{I}_x$ is the $x \times x$ identity matrix.

Let's consider again the two matrices $\mathbf{A}$ and $\mathbf{B}$, where $\mathbf{A}$ is the same as before, and in $\mathbf{B}$ the last column has been deleted. The computation of their Kronecker sum is:

$$\mathbf{A} \otimes \mathbf{Id}_2 = \begin{pmatrix} a_{00} & 0 & a_{01} & 0 \\ 0 & a_{00} & 0 & a_{01} \\ a_{10} & 0 & a_{11} & 0 \\ 0 & a_{10} & 0 & a_{11} \end{pmatrix}$$

$$\mathbf{Id}_1 \otimes \mathbf{B} = \begin{pmatrix} b_{00} & b_{01} & 0 & 0 \\ b_{10} & b_{11} & 0 & 0 \\ 0 & 0 & b_{00} & b_{01} \\ 0 & 0 & b_{10} & b_{11} \end{pmatrix}$$

$$\mathbf{D} = \begin{pmatrix} a_{00} + b_{00} & b_{01} & a_{01} & 0 \\ b_{10} & a_{00} + b_{11} & 0 & a_{01} \\ a_{10} & 0 & a_{11} + b_{00} & b_{01} \\ 0 & a_{10} & b_{10} & a_{11} + b_{11} \end{pmatrix}$$

The generalization to K matrices is straightforward. Let $\mathbf{A} = \bigoplus_{k=1}^{K} \mathbf{A}^k$ denote the Kronecker sum of K matrices $\mathbf{A}^k \in \mathbb{R}^{n_k \times n_k}$:

$$\mathbf{A} = \sum_{k=1}^{K} \mathbf{I}_{n_1} \otimes \cdots \otimes \mathbf{I}_{n_{k-1}} \otimes \mathbf{A}^k \otimes \mathbf{I}_{n_{k+1}} \otimes \cdots \otimes \mathbf{I}_{n_K} = \sum_{k=1}^{K} \mathbf{I}_{n_1^{k-1}} \otimes \mathbf{A}^k \otimes \mathbf{I}_{n_{k+1}^K}.$$

2.1   Kronecker Operators and Independent Composition of Markov Chains

Let us consider two independent Markov chains $\mathcal{M}_1$ and $\mathcal{M}_2$, and assume that we know the set of reachable states of the two, $\mathcal{S}^1$ and $\mathcal{S}^2$. Assuming that the two are finite, of cardinality n_1 and n_2 respectively, it is a straightforward task to associate a number from 1 to n_1 (n_2) to each state of the first (second) model.

If we now consider the global model $\mathcal{M}$ obtained by the parallel and independent composition of the two models, then a global state is a pair $\bar{\imath} = (i_1, i_2)$, and, if n_1 and n_2 are finite, it is straightforward to associate to $\bar{\imath}$ the number $i = i_1 \cdot n_2 + i_2$.

In general, if we have K models $\mathcal{M}_k$, each of state space $\mathcal{S}^k$, of n_k states each, then the state space of the model $\mathcal{M}$, obtained by the independent parallel composition of the K models, is the Cartesian product of the state spaces:

$$\mathcal{S} = \mathcal{S}^1 \times \ldots \mathcal{S}^K$$

and the number of states is:

$$n = |\mathcal{S}| = \prod_{k=1}^{K} n_k$$

Again, each state is either the number i or as the vector $\bar{\imath}$ of size K.

In case of square matrices, if $\mathbf{A}$ and $\mathbf{B}$ are interpreted as state transition matrices of two discrete time Markov chains, it is immediate to recognize (see Davio in [15]) that $\mathbf{C} = \mathbf{A} \otimes \mathbf{B}$ is the transition probabilities matrix of the process obtained as independent composition of the two original processes. Indeed, for the global system to move from $\bar{\imath} = (i_1, \ldots, i_K)$ to $\bar{\jmath} = (j_1, \ldots, j_K)$ each component k has to move from i_k to j_k at the same time step, and this indeed happens with a probability that is the product of the probabilities of the local moves.

Again if we consider $\mathbf{A}$ and $\mathbf{B}$ as the infinitesimal generator of two continuous time Markov chains, then $\mathbf{D} = \mathbf{A} \oplus \mathbf{B}$ is the infinitesimal generator of the process obtained by independent composition of the two original ones. Indeed the behaviour of the independent composition of 2 CTMC is the sum of two terms: the first CTMC moves while, at the same time, the second one does not move ($\mathbf{A} \otimes \mathbf{I}_p$) or viceversa ($\mathbf{I}_n \otimes \mathbf{B}$).

Observe that a system obtained by the independent composition of K CTMCs, can move from a state $\bar{\imath} = (i_1, \ldots, i_K)$ to a state $\bar{\jmath} = (j_1, \ldots, j_K)$ only if it exists one and only one index k such that $i_k \neq j_k$, since we are in a continuous time environment, and indeed the Kronecker sum produces a matrix in which all transition rates among states that differ by more than a single component are set equal to zero.

2.2   Vector-Matrix Multiplication

An aspect that deserves attention is that even if the infinitesimal generator of the global system is rewritten using only the infinitesimal generators of the component systems, nevertheless an evaluation of the expression leads to a matrix of size equal to the product of the size of the components. The interesting point is that the expression need not to be evaluated explicitly: indeed each time that an element a_{ij} of $\mathbf{A}$ is used, we only need to compute its value according to the Kronecker product definition, which requires $K-1$ additional multiplications.

A different method, called shuffle method, has been instead explained in [15]: the $\mathbf{A}^k$ matrices are considered sequentially, one at a time, exploiting the equality [15]:

$$\mathbf{D} = \bigotimes_{k=1}^{K} \mathbf{A}^k = \prod_{k=1}^{K} \mathbf{S}_{(n_1 \cdots n_k, n_{k+1} \cdots n_K)} \cdot (\mathbf{I}_{\bar{n}_k} \otimes \mathbf{A}^k) \cdot \mathbf{S}^T_{(n_1 \cdots n_k, n_{k+1} \cdots n_K)} \qquad (2)$$

where $\bar{n}_k = n_1 \cdot \ldots \cdot n_{k-1} \cdot n_{k+1} \cdot \ldots \cdot n_K$, and $\mathbf{S}_{(a,b)} \in \{0,1\}^{a \cdot b \times a \cdot b}$ is the matrix describing an (a,b) perfect shuffle permutation:

$$(\mathbf{S}_{(a,b)})_{i,j} = \begin{cases} 1 \text{ if } j = (i \bmod a) \cdot b + (i \text{ div } a) \\ 0 \text{ otherwise} \end{cases}.$$

Complexity plays a relevant role in Kronecker algebra based method: while the storage advantages are obvious (order K matrices of size $n_k \times n_k$ instead of one of size $n \times n$ with $n = \prod_{k=1}^{K}$), the time advantages/disadvantages are less intuitive. A discussion of the complexity of the shuffle based algorithms, for the full storage case, can be found in [36], while [6] presents alternative algorithms for the matrix vector multiplication and compares the time complexity of the new algorithms and of shuffle in the case of both full and sparse storage. We report in the following the results of the comparison in [6], for the computation of $\mathbf{x} \cdot \mathbf{A}$, where $\mathbf{A} = \bigotimes_{k=1}^{K} \mathbf{A}^k$. Each matrix $\mathbf{A}^k$ is a square matrix of size $n_k \times n_k$, that is to say with $(n_k)^2$ elements, while we indicate with $\eta[A^k]$ the number of non zero elements of $\mathbf{A}^k$; by definition of Kronecker product, $\mathbf{A}$ has $n = \prod_{k=1}^{K} n_k$ elements, with $\eta[\mathbf{A}] = \prod_{k=1}^{K} \eta[\mathbf{A}^k]$ non zeros.

For the complexity we consider three cases: *ordinary*, where matrix A is computed and stored in memory, *Shuffle*, that uses the shuffle based equations presented above, and *Row*, that is a straightforward algorithm (presented in figure 5.3 in [6]) that does not store $\mathbf{A}$, but that computes each element of the matrix "on the fly" when it is needed, which implies in theory K additional multiplications per element, but the algorithm optimizes this computation by memorizing in K additional variables partial computations (prefixes of the products).

If the matrices $\mathbf{A}^k$ are stored in full storage, the complexity is

- *Ordinary* $\hspace{6cm} O\left(n^2\right)$

- *Shuffle* $\hspace{2cm} O\left(\sum_{k=1}^{K} \frac{N}{n_k} \cdot (n_k)^2\right) = O\left(n \cdot \sum_{k=1}^{K} n_k\right)$

- *Row* $\hspace{6.5cm} O\left(n^2\right)$

Observe that *Row* effectively amortizes the K additional multiplications if the matrix is full, so that there is basically no additional time overhead for the reduced space complexity, while *Shuffle* performs better than the ordinary multiplication, so that the saving in space is coupled with a saving in time.

If the matrices $\mathbf{A}^k$ are stored in sparse storage and are "not too sparse":

- *Ordinary*: $\hspace{10cm} O(\eta(\mathbf{A}))$

- *Shuffle*: $\hspace{8cm} O\left(n \cdot \sum_{k=1}^{K} \frac{\eta(\mathbf{A}^k)}{n_k} \right)$

- *Row*: $\hspace{3cm} O\left(\sum_{k=1}^{K} \prod_{l=1}^{k} \eta(\mathbf{A}^l) \right) = O\left(\prod_{k=1}^{K} \eta(\mathbf{A}^k) \right) = O(\eta(\mathbf{A}))$

Depending on the sparsity of the matrix, *Shuffle* can still do better than the other algorithms, but if the matrices $\mathbf{A}^k$ are *ultrasparse*, that is to say, most rows have no or one nonzero, which implies that $\eta(\mathbf{A}^k) \approx n_k$, then the complexity of *Shuffle* and *Row* becomes:

$$O\left(K \cdot n \right) = O\left(K \cdot \prod_{k=1}^{K} \eta(\mathbf{A}^k) \right) = O(K \cdot \eta(\mathbf{A}))$$

that is to say, both *Shuffle* and *Row* pay an additional K overhead in time, with respect to a straightforward multiplication, for the saving in space.

A particular case of multiplication is the one used for the Kronecker sum, since, by definition of sum, all matrices in the product are identities, but one, $\mathbf{A}^k$. Specific algorithms have been designed for this particular case, but, independently on whether *Shuffle* or *Row* is used, the resulting complexity is:

$$O\left(n \cdot \frac{\eta(\mathbf{A}^k)}{n_k} \right)$$

and, since there are K such products for a Kronecker sum, the total complexity for the vector-matrix multiplication is

$$\sum_{k=1}^{K} O\left(n \cdot \frac{\eta(\mathbf{A}^k)}{n_k} \right)$$

3   Kronecker Algebra and Petri Nets

An obvious application of Kronecker operators to Petri nets is to use them to define the CTMC of a model obtained from the independent composition of K GSPNs. The example we present here is due to Ciardo, in our joint tutorial presentation in [11].

Given K nets $\mathcal{N}^1, \ldots, \mathcal{N}^K$, let

- $\mathcal{S}^k$ be the local tangible reachability set for model k.
- $n_k = |\mathcal{S}^k|$.

 – $\widehat{\mathcal{S}} = \mathcal{S}^1 \times \cdots \times \mathcal{S}^K$ is the *potential state space* (also called *product* state space.

The model obtained by *parallel composition* of the K models

$$\mathcal{N} = \mathcal{N}^1 \; || \; \cdots \; || \; \mathcal{N}^K$$

has tangible reachability set $\mathcal{S} = \widehat{\mathcal{S}}$, and

$$|\mathcal{S}| = |\widehat{\mathcal{S}}| = n = \prod_{k=1}^{K} n_k$$

For $\mathcal{N}$ we can compute the rate matrix $\mathbf{R}$ (the matrix obtained from the infinitesimal generator deleting to zero all the diagonal elements) from the rate matrix of the components $\mathbf{R}^k$ as:

$$\mathbf{R} = \bigoplus_{k=1}^{K} \mathbf{R}^k$$

that follows from the analogous result for independent composition of CTMCs.

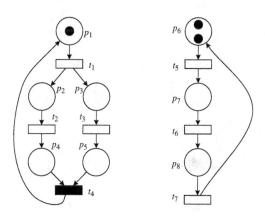

Fig. 1. Two independent GSPNs

An example of the rate matrix computation is now presented using the GSPNs of Figure 1: $\mathcal{N}^1$ on the left, and $\mathcal{N}^2$ on the right. $\mathcal{N}^1$ has 4 states numbered from 0 to 3:

$$
\begin{aligned}
p_1 &\rightarrow 0 \\
p_2 + p_3 &\rightarrow 1 \\
p_3 + p_4 &\rightarrow 2 \\
p_2 + p_5 &\rightarrow 3
\end{aligned}
$$

while $\mathcal{N}^2$ has 6 states numbered from 0 to 5:

$$
\begin{aligned}
2 \cdot p_6 &\rightarrow 0 \\
p_6 + p_7 &\rightarrow 1 \\
2 \cdot p_7 &\rightarrow 2 \\
p_6 + p_8 &\rightarrow 3 \\
p_7 + p_8 &\rightarrow 4 \\
2 \cdot p_8 &\rightarrow 5
\end{aligned}
$$

The GSPN $\mathcal{N} = \mathcal{N}^1 \,||\mathcal{N}^2$ has 24 states, with

$$
n = n_1 \cdot n_2 = |\widehat{\mathcal{S}}| = |\mathcal{S}| = 24
$$

For net $\mathcal{N}$ we can compute the rate matrix $\mathbf{R}$ as:

$$
\mathbf{R} = \mathbf{R}^1 \oplus \mathbf{R}^2
$$

where $\mathbf{R}^i$ is the rate matrix of $\mathcal{N}^i$.

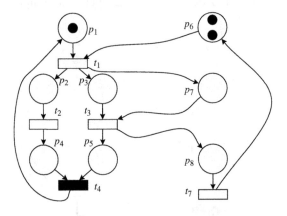

Fig. 2. Two dependent GSPNs

What happens if there are instead dependencies among nets, in the sense that the change of state in one GSPN is conditioned on the current (sub)state of the other? An example of such dependencies is shown in the net $\mathcal{N}$ of Figure 2 where it is immediate to recognize nets $\mathcal{N}^1$ and $\mathcal{N}^2$ of Figure 1, and the dependency is obtained by imposing a synchronization between transition t_1 of $\mathcal{N}^1$ and t_5 of $\mathcal{N}^2$ and between t_3 of $\mathcal{N}^1$ and t_6 of $\mathcal{N}^2$. This net belongs to the class called "Superposed Generalized Stochastic Petri Nets" (SGSPN) [17], that is to say a set of GSPNs synchronized on transitions.

A first consequence of the dependencies among the two nets is that the Cartesian product $\widehat{\mathcal{S}} = \mathcal{S}^1 \times \mathcal{S}^2$ of the state spaces of the components considered in isolation is now a superset of the reachability set $\mathcal{S}$ of $\mathcal{N}$, that is to say $\mathcal{S} \subseteq \widehat{\mathcal{S}}$. We shall use n for the cardinality of $\mathcal{S}$ and $\hat{n}$ for the cardinality of $\widehat{\mathcal{S}}$: of the 24 states of $\widehat{\mathcal{S}}$ only 10 are reachable.

A second effect of the dependencies is that $\mathbf{R}^1 \oplus \mathbf{R}^2$ does not correctly describe the rate matrix $\mathbf{R}$ of $\mathcal{N}$: indeed the effect of the synchronizations is to eliminate the contributions due to the firings of t_1 in any state in which net $\mathcal{N}^1$ is not in a (local) state in which transition t_5 is enabled, and to have an additional contribution due to the simultaneous change of (local) states. With respect to the rate matrix obtained as $\mathbf{R}^1 \oplus \mathbf{R}^2$, we have therefore to remove and add contributions: removal is realized by setting to zero the contributions in the component matrices $\mathbf{R}^k$, while the addition is realized summing matrices of size $\hat{n}$ that are also expressed in Kronecker form.

The resulting expression is [17,23]:

$$\mathbf{R} = \widehat{\mathbf{R}}_{\mathcal{S},\mathcal{S}} = \left(\mathbf{R}^1 \oplus \mathbf{R}^2 + w(t_1)\mathbf{B}^{1,t_1} \otimes \mathbf{B}^{2,t_1} + w(t_3)\mathbf{B}^{1,t_3} \otimes \mathbf{B}^{2,t_3} \right)_{\mathcal{S},\mathcal{S}}$$

$\mathbf{R}^k$ is the rate matrix of component k, where all contributions due to synchronization transitions have been deleted, $\mathbf{w}(t)$ is the rate of transition t, and $\mathbf{B}^{k,t}$ is the matrix that describes the reachability in component k due to the firing of t. The notation $\widehat{\mathbf{R}}_{\mathcal{S},\mathcal{S}}$ indicates the matrix obtained by selecting only rows and columns corresponding to reachable state (states in $\mathcal{S}$), since, indeed, the matrix resulting from the Kronecker sum and product have the same size as $\widehat{\mathcal{S}}$.

The formula requires the construction of the following matrices (where, for brevity, we have indicated with λ_i the rate $\mathbf{w}(t_i)$):

$$\mathbf{R}^1 = \begin{bmatrix} & 0 & \\ & & \lambda_2 & 0 \\ & \lambda_2 & & \end{bmatrix} \quad \mathbf{R}^2 = \begin{bmatrix} & 0 & & \\ & & 0 & 0 & \\ & \lambda_7 & & & 0 \\ & & \lambda_7 & & 0 \\ & & & \lambda_7 & \end{bmatrix}$$

$$\mathbf{w}(t_1) \cdot \mathbf{B}^{1,t_1} = \begin{bmatrix} & \lambda_1 & \\ & & 0 & 0 \\ 0 & & & \\ 0 & & & \end{bmatrix} \quad \mathbf{w}(t_1) \cdot \mathbf{B}^{2,t_1} = \begin{bmatrix} \lambda_1 & & & \\ & \lambda_1 & 0 & \\ & & & 0 \\ 0 & & & \lambda_1 \\ & & 0 & \end{bmatrix}$$

$$\mathbf{w}(t_3) \cdot \mathbf{B}^{1,t_3} = \begin{bmatrix} & 0 & \\ & & 0 & \lambda_3 \\ \lambda_3 & & & \\ 0 & & & \end{bmatrix} \quad \mathbf{w}(t_3) \cdot \mathbf{B}^{2,t_3} = \begin{bmatrix} & 0 & & \\ & & 0 & \lambda_3 & \\ & & & \lambda_3 & \\ 0 & & & & \lambda_3 \\ & & 0 & & \end{bmatrix}$$

And the evaluation of the formula for $\widehat{\mathbf{R}}$ leads to: $\widehat{\mathbf{R}} =$

$$
\begin{bmatrix}
 & \lambda_1 & \\
 & \quad \lambda_1 & \\
\lambda_7 & \lambda_1 & \\
\quad \lambda_7 & & \\
\quad\quad \lambda_7 & & \\
 & \lambda_2 & \\
 & \quad \lambda_2 & \lambda_3 \\
 & \quad\quad \lambda_2 \; \lambda_2 & \quad \lambda_3 \\
 \lambda_7 & \quad\quad \lambda_2 \; \lambda_2 & \\
 \quad \lambda_7 & \quad\quad\quad\quad \lambda_2 & \lambda_3 \\
 \quad\quad \lambda_7 & & \\
\lambda_3 & & \\
\quad \lambda_3 & \lambda_7 & \\
\quad\quad \lambda_3 & \quad \lambda_7 & \\
 & \quad\quad \lambda_7 & \\
\lambda_2 & & \\
\quad \lambda_2 & & \\
\quad \lambda_2 \; \lambda_2 & & \lambda_7 \\
\quad\quad\quad \lambda_2 & & \quad \lambda_7 \\
\quad\quad\quad\quad \lambda_2 & & \quad\quad \lambda_7 \\
\end{bmatrix}
$$

By selecting only rows and columns in $\mathcal{S}$ we get the correct rate matrix $\mathbf{R}$:

$$
\mathbf{R} =
\begin{bmatrix}
 & \lambda_1 & & & \\
 & & \lambda_2 \; \lambda_3 & & \\
 & & & \lambda_3 & \\
 & & & \lambda_2 \; \lambda_7 & \\
 & & & & \lambda_1 \\
\lambda_7 & & & & \\
\lambda_2 & & & & \\
 & \lambda_7 & & \lambda_2 \; \lambda_3 & \\
 & \quad \lambda_7 & & & \lambda_3 \\
 & & \lambda_7 & & \lambda_2 \\
 & & \quad \lambda_7 & & \\
\end{bmatrix}
$$

In general, given a GSPN $\mathcal{N}$, obtained from K GSPNs $\mathcal{N}^k$, synchronized other a set $\mathcal{T}^*$ of timed transitions, the expression for the rate matrix is as follows [17,23]:

$$
\mathbf{R} = \widehat{\mathbf{R}}_{\mathcal{S},\mathcal{S}} = \left(\bigoplus_{k=1}^{K} \mathbf{R}^k + \sum_{t\in\mathcal{T}^*} \mathbf{w}(t) \cdot \bigotimes_{k=1}^{k} \mathbf{B}^{k,t} \right)_{\mathcal{S},\mathcal{S}}
$$

where $\mathbf{R}^k$ is the rate matrix of component k, where all contributions due to transitions in $\mathcal{T}^*$ have been deleted, $\mathbf{w}(t)$ is the rate of transition t, and $\mathbf{B}^{k,t}$ is the matrix that describes the reachability in component k due to the firing of t.

When performing a matrix-vector multiplication for steady state or transient analysis [26], the vector is sized according to $\widehat{\mathcal{S}}$, but, even if there are non reachable (spurious) states in $\widehat{\mathcal{S}}$, the Kronecker algebra approach leads to exact solution [34,17], if a non-zero initial probability is assigned only to reachable states. Nevertheless the storage and computational complexity may be increased in practice to the point that the advantages of the technique are lost, unless specific counter-measures are taken, as will be discussed in the next section.

4   Problems and Some Solutions

In this section we summarize, with brief explanations and pointers to the literature, the major problems that the research community had, and has, to face to make the Kronecker approach effective in a larger context than SGSPN and with a more efficient solution process.

4.1 $\widehat{\mathcal{S}}$ versus $\mathcal{S}$: Potential versus Actual

In general the distance between $\mathcal{S}$ and the $\widehat{\mathcal{S}}$ can be such that all the space saving due to the use of a Kronecker expression instead than the explicit storage of **R** may be lost due to the size of the probability vector. A number of techniques have been devised, we discuss here two orthogonal ones: limiting $\widehat{\mathcal{S}}$ using the so-called "abstract views", and using ad-hoc, efficient data structures that allow to work directly with $\mathcal{S}$.

Using abstract views. Figure 3 shows a GSPN $\mathcal{S}$ that can be considered as the composition of two GSPNs $\mathcal{S}^1$ and $\mathcal{S}^2$ over three common transitions $T1, T2$ and $T3$. Places whose names start with letter a define component $\mathcal{S}^1$, and those starting with b define $\mathcal{S}^2$. We assume that there is a sequence of n places and transitions between $b21$ and $b2n$, and of m places and transitions between $a31$ and $a3m$. $\mathcal{S}^1$ has therefore $m + 2$ states, $\mathcal{S}^2$ has $n + 2$. A product state space $\widehat{\mathcal{S}}$ can then be defined as

$$\widehat{\mathcal{S}} = \mathcal{S}^1 \times \mathcal{S}^2$$

and it is straightforward to observe that $\mathcal{S} \subseteq \widehat{\mathcal{S}}$, since $\widehat{\mathcal{S}}$ has $(m + 2) \cdot (n + 2)$ states, but the reachability set of $\mathcal{S}$ has only $m + n + 1$.

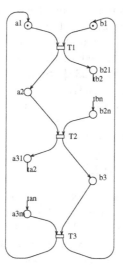

Fig. 3. Explaining abstract views.

To limit the number of spurious states an abstract description of the system can be used to appropriately pre-select the subsets of the states that should enter in the Cartesian product.

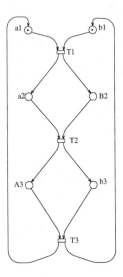

Fig. 4. Explaining abstract views.

For example, we can consider the net $\mathcal{S}^a$ of Figure 4 as an abstract representation of the one in Figure 3, with place $B2$ "summarizing" the structure of places $b21, \ldots, b2n$ and $A3$ "summarizing" the structure of places $a31, \ldots, a3m$. $\mathcal{S}^a$ has three reachable states: $z_1 = (a1, b1)$, $z_2 = (a2, B2)$, and $z_3 = (A3, b3)$. The states of $\mathcal{S}^1$ and $\mathcal{S}^2$ can be partitioned according to the states of $\mathcal{S}^a$: the $m + 2$ states of $\mathcal{S}^1$ are partitioned in three equivalence classes: $\mathcal{S}^1_{z_1} = \{a1\}$, $\mathcal{S}^1_{z_2} = \{a2\}$, and $\mathcal{S}^1_{z_3} = \{a31, \ldots, a3n\}$. Similarly, for $\mathcal{S}^2$ we get: $\mathcal{S}^2_{z_1} = \{b1\}$, $\mathcal{S}^2_{z_2} = \{b21, \ldots, b2n\}$, and $\mathcal{S}^2_{z_3} = \{b3\}$.

The restricted product state space RPS can then be built as:

$$\text{RPS} = \biguplus_{z \in \mathcal{S}^a)} \mathcal{S}^1_z \times \mathcal{S}^2_z =$$

$$\{a1\} \times \{b1\} \cup \{a2\} \times \{b21, \ldots, b2n\} \ \cup \ \{a31, \ldots, a3n\} \times \{b3\}$$

where $\biguplus$ is the *disjoint* set union. Note that in this case we obtain a precise characterization of the state space, but the union of Cartesian products can in general produce a superset of the reachable state space, depending on how precise is the abstract representation. We refer to methods based on a construction of a restricted state space as *abstract*, or *two levels*.

Since the state space is no longer the Cartesian product of sets of local state spaces, but the union of Cartesian products, we cannot expect to have for the infinitesimal generator a Kronecker expression as simple as before: we can build a matrix $\widehat{\mathbf{R}}$, supermatrix of the rate matrix $\mathbf{R}$, of size $|RPS| \times |RPS|$, and then consider it as block structured according to the states of $\mathcal{S}^a$. Each block will thus refer to a set of states obtained by a Cartesian product, and a Kronecker expression for each block can be derived. A similar method works for the reachability graph matrix.

Working with $\mathcal{S}$ only. Instead of trying to reduce the distance between $\mathcal{S}$ and $\widehat{\mathcal{S}}$ it is possible to size the vector used for the vector matrix multiplication according to n instead that $\hat{n}$. This is possible since it is known that the vector entries corresponding to non reachable states are always equal to zero. But the problem to be faced is then to multiply a vector by a matrix of different size, considering that it is not possible any longer to use the mixed base transformation to translate a vector index i into the K element vector needed for the multiplication, and, of course, viceversa. The problem has been solved using a tree-like data structure with K levels, with the lowest level pointing to the vector to be multiplied. A path in the tree from root to leaf correspond to a reachable state and to an efficient computation of the index in the vector of size $\mathcal{S}$. Observe that such a tree structure contains a description of the reachability set, while the reachability graph can be derived from the Kronecker expression.

Several variations of it have been studied in the thesis of Kemper [24] and by Ciardo and Miner in [12], until a very efficient representation based on multivalued decision diagram was devised and implemented in the thesis of Miner [30]. The efficiency is so high in many practical cases that enormous state spaces can be stored, but of course no vector matrix multiplication can be performed, since even the vector does not fit in memory (although the reachability set does!): nevertheless the efficient data structure can be use to effectively prove temporal logic properties using model checking [10].

4.2   Only SGSPN?

Up to know we have discussed only results in the field of SGSPN, that is to say the net must be seen as a set of GSPNs that interact by synchronization over *timed* transitions, and for each GSPN we must be able to generate a finite state space in isolation and the corresponding rate matrix. These limitations have mostly been overcome, and we discuss some of the solutions in the following.

Asynchronous composition. Asynchronous composition of model is as natural as the synchronous one, so a certain attention has been devoted to it. Buchholz was the first one to consider asynchronous composition, in the context of marked graphs [7], and in that of hierarchical Petri nets and queueing networks [5]. To solve the problem connected to the computation of the state spaces in isolation he introduced the concept of "high level view", that we have presented above in the context of SGSPN. The work on marked graph was adapted by Campos, Donatelli and Silva to the larger class of DSSP in [8], and, finally to the general case of GSPNs that interact through places (instead of transitions) in [9].

But two levels methods may work even if there is no high level model given: the work in [4] shows how to build automatically a two levels hierarchy starting from any state space, while a more recent improvement [3] shows how this can be done at the net level.

Stochastic well-formed nets. The work in [32] shows how the Kronecker approach can be extended to the case of high level nets. In particular they show that it is not possible to use the aggregated symbolic reachability graph of the components in isolation for the Kronecker expression, since they may not correctly take into account the effect of the synchronization activities, but that each net in isolation should include some information on the other nets with which is going to be synchronized.

Synchronizing over immediate. All the initial works on SGSPN pose the limitation that modules should synchronize on timed transitions, a restriction that has been removed by Ciardo and Tilgner in [13], and by Donatelli and Kemper in [18].

Kronecker and Phase-type distributions. Phase-type distributions in queueing networks and stochastic Petri nets have the effect of enlarging the state space in a rather regular, almost Cartesian-like, fashion. This fact has been exploited in queueing networks by Beounes in [1] and in [22] for Petri nets, using a flat approach inspired by the classical work on SGSPN, that leads to a high number of non-reachable states, while a more elaborated method, based on abstract views consisting of the enabling degree of phase type transitions, has been proposed in [19], that allows a precise characterization of the state space (no spurious states), and it has been followed also for the discrete phase-type case in [2].

4.3   Working without Components

What happen if the net does not have an explicit structure since it has not been obtained by composition, and/or no information on partitioning of the net into components with finite state space is available? It is possible to show that the modularity unit can be as fine as places (each place is a module) as it has been shown in [13] and [9], although the improvements in space and time may be lost in some cases: the method can therefore be applied to a general GSPN, but its efficacy in reducing the solution cost may be very low, or even result in a worsening of the problem. Despite the fact that the modularity can be as fine as desired, it is nevertheless necessary to be able to generate the local state spaces, or a superset of them, in isolation. If the modularity is at place level, a bound on the number of tokens in each place is enough to generate local state spaces, and it is well-known that this bound can be computed if the net is covered by P-invariants.

4.4   The Vector Bottleneck

We have discussed above of the possibility of building and storing in compact form, using multivalued decision diagrams, very huge state space: but what are the advantages for performance evaluation? Indeed for the computation of the performance indices is necessary to have the vector of steady state (or transient) probabilities, that has as many entries as there are states, so it may appear that there is no real advantage in building state spaces larger than a few million states (the current limit on a reasonably equipped PC nowadays). There are two possible solutions here: to limit our desires, considering approximations instead of exact results, as it has been done in [31], or to devise a method to store also the vector in an efficient manner (and no results are known up to now to the best of the author's knowledge).

5   Is/Was It Worth the Effort?

It is my opinion that the approach was worth the effort put into it for at least three reasons.

1. A contribution of Kronecker based method to the Petri net field has been that of thinking of global states in terms of product of local states, and to carry on this association also at the matrix level, for both Markov chain and reachability graph generation.
2. The transition from the potential state space to multilevel tree and then to multivalued decision diagram for the representation of the state space of a net can also be considered a very important result/effect of the application of Kronecker algebra to nets.
3. The compositional approach was traditionally seen in nets more as a descriptive mechanism, than as a mean to reach an efficient solution: indeed Kronecker based approach have helped in changing this viewpoint.

Despite the above positive "theoretical" points, the impact of the Kronecker based approach for the study of models coming from the real world is still very limited. There are may be two major reasons for this. The complexity results available at the beginning only referred to the case of matrices memorized using full storage scheme, and it was not at all obvious what the complexity actually was using sparse scheme: the work in [6] clearly removes this limitation, although the abstract view case is not discussed. Absence of a clear statement on practical complexity as well as a certain use of somewhat cumbersome notation, has indeed limited the widespreading of the technique. The other reason is indeed the availability of tools, or, better, the availability of Kronecker-based solution in general purpose tools. Plateau implemented a first version of the solution of Stochastic Automata Networks in [35], while Kemper has implemented in a tool the solution of SGSPN models [27]. The SGSPN solution module is integrated into the environment Toolbox [28,29] that allows also the solution of hierarchical queueing Petri nets using the two level methods due to Buchholz. SMART [14] is the tool by Ciardo and Miner that implements Kronecker based methods using the very efficient representation of the state space based on multivalued decision diagrams, unfortunately the tool is not yet distributed to the large public, since the (in my opinion right) choice was taken to deliver the tool only when it is general purpose enough to support all the activities of analysis (including simulation), to allow its use from people that are not Kronecker experts, and when a reasonable manual is available.

As a final remark we can say that the success of the Kronecker-based research line could be measured in the near feature by the number of improvements that will be hidden deep in the tools, that is to say, how many of the results summarized in this paper will be made usable to a public with a knowledge in model verification and evaluation, but no specific knowledge on Kronecker algebra.

Acknowledgements. I would like to thank all the colleagues with which I have shared my research effort in the field of Kronecker algebra and Petri nets, in particular Peter Buchholz, Gianfranco Ciardo, and Peter Kemper since our joint work on complexity was a true and deep exchange of knowledge and experience for, I think, all of us. A special thank goes to Ciardo for allowing me to re-use his latex effort for the example that he had prepared for our joint tutorial, and to Gianfranco Balbo for telling me, already many years ago, when I was a PhD student: "I have listened to an interesting presentation by Plateau on Kronecker algebra: why don't you take a look to it?".

References

1. C. Beounes. Stochastic Petri net modeling for dependability evaluation of complex computer system. In *Proc. of the International Workshop on Timed Petri nets*, Torino, Italy, July 1985. IEEE-CS Press.
2. M. Scarpa and A. Bobbio. Kronecker representation of stochastic Petri nets with discrete PH distributions. In *Proceedings of International Computer Performance and Dependability Symposium - IPDS98*, IEEE Computer Society Press, 52-61, 1998.
3. P. Buchholz, and P. Kemper. On generating a hierarchy for GSPN analysis *ACM Performance Evaluation Review*, Vol. 26 (2), 1998, pages 5-14.
4. P. Buchholz. Hierarchical structuring of Superposed GSPN. In *Proc. of the 7th Intern. Workshop on Petri Nets and Performance Models*, pages 11–90, Saint Malo, France, June 1997. IEEE-CS Press.
5. P. Buchholz. A hierarchical view of GCSPN's and its impact on qualitative and quantitative analysis. *Journal of Parallel and Distributed Computing*, 15(3):207–224, July 1992.
6. P. Buchholz, G. Ciardo, S Donatelli, and P. Kemper. Complexity of memory-efficient Kronecker operations with applications to the solution of Markov models. *INFORMS Journal on Computing*, vol. 12, n.3, summer 2000.
7. P. Buchholz and P. Kemper. Numerical analysis of stochastic marked graphs. In *Proc. 6th Intern. Workshop on Petri Nets and Performance Models*, pages 32–41, Durham, NC, USA, October 1995. IEEE-CS Press.

8. J. Campos, S. Donatelli, and M. Silva. Structured solution of stochastic DSSP systems. In *Proc. of the 7th Intern. Workshop on Petri Nets and Performance Models*, pages 91–100, Saint Malo, France, June 1997. IEEE-CS Press.

9. J. Campos, S. Donatelli, and M. Silva. Structured solution of Asynchronously Communicating Stochastic Modules. In *IEEE Transactions on Software Engineering*, 25(2), April 1999.

10. G. Ciardo, G. Luettgen, and Siminiceanu. Efficient symbolic state-space construction for asynchronous systems. In *Proceedings of the 21st International Conference on Application and Theory of Petri Nets*, Lecture Notes in Computer Science 1825, pages 103-122, June 2000. Springer-Verlag.

11. G. Ciardo and S. Donatelli. Kronecker operators and Markov chain solution. Tutorial presentation at the *joint ACM SIGMETRICS and PERFORMANCE 1998 conference*, June 1998, Wisconsin (USA).

12. G. Ciardo and A. S. Miner. Storage alternatives for large structured state spaces. In R. Marie, B. Plateau, M. Calzarossa, and G. Rubino, editors, *Proc. 9th Int. Conf. on Modelling Techniques and Tools for Computer Performance Evaluation*, LNCS 1245, pages 44–57, Saint Malo, France, June 1997. Springer-Verlag.

13. G. Ciardo and M. Tilgner. On the use of Kronecker operators for the solution of generalized stochastic Petri nets. *ICASE Report* 96-35, Institute for Computer Applications in Science and Engineering, Hampton, VA, May 1996.

14. G. Ciardo and A. S. Miner. SMART: Simulation and Markovian Analyzer for Reliability and Timing. In *Proc. IEEE International Computer Performance and Dependability Symposium (IPDS'96)*, Urbana-Champaign, IL, USA. Sept. 1996. IEEE Comp. Soc. Press.

15. M. Davio. Kronecker products and shuffle algebra. *IEEE Transactions on Computers*, 30(2):116–125, 1981.

16. S. Donatelli. Superposed stochastic automata: a class of stochastic Petri nets with parallel solution and distributed state space. *Performance Evaluation*, 18:21–36, 1993.

17. S. Donatelli. Superposed generalized stochastic Petri nets: definition and efficient solution. In R. Valette, editor, *Proc. of the 15th Intern. Conference on Applications and Theory of Petri Nets*, volume 815 of *Lecture Notes in Computer Science*, pages 258–277. Springer-Verlag, Berlin Heidelberg, 1994.

18. S. Donatelli and P. Kemper. Integrating synchronization with priority into a Kronecker representation *Performance evaluation*, 44 (1-4), 2001.

19. S. Donatelli, S. Haddad, and P. Moreaux. Structured characterization of the Markov Chain of phase-type SPN. In *Proc. 10th Int. Conf. on Modelling Techniques and Tools for Computer Performance Evaluation*, Palma de Mallorca, September 98; LNCS 1469, Springer-Verlag.

20. P. Fernandes, B. Plateau, and W. J. Stewart. Efficient descriptor-vector multiplication in stochastic automata networks. *Journal of the ACM*, 45(3), 1998.

21. P. Fernandes, B. Plateau, and W. J. Stewart. Numerical issues for stochastic automata networks. *INRIA research report* no 2938, July 1996 (available by ftp from ftp.inria.fr).

22. S. Haddad, P. Moreaux, and G. Chiola. Efficient handling of phase-type distributions in generalized stochastic Petri nets. In *Proc. of the 18th International Conference on Application and Theory of Petri Nets*, number 1248 in LNCS, pages 175–194, Toulouse, France, June 23–27 1997. Springer–Verlag.

23. P. Kemper. Numerical analysis of superposed GSPN. *IEEE Transactions on Software Engineering*, 22(4):615–628, September 1996.

24. P. Kemper. Superposition of generalized stochastic Petri nets and its impact on performance analysis. *PhD thesis*, Universität Dortmund, 1996.

25. P. Kemper. Reachability analysis based on structured representations In *Proc. 17th International Conference Application and Theory of Petri Nets*, Osaka (JP), June 1996, pp. 269 - 288, LNCS 1091. Springer, 1996.

26. P. Kemper. Transient analysis of superposed GSPNs. *IEEE Trans. on Software Engineering*, 25(2), March/April 1999. Revised and extended version of a paper with same title in *7-th International Conference on Petri Nets and Performance Models - PNPM97*, pages 101–110. IEEE Computer Society, 1997.

27. P. Kemper. SupGSPN Version 1.0 - an analysis engine for superposed GSPNs. *Technical report*, Universität Dortmund, 1997.

28. P. Kemper F. Bause, P. Buchholz. A toolbox for functional and quantitative analysis of deds. *Technical Report* 680, Universität Dortmund, 1998.

29. P. Kemper F. Bause, P. Buchholz. A toolbox for functional and quantitative analysis of DEDS. Short paper at PERFORMANCE TOOLS'98, *10th International Conference on Modelling Techniques and Tools for Computer Performance Evaluation* Palma de Mallorca, Spain, 1998; LNCS 1469, Springer-Verlag.

30. A.S. Miner. Superposition of generalized stochastic Petri nets and its impact on performance analysis. *PhD thesis*, The college of William and Mary, Williamsburg (USA), 2000.

31. A. Miner, G. Ciardo and S. Donatelli. Using the exact state space of a Markov model to compute approximate stationary measures. In J. Kurose and P. Nain, editors, *Proceedings of the 2000 ACM SIGMETRICS Conference on Measurement and Modeling of Computer Systems*, pages 207-216, June 2000. ACM Press.

32. S. Haddad and P. Moreaux. Asynchronous Composition of High Level Petri Nets: A Quantitative Approach. In *Proc. of the 17th Intern. Conference on Applications and Theory of Petri Nets*, June 1996, LNCS 1091, Springer-Verlag.

33. B. Plateau and K. Atif. Stochastic automata network for modeling parallel systems. *IEEE Transactions on Software Engineering*, 17(10):1093–1108, 1991.

34. B. Plateau. On the stochastic structure of parallelism and synchronization models for distributed algorithms. In *Proc. 1985 ACM SIGMETRICS Conference*, pages 147–154, Austin, TX, USA, August 1985. ACM Press.

35. B. Plateau. PEPS: A package for solving complex Markov models of parallel systems. In R. Puigjaner and D. Poiter, editors, *Modeling techniques and tools for computer performance evaluation*, pages 291–306. Plenum Press, New York and London, 1990.

36. W. J. Stewart. Introduction to the Numerical Solution of Markov Chains. Princeton University Press, 1994.

The Flux of Interaction

Robin Milner

University of Cambridge, The Computer Laboratory,
New Museums Site, Cambridge CB2 3QG
Robin.Milner@cl.cam.ac.uk

Abstract. A graphical model of interactive systems called *bigraphs* is introduced, resting on the orthogonal treatment of *connectivity* and *locality*. The model will be shown to underlie several calculi for mobile systems, in particular the π-calculus and the ambient calculus. Its core behavioural theory will be outlined.

Lecture Summary

The lecture will be about a simple graphical model for mobile computing.

Graphical or geometric models of computing are probably as old as the stored-program computer, possibly older. I do not know when the first flowchart was drawn. Though undeniably useful, flowcharts were denigrated because vital notions like parametric computing –the *procedure*, in Algol terms– found no place in them. But a graphical reduction model was devised by Wadsworth [16] for the lambda calculus, the essence of parametric (functional) computing. Meanwhile, Petri nets [13] made a breakthrough in understanding synchronization and concurrent control flow. Later, the chemical abstract machine (Cham) [2] –employing chemical analogy but clearly a spatial concept– clarified and generalised many features of process calculi.

Before designing CCS, I defined *flowgraphs* [9] as a graphical presentation of *flow algebra*, an early form of what is now called structural congruence; it represented the *static* geometry of interactive processes. The pi calculus and related calculi are all concerned with a form of mobility; they all use some form of structural congruence, but are also informed by a kind of *dynamic* geometrical intuition, even if not expressed formally in those terms.

There are now many such calculi and associated languages. Examples are the pi calculus [11], the fusion calculus [12], the join calculus [5], the spi calculus [1], the ambient calculus [3], Pict [14], nomadic Pict [17], explicit fusions [6]. While these calculi were evolving, in the action calculus project [10] we tried to distill their shared mobile geometry into the notion of *action graph*. This centred around a notion of *molecule*, a node in which further graphs may nest. All action calculi share this kind of geometry, and are distinguished only by a *signature* (a set of molecule types) and a set of *reaction rules*. The latter determine what configurations of molecules can react, and the contexts in which these reactions can take place.

J.-M. Colom and M. Koutny (Eds.): ICATPN 2001, LNCS 2075, pp. 19–22, 2001.
© Springer-Verlag Berlin Heidelberg 2001

Such a framework does not necessarily help in designing and analysing a calculus for a particular purpose. It becomes useful when it supplies non-trivial theory relevant to all, or a specific class of, calculi. Most process calculi are equipped with a behavioural theory – often a labelled transition system (LTS), or a reaction (= reduction) relation, together with a trace-based or (bi)simulation-based behavioural preorder or equivalence. Developing this theory is often hard work, especially proving that the behavioural relation is preserved by (some or all) contexts. Recently [8] we have defined a simple categorical notion of *reactive system*, and shown that under certain conditions an LTS may be uniformly *derived* for it, in such a way that various behavioural relations –including the failures preorder and bisimilarity– will automatically be congruential (i.e. preserved by contexts). We have also shown [4] that a substantial class of action calculi satisfy the required conditions. Thus we approach a non-trivial general theory for those calculi which fit the framework, as many do.

This work has encouraged us to base the theory on a simpler notion: a *bigraph*. Here is an example:

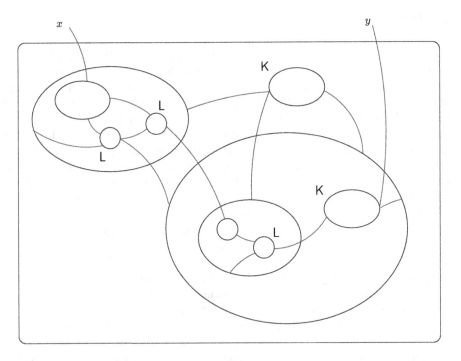

These graphs are a generalisation of Lafont's interaction nets [7]. They consist just of nodes (with many ports) and edges, but with a *locality* –i.e. a forest structure– imposed upon nodes quite independently of the edge wiring. This notion has grown out of action calculi but is also inspired by the Cham of Berry and Boudol [2], the ambient calculus of Cardelli and Gordon [3], the language Nomadic Pict of Sewell and Wojciechowski [17], and the fusion concept of Parrow and Victor [12] further developed by Gardner and Wischik [6]. Graphs with such additional structure are widely useful, and are studied in a recent Handbook edited by Rozenberg [15]. The intuition of bigraphs is that *nodes* have locality, *wires* (per se) don't. A node and its (nodal) contents can be an ambient, a physical location, a λ-abstraction, a program script, an administrative region, ... A node without contents can be a date constructor, a cryptographic key, a merge or copy node, a message envelope, ...

In the lecture I shall outline the basic behavioural theory of bigraphs. I shall show how it leads to congruential behavioural relations for a wide class of calculi for mobile systems. For example, I hope to compare the notion of bisimulation which it generates with that originally defined for the π-calculus [11]. This is work in progress.

Acknowledgement. I would like to thank my colleagues Luca Cattani, Philippa Gardner, Jamey Leifer and Peter Sewell for their co-operation, inspiration and patience.

References

1. Abadi, M. and Gordon, A.D. (1997), A calculus for cryptographic protocols: the spi calculus. Proc. 4th ACM Conference on Computer and Communications Security, ACM Press, 36-47.
2. Berry, G. and Boudol, G. (1992), The chemical abstract machine. Journal of Theoretical Computer Science, Vol 96, pp. 217–248.
3. Cardelli, L. and Gordon, A.D. (2000), Mobile ambients. Foundations of System Specification and Computational Structures, LNCS 1378, 140–155.
4. Cattani, G.L., Leifer, J.J. and Milner, R. (2000),. Contexts and Embeddings for closed shallow action graphs. University of Cambridge Computer Laboratory, Technical Report 496. [Submitted for publication. Available at http://www.cam.cl.ac.uk/users/jjl21 .]
5. Fournet, C. and Gonthier, G. (1996), The reflexive Cham and the join calculus. Proc. 23rd Annual ACM Symposium on Principles of Programming Languages, Florida, pp. 372–385.
6. Gardner, P.A. and Wischik, L.G. (2000), Explicit fusions. Proc. MFCS 2000. LNCS 1893.
7. Lafont, Y. (1990), Interaction nets. Proc. 17th ACM Symposium on Principles of Programming Languages (POPL 90), pp. 95–108.
8. Leifer, J.J. and Milner, R. (2000), Deriving bisimulation congruences for reactive systems. Proc. CONCUR2000. [Available at http://www.cam.cl.ac.uk/users/jjl21.]
9. Milner, R. (1979), Flowgraphs and Flow Algebras. Journal of ACM, 26,4,1979, pp. 794–818.
10. Milner, R. (1996), Calculi for interaction. Acta Informatica 33, 707–737.
11. Milner, R., Parrow, J. and Walker D. (1992), A calculus of mobile processes, Parts I and II. Journal of Information and Computation, Vol 100, pp. 1–40 and pp. 41–77.
12. Parrow, J. and Victor, B. (1998), The fusion calculus: expressiveness and symmetry in mobile processes. Proc. LICS'98, IEEE Computer Society Press.
13. Petri, C.A. (1962), Fundamentals of a theory of asynchronous information flow. Proc. IFIP Congress '62, North Holland, pp. 386–390.

14. Pierce, B.C. and Turner, D.N. (2000), Pict: A programming language based on the pi-calculus. In *Proof, Language and Interaction: Essays in Honour of Robin Milner*, ed. G.D.Plotkin, C.P.Stirling and M.Tofte, MIT Press, pp. 455–494.
15. G. Rozenberg (ed.) (1997), Handbook of Graph Grammars and Computing by Graph Transformation, Volume 1: Foundations, World Scientific.
16. Wadsworth, C.P. (1971), Semantics and pragmatics of the lambda-calculus. Dissertation, Oxford University.
17. Wojciechowski, P.T. and Sewell, P. (1999), Nomadic Pict: Language and infrastructure design for mobile agents. Proc. ASA/MA '99, Palm Springs, California.

Towards a Notion of
Distributed Time for Petri Nets
Extended Abstract

Mogens Nielsen[1], Vladimiro Sassone[2*], and Jiří Srba[3**]

[1,3] **BRICS**[***], Dept. of Computer Science, University of Aarhus, DK
{mn,srba}@brics.dk

[2] University of Sussex, UK
vs@susx.ac.uk

Abstract. We set the ground for research on a timed extension of Petri
nets where time parameters are associated with tokens and arcs carry
constraints that qualify the age of tokens required for enabling. The
novelty is that, rather than a single global clock, we use a set of unre-
lated clocks — possibly one per place — allowing a local timing as well
as distributed time synchronisation. We give a formal definition of the
model and investigate properties of local versus global timing, including
decidability issues and notions of processes of the respective models.

1 Introduction

Verification of finite state systems has been an important area with successful applications to e.g. communication protocols, hardware structures, mobile
phones, hi-fi equipment and many others. For systems that operate for example on data from unbounded domains, new methods must be proposed since
they are not finite state any more and model/equivalence checking is usually
more difficult. Recently algorithmic methods have been developed for process
algebras generating infinite state systems [Mol96,BE97], timed process algebra [Yi90], Petri nets [Jan90], lossy vector addition systems [BM99], counter machines [Jan97,AC98], real time systems [ACD90,AD90,AD94,LPY95] and many
others. In particular, the idea to equip automata with real time appeared to be
very fruitful and there are even automatic verification tools for such systems as
UPPAAL [LPY97] and KRONOS [BDM+98].

The main idea behind timed automata is to equip a standard automaton
with a number of synchronous clocks, and to allow transitions (a) to be conditioned on clock values, and (b) to affect (reset) clocks. One of the objections
to this formalism is the assumption of perfect synchrony between clocks. For

---

[*] Author partly supported by MUST project *TOSCA*.
[**] Author partly supported by the GACR, grant No. 201/00/0400.
[***] Basic Research in Computer Science, Centre of the Danish National Research Foun-
dation.

J.-M. Colom and M. Koutny (Eds.): ICATPN 2001, LNCS 2075, pp. 23–31, 2001.

many applications this assumption is justified, but for others this is an unrealistic assumption. It is easy to imagine systems which are geographically highly distributed where this is the case, but also within hardware design the issue has been addressed, e.g. within work on so-called Globally Asynchronous Locally Synchronous (GALS) systems [MHKEBLTP98].

We are looking for a formalism in which to model such systems. Petri nets seem to be a natural starting point, since one of the virtues of nets is the explicit representation of locality.

Several models that take time features into account have been presented in the literature (for a survey see [Bow96,Wan98]). For example *timed transitions Petri nets* were proposed by Ramchandani [Ram73]. Here each transition is annotated with its firing duration. Another model where time parameters are associated to the places is called *timed places Petri nets*, introduced by Sifakis [Sif77]. We will analyse *timed-arc Petri nets* [BLT90,Han93], a time extension of Petri nets where time (age) is associated to tokens and transitions are labelled by time intervals, which restrict the age of tokens that can be used to fire the transition. In this model, time is considered to be *global*, i.e., all tokens grow older with the same speed. In spite of the fact that reachability is decidable for ordinary Petri nets [May81], reachability for global timed-arc Petri nets is undecidable [RGdFE99]. On the other hand, coverability is decidable for global timed-arc Petri nets [RdFEA00,AN01]. It is also known that the model offers 'weak' expressiveness, in the sense that it cannot simulate Turing machines [BC89].

We suggest a new model where time elapses in a place independently on other places, taking the view that places represent "localities". We generalise this idea of local clocks in such a way that we allow to define an equivalence relation on places such that two places must synchronise if and only if they are in the same equivalence class. We call this model *distributed timed-arc Petri nets*. As special instances we get *local* timed-arc Petri nets (LT nets) where no places are forced to synchronise, and *global* timed-arc Petri nets (GT nets) with full synchronisation. There is yet another motivation for considering LT nets, namely that they seem to be a weaker model than the original one with global time and some properties could be algorithmically verified. We investigate here to what extent this hope is justified.

2 Distributed Timed-Arc Petri Nets

In this section we define formally the model and we consider both continuous and discrete time.

Definition 1 (Distributed timed-arc Petri net).
A distributed timed-arc Petri net (DTAPN) is a tuple $N = (P, T, F, c, E, D)$, where

- *P is a finite set of* places,
- *T is a finite set of* transitions *such that $T \cap P = \emptyset$,*

- $F \subseteq (P \times T) \cup (T \times P)$ *is a* flow relation,
- $c : F|_{P \times T} \to D \times (D \cup \{\infty\})$ *is a* time constraint *on transitions such that for each arc* $(p,t) \in F$ *if* $c(p,t) = (t_1, t_2)$ *then* $t_1 \leq t_2$,
- $E \subseteq P \times P$ *is an equivalence relation on places* (synchronisation relation)
- $D \in \{\mathbb{R}_0^+, \mathbb{N}\}$ *is either* continuous *or* discrete *time.*

Let $x \in D$ and $c(p,t) = (t_1, t_2)$. We write $x \in c(p,t)$ whenever $t_1 \leq x \leq t_2$. We also define $\bullet t = \{p \mid (p,t) \in F\}$ and $t^\bullet = \{p \mid (t,p) \in F\}$.

Definition 2 (Marking).
Let $N = (P,T,F,c,E,D)$ *be a DTAPN. A marking* M *is a function*

$$M : P \to \mathcal{B}(D)$$

where $\mathcal{B}(D)$ *denotes the set of finite multisets on* D.

Each place is thus assigned a certain number of tokens, and each token is annotated with a real (natural) number (*age*). Let $x \in \mathcal{B}(D)$ and $a \in D$. We define $x \triangleleft\!+ a$ in such a way that we add the value a to every element of x, i.e., $x \triangleleft\!+ a = \{b + a \mid b \in x\}$. As *initial markings* we allow only markings with all tokens of age 0.

Definition 3 (Marked DTAPN).
A marked DTAPN *is a pair* (N, M) *where* N *is a distributed timed-arc Petri net and* M *is an initial marking.*

Let us now define the dynamics of DTAPNs. We introduce two types of transition rules: *firing* of a transition and *time-elapsing*.

Definition 4 (Transition rules).
Let $N = (P,T,F,c,E,D)$ be a DTAPN, M a marking and $t \in T$.

- We say that t is *enabled* by M iff $\forall p \in {}^\bullet t. \ \exists x \in M(p). \ x \in c(p,t)$.
- If t is enabled by M then it can be *fired*, producing a marking M' such that:

$$\forall p \in P. \ M'(p) = \Big(M(p) \smallsetminus C^-(p,t) \Big) \cup C^+(t,p)$$

where C^- and C^+ are chosen to satisfy the following equations (note that there may be more possibilities and that all the operations are on multisets):

$$C^-(p,t) = \begin{cases} \{x\} & \text{if } p \in {}^\bullet t \wedge x \in M(p) \wedge x \in c(p,t) \\ \emptyset & \text{otherwise} \end{cases}$$

$$C^+(t,p) = \begin{cases} \{0\} & \text{if } p \in t^\bullet \\ \emptyset & \text{otherwise.} \end{cases}$$

Then we write $M[t\rangle M'$. Note that the new tokens added to places $t^\bullet$ are of the initial age 0.

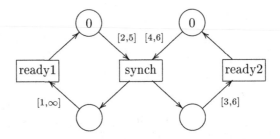

Fig. 1. Example of time synchronisation for GT nets

- We define a *time-elapsing* transition ϵ, for $\epsilon\colon P/E \to D$, as follows, where $[p]_E$ denotes the E-equivalence class of p:

$$M[\epsilon\rangle M' \quad \text{iff} \quad \forall p \in P.\ M'(p) = M(p) \lessdot+ \epsilon([p]_E).$$

We write $M \longrightarrow M'$ iff either $M[t\rangle M'$ or $M[\epsilon\rangle M'$ for some t or ϵ.

In particular we can consider the following two classes of DTAPNs. The first one requires an absolute synchronisation and was studied in the past, while the other one is a new model — completely asynchronous.

- *Global timed-arc Petri nets (GT nets):* $E = P \times P$.
- *Local timed-arc Petri nets (LT nets):* $E = \Delta_P = \{(p, p) \mid p \in P\}$.

3   Examples

In this section we present three examples of timed-arc Petri nets in order to demonstrate the usefulness of GT nets, LT nets and the general model of distributed timed-arc Petri nets. Let us first consider an example of a GT net. Figure 1 gives its graphical representation.

Places are drawn as circles and squares represent transitions with given names. The flow relation is present in form of arcs and every arc from a place to a transition contains a time interval. In the initial marking a pair of tokens of age 0 is present in the upper two places of the picture. An interesting transition is named 'synch'. This is an example of *time synchronisation*, in the sense that in order to fire this transition from the initial marking, there must be some time-elapsing step by 4 or 5 time units. If the net is considered with continuous time also any ϵ-elapsing step is possible for $4 \leq \epsilon(P) \leq 5$. Then we can fire the transition 'synch'. Whenever we want to fire this transition again, the age of tokens in the places from •synch must by synchronised in a similar fashion. Observe that the system can easily deadlock since tokens in places may become *dead*, i.e., they are too old to be useful for firing a transition.

Let us have a look at Figure 2 now. This example is to demonstrate a simple producer/consumer system with continuous time. The net is considered with a local time and whenever a time constraint is missing on an arc, we implicitly

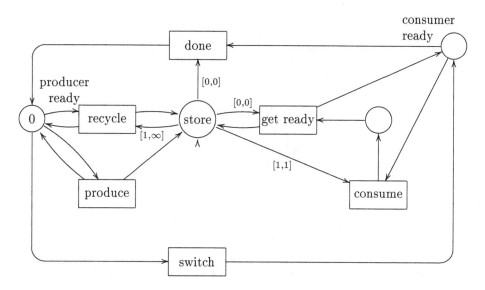

Fig. 2. Producer/consumer example for LT nets

assume that this constraint is irrelevant, i.e., it is of the form $[0, \infty]$. Thus the
only interesting place where time-parameters are of importance is 'store'. The
other places are just control ones and time can elapse completely independently
there. From the initial marking we can fire a transition 'produce' which adds a
number of products (tokens) into 'store'. If time elapses during the production
process, products of several ages can appear in 'store'. By firing the transition
'switch' we add one token of age 0 into 'store' and one into the place 'consumer
ready'. Now producer is active and he can consume products of age 1 from
'store'. Notice that no time elapsing step is allowed, otherwise there is no token
of age 0 in 'store' and the transition 'get ready' cannot be fired. When consumer
consumed all the products he wanted, a transition 'done' is performed (again
checking that there is still the control token of age 0 in 'store') and producer
becomes active again. Since consumer is not forced to consume all the products
of age 1, it can be the case that products that are too old appear in 'store',
however, they can be recycled by firing the transition 'recycle'. The example in
Figure 2 demonstrates that LT nets are not so weak as they may look. First, it
shows that they allow to implement a potentially *infinite timed-queue* in a place
— in our example in the place 'store'. Second, a mechanism is sketched how to
restrict a time-elapsing step by means of a control token — in our case this token
is added by the transition 'switch'.

The last example we will consider is a *Fischer's protocol for mutual exclusion*.
Fischer's protocol was suggested by Schneider, Bloom and Marzullo in [SBM92]
for testing real-time systems and successfully verified using GT nets by Abdulla
and Nylen [AN01]. Figure 3 is taken from the paper [AN01] and it demonstrates
a running code for a process i. The idea is that we have potentially infinitely

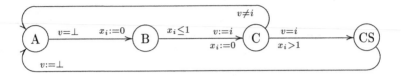

Fig. 3. Fischer's protocol for mutual exclusion

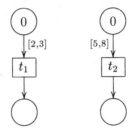

Fig. 4. Dependent transitions in a GT net and independent in an LT net

many processes, each of them running the previously mentioned code. Processes operate on a common shared variable v, A is the initial state and each process has got its own clock x_i. All the clocks are globally synchronised. Our aim is to show that this protocol is correct in the sense that at most one process can enter the critical section CS.

Fischer's protocol can be easily modelled in the GT net formalism as was shown in [AN01]. Synchronisation between places B and C is essential for the mutual exclusion property. However, in this example no behaviour of processes in the critical section is considered. So the protocol only insures safe scheduling mechanism. Assume that we have another GT net N that models the process behaviour in the critical section. If we want e.g. to put the control mechanism together with N and still separate their time-parameters, one solution is to define it as *distributed timed-arc Petri net*. The places in the control mechanism will belong to one equivalence class and the places of N will belong to the other equivalence class. Thus we obtain a complete time-independence between the scheduling process and the process behaviour in the critical section.

4   Investigating DTAPNs

We aim at providing a common ground on which to assess relative expressiveness of GT nets and LT nets. One attempt is to formalise a notion of processes of DTAPNs. The standard notion of processes of P/T nets [GR83] lends itself more readily to LT nets than to GT nets, as illustrated by the net of Figure 4.

Were this net an ordinary, untimed net, we could safely think of the transitions t_1 and t_2 as being completely independent. The situation is not so neat when we consider the time constraints. If we interpret the net as a GT net, i.e., we take the time to be global, after firing t_2, the transition t_1 cannot possibly fire anymore. So, even if there are no static connections between t_1 and t_2, time constrains do not allow to consider them as totally independent. If instead we consider the net under the local time interpretation, t_1 and t_2 are again independent, as they cannot affect each other's enabledness.

We study DTAPN processes in order to establish their properties with respect to timed firing sequences and to be able to prove results assessing the relative expressiveness of LT versus GT nets.

Another attempt is to consider various decidability questions. Ruiz, Gomez and Escrig recently proved in [RGdFE99] that reachability is undecidable for GT nets. Their proof does not imply undecidability for LT nets, because it relies on synchronised places. In principle, it may seem that the model of LT nets is less powerful than the one of GT nets.

Nevertheless, we demonstrate that reachability for LT nets is undecidable as well. The proof is based on a reduction from the halting problem of Minsky machine with two counters. Notice that this contrasts with the result by Mayr [May81] stating the decidability of reachability for ordinary Petri nets. The reachability problem for local timed-arc Petri nets can be formulated as follows.

Problem: Reachability for LT nets.
Instance: A marked LT net (N, M) and a final marking M'.
Question: $M \longrightarrow^* M'$?

Theorem 1. *Reachability for LT nets is undecidable.*

On the other hand, we only need to restrict the class of considered nets very little in order to get the expected difference between local and global timed nets. Say that a marking is *simple* if each place contains at most one token, and that a marked DTAPN is *simple* if the initial marking is simple.

Theorem 2. *Reachability is decidable for simple LT nets, but undecidable for simple GT nets.*

The coverability problem for GT nets was shown to be decidable — for discrete time in [RdFEA00] and for continuous time in [AN01]. By modifying these results we get that coverability is decidable even for DTAPNs. The problem is defined as follows.

Problem: Coverability for DTAPNs.
Instance: A marked DTAPN (N, M) and a final marking M'.
Question: $\exists M''.\ M \longrightarrow^* M'' \wedge \forall p \in P.\ M'(p) \subseteq M''(p)$?

Theorem 3. *Coverability for DTAPNs is decidable.*

5   Conclusion

We have introduced a Petri net model aimed at capturing the ideas behind the Globally Asynchronous Locally Synchronous paradigm, and provided some initial results on our model. However, we believe there are many interesting problems to be addressed in the future for such models.

References

[AC98]   P. A. Abdulla and K. Cerans. Simulation is decidable for one-counter nets. In *Proceedings of 9th International Conference on Concurrency Theory (CONCUR'98)*, volume 1466 of *LNCS*, pages 253–268, 1998.

[ACD90]   R. Alur, C. Courcoubetis, and D. Dill. Model-checking for real-time systems. *5th Symp. on Logic in Computer Science (LICS 90)*, pages 414–425, 1990.

[AD90]   R. Alur and D. Dill. Automata for modelling real-time systems. In *Proc. of Int. Colloquium on Algorithms, Languages and Programming*, volume 443 of *LNCS*, pages 322–335, 1990.

[AD94]   R. Alur and D. Dill. Automata for Modelling Real-Time Systems. *Theoretical Computer Science*, 126(2):183–236, 1994.

[AN01]   P.A. Abdulla and A. Nylen. BQOs and timed Petri nets. In *International Conference on Application and Theory of Petri Nets (ICATPN 2001)*, 2001.
            http://www.docs.uu.se/~parosh/publications/publications.shtml.

[BC89]   T. Bolognesi and P. Cremonese. The weakness of some timed models for concurrent systems. Technical Report CNUCE C89-29, CNUCE–C.N.R., 1989.

[BDM⁺98]   M. Bozga, C. Daws, O. Maler, A. Olivero, S. Tripakis, and S. Yovine. Kronos: A model-checking tool for real-time systems. In *Proceedings of CAV'98*, volume 1427 of *LNCS*, pages 546–550. Springer–Verlag, 1998.

[BE97]   O. Burkart and J. Esparza. More infinite results. *Bulletin of the European Association for Theoretical Computer Science*, 62:138–159, June 1997. Columns: Concurrency.

[BLT90]   T. Bolognesi, F. Lucidi, and S. Trigila. From timed Petri nets to timed LOTOS. In *Proceedings of the IFIP WG 6.1 Tenth International Symposium on Protocol Specification, Testing and Verification (Ottawa 1990)*, pages 1–14. North-Holland, Amsterdam, 1990.

[BM99]   A. Bouajjani and R. Mayr. Model checking lossy vector addition systems. In *Annual Symposium on Theoretical Aspects of Computer Science (STACS'99)*, volume 1563 of *LNCS*, pages 323–333. Springer-Verlag, 1999.

[Bow96]   Fred D.J. Bowden. Modelling time in Petri nets. In *Proceedings of the Second Australia-Japan Workshop on Stochastic Models*, 1996.
            http://www.itr.unisa.edu.au/~fbowden/pprs/stomod96/.

[GR83]   U. Goltz and W. Reisig. The non-sequential behaviour of Petri nets. *Information and Computation*, 57:125–147, 1983.

[Han93]   H.M. Hanisch. Analysis of place/transition nets with timed-arcs and its application to batch process control. In *Application and Theory of Petri Nets*, volume 691 of *LNCS*, pages 282–299, 1993.

[Jan90]   P. Jancar. Decidability of a temporal logic problem for Petri nets. *Theoretical Computer Science*, 74(1):71–93, 1990.

[Jan97]   P. Jancar. Bisimulation equivalence is decidable for one-counter processes. In *Automata, Languages and Programming, 24th International Colloquium (ICALP'97)*, volume 1256 of *LNCS*, pages 549–559. Springer-Verlag, 1997.

[LPY95]   K.G. Larsen, P. Pettersson, and W. Yi. Model-checking for real-time systems. In *Proc. of Fundamentals of Computation Theory*, number 965 in LNCS, pages 62–88, 1995.

[LPY97]   K.G. Larsen, P. Pettersson, and W. Yi. UPPAAL in a Nutshell. *Int. Journal on Software Tools for Technology Transfer*, 1(1–2):134–152, 1997.

[May81]   E.W. Mayr. An algorithm for the general Petri net reachability problem (preliminary version). In *Proc. 13th Ann. ACM Symposium on Theory of Computing*, pages 238–246. Assoc. for Computing Machinery, 1981.

[MHKEBLTP98]   T. Meincke, A. Hemani, S. Kumar, P. Ellervee, J. Berg, D. Lindqvist, H. Tenhunen, A. Postula. Evaluating Benefits of Globally Asynchronous Locally Synchronous VLSI Architecture). In *Proc. 16th Norchip*, pages 50–57, 1998.

[Mol96]   F. Moller. Infinite results. In *Proceedings of CONCUR'96*, volume 1119 of *LNCS*, pages 195–216. Springer-Verlag, 1996.

[Ram73]   C. Ramchandani. *Performance Evaluation of Asynchronous Concurrent Systems by Timed Petri Nets*. PhD thesis, Massachusetts Institute of Technology, Cambridge, 1973.

[RdFEA00]   V. Valero Ruiz, D. de Frutos Escrig, and O. Marroquin Alosno. Decidability of properties of timed-arc Petri nets. In *ICATPN 2000*, volume 1825, pages 187–206, 2000.

[RGdFE99]   V. Valero Ruiz, F. Cuartero Gomez, and D. de Frutos Escrig. On non-decidability of reachability for timed-arc Petri nets. In *Proceedings of the 8th Int. Workshop on Petri Net and Performance Models (PNPM'99)*, pages 188–196, 1999.

[SBM92]   F. B. Schneider, B. Bloom, and K. Marzullo. Putting time into proof outlines. In *Proceedings REX Workshop on Real-Time: Theory in Practice*, volume 600 of *LNCS*, pages 618–639. Springer-Verlag, 1992.

[Sif77]   J. Sifakis. Use of Petri nets for performance evaluation. In *Proceedings of the Third International Symposium IFIP W.G. 7.3., Measuring, modelling and evaluating computer systems (Bonn-Bad Godesberg, 1977)*, pages 75–93. Elsevier Science Publishers, Amsterdam, 1977.

[Wan98]   J. Wang. *Timed Petri Nets, Theory and Application*. Kluwer Academic Publishers, 1998.

[Yi90]   Wang Yi. Real–time behaviour of asynchronous agents. In *Proceedings of the International Conference on Concurrency Theory (CONCUR'90)*, number 458 in LNCS. Springer–Verlag, 1990.

Identifying Commonalities and Differences in Object Life Cycles Using Behavioral Inheritance

Wil M.P. van der Aalst[1] and Twan Basten[2]

[1] Dept. of Technology Management and Dept. of Computing Science, Eindhoven University of Technology, The Netherlands, `w.m.p.v.d.aalst@tue.nl`
[2] Dept. of Electrical Engineering, Eindhoven University of Technology, The Netherlands, `a.a.basten@tue.nl`

Abstract. The behavioral-inheritance relations of [7,8] can be used to compare the life cycles of objects defined in terms of Petri nets. They yield partial orders on object life cycles (OLCs). Based on these orders, we define concepts such as the greatest common divisor and the least common multiple of a set of OLCs. These concepts have practical relevance: In component-based design, workflow management, ERP reference models, and electronic-trade procedures, there is a constant need for identifying commonalities and differences in OLCs. Our results provide the theoretical basis for comparing, customizing, and unifying OLCs.

Keywords: Petri nets, inheritance, lattices, partial orders, object-oriented methods, workflow management.

1 Introduction

For several years, we have been working on notions of *inheritance of behavior* [7, 8]. Inheritance is a key issue in object-oriented design [10]. It allows for the definition of subclasses that inherit features of some superclass. Inheritance is well defined for static properties of classes such as attributes and methods. However, there is no general agreement on the meaning of inheritance when considering the dynamic behavior of objects, captured by their life cycles. In our work, we use Petri nets [18,19] for defining object life cycles (OLCs); they allow for a graphical representation with an explicit representation of object states. In [7, 8], four behavioral-inheritance notions have been defined, based on the principle that by blocking and/or hiding methods of a subclass the resulting behavior should match the behavior of the superclass.

We have applied the behavioral-inheritance concepts in different domains ranging from workflow management [4] and electronic commerce [3] to object-oriented methods [7,8] and component-based software architectures [5]. In each of these applications, objects are designed and compared. The objects of interest can be insurance claims, orders, bank accounts, hardware modules, or software components. One thing they have in common is that they have a *life cycle*. The inheritance notions have been used as a basis for the comparison of these

J.-M. Colom and M. Koutny (Eds.): ICATPN 2001, LNCS 2075, pp. 32–52, 2001.

OLCs. The applications revealed a new and intriguing question: Given a set of OLCs, what do these OLCs have in common? In this paper, we provide some fundamental results that can be used to answer this question.

Consider a set of OLCs that are variants of some process (a workflow or trade procedure, or the control flow in a hardware or software component). Each of the inheritance relations yields a partial order that can be used to reason about a common super- or subclass of the variants. The *Greatest Common Divisor* (GCD) is the common superclass which preserves as much information about the OLCs as possible, i.e., it is not possible to construct a more detailed OLC which is also a superclass of all variants. The GCD describes the behavior all variants agree on. The *Least Common Multiple* (LCM) is the most compact OLC which is still a subclass of all variants. For each of the application domains mentioned, there is a clear use for such concepts. Consider for example two similar software components. The GCD can be used to deduce what both components have in common; the LCM can be used to construct a generic component which can be used to replace the other two. Another example is the use of ad-hoc workflows. Workflow management systems such as InConcert (TIBCO) allow for case-specific variants of a workflow process. Both the GCD and the LCM of these variants can be used to generate meaningful management information [4].

In this paper, we define the concepts of GCDs and LCMs based on the four inheritance relations mentioned earlier. Since none of them forms a (complete) lattice, a restrictive definition leads to situations where there is no GCD (LCM) and a more liberal definition leads to situations where there are multiple GCDs (LCMs). Both situations are undesirable. We tackle this problem by giving both more restrictive and more liberal definitions. For the latter, we use the terms *Maximal Common Divisor* (MCD) and *Minimal Common Multiple* (MCM). We use the Dedekind-MacNeille [17] completion to turn an inheritance partial order into a complete lattice with virtual nodes. In such a lattice, each set of variants has a GCD and an LCM. However, they may correspond to a so-called virtual OLC. Although a virtual OLC cannot be represented by a single Petri net, it provides meaningful information on commonalities and differences.

The paper is organized as follows. Section 2 introduces preliminaries. The behavioral-inheritance concepts are given in Section 3. The other sections deal with GCDs, LCMs, MCDs, and MCMs. Section 4 studies these notions in the context of life-cycle inheritance, the most general form of inheritance. In Section 5, the results are extended to the three other notions of inheritance. Section 6 uses the Dedekind-MacNeille completion to guarantee the existence of GCDs and LCMs. We conclude with some remarks on the application of our results.

2   Preliminaries

This section introduces the techniques used in the remainder. Standard definitions for Petri nets are given. Moreover, more advanced concepts such as branching bisimilarity and OLCs are presented.

2.1   Labeled Place/Transition Nets

We define a variant of the classic Petri-net model, namely labeled Place/Transition nets. For an elaborate introduction to Petri nets, the reader is referred to [12, 18,19]. Let L be some set of *action labels*. These labels correspond to methods when modeling OLCs.

Definition 2.1. (Labeled P/T-nets)[1] An L-labeled Place/Transition net, or simply labeled P/T-net, is a tuple (P, T, F, ℓ) where:

1. P is a finite set of *places*,
2. T is a finite set of *transitions* such that $P \cap T = \emptyset$,
3. $F \subseteq (P \times T) \cup (T \times P)$ is a set of directed arcs, called the *flow relation*, and
4. $\ell : T \to L$ is a *labeling function*.

A *marked*, L-labeled P/T-net is a pair (N, s), where $N = (P, T, F, \ell)$ is an L-labeled P/T-net and where s is a bag over P denoting the *marking* of the net. The set of all marked, L-labeled P/T-nets is denoted $\mathcal{N}$.

A marking is a *bag* over the set of places P, i.e., it is a function from P to the natural numbers. We use square brackets for the enumeration of a bag, e.g., $[a^2, b, c^3]$ denotes the bag with two a-s, one b, and three c-s. The sum of two bags $(X + Y)$, the difference $(X - Y)$, the presence of an element in a bag $(a \in X)$, and the notion of subbags $(X \le Y)$ are defined in a straightforward way and they can handle a mixture of sets and bags.

Transition labeling is needed for two reasons. First, a P/T-net modeling an OLC may contain several transitions referring to a single method (identified by the label) in the OLC. Second, we use transition labels as a mechanism to abstract from (internal) methods. For simplicity, we assume that transition labels are identical to transition identifiers unless explicitly stated otherwise.

Let $N = (P, T, F, \ell)$ be a labeled P/T-net. Elements of $P \cup T$ are called *nodes*. A node x is an *input node* of another node y iff there is a directed arc from x to y (i.e., xFy). Node x is an *output node* of y iff yFx. For any $x \in P \cup T$, $\bullet x = \{y \mid yFx\}$ and $x\bullet = \{y \mid xFy\}$; the superscript N may be omitted if clear from the context.

The dynamic behavior of marked, labeled P/T-nets is defined by a *firing rule*.

Definition 2.2. (Firing rule) Let $(N = (P, T, F, \ell), s)$ be a marked, labeled P/T-net. Transition $t \in T$ is *enabled*, denoted $(N, s)[t\rangle$, iff $\bullet t \le s$. The *firing rule* $\_[\_\rangle\_ \subseteq \mathcal{N} \times L \times \mathcal{N}$ is the smallest relation satisfying for any $(N = (P, T, F, \ell), s) \in \mathcal{N}$ and any $t \in T$, $(N, s)[t\rangle \Rightarrow (N, s)[\ell(t)\rangle (N, s - \bullet t + t\bullet)$.

A transition firing is also referred to as an *action*.

---

[1] In the literature, the class of Petri nets introduced in Definition 1 is sometimes referred to as the class of (labeled) *ordinary* P/T-nets to distinguish it from the class of Petri nets that allows more than one arc between a place and a transition.

Definition 2.3. (Reachable markings) Let (N, s_0) be a marked, labeled P/T-net in $\mathcal{N}$. A marking s is *reachable* from the initial marking s_0 iff there exists a sequence of enabled transitions whose firing leads from s_0 to s. The set of reachable markings of (N, s_0) is denoted $[N, s_0\rangle$.

Definition 2.4. (Connectedness) A net $N = (P, T, F, \ell)$ is *weakly connected*, or simply *connected*, iff, for every two nodes x and y in $P \cup T$, $x(F \cup F^{-1})^* y$, where R^{-1} is the inverse and R^* the reflexive and transitive closure of a relation R. Net N is *strongly connected* iff, for every two nodes x and y, $xF^* y$.

We assume that all nets are weakly connected and have at least two nodes.

Definition 2.5. (Boundedness, safeness) A marked net $(N = (P, T, F, \ell), s)$ is *bounded* iff the set of reachable markings $[N, s\rangle$ is finite. It is *safe* iff, for any $s' \in [N, s\rangle$ and any $p \in P$, $s'(p) \leq 1$. Note that safeness implies boundedness.

Definition 2.6. (Dead transitions, liveness) Let $(N = (P, T, F, \ell), s)$ be a marked, labeled P/T-net. A transition $t \in T$ is *dead* in (N, s) iff there is no reachable marking $s' \in [N, s\rangle$ such that $(N, s')[t\rangle$. (N, s) is *live* iff, for every reachable marking $s' \in [N, s\rangle$ and $t \in T$, there is a reachable marking $s'' \in [N, s'\rangle$ such that $(N, s'')[t\rangle$. Note that liveness implies the absence of dead transitions.

2.2 Branching Bisimilarity

To formalize the inheritance concepts in this paper, we need a notion of equivalence. We choose *branching bisimilarity* [13] as the standard equivalence relation. The notion of a *silent action* is pivotal to branching bisimilarity. Silent actions, denoted with the label τ, are actions that cannot be observed. Thus, only the firings of transitions of a P/T-net with a label different from τ are observable.

We distinguish *successful termination* from *deadlock*. A *termination predicate* $\downarrow \subseteq \mathcal{N}$ defines in what states a marked net can terminate successfully. A marked net that cannot perform any actions or terminate successfully is in a *deadlock*.

We need two auxiliary definitions: (1) a relation expressing that a marked net can evolve via zero or more τ actions into another marked net; (2) a predicate expressing that a marked net can terminate via zero or more τ actions.

Definition 2.7. The relation $\_ \Longrightarrow \_ \subseteq \mathcal{N} \times \mathcal{N}$ is defined as the smallest relation satisfying, for any $p, p', p'' \in \mathcal{N}$, $p \Longrightarrow p$ and $(p \Longrightarrow p' \wedge p' [\tau\rangle p'') \Rightarrow p \Longrightarrow p''$. The predicate $\Downarrow \_ \subseteq \mathcal{N}$ is defined as the smallest set of marked, labeled P/T-nets satisfying, for any $p, p' \in \mathcal{N}$, $\downarrow p \Rightarrow \Downarrow p$ and $(\Downarrow p \wedge p' [\tau\rangle p) \Rightarrow \Downarrow p'$.

For any two marked, L-labeled P/T-nets $p, p' \in \mathcal{N}$ and action $\alpha \in L$, $p [(\alpha)\rangle p'$ is an abbreviation of $(\alpha = \tau \wedge p = p') \vee p [\alpha\rangle p'$. Thus, $p [(\tau)\rangle p'$ means that zero τ actions are performed, when the first disjunct is satisfied, or that one τ action is performed, when the second disjunct is satisfied. For any observable action $a \in L \setminus \{\tau\}$, the first disjunct of the predicate can never be satisfied. Hence, $p [(a)\rangle p'$ simply equals $p [a\rangle p'$, meaning that a single a action is performed.

Definition 2.8. (Branching bisimilarity) A binary relation $\mathcal{R} \subseteq \mathcal{N} \times \mathcal{N}$ is called a *branching bisimulation* if and only if, for any $p, p', q, q' \in \mathcal{N}$ and $\alpha \in L$,

1. $p\mathcal{R}q \wedge p\,[\alpha\rangle\,p' \Rightarrow (\exists q', q'' : q', q'' \in \mathcal{N} : q \Longrightarrow q'' \wedge q''\,[(\alpha)\rangle\,q' \wedge p\mathcal{R}q'' \wedge p'\mathcal{R}q')$,
2. $p\mathcal{R}q \wedge q\,[\alpha\rangle\,q' \Rightarrow (\exists p', p'' : p', p'' \in \mathcal{N} : p \Longrightarrow p'' \wedge p''\,[(\alpha)\rangle\,p' \wedge p''\mathcal{R}q \wedge p'\mathcal{R}q')$,
3. $p\mathcal{R}q \Rightarrow (\downarrow p \Rightarrow \Downarrow q \wedge \downarrow q \Rightarrow \Downarrow p)$.

Two marked, labeled P/T-nets are called *branching bisimilar*, denoted $p \sim_b q$, if and only if there exists a branching bisimulation $\mathcal{R}$ such that $p\mathcal{R}q$.

Branching bisimilarity is an equivalence relation on $\mathcal{N}$, i.e., $\sim_b$ is reflexive, symmetric, and transitive (see [7] for a detailed proof).

2.3   Object Life Cycles

Petri nets allow for the graphical representation of OLCs with an explicit representation of states and a clear definition of the initial (object creation) and final (object termination) state. OLCs correspond to the diagrams used in object-oriented methods (e.g., statechart diagrams in UML [10]), process definitions used by workflow management systems [14,4], reference models used in ERP systems (e.g., EPCs used by SAP [15]), and trade procedures as defined in [16].

Definition 2.9. (Object life cycle) Let $N = (P, T, F, \ell)$ be an L-labeled P/T-net and $\bar{t}$ a fresh identifier not in $P \cup T$. N is an *object life cycle* (OLC) iff:

1. *object creation*: P contains an input place i such that $\bullet i = \emptyset$,
2. *object completion*: P contains an output place o such that $o\bullet = \emptyset$,
3. *connectedness*: $\bar{N} = (P, T \cup \{\bar{t}\}, F \cup \{(o, \bar{t}), (\bar{t}, i)\}, \ell \cup \{(\bar{t}, \tau)\})$ is strongly connected,
4. *safeness*: $(N, [i])$ is safe,
5. *proper completion*: for any marking $s \in [N, [i]\rangle$, $o \in s$ implies $s = [o]$,
6. *option to complete*: for any marking $s \in [N, [i]\rangle$, $[o] \in [N, s\rangle$, and
7. *absence of dead methods*: $(N, [i])$ contains no dead transitions.

The set of all OLCs is denoted $\mathcal{O}$.

An OLC satisfies seven requirements. First, an OLC has one place i without any input transitions. A token in i corresponds to an object which is created, i.e., at the beginning of its life cycle. Second, an OLC has one place o without output transitions. A token in o corresponds to an object that is destroyed. Third, an OLC should not have "dangling" transitions and/or places. Thus, every node of an OLC should be located on a path from i to o. This requirement corresponds to strongly connectedness if o is connected to i via an additional transition $\bar{t}$. The net $\bar{N}$ used to formulate the connectedness constraint is called the *short-circuited* net. The label of the new transition is not important and simply set to τ. The fourth requirement says that an OLC is safe. This is a reasonable assumption since places in an OLC correspond to conditions which are either true (marked by a token) or false (empty). The fifth requirement states that the moment a token is put into o all the other places should be empty, which corresponds to the completion of an OLC without leaving dangling references. The sixth requirement states that starting from the initial marking $[i]$ it is always possible to reach the marking with one token in o, which means that it is always feasible to complete the OLC. The last requirement implies that for each transition there is a scenario in which the transition is performed.

The notion of an OLC is strongly related to the notion of a sound workflow net [2,4]. A workflow net also describes the life cycle of one object, often called a *case* or a *workflow instance*. The applicability of the results in this paper transcends workflow management. Therefore, we prefer the term object life cycle (OLC). Since transition labels in OLCs correspond to methods, we also use the term "method" implicitly for transitions.

The last four requirements in Definition 2.9 coincide with liveness and safeness of the short-circuited net [1]. Thus, we can use standard techniques for checking the life-cycle requirements. Our tool *Woflan* [20] has been specifically designed to analyze the requirements stated in Definition 2.9.

We introduced branching bisimilarity as the standard equivalence. Recall that branching bisimilarity distinguishes successful termination and deadlock. An OLC can only terminate successfully in marking $[o]$.

Definition 2.10. The class of marked, labeled P/T-nets $\mathcal{N}$ is equipped with the following termination predicate: $\downarrow = \{(N, [o]) \mid N \in \mathcal{O}\}$.

Definition 2.11. (Behavioral equivalence of OLCs) For OLCs N_0 and N_1 in $\mathcal{O}$, $N_0 \cong N_1$ if and only if $(N_0, [i]) \sim_b (N_1, [i])$.

The set of *visible* actions that an OLC can perform is called the *alphabet* of the OLC. Since an OLC does not have any dead transitions, its alphabet simply is the set of its transition labels excluding silent action τ.

Definition 2.12. (Alphabet) For any OLC $N = (P, T, F, \ell)$ in $\mathcal{O}$, its alphabet $\alpha(N)$ equals $\{\ell(t) \mid t \in T \wedge \ell(t) \neq \tau\}$.

3   Inheritance

In this section, we define four *behavioral*-inheritance relations for OLCs. For a detailed motivation and an overview of related work, we refer to [8]. Consider two OLCs x and y. When is x a subclass of y? Intuitively, one could say that x is a subclass of y iff x can do what y can do. Clearly, all methods of y should also be present in x. Moreover, x will typically add new methods. Therefore, it is reasonable to demand that x can do what y can do with respect to the methods present in y. With respect to new methods (i.e., methods present in x but not in y), there are basically two mechanisms which can be used. The first one simply disallows the execution of any new methods.

> If it is not possible to distinguish the behaviors of x and y when only methods of x that are also present in y are executed, then x is a subclass of y.

This definition conforms to *blocking* methods new in x. The resulting inheritance concept is called *protocol inheritance*; x inherits the protocol of y.

Another mechanism would be to allow for the execution of new methods but to consider only the effects of old ones.

> If it is not possible to distinguish the behaviors of x and y when arbitrary methods of x are executed but when only the effects of methods that are also present in y are considered, then x is a subclass of y.

This inheritance notion is called *projection inheritance*; it conforms to *hiding* methods new in x. This can be achieved by renaming these methods to the silent action τ.

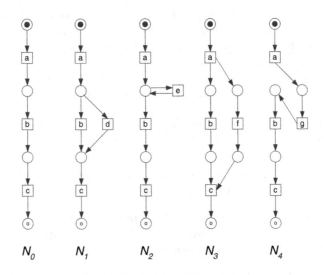

Fig. 1. Five object life cycles.

Although the distinction between the two inheritance mechanisms may seem subtle, the corresponding inheritance notions are quite different. To illustrate this, we use the five OLCs of Figure 1. N_0 corresponds to a sequential OLC consisting of three methods a, b, and c. Each of the other OLCs extends N_0 with one additional method. In N_1, method d can be executed instead of b. N_1 is a subclass of N_0 under protocol inheritance; if d is blocked, N_1 is equivalent to N_0. N_1 is not a subclass of N_0 under projection inheritance, because it is possible to skip method b by executing the (hidden) method d. In N_2, method e can be executed arbitrarily many times between a and b. N_2 is a subclass of N_0 under protocol inheritance; if e is blocked, then N_2 equals N_0. N_2 is also a subclass of N_0 under projection inheritance; if every execution of e is hidden, then N_2 is equivalent (as defined in Definition 2.11) to N_0. In OLC N_3, method f is executed in parallel with b. N_3 is not a subclass of N_0 under protocol inheritance; if f is blocked, then c cannot be executed. However, N_3 is a subclass of N_0 under projection inheritance. If one hides the newly-added method f, one cannot distinguish N_3 and N_0. Method g is inserted between a and b in the remaining OLC N_4. N_4 is not a subclass of N_0 under protocol inheritance; if g is blocked, the OLC deadlocks after executing a. However, N_4 is a subclass of N_0 under projection inheritance. If one hides g, one cannot observe any differences between the behaviors of N_4 and N_0.

The two mechanisms (i.e., blocking and hiding) result in orthogonal inheritance notions. We also consider combinations of the two. An OLC is a subclass of another OLC under *protocol/projection inheritance* iff both by hiding the new methods and by blocking the new methods one cannot detect any differences, i.e., it is a subclass under both protocol and projection inheritance. In Figure 1, N_2 is a subclass of N_0 under protocol/projection inheritance. The two mechanisms can also be used to obtain a more general form of inheritance. An OLC is a subclass of another OLC under *life-cycle inheritance* iff by blocking some newly-added methods and by hiding some others one cannot distinguish them. All OLCs in Figure 1 are subclasses of N_0 under life-cycle inheritance.

To formalize the inheritance relations, we define two operators on P/T-nets: *encapsulation* for blocking and *abstraction* for hiding methods. They are inspired by the encapsulation and abstraction operators from process algebra [6].[2]

Definition 3.1. (Encapsulation) Let $N = (P, T_0, F_0, \ell_0)$ be an L-labeled P/T-net. For any $H \subseteq L \setminus \{\tau\}$, the encapsulation operator ∂_H is a function that removes from a given P/T-net all transitions with a label in H. Formally, $\partial_H(N) = (P, T_1, F_1, \ell_1)$ such that $T_1 = \{t \in T_0 \mid \ell_0(t) \notin H\}$, $F_1 = F_0 \cap ((P \times T_1) \cup (T_1 \times P))$, and $\ell_1 = \ell_0 \cap (T_1 \times L)$.

Note that removing transitions from an OLC as defined in Definition 2.9 might yield a P/T-net that is no longer an OLC.

Definition 3.2. (Abstraction) Let $N = (P, T, F, \ell_0)$ be an L-labeled P/T-net. For any $I \subseteq L \setminus \{\tau\}$, the abstraction operator τ_I is a function that renames all transition labels in I to the silent action τ. Formally, $\tau_I(N) = (P, T, F, \ell_1)$ such that, for any $t \in T$, $\ell_0(t) \in I$ implies $\ell_1(t) = \tau$ and $\ell_0(t) \notin I$ implies $\ell_1(t) = \ell_0(t)$.

The formal definitions of the four inheritance relations are slightly more general than the informal definitions given above: An OLC is a subclass of another OLC if and only if there exists *some* set of methods such that encapsulating or hiding these methods in the first OLC yields the other OLC. Not requiring that the methods being encapsulated or hidden must be *exactly* the newly-added methods can sometimes be convenient. In [7,8], it is shown that the formal and informal definitions are equivalent. Recall Definition 2.8 (branching bisimilarity, $\sim_b$).

Definition 3.3. (Inheritance relations)

1. *Protocol inheritance*: For any OLCs N_0 and N_1 in $\mathcal{O}$, OLC N_1 is a subclass of N_0 under protocol inheritance, denoted $N_1 \leq_{pt} N_0$, iff there is an $H \subseteq L \setminus \{\tau\}$ such that $(\partial_H(N_1), [i]) \sim_b (N_0, [i])$.

---

[2] Note that the terms "abstraction" and "encapsulation" in process algebra have a different meaning than the same terms in object-oriented design. In this paper, they always refer to the process-algebraic concepts.

2. *Projection inheritance*: For any OLCs N_0 and N_1 in $\mathcal{O}$, OLC N_1 is a subclass of N_0 under projection inheritance, denoted $N_1 \leq_{pj} N_0$, iff there is an $I \subseteq L \setminus \{\tau\}$ such that $(\tau_I(N_1), [i]) \sim_b (N_0, [i])$.

3. *Protocol/projection inheritance*: For any OLCs N_0 and N_1 in $\mathcal{O}$, OLC N_1 is a subclass of N_0 under protocol/projection inheritance, denoted $N_1 \leq_{pp} N_0$, iff there is an $H \subseteq L \setminus \{\tau\}$ such that $(\partial_H(N_1), [i]) \sim_b (N_0, [i])$ *and* an $I \subseteq L \setminus \{\tau\}$ such that $(\tau_I(N_1), [i]) \sim_b (N_0, [i])$.

4. *Life-cycle inheritance*: For any OLCs N_0 and N_1 in $\mathcal{O}$, N_1 is a subclass of N_0 under life-cycle inheritance, denoted $N_1 \leq_{lc} N_0$, iff there are an $I \subseteq L \setminus \{\tau\}$ and an $H \subseteq L \setminus \{\tau\}$ such that $I \cap H = \emptyset$ and $(\tau_I \circ \partial_H(N_1), [i]) \sim_b (N_0, [i])$.

Note that for life-cycle inheritance the new methods are partitioned into two sets H and I: methods that are blocked by means of the operator ∂_H and methods that are hidden by means of τ_I. It is easy to see that protocol/projection inheritance implies both protocol and projection inheritance. Moreover, both protocol and projection inheritance imply life-cycle inheritance. However, life-cycle inheritance does not imply protocol or projection inheritance.

The inheritance relations have a number of desirable properties. First, they are preorders (i.e., they are reflexive and transitive; see Property 6.19 in [8]). Furthermore, if one OLC is a subclass of another OLC under any of the four inheritance relations and vice versa, then the two OLCs are equivalent as defined in Definition 2.11 (i.e., the two OLCs are branching bisimilar; see Property 6.21 in [8]). In other words, the four inheritance relations are anti-symmetric. A relation that is reflexive, anti-symmetric, and transitive is a partial order.

Property 3.4. Assuming $\cong$, as defined in Definition 2.11, as the equivalence on OLCs, $\leq_{lc}$, $\leq_{pt}$, $\leq_{pj}$, and $\leq_{pp}$ are partial orders.

4   GCDs and LCMs under Life-Cycle Inheritance

Each of the four notions of inheritance provides a partial ordering on OLCs. This inspired us to investigate whether it is possible to define the notions of a *Greatest Common Divisor* (GCD) and a *Least Common Multiple* (LCM) for sets of OLCs. In this section, we restrict ourselves to life-cycle inheritance (Definition 3.3-4). In Section 5, we consider the other three inheritance notions. We use the term *variant* for an OLC in a set of OLCs. The idea is that the GCD should capture the commonality of the variants, i.e., the part where they agree on. The LCM should capture all possible behaviors of all the variants. Consider for example the five OLCs of Figure 1. The GCD of these OLCs should be N_0. All the OLCs execute a, b, and c in sequential order. Each of the five variants is a subclass of N_0 and it is not possible to find a different OLC that is also a superclass of N_0 through N_4 and at the same time a subclass of N_0. Figure 2 shows $N_{GCD} = N_0$ as the GCD of the five OLCs of Figure 1. It also shows the OLC N_{LCM}. N_{LCM} is a subclass of each of the five variants considered. Moreover, it is not possible to find a different OLC which is also a subclass of N_0 through N_4 and at the same time a superclass of N_{LCM}. Thus, N_{LCM} is a good choice for the LCM of

N_0 through N_4. Any sequence of transition firings generated by one of the five OLCs can also be generated by N_{LCM} after the appropriate abstraction.

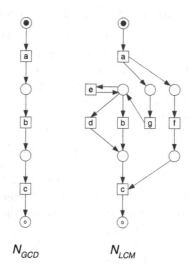

$$N_{GCD} \qquad N_{LCM}$$

Fig. 2. The GCD and the LCM of the five OLCs shown in Figure 1.

To formalize the GCD and LCM concepts, we need some partial-order theory.

Definition 4.1. (Lattices) Let $(Q, \leq)$ be a partial order; let $S \subseteq Q$ and $q \in Q$.

1. q is an *upper bound* of S iff $s \leq q$ for all $s \in S$;
2. $S^\uparrow = \{x \in Q \mid (\forall s : s \in S : s \leq x)\}$ is the set of all upper bounds of S;
3. q is a *lower bound* of S iff $q \leq s$ for all $s \in S$;
4. $S^\downarrow = \{x \in Q \mid (\forall s : s \in S : x \leq s)\}$ is the set of all lower bounds of S;
5. q is the *least upper bound* (lub) of S iff q is an upper bound of S and $q \leq s$ for all $s \in S^\uparrow$;
6. q is the *greatest lower bound* (glb) of S iff q is a lower bound of S and $s \leq q$ for all $s \in S^\downarrow$;
7. $(Q, \leq)$ is a *lattice* iff any pair of elements in Q has a lub and a glb;
8. $(Q, \leq)$ is a *complete* lattice iff any subset of Q has a lub and a glb.

We are not interested in distinguishing OLCs that are branching bisimilar. That is, we consider equivalence classes of OLCs under behavioral equivalence (Definition 2.11). The set of all equivalence classes is denoted $\mathcal{O}_{/\cong}$. We can lift life-cycle inheritance ($\leq_{lc}$) to $\mathcal{O}_{/\cong}$ resulting in the partial order $(\mathcal{O}_{/\cong}, \leq_{lc})$. For convenience, we refer to elements of $\mathcal{O}_{/\cong}$ as OLCs.

Definition 4.2. (MCD/GCD, MCM/LCM) Consider the inheritance partial order $(\mathcal{O}_{/\cong}, \leq_{lc})$. Let $S \subseteq \mathcal{O}_{/\cong}$ be a set of OLCs.

1. OLC N is a *Maximal Common Divisor* (MCD) of S iff (a) it is an upper bound of S (i.e., $N \in S^\uparrow$) and (b) for all $N' \in S^\uparrow$, $N' \leq_{lc} N$ implies that N' equals N (i.e., it is minimal in $S^\uparrow$).

2. OLC N is the *Greatest Common Divisor* (GCD) of S iff it is the lub of S.
3. OLC N is a *Minimal Common Multiple* (MCM) of S iff (a) $N \in S^{\downarrow}$ and (b) for all $N' \in S^{\downarrow}$, $N \leq_{lc} N'$ implies that N' equals N.
4. OLC N is the *Least Common Multiple* (LCM) of S iff it is the glb of S.

Note that the notions of an MCD and a GCD (an MCM and an LCM) coincide if $(\mathcal{O}_{/\cong}, \leq_{lc})$ is a complete lattice. On a first reading, Definition 4.2 might be counterintuitive: An MCD is required to be a *super*class of the OLCs in S, whereas an MCM is a *sub*class of the OLCs in S. It may be more intuitive to consider the size of the OLCs as determined by the numbers of their methods. If N_{MCD} is an MCD of two OLCs N_0 and N_1, then N_{MCD} typically contains *fewer* methods than N_0 and N_1, which conforms to the intuitive notion of an MCD. Similarly, if N_{MCM} is an MCM of N_0 and N_1, then N_{MCM} typically contains *more* methods than N_0 and N_1. Moreover, although it is straightforward to show that any MCM is a subclass under life-cycle inheritance of any MCD ($\leq_{lc}$ is transitive (Property 3.4)), an MCM is typically larger than an MCD in terms of their numbers of methods. Consider for example the OLCs of Figure 2. By Definition 4.2, N_{GCD} is an MCD of the OLCs of Figure 1 and N_{LCM} is an MCM of these OLCs. Although $N_{LCM} \leq_{lc} N_{GCD}$, N_{LCM} has more methods than N_{GCD}.

Definition 4.2 raises two interesting questions:

1. Has any set of OLCs always at least one MCD and at least one MCM?
2. Has any set of OLCs a GCD and an LCM (i.e., is $(\mathcal{O}_{/\cong}, \leq_{lc})$ a complete lattice)?

We show that the answer to the first question is (almost always) affirmative. Unfortunately, the answer to the second question is negative.

The following two properties are needed. The first one is straightforward.

Property 4.3. Let N_{τ} be the OLC containing one method labeled τ: $N_{\tau} = (\{i, o\}, \{\tau\}, \{(i, \tau), (\tau, o)\}, \{(\tau, \tau)\})$. N_{τ} is a superclass under life-cycle inheritance of any OLC, i.e., it is an upper bound of $\mathcal{O}_{/\cong}$ in $(\mathcal{O}_{/\cong}, \leq_{lc})$.

A set of *totally* ordered (according to $\leq_{lc}$) OLCs is called a *chain*.

Property 4.4. Let N_0 and N_1 be two OLCs in $\mathcal{O}_{/\cong}$ such that $N_0 \leq_{lc} N_1$. There is no *infinite* chain $N^0 \leq_{lc} N^1 \leq_{lc} \ldots$ of *different* OLCs $N^0, N^1, \ldots \in \mathcal{O}_{/\cong}$ such that $N_0 \leq_{lc} N^0 \leq_{lc} N^1 \leq_{lc} \ldots \leq_{lc} N_1$.

Proof. Let N and N' be two OLCs with $N \leq_{lc} N'$. The following three observations are important. First, $\alpha(N') \subseteq \alpha(N)$. Second, if N and N' are different, then $\alpha(N') \subset \alpha(N)$. Third, $\alpha(N) \setminus \alpha(N')$ is finite.

Let $N^0 \leq_{lc} N^1 \leq_{lc} \ldots$ be an infinite chain of different OLCs $N^0, N^1, \ldots$ such that $N_0 \leq_{lc} N^0 \leq_{lc} N^1 \leq_{lc} \ldots \leq_{lc} N_1$. It follows from the first two of the above observations that $\alpha(N_1) \subseteq \ldots \subset \alpha(N^1) \subset \alpha(N^0) \subseteq \alpha(N_0)$. The third observation above states that $\alpha(N_0) \setminus \alpha(N_1)$ is finite, yielding a contradiction. $\square$

It follows immediately from the previous two properties that any non-empty set of OLCs has an MCD. The empty set does not have an MCD because $\mathcal{O}_{/\cong}$

is infinite and does not have minimal elements. Any finite set of OLCs has an MCM. First, OLC N_τ of Property 4.3 is an MCM of the empty set. Second, consider a non-empty finite set $\{N_0, N_1, \ldots, N_{n-1}\}$ of n OLCs. Let N_∂ be the OLC that is constructed from the variants as follows. The source place i of N_∂ has n output transitions, one for each variant. Each of them has a unique method label that does not occur in the alphabets of any of the variants. The source place of each variant is given a new identifier and connected as an output place to one of the n new transitions. In this way, the new transitions act as guards for the n original variants. The sink places of the n variants are simply fused together, yielding the sink place o of N_∂. Clearly, N_∂ is a subclass of each variant; by blocking all new transitions except one which is hidden, one obtains an OLC branching bisimilar to one of the variants. Based on Property 4.4, we may conclude that an MCM of the n variants exists. The above deliberations lead to the following theorem, answering the first question posed above.

Theorem 4.5. (Existence of an MCD and an MCM) Let $S \subseteq \mathcal{O}_{/\cong}$ be a set of OLCs. If S is *non-empty*, it has an MCD; if S is *finite*, it has an MCM.

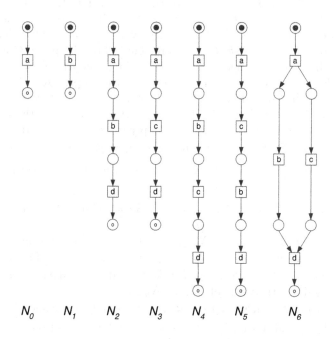

Fig. 3. Seven object life cycles.

As already mentioned, the answer to the second question posed above is negative. A set of OLCs may have two or more different MCDs, which means that it has no GCD. Similarly, a set of OLCs may have two or more different MCMs and, thus, no LCM. Consider OLCs N_4 and N_5 of Figure 3. They have at least two MCDs. It is easy to verify that both OLC N_2 and OLC N_3 are MCDs of N_4

and N_5. Each one is a superclass of both N_4 and N_5 and, in both cases, there is not a smaller (according to $\leq_{lc}$) candidate. Similarly, the two OLCs N_2 and N_3 in Figure 3 have more than one MCM. Each of the OLCs N_4, N_5, and N_6 is an MCM of N_2 and N_3. Note that hiding method c in any of the OLCs N_4, N_5, and N_6 yields an OLC equivalent to N_2. Hiding method b in any of the OLCs N_4, N_5, and N_6 yields an OLC equivalent to N_3. Clearly, in each case, there is no larger candidate.

Based on the examples in Figure 3, we conclude that a given set of OLCs can have several MCDs and MCMs. The reason that there is not a GCD for N_4 and N_5 is that they do agree on the presence of the methods b and c, whereas they do not agree on their ordering. The reason that there is not an LCM for N_2 and N_3 is that there are several ways to add methods b and c to a common subclass. However, in many situations, there is one unique MCD, which is therefore the GCD, and one unique MCM, the LCM. For example, the five variants shown in Figure 1 have a GCD and an LCM, namely the nets N_{GCD} and N_{LCM} of Figure 2, respectively. There are situations where it is quite easy to pinpoint the GCD and/or the LCM of a set of OLCs. If the set forms a *chain*, i.e., the OLCs are totally ordered according to the $\leq_{lc}$ relation, then the least element is the LCM and the greatest element is the GCD. Second, if one OLC is a superclass of all the other OLCs, then this variant is the GCD. Note that the five OLCs of Figure 1 satisfy this requirement. Third, if one OLC is a subclass of all the other variants, then this OLC is the LCM. Fourth, if two variants have no methods in common, then the GCD equals the empty OLC N_τ of Property 4.3. Finally, if the OLCs have nothing in common (i.e., with respect to internal places, transitions, and labels) and always start with a real method (i.e., a non-τ-labeled transition), then the LCM is simply the union of all OLCs (where the union means the element-wise union of the tuples defining the OLCs).

Property 4.6. Let $N_0, N_1, \ldots, N_{n-1}$, with n a positive natural number, be n OLCs.

1. If $N_0 \leq_{lc} N_1 \leq_{lc} \cdots \leq_{lc} N_{n-1}$, then N_0 is the LCM and N_{n-1} is the GCD of $N_0, \ldots, N_{n-1}$.
2. If, for all k with $0 \leq k < n$, $N_k \leq_{lc} N_0$, then N_0 is the GCD of $N_0, \ldots, N_{n-1}$.
3. If, for all k with $0 \leq k < n$, $N_0 \leq_{lc} N_k$, then N_0 is the LCM of $N_0, \ldots, N_{n-1}$.
4. If, for some j and k with $0 \leq j < k < n$, $\alpha(N_j) \cap \alpha(N_k) = \emptyset$, then N_τ of Property 4.3 is the GCD of $N_0, \ldots, N_{n-1}$.
5. If, for all j and k with $0 \leq j < k < n$, $\alpha(N_j) \cap \alpha(N_k) = \emptyset$ and $(P_j \cup T_j) \cap (P_k \cup T_k) = \{i, o\}$ and, for all k with $0 \leq k < n$ and all transitions $t \in i \overset{N_k}{\bullet}$, t has a label different from τ, then $N_\partial = \bigcup_{0 \leq k < n} N_k$ is the LCM of $N_0, \ldots, N_{n-1}$. (Note the similarity between N_∂ in this property and N_∂ as defined before Theorem 4.5.)

Proof. The first three properties follow immediately from Definitions 4.1 and 4.2.

To prove the fourth property, let N' be an arbitrary superclass of $N_0, N_1, \ldots, N_{n-1}$. Consider two variants N_j and N_k, with $0 \leq j < k < n$, such that $\alpha(N_j) \cap$

$\alpha(N_k) = \emptyset$. Since $N_j \leq_{lc} N'$, it follows that $\alpha(N') \subseteq \alpha(N_j)$; similarly, $\alpha(N') \subseteq \alpha(N_k)$. Hence, it follows that $\alpha(N') \subseteq \alpha(N_j) \cap \alpha(N_k) = \emptyset$, which means that $\alpha(N') = \emptyset$. Consequently, N' equals N_τ, which means that N_τ is the GCD of the set of variants N_0 through N_{n-1}.

To prove the last property, we first show that N_∂ is a subclass of each of the variants. Consider a variant N_k, for some k with $0 \leq k < n$. Since for all j with $0 \leq j < n$ and $j \neq k$, $\alpha(N_j) \cap \alpha(N_k) = \emptyset$, $(P_j \cup T_j) \cap (P_k \cup T_k) = \{i, o\}$, and all transitions $t \in i \overset{N_j}{\bullet}$ have a label different from τ, blocking all transitions in $i \overset{N_\partial}{\bullet} \setminus i \overset{N_k}{\bullet}$ in N_∂, yields a marked net branching bisimilar to N_k. Hence, $N_\partial \leq_{lc} N_k$, which means that it is a subclass of all n variants. Second, we prove that any OLC N' that is a subclass of all the variants is also a subclass of N_∂. Assume that N is a subclass of all variants. Let, for all k with $0 \leq k < n$, I_k and H_k be sets of method labels such that $(\tau_{I_k} \circ \partial_{H_k}(N'), [i]) \sim_b (N_k, [i])$ (see Definition 3.3-4 (Life-cycle inheritance)). Let $I = \bigcup_{0 \leq k < n} I_k$ and $H = \bigcup_{0 \leq k < n} H_k$. Clearly, $(\tau_I \circ \partial_H(N'), [i]) \sim_b (N_\partial, [i])$, because each label in H or I appears in the alphabet of precisely one of the n variants. Hence, $N' \leq_{lc} N_\partial$. Combining the results derived so far yields that N_∂ is the LCM of the set of variants N_0 through N_{n-1}. $\square$

5   How about the Other Three Notions of Inheritance?

The results presented in Section 4 are restricted to life-cycle inheritance. In this section, we explore the other three notions of inheritance. First, we define the concepts MCD, MCM, GCD, and LCM for each of the four notions of inheritance.

Definition 5.1. (MCD$_*$, MCM$_*$, GCD$_*$, and LCM$_*$) Let S be some set of OLCs in $\mathcal{O}_{/\cong}$. OLC N is an MCD$_*$, MCM$_*$, GCD$_*$, or LCM$_*$ of S in $(\mathcal{O}_{/\cong}, \leq_*)$ with $* \in \{pt, pj, pp, lc\}$ iff the corresponding requirement stated in Definition 4.2 holds with respect to the corresponding notion of inheritance.

Note that MCD$_{lc}$, MCM$_{lc}$, GCD$_{lc}$, and LCM$_{lc}$ coincide with the concepts introduced in Section 4. As an example of this definition, consider the five variants of Figure 1. It is easy to see that N_0 is the GCD$_{pj}$ of $\{N_2, N_3, N_4\}$, the GCD$_{pt}$ of $\{N_1, N_2\}$, and the GCD$_{pp}$ of $\{N_0, N_2\}$.

The questions raised in previous section arise again: Do MCD$_*$, MCM$_*$, GCD$_*$, and LCM$_*$ exist for $* \in \{pt, pj, pp, lc\}$?

In Theorem 4.5, it has been shown that any non-empty set of variants always has an MCD$_{lc}$. Properties 4.3 and 4.4 carry over to projection inheritance. Thus, we arrive at the following theorem.

Theorem 5.2. (Existence of MCD$_{pj}$) Any *non-empty* set of OLCs $S \subseteq \mathcal{O}_{/\cong}$ has an MCD$_{pj}$ in $(\mathcal{O}_{/\cong}, \leq_{pj})$.

An MCD$_{pj}$ of the five variants of Figure 1 is the sequential OLC containing just the methods a and c. Note that N_0 is not an MCD$_{pj}$, because in N_1 it is possible to bypass b, i.e., N_{GCD} of Figure 2 is not an MCD under projection inheritance.

Unfortunately, MCD$_{pt}$, MCD$_{pp}$, MCM$_{pj}$, MCM$_{pt}$, and MCM$_{pp}$ are not guaranteed to exist. We use the variants shown in Figure 3 to give counterexamples.

Consider N_0 and N_1. There is no MCD_{pt} for these two variants. Suppose that N is an MCD_{pt} of the set $\{N_0, N_1\}$. N should be a superclass of both N_0 and N_1 under protocol inheritance. This implies that the alphabet of N is a subset of the intersection of the alphabets of N_0 and N_1. Since the alphabets of these two variants are disjoint, the alphabet of N is the empty set. There is just one OLC (modulo branching bisimilarity) that has the empty alphabet. This is OLC N_τ of Property 4.3. However, N_0 is not a subclass of N_τ with respect to protocol inheritance because encapsulating method a does not yield N_τ. (In fact, encapsulating a does not yield an OLC.) Therefore, there cannot be an MCD_{pt} for OLCs N_0 and N_1 of Figure 3. It follows immediately from the definition of protocol/projection inheritance (Definition 3.3-3) that this example also implies that MCD_{pp} does not always exist.

To prove that there may be sets of OLCs for which there is no MCM_{pj}, we use the variants N_4 and N_5 of Figure 3. Suppose that N is a subclass of both N_4 and N_5 under projection inheritance. The alphabet of N will include $\{a, b, c, d\}$. Let I be the set of methods in N but not in N_4 and N_5, i.e., $I = \alpha(N) \setminus \{a, b, c, d\}$. By the definition of projection inheritance, we find that $(\tau_I(N), [i]) \sim_b (N_4, [i])$ and $(\tau_I(N), [i]) \sim_b (N_5, [i])$. Hence, since $\sim_b$ is an equivalence, $N_4 \cong N_5$; that is, the two variants are equivalent modulo branching bisimilarity. Clearly, this is a contradiction. Therefore, N_4 and N_5 cannot have a common subclass under projection inheritance. As a result, they have no MCM_{pj}. It follows immediately from the definition of protocol/projection inheritance that there is also no MCM_{pp} for N_4 and N_5.

It remains to be shown that an MCM_{pt} does not necessarily exist. Consider the set S of OLCs $\{N_3, N_4\}$. Assume that N is a subclass of N_3 and N_4 under protocol inheritance. Clearly, $\alpha(N)$ is a superset of $\{a, b, c, d\}$. Let H be $\alpha(N) \setminus \{a, b, c, d\}$; that is, H contains the methods added to N_4 to obtain N, whereas $H \cup \{b\}$ contains the methods added to N_3 to obtain N. It follows from the definition of protocol inheritance that $(\partial_H(N), [i]) \sim_b (N_4, [i])$ and that $(\partial_{H \cup \{b\}}(N), [i]) \sim_b (N_3, [i])$. The definition of branching bisimilarity implies that $(N_3, [i]) \sim_b (\partial_{H \cup \{b\}}(N), [i]) \sim_b (\partial_{\{b\}}(\partial_H(N)), [i]) \sim_b (\partial_{\{b\}}(N_4), [i])$. The latter is the process that can only execute an a and then deadlocks. This is clearly not branching bisimilar to N_3. Hence, we have again a contradiction, showing that N_3 and N_4 cannot have a common subclass under protocol inheritance. This, in turn, implies that $\{N_3, N_4\}$ does not have an MCM_{pt}.

The counterexamples given show that MCD_{pt}, MCD_{pp}, MCM_{pj}, MCM_{pt}, and MCM_{pp} are not guaranteed to exist. Consequently, also GCD_{pt}, GCD_{pp}, LCM_{pj}, LCM_{pt}, and LCM_{pp} may not exist for a given set of variants. In the previous section, it has already been shown that GCD_{lc} and LCM_{lc} do not need to exist. An argument similar to the one used in the previous section shows that N_4 and N_5 in Figure 3 do not have a GCD_{pj}. Thus, also GCD_{pj} does not necessarily exist.

It remains to generalize Property 4.6 to the other notions of inheritance. The proof is omitted because it is similar to the proof of Property 4.6.

Property 5.3. Let $N_0, N_1, \ldots, N_{n-1}$, with n some positive natural number, be n OLCs and let $* \in \{pt, pj, pp, lc\}$.

1. If $N_0 \leq_* N_1 \leq_* \ldots \leq_* N_{n-1}$, then N_0 is the LCM$_*$ and N_{n-1} is the GCD$_*$ of $N_0, \ldots, N_{n-1}$.
2. If, for all k with $0 \leq k < n$, $N_k \leq_* N_0$, then N_0 is the GCD$_*$ of $N_0, \ldots, N_{n-1}$.
3. If, for all k with $0 \leq k < n$, $N_0 \leq_* N_k$, then N_0 is the LCM$_*$ of $N_0, \ldots, N_{n-1}$.
4. If, for some j and k with $0 \leq j < k < n$, $\alpha(N_j) \cap \alpha(N_k) = \emptyset$, then N_τ of Property 4.3 is the GCD$_{pj}$ and the GCD$_{lc}$ of $N_0, \ldots, N_{n-1}$.
5. If, for all j and k with $0 \leq j < k < n$, $\alpha(N_j) \cap \alpha(N_k) = \emptyset$ and $(P_j \cup T_j) \cap (P_k \cup T_k) = \{i, o\}$ and, for all k with $0 \leq k < n$ and all transitions $t \in i\overset{N_k}{\bullet}$, t has a label different from τ, then $N_\partial = \bigcup_{0 \leq k < n} N_k$ is the LCM$_{pt}$ and the LCM$_{lc}$ of $N_0, \ldots, N_{n-1}$.

6   Virtual OLCs and the Dedekind-MacNeille Completion

As we have seen, each of the four inheritance relations provides a partial order on OLCs but none of these orders is a (complete) lattice. If the inheritance relations would have been complete lattices, there would have been a GCD and an LCM for any set of OLCs under any of the inheritance relations. It does not make any sense to try to modify the inheritance relations into lattices. The four relations have been carefully chosen and any attempt to transform them into lattices would reduce their applicability. If a set of OLCs has no GCD/LCM, one could settle for an MCD/MCM. However, also the MCD/MCM do not always exist, particularly for the more restrictive forms of inheritance. Fortunately, the *Dedekind-MacNeille completion* [17,11] can be used to extend the inheritance partial orders to complete lattices. The Dedekind-MacNeille completion provides the smallest complete lattice that embeds a given partial order.

We illustrate the concepts of this section using the seven OLCs shown in Figure 4(a). Transitions without a label correspond to τ-labeled transitions (i.e., silent steps). Figure 4(b) shows the ordering relations between these OLCs under life-cycle inheritance. An OLC N is a superclass of OLC N' (i.e., $N' \leq_{lc} N$) if and only if there is a path of downward going lines from N to N'. The unconnected line segments illustrate that the seven depicted OLCs form only a part of the larger partial order $(\mathcal{O}_{/\cong}, \leq_{lc})$. Note that each element in the partial order in fact corresponds to an equivalence class of OLCs modulo branching bisimilarity.

Consider the set S of OLCs $\{N_0, N_1, N_2\}$. The elements of S are all upper bounds of the OLC sets $S_0 = \{N_3, N_4\}$ and $S_1 = \{N_3, N_4, N_5, N_6\}$; it is not difficult to see that $S_0^\uparrow = S_1^\uparrow = S$ (see Definition 4.1). S_0 and S_1 have no lub, because N_1 and N_2 are incomparable under life-cycle inheritance. In terms of Definition 4.2 (MCD/GCD, MCM/LCM), N_1 and N_2 are MCDs of S_0 and S_1, but N_0 is not; moreover, S_0 and S_1 have no GCD. Similarly, N_3, N_4, N_5, and N_6 are lower bounds and MCMs of the OLCs in S, whereas S has no glb or LCM. The reason for all this is that N_3, N_4, N_5, and N_6 agree on the presence of methods a and b but not on their ordering.

Essential in the Dedekind-MacNeille completion is the notion of *cuts*.

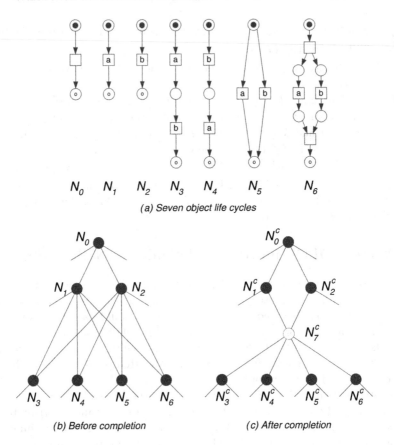

Fig. 4. Seven OLCs and their ordering under life-cycle inheritance before and after completion.

Definition 6.1. (Cut) Let $(Q, \leq)$ be a partial order and $A, B \subseteq Q$. (A, B) is a cut of Q if and only if $A^\uparrow = B$ and $A = B^\downarrow$.

Consider Figure 4(b). It is easy to see that $(\{N_3, N_4, N_5, N_6, \ldots\}, \{N_0, N_1, N_2\})$ is a cut, where the dots represent the subclasses of N_3, N_4, N_5, and N_6 not shown in the figure; the pair $(\{N_3, N_4, N_5, N_6, \ldots\}, \{N_0, N_1\})$ is not a cut.

We need one more definition to formalize the Dedekind-MacNeille completion.

Definition 6.2. (Order-isomorphy) Partial orders $(Q, \leq)$ and $(Q', \leq')$ are *order-isomorphic* iff there exists a bijective function $\phi : Q \to Q'$ such that, for any $x, y \in Q$, $x \leq y$ iff $\phi(x) \leq' \phi(y)$.

Theorem 6.3. (Dedekind-MacNeille completion [17,11]) Let $(Q, \leq)$ be a partial order. Let $(Q^c, \leq^c)$ be the partial order with Q^c the set of all cuts of Q and $\leq^c$ the ordering such that, for any (A_1, B_1) and (A_2, B_2) in Q^c, $(A_1, B_1) \leq^c (A_2, B_2)$ iff $A_1 \subseteq A_2$. Order $(Q^c, \leq^c)$ is the smallest complete lattice containing an ordered subset that is order-isomorphic with $(Q, \leq)$.

An element (A, B) of Q^c corresponds to an element q of Q iff $A \cap B = \{q\}$; it has no corresponding element in Q iff $A \cap B = \emptyset$. If $S^c = \{(A, B) \in Q^c \mid A \cap B \neq \emptyset\}$, then $(S^c, \leq^c \cap (S^c \times S^c))$ is order-isomorphic with $(Q, \leq)$.

The construction of lattice $(Q^c, \leq^c)$ is known as the Dedekind-MacNeille completion. $(Q^c, \leq^c)$ is order-isomorphic with $(Q, \leq)$ if $(Q, \leq)$ is already a complete lattice. The cuts corresponding to elements of Q are called *concrete* elements; other cuts are called *virtual* elements. The Dedekind-MacNeille completion can be applied to the four inheritance partial orders.

Definition 6.4. (Dedekind-MacNeille completion) For $* \in \{pt, pj, pp, lc\}$, $(\mathcal{O}_*^c, \leq_*^c)$ is the Dedekind-MacNeille completion of partial order $(\mathcal{O}_{/\cong}, \leq_*)$.

If we apply the Dedekind-MacNeille completion to the partial order $(\mathcal{O}_{/\cong}, \leq_{lc})$, we obtain the complete lattice (partially) shown in Figure 4(c). Elements N_0^c through N_7^c are cuts: For example, $N_0^c = (\{N_0, N_1, N_2, N_3, N_4, N_5, N_6, \ldots\}, \{N_0\})$, $N_1^c = (\{N_1, N_3, N_4, N_5, N_6, \ldots\}, \{N_0, N_1\})$, $N_4^c = (\{N_4, \ldots\}, \{N_0, N_1, N_2, N_4\})$, and $N_7^c = (\{N_3, N_4, N_5, N_6, \ldots\}, \{N_0, N_1, N_2\})$. The black nodes in the completion of Figure 4(c) are concrete; the corresponding OLCs in Figure 4(b) can be obtained as explained in Theorem 6.3: for example, for cut N_1^c, $\{N_1, N_3, N_4, N_5, N_6, \ldots\} \cap \{N_0, N_1\}) = \{N_1\}$. Node N_7^c is virtual; it does not correspond to an OLC: $\{N_3, N_4, N_5, N_6, \ldots\} \cap \{N_0, N_1, N_2\} = \emptyset$.

Theorem 6.5. Consider the Dedekind-MacNeille completion $(\mathcal{O}_*^c, \leq_*^c)$, with $* \in \{pt, pj, pp, lc\}$. Let $S \subseteq \mathcal{O}_{/\cong}$ be some set of OLCs; let $S^c \subseteq \mathcal{O}_*^c$ be the set of corresponding elements in $\mathcal{O}_*^c$.

1. Let N_{GCD}^c be the lub of S^c in $(\mathcal{O}_*^c, \leq_*^c)$. If N_{GCD}^c is virtual, then S has no GCD$_*$ in $(\mathcal{O}_{/\cong}, \leq_*)$; if N_{GCD}^c is concrete, then the corresponding element N_{GCD} in $\mathcal{O}_{/\cong}$ is the GCD$_*$ of S in $(\mathcal{O}_{/\cong}, \leq_*)$.

2. Let N_{LCM}^c be the glb of S^c in $(\mathcal{O}_*^c, \leq_*^c)$. If N_{LCM}^c is virtual, then S has no LCM$_*$ in $(\mathcal{O}_{/\cong}, \leq_*)$; if N_{LCM}^c is concrete, then the corresponding element N_{LCM} in $\mathcal{O}_{/\cong}$ is the LCM$_*$ of S in $(\mathcal{O}_{/\cong}, \leq_*)$.

Proof. It follows directly from Theorem 6.3 and Definitions 4.2 and 6.4.   □

Theorem 6.5 illustrates that the Dedekind-MacNeille completion can be used to construct a virtual GCD$_*$ or LCM$_*$ for a set of OLCs if *and only if* it has no concrete GCD$_*$/LCM$_*$. A virtual GCD$_*$/LCM$_*$ cannot be drawn as an ordinary P/T-net. However, it can be expressed in terms of concrete OLCs. Consider again the virtual OLC $N_7^c = (\{N_3, N_4, N_5, N_6, \ldots\}, \{N_0, N_1, N_2\})$ of Figure 4(c). Let A and B be the first and second element of N_7^c, respectively. Sets A and B are the sets of all OLCs corresponding to the *concrete* lower bounds and *concrete* upper bounds of N_7^c in the completion, respectively. Note that the maximal elements of A and the minimal elements of B correspond to MCMs of B and MCDs of A, respectively. Virtual OLC N_7^c provides a good characterization of the GCD of N_3, N_4, N_5, and N_6, and of the LCM of N_0, N_1, and N_2. Algorithms for computing the Dedekind-MacNeille completion, see for example [9], can be used to compute (virtual) GCDs and LCMs.

7 Applications and Conclusion

We have focused on the theoretical foundations for GCDs and LCMs of OLCs based on various notions of inheritance. The results are not only intriguing from a theoretical point of view. They have many applications. In component-based software development, workflow management, ERP reference models, and electronic-trade procedures, there is a constant need for identifying commonalities and differences. To conclude this paper, we discuss some of these applications.

Object-oriented methods such as *UML* [10] emphasize reuse and offer various inheritance notions. However, there is no agreement on the meaning of inheritance when considering the dynamic behavior of objects. The inheritance relations in this paper focus on dynamics [8]. One application of the GCD and LCM notions in the context of UML is the following. UML allows for the specification of sequence and collaboration diagrams. Both types of diagrams are used to describe use cases and typically describe one of many possible scenarios. A scenarios is easily translated to an OLC. The GCD of the resulting set of OLCs provides a succinct OLC capturing the behavior all scenarios agree on. The LCM of the set captures all possible behaviors generated by any of the scenarios.

Projection inheritance has been applied in the context of *component-based software architectures* [5]. One of the central issues when dealing with components is the question whether a component "fits." The framework of [5] focuses on the external behavior of a component. The question whether a component "fits" is easily expressed using inheritance. The application potential of the GCD and the LCM of a set of components is promising. The GCD can be used to deduce commonalities for a given set of similar components. The LCM can be used to construct the smallest component that can replace any of the components.

From a conceptual viewpoint, a *workflow procedure* is very similar to an OLC. Workflow management systems are driven by models that describe the life cycle of a case (e.g., insurance claim, order, or tax declaration) [1,2,14]. We applied the inheritance notions in the context of workflow change [4]. Using a number of construction rules, we can construct subclasses of a given workflow (i.e., correctness-by-construction). These rules allow for the automatic migration of cases from sub- to superclass and vice versa. A problem of workflow management systems supporting multiple variants of a workflow (e.g., InConcert (TIBCO) and Ensemble (Filenet)) is the lack of aggregated management information. Using the techniques of this paper, we can calculate the GCD and LCM of a set of variants. These variants may be the result of ad-hoc or evolutionary workflow changes. By migrating the status of every case residing in any of the variants to the GCD and/or LCM, one obtains aggregated management information, i.e., one diagram containing condensed information on the work in progress.

The applicability of the techniques presented in this paper is not limited to workflow within one organization. Especially *interorganizational workflows* [3] and *electronic-trade procedures* [16] can benefit from notions such as the GCD and the LCM. In [3], the notion of a view is introduced. A view is the workflow as seen by one of the parties involved. The GCD of all views is the contract all parties should agree upon. The LCM is the actual workflow being executed. The interested reader is referred to a technical report for more details [3].

Enterprise Resource Planning (ERP) systems such as SAP, Baan, Peoplesoft, and JD Edwards use *reference models* to describe and enact "best practices," i.e., proven business process models are used to drive these systems. Whenever an ERP system is installed, a considerable amount of customization is needed to adapt either the business processes inside the enterprise to the ERP system or vice versa. To determine the amount of customization, the reference model needs to be compared to the desired or actual business process. The GCD can be used to determine the commonalities between both processes and is a good predictor for the customization efforts required.

Another application of GCDs and LCMs is the *unification of procedures* in Europe. Consider for example labor mobility in Europe; a harmonization of national procedures with respect to health insurance, pensions, and so on is needed so that people can move from one EU country to another without bureaucratic confusion. Another example is the unification of financial processes resulting from the introduction of the Euro.

The applications briefly introduced in this final section show the relevance of the questions tackled in this paper. It remains for future work to study these applications in more detail.

Acknowledgment. We thank the anonymous referees for their useful comments.

References

1. W.M.P. van der Aalst. Verification of Workflow Nets. In P. Azéma and G. Balbo, editors, *Application and Theory of Petri Nets 1997*, Lecture Notes in Computer Science 1248, pages 407–426. Springer, Berlin, Germany, 1997.
2. W.M.P. van der Aalst. The Application of Petri Nets to Workflow Management. *The Journal of Circuits, Systems and Computers*, 8(1):21–66, 1998.
3. W.M.P. van der Aalst. Inheritance of Interorganizational Workflows: How to Agree to Disagree Without Loosing Control? BETA Working Paper Series, WP 46, Eindhoven University of Technology, The Netherlands, 2000.
4. W.M.P. van der Aalst and T. Basten. Inheritance of Workflows: An approach to tackling problems related to change. To appear in *Theoretical Computer Science*.
5. W.M.P. van der Aalst, K.M. van Hee, and R.A. van der Toorn. Component-Based Software Architectures: A Framework Based on Inheritance of Behavior. To appear in *Science of Computer Programming*.
6. J.C.M. Baeten and W.P. Weijland. *Process Algebra*. Cambridge Tracts in Theoretical Computer Science 18. Cambridge University Press, Cambridge, UK, 1990.
7. T. Basten. *In Terms of Nets: System Design with Petri Nets and Process Algebra*. PhD thesis, Eindhoven University of Technology, The Netherlands, 1998.
8. T. Basten and W.M.P. van der Aalst. Inheritance of Behavior. *Journal of Logic and Algebraic Programming*, 47(2):47–145, 2001.

9. K. Bertet, M. Morvan, and L. Nourine. Lazy MacNeille Completion of a Partial Order. In G. Mineau and A. Fall, editors, *Proc. of the 2nd Int. Symp. on Knowledge Retrieval, Use and Storage for Efficiency, KRUSE '97*, pages 72–81, 1997

10. G. Booch, J. Rumbaugh, and I. Jacobson. *The Unified Modeling Language User Guide*. Addison-Wesley, Reading, MA, 1998.

11. B.A. Davey and H.A. Priestley. *Introduction to Lattices and Order*. Cambridge University Press, Cambridge, UK, 1990.

12. J. Desel and J. Esparza. *Free Choice Petri Nets. Cambridge Tracts in Theoretical Computer Science* 40. Cambridge University Press, Cambridge, UK, 1995.

13. R.J. van Glabbeek and W.P. Weijland. Branching Time and Abstraction in Bisimulation Semantics. *Journal of the ACM*, 43(3):555–600, 1996.

14. S. Jablonski and C. Bussler. *Workflow Management: Modeling Concepts, Architecture, and Implementation*. Int. Thomson Computer Press, London, UK, 1996.

15. G. Keller and T. Teufel. *SAP R/3 Process Oriented Implementation*. Addison-Wesley, Reading, MA, 1998.

16. R.M. Lee. Distributed Electronic Trade Scenarios: Representation, Design, Prototyping. *International Journal of Electronic Commerce*, 3(2):105–120, 1999.

17. H.M. MacNeille. Partially ordered sets. *Transactions of the American Mathematical Society*, 42:416–460, 1937.

18. T. Murata. Petri Nets: Properties, Analysis and Applications. *Proceedings of the IEEE*, 77(4):541–580, 1989.

19. W. Reisig and G. Rozenberg, editors. *Lectures on Petri Nets I: Basic Models, Lecture Notes in Computer Science* 1491. Springer, Berlin, Germany, 1998.

20. H.M.W. Verbeek and W.M.P. van der Aalst. Woflan 2.0: A Petri-net-based Workflow Diagnosis Tool. In M. Nielsen and D. Simpson, editors, *Application and Theory of Petri Nets 2000, Lecture Notes in Computer Science* 1825, pages 475–484. Springer, Berlin, Germany, 2000. http://www.tm.tue.nl/it/woflan.

Timed Petri Nets and BQOs

Parosh Aziz Abdulla and Aletta Nylén

Department of Computer Systems, Uppsala University
P.O. Box 337, SE-751 05 Uppsala, Sweden
{parosh, aletta}@docs.uu.se

Abstract. We consider (unbounded) *Timed Petri Nets (TPNs)* where each token is equipped with a real-valued clock representing the "age" of the token. Each arc in the net is provided with a subinterval of the natural numbers, restricting the ages of the tokens travelling the arc. We apply a methodology developed in [AN00], based on the theory of *better quasi orderings (BQOs)*, to derive an efficient constraint system for automatic verification of safety properties for TPNs. We have implemented a prototype based on our method and applied it for verification of a parametrized version of Fischer's protocol.

1 Introduction

One of the most widely used techniques for automatic verification of programs is that of *model checking* [CES86,QS82]. A major current challenge is to extend the applicability of model checking to the context of infinite-state systems. A program may be infinite-state since it operates on unbounded data structures, e.g. timed automata [ACD90], hybrid automata [Hen95], data-independent systems [JP93,Wol86], relational automata [Čer94], counter machines [Jan97, AČ98], pushdown processes [BS95], lossy channel systems [AJ96], completely specified protocols [Fin94] etc. A program may also be infinite-state since it has an infinite control part, e.g. Petri nets [Esp95,JM95], and parameterized systems [GS92,AJ98b,KMM⁺97], in which the topology of the system is parameterized by the number of processes inside the system. Petri nets are one of the most widely used models for analysis and verification of concurrent systems. Furthermore, several classes of *Timed Petri Nets (TPNs)* have been introduced in the literature for studying the behaviours of real-time systems; e.g. [RP85,MF76, BD91,GMMP91] (also see [Bow96] for a survey).

In this paper we consider verifying coverability properties of TPNs. In our model, each token in a TPN has an "age" which is represented by a real number. A marking of the net is therefore a mapping which assigns a bag of real numbers to each place. The bag represents the numbers and ages of the tokens in the corresponding place. Each arc of the net is equipped with an interval defined by two natural numbers. A transition may fire only if its input places have tokens with ages satisfying the intervals of the corresponding arcs. Tokens generated by transitions will have ages in the intervals of the output arcs. Furthermore,

J.-M. Colom and M. Koutny (Eds.): ICATPN 2001, LNCS 2075, pp. 53–70, 2001.

we assume a lazy (non-urgent) behaviour of the TPN. This means that transitions may be delayed, even if that implies that some transitions become disabled because their input tokens become too old. Observe that TPNs cannot be modelled within the context of real-time automata [AD90], as the latter operate on a finite number of clocks. In fact TPNs are infinite in two dimensions; they have an unbounded number of tokens and each token has a real-valued clock.

An instance of the coverability problem consists of an initial marking, and a upward closed set of *bad markings*. Intuitively, we do not want the bad markings to occur during the execution of the TPN, and therefore we are interested in showing that no bad marking is reachable from the initial marking. Using standard techniques [VW86,GW93], we can reduce several classes of safety properties for TPNs into the coverability problem.

To solve the coverability problem, we apply an instance of a general algorithm described in [AČJYK96a,AJ98a] for reachability analysis of infinite-state systems. We use a symbolic representation, called *existential zones* for representing (infinite) upward closed sets of markings. We perform a fixpoint iteration, in which we generate existential zones characterizing the set of markings from which a bad marking is reachable within j steps, for increasing values of j.

A main issue when using such an algorithm is to show that the fixpoint iteration always terminates. Applying the method of [AČJYK96a,AJ98a] to existential zones, we can show that the termination of our algorithm is guaranteed if we show that existential zones are *well quasi-ordered*, i.e., for each infinite sequence of zones $Z_0, Z_1, Z_2, \ldots$, there are i and j with $i < j$ where Z_j characterizes a set of markings which is a subset of the set of markings characterized by Z_i. To show the well quasi-ordering of existential zones, we follow the methodology of [AN00], and show that existential zones in fact satisfy a stronger property than well quasi-ordering, namely that they are *better quasi-ordered*. It is worth noting that the well quasi-ordering of existential zones is not possible to show with the framework of [AČJYK96a,AJ98a]. Thus, model checking of TPNs provides a strong evidence that better quasi-orderings are more suitable to use in the context of symbolic model checking than well quasi-orderings.

Based on our algorithm, we have implemented a prototype for automatic verification of safety properties for TPNs. We have used the tool for verification of a parameterized version of Fischer's protocol with encouraging results.

Related Work. Existential zones are variants of another symbolic representation namely that of *zones*. Zones are used in the design of existing tools for verification of real-time systems, such as KRONOS [Yov97] and UPPAAL [LPY97]. However, zones characterize finite sets of clocks, and therefore cannot be used to analyze TPNs.

In [AJ98b] we consider a model close to TPNs, namely *timed networks*. A timed network consists of an arbitrary number of timed processes, each with a single real-valued clock. However, in [AJ98b] we use *existential regions* for verification of timed networks. Existential regions are related to *regions* in the same manner as existential zones are related to zones. In the same manner as regions are less efficient than zones, existential regions are far less efficient than existential zones and explode even on very small applications. In fact, an (existential) zone is the union of a (often large) number of (existential) regions, and therefore (existential) zones offer a much more compact representation of the state space.

Most earlier work on studying decidability issues for TPNs, e.g. [RP85,BD91,GMMP91,RGdFE99] either report undecidability results or decidability under the assumption that the TPN is bounded. A work closely related to ours is [dFERA00]. The authors consider the coverability problem for a class of TPNs similar to our model. The main difference is that in [dFERA00], it is assumed that the ages of the tokens are natural numbers. Furthermore, it is not evident how efficient the constraint system is in practical applications.

In this paper, we consider only lazy TPNs. In fact it can be shown [JLL77] that very simple classes of TPNs with urgent behaviours can simulate two-counter machines, and hence almost all verification problems are undecidable for them. This is not a problem when checking coverability since the set of transitions of an urgent TPN is a subset of the set of transitions of the corresponding lazy TPN. This means that if a set of markings is not reachable in the lazy TPN, then it is certainly not reachable in the urgent TPN.

Outline. In the next section we introduce timed Petri nets. In Section 3 we give an overview of our reachability algorithm. A constraint system which we call existential zones is introduced in Section 4 and in the following section we define an entailment relation on existential zones. In Section 6 we show how *Pre* is computed and in Section 7 we prove that the reachability algorithm terminates. Section 8 introduces existential DDDs, the constraint system used in our experimental work which is presented in Section 9.

2   Timed Petri Nets

We consider *Timed Petri Nets (TPNs)* where each token is equipped with a real-valued clock representing the "age" of the token. The firing conditions of a transition include the usual ones for Petri nets. Furthermore, each arc between a place and a transition is labeled with a subinterval of the natural numbers. When a transition is fired, the tokens removed from the input places of the transition and the tokens added to the output places should have ages lying in the intervals of the corresponding arcs. We let $\mathcal{N}$, $\mathcal{Z}$, and $\mathcal{R}^{\geq 0}$ denote the sets of natural numbers, integers, and nonnegative reals respectively. For a set A, we define the set $Bags(A)$ of *bags* over A to be the set of mappings from A to $\mathcal{N}$. Sometimes we write bags as lists, so e.g. $(2.4, 5.1, 5.1, 2.4, 2.4)$ represents a bag B over $\mathcal{R}^{\geq 0}$ where $B(2.4) = 3$, $B(5.1) = 2$ and $B(x) = 0$ for $x \neq 2.4, 5.1$. We may also write B as $(2.4^3, 5.1^2)$. For bags B_1 and B_2 over a set A, we say that $B_1 \leq B_2$ if $B_1(a) \leq B_2(a)$ for each $a \in A$. We define $B_1 + B_2$ to be the bag B where $B(a) = B_1(a) + B_2(a)$, and (assuming $B_1 \leq B_2$) we define $B_2 - B_1$ to be the bag B where $B(a) = B_2(a) - B_1(a)$, for each $a \in A$. We use $\emptyset$ to denote the empty bag, i.e., $\emptyset(a) = 0$ for each $a \in A$.

We use a set $Intrv$ of $intervals$ of the form $[a : b]$, where $a \in \mathcal{N}$ and $b \in \mathcal{N} \cup \{\infty\}$. For $x \in \mathcal{R}^{\geq 0}$, we write $x \in [a : b]$ to denote that $a \leq x \leq b$.

A $Timed$ $Petri$ Net (TPN) is a tuple $N = (P, T, In, Out)$ where P is a finite set of $places$, T is a finite set of $transitions$ and $In, Out : T \times P \mapsto Bags(Intrv)$. If $In(t, p)(\mathcal{I}) \neq \emptyset$ ($Out(t, p)(\mathcal{I}) \neq \emptyset$), for some interval $\mathcal{I}$, we say that p is an $input$ $(output)$ $place$ of t.

A $marking$ M of N is a finite bag over $P \times \mathcal{R}^{\geq 0}$. The marking M defines numbers and ages of the tokens in each place in the net. That is, $M(p, x)$ defines the number of tokens with age x in place p. For example, if $M = ((p_1, 2.5), (p_1, 1.3), (p_2, 4.7), (p_2, 4.7))$, then, in the marking M, there are two tokens with ages 2.5 and 1.3 in p_1, and two tokens each with age 4.7 in the place p_2. Abusing notation, we define, for each place p, a bag $M(p)$ over $\mathcal{R}^{\geq 0}$, where $M(p)(x) = M(p, x)$. Notice that untimed Petri nets are a special case in our model where all intervals are of the form $[0 : \infty]$.

We define two types of transition relations on markings. A $timed$ $transition$ increases the age of all tokens by the same real number. Formally $M_1 \longrightarrow_T M_2$ if M_1 is of the form $((p_1, x_1), \ldots, (p_n, x_n))$, and there is $\delta \in \mathcal{R}^{\geq 0}$ such that $M_2 = ((p_1, x_1 + \delta), \ldots, (p_n, x_n + \delta))$.

We define the set of $discrete$ $transitions$ $\longrightarrow_d$ as $\bigcup_{t \in T} \longrightarrow_t$, where $\longrightarrow_t$ represents the effect of firing the transition t. More precisely, we define $M_1 \longrightarrow_t M_2$ if, for each place p with $In(t, p) = (\mathcal{I}_1, \ldots, \mathcal{I}_m)$ and $Out(t, p) = (\mathcal{J}_1, \ldots, \mathcal{J}_n)$, there are bags $B_1 = (x_1, \ldots, x_m)$ and $B_2 = (y_1, \ldots, y_n)$ over $\mathcal{R}^{\geq 0}$, such that the following holds.

- $B_1 \leq M_1(p)$.
- $x_i \in \mathcal{I}_i$, for $i : 1 \leq i \leq m$.
- $y_i \in \mathcal{J}_i$, for $i : 1 \leq i \leq n$.
- $M_2(p) = M_1(p) - B_1 + B_2$.

Intuitively, a transition t may be fired only if for each incoming arc to the transition, there is a token with the "right" age in the corresponding input place. This token will be removed from the input place when the transition is fired. Furthermore, for each outgoing arc, a token with an age in the interval will be added to the output place. We define the relation $\longrightarrow$ to be $\longrightarrow_T \cup \longrightarrow_d$, and define $\stackrel{*}{\longrightarrow}$ to be the reflexive transitive closure of $\longrightarrow$. For markings M_1 and M_2, we say that M_1 is $reachable$ from M_2 if $M_1 \stackrel{*}{\longrightarrow} M_2$. For a marking M and a set of markings M, we write $M \stackrel{*}{\longrightarrow}$ M to denote that there is a $M' \in$ M such that $M \stackrel{*}{\longrightarrow} M'$.

For set M of markings we let $Pre(\mathsf{M})$ denote the set $\{M; \exists M' \in \mathsf{M}. M \longrightarrow M'\}$, i.e., $Pre(\mathsf{M})$ is the set of markings from which we can reach a marking in M through the application of a single (timed or discrete) transition.

A set M of markings is said to be $upward$ $closed$ if it is the case that $M \in$ M and $M \leq M'$ imply $M' \in$ M.

Coverability. The *coverability problem* is defined as follows.

Instance A TPN N, a marking M_{init} of N, and an upward closed set M_{fin} of markings of N.

Question $M_{init} \xrightarrow{*} M_{fin}$?

Using standard techniques [VW86,GW93], we can show that checking several classes of safety properties for TPNs can be reduced to the coverability problem.

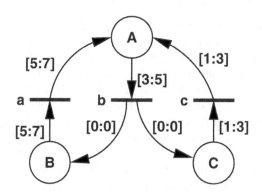

Fig. 1. A small timed Petri net.

Example. Figure 1 shows an example of a TPN where $P = \{A, B, C\}$ and $T = \{a, b, c\}$. For instance, $In(a) = ((B, [5 : 7]))$ and $Out(b) = ((B, [0 : 0]), (C, [0 : 0]))$. The initial marking of this net is the marking $M_{init} = ((A, 0.0))$ with only one token with age 0 in place A.

Remark 1. For simplicity of presentation we use only non-strict inequalities. All the results can be generalized in a straightforward manner to include the more general case, where we also allow strict inequalities.

Remark 2. Notice that, in our definition of the operational behaviour of TPNs, we assume a lazy (non-urgent) behaviour of the net. This means that we may choose to "let time pass" instead of firing enabled transitions, even if that makes transitions disabled due to some of the needed tokens becoming "too old". Tokens that are too old to participate in firing transitions are usually called dead tokens. In an urgent TPN, timed transitions that cause dead tokens are not allowed. This means that the set of transitions of an urgent TPN is a subset of the set of transitions of the corresponding lazy TPN. Therefore, if a set of markings is not reachable in the lazy TPN it is not reachable in the urgent TPN either. In other words safety properties that hold for the lazy TPN also hold for the urgent TPN.

3   Overview of the Verification Algorithm

We give an overview of our algorithm for solving the coverability problem. The main ingredients of the algorithm are

- an instance of a symbolic algorithm described in [AČJYK96b] for checking reachability properties of infinite-state systems
- an application of a methodology based on the theory of *better quasi orderings* described in [AN00] for designing efficient data structures in the implementation of the above symbolic algorithm.

We use a symbolic representation of markings called *existential zones* (Section 4), where each zone Z characterizes an upward closed set of markings $[\![Z]\!]$. The coverability algorithm operates on finite sets of zones. Intuitively, a finite set Z of zones represents the union of the interpretations of its members, i.e., $[\![\mathsf{Z}]\!] = \bigcup_{Z \in \mathsf{Z}}[\![Z]\!]$. Given an instance of the coverability problem (Section 2), defined by M_{init} and a zone Z_0 such that $[\![Z_0]\!] = \mathsf{M}_{fin}$, the symbolic algorithm consists of performing a fixpoint iteration, generating a sequence $\mathsf{Z}_0, \mathsf{Z}_1, \mathsf{Z}_2, \ldots$, where each Z_i is a finite set of existential zones. The set Z_0 is defined to be the singleton $\{Z_0\}$. We define Z_{i+1} to be $\mathsf{Z}_i \cup Pre(\mathsf{Z}_i)$, where $[\![Pre(\mathsf{Z}_i)]\!] = Pre([\![\mathsf{Z}_i]\!])$. In other words, $Pre(\mathsf{Z}_i)$ characterizes exactly the markings from which we can reach a marking in $[\![\mathsf{Z}_i]\!]$ through the application of a single step of the transition relation. In Section 6, we show that the set $Pre(\mathsf{Z}_i)$ exists and is computable. Observe that Z_i characterizes the set of markings from which we can reach M_{fin} in i or fewer steps. We also notice that the elements of the sequence denote larger and larger sets of markings, i.e., $[\![\mathsf{Z}_0]\!] \subseteq [\![\mathsf{Z}_1]\!] \subseteq [\![\mathsf{Z}_2]\!] \subseteq \cdots$. This implies that the procedure of generating new elements of the sequence can be terminated when we reach a point j, where $[\![\mathsf{Z}_j]\!] \supseteq [\![\mathsf{Z}_{j+1}]\!]$. In such a case we have reached the fixpoint, and Z_j characterizes the set of all markings from which M_{fin} is reachable. Consequently, the reachability of M_{fin} from M_{init} is equivalent to whether $M_{init} \in [\![\mathsf{Z}_j]\!]$. In Section 5 (Lemma 3), we show that the relation $[\![\mathsf{Z}_{j+1}]\!] \subseteq [\![\mathsf{Z}_j]\!]$ is decidable, and in Section 4 (Lemma 1), we show that the relation $M_{init} \in [\![\mathsf{Z}_j]\!]$ is decidable. One key issue is to show that the symbolic algorithm always terminates. In [AČJYK96b] we show that, for any constraint system, the termination of the algorithm is guaranteed if the constraint system satisfies a certain property, namely that the constraint system is *well quasi-ordered*. In Section 7 we show well quasi-ordering of existential zones. We do that by applying the methodology developed in [AN00]. More precisely, we show that existential zones satisfy a stronger property than well quasi-ordering; namely that they are *better quasi-ordered*. Better quasi-ordering of existential zones follows from the fact that they can be derived starting from finite domains and then then repeatedly applying the operations of building sets, bags, strings, and taking unions.

4   Existential Zones

In this section we introduce a constraint system called *existential zones*. Intuitively, an existential zone characterizes a upward closed set of markings. An existential zone Z represents minimal conditions on markings. More precisely, Z specifies a minimum number of tokens which should be in the marking, and then imposes certain conditions on these tokens. The conditions are formulated as specifications of the places in which the tokens should reside and restrictions on their ages. The age restrictions are stated as bounds on values of clocks, and bounds on differences between values of pairs of clocks. A marking M which satisfies Z should have at least the number of tokens specified by Z. Furthermore, the places and ages of these tokens should satisfy the conditions imposed by Z. In such a case, M may have any number of additional tokens (whose places and ages are irrelevant for the satisfiability of the zone by the marking).

For a natural number n, we let n^* denote the set $\{0, 1, 2, \dots, n\}$, and let n^+ denote the set $\{1, 2, \dots, n\}$. We assume a TPN (P, T, In, Out).

An *existential zone* Z is a triple $(m, \bar{P}, D)$, where m is an natural number, $\bar{P}$ (called a *placing*) is a mapping $\bar{P} : m^+ \rightarrow P$, and D (called a *difference bound matrix*) is a mapping $D : m^* \times m^* \rightarrow \mathcal{N} \cup \{\infty\}$. Intuitively, m defines the minimum number of tokens in the marking, $\bar{P}$ maps each token to a place, and D defines restrictions on the ages of the tokens in forms of bounds on clock values and on differences between clock values. Difference bound matrices, or DBMs, are widely used in verification of timed automata, e.g., [Dil89,LPY95].

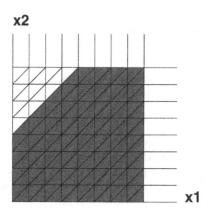

Fig. 2. Example of restrictions on ages of tokens.

Consider the example from Section 2. Assume that we are interested in checking the coverability of markings with at least two tokens, one in place B and one in place C, such that the ages of the tokens are at most 8 and the token in B is at most 4 time units older than the one in C. The markings satisfying these constraints can be described by the existential zone $Z = (2, \bar{P}, D)$ where $\bar{P}(1) = B$, $\bar{P}(2) = C$ and D is described by the following table where eg.

$D(0,i) = 0$ and $D(2,1) = 4$.

$$D = \begin{array}{c|ccc} & 0 & 1 & 2 \\ \hline 0 & - & 0 & 0 \\ 1 & 8 & - & 8 \\ 2 & 8 & 4 & - \end{array}$$

Figure 2 shows an illustration of the age restrictions of Z.

Consider a marking $M = ((p_1, x_1), \ldots, (p_n, x_n))$ and an injection $h : m^+ \to n^+$ (called a *witness*). We say that M *satisfies* Z *with respect to* h, written $M, h \models Z$, if the following conditions are satisfied.

- $\bar{P}(i) = p_{h(i)}$, for each $i : 1 \le i \le m$.
- $x_{h(j)} - x_{h(i)} \le D(j,i)$, for each $i,j \in m^+$ with $i \ne j$.
- $x_{h(i)} \le D(i,0)$ and $-D(0,i) \le x_{h(i)}$, for each $i \in m^+$.

We say that M *satisfies* Z, written $M \models Z$, if $M, h \models Z$ for some h. Notice that if M satisfies Z then $m \le n$ (since h is an injection), i.e., M has at least the number of tokens required by Z, and furthermore, the places and ages of the tokens satisfy the requirements of Z. We define $[\![Z]\!] = \{M; \ M \models Z\}$. Notice that the value of $D(i,i)$ is irrelevant for the satisfiability of Z.

Membership. From the above definitions the following lemma is straightforward.

Lemma 1. *For an existential zone Z and a marking M, it is decidable whether $M \models Z$.*

Upward Closedness. We observe that Z defines a number of minimal requirements on M, in the sense that M should contain at least m tokens whose places and ages are constrained by the functions $\bar{P}$ and D respectively. This means the set $[\![Z]\!]$ is upward closed since $M \models Z$ and $M \le M'$ implies $M' \models Z$.

Normal and Consistent Existential Zones. An existential zone $Z = (m, \bar{P}, D)$ is said to be *normal* if for each $i,j,k \in m^*$, we have $D(j,i) \le D(j,k) + D(k,i)$. It is easy to show the following.

Lemma 2. *For each existential zone Z there is a unique (up to renaming of the index set) normal existential zone, written $\tilde{Z}$, such that $[\![\tilde{Z}]\!] = [\![Z]\!]$.*

This means that we can assume without loss of generality that all existential zones we work with are normal.

An existential zone Z is said to be *consistent* if $[\![Z]\!] \ne \emptyset$.

5   Entailment

Given zones Z_1 and Z_2, we say that Z_1 is *entailed* by Z_2, written $Z_1 \preceq Z_2$, if $[\![Z_2]\!] \subseteq [\![Z_1]\!]$.

We reduce checking entailment between existential zones into validity of formulas in a logic which we here call *Difference Bound Logic (DBL)*. The atomic formulas are either of the form $v \leq c$ or of the form $v - u \leq c$, where v and u are variables interpreted over $\mathcal{R}^{\geq 0}$ and $c \in \mathcal{N}$. Furthermore the set of formulas is closed under the propositional connectives. It is easy to see that validity of DBL-formulas is NP-complete.

Suppose that we are given two existential zones $Z_1 = (m_1, \bar{P}_1, D_1)$ and $Z_2 = (m_2, \bar{P}_2, D_2)$. We translate the relation $Z_1 \preceq Z_2$ into validity of a DBL-formula F as follows. We define the set of free variables in F to be $\{v_i; \ i \in m_2{}^+\}$. Let H be the set of injections from $m_1{}^+$ to $m_2{}^+$ such that $h \in H$ if and only if $\bar{P}_1(i) = \bar{P}_2(h(i))$ for each $i \in m_1{}^+$. We define $F = \left(F_1 \implies \left(\bigvee_{h \in H} F_2\right)\right)$, where $F_1 = F_{11} \wedge F_{12} \wedge F_{13}$, and $F_2 = F_{21} \wedge F_{22} \wedge F_{23}$, and

- $F_{11} = \bigwedge_{i,j \in m_2{}^+, j \neq i} (v_j - v_i \leq D_2(j, i))$.
- $F_{12} = \bigwedge_{i \in m_2{}^+} (v_i \leq D_2(i, 0))$.
- $F_{13} = \bigwedge_{i \in m_2{}^+} (-D_2(0, i) \leq v_i)$.
- $F_{21} = \bigwedge_{i,j \in m_1{}^+, j \neq i} (v_{h(j)} - v_{h(i)} \leq D_1(h(j), h(i)))$.
- $F_{22} = \bigwedge_{i \in m_1{}^+} (v_{h(i)} \leq D_1(h(i), 0))$.
- $F_{23} = \bigwedge_{i \in m_1{}^+} (-D_1(0, h(i)) \leq v_{h(i)})$.

This gives the following.

Lemma 3. *The entailment relation is decidable for existential zones.*

Notice that in contrast to zones for which entailment can be checked in polynomial time, the entailment relation for existential zones can be checked only in nondeterministic polynomial time (as we have to consider exponentially many witnesses). This is the price we pay for working with an unbounded number of clocks. On the other hand, when using zones, the size of the problem grows exponentially with the number of clocks inside the system.

6   Computing Predecessors

We define a function *Pre* such that for a zone Z, the value of $Pre(Z)$ is a finite set $\{Z_1, \ldots, Z_m\}$ of zones. The set $Pre(Z)$ characterizes the set of markings from which we can reach a marking satisfying Z through the performance of a single discrete or timed transition. In other words $Pre[\![Z]\!] = [\![Z_1]\!] \cup \cdots \cup [\![Z_m]\!]$. We define $Pre = Pre_D \cup Pre_\delta$, where Pre_D corresponds to firing transitions backwards and Pre_δ corresponds to running time backwards.

We define $Pre_D = \cup_{t \in T} Pre_t$, where Pre_t characterizes the effect of running the transition t backwards. To define Pre_t, we need the following operations on zones. In the rest of the section we assume a normal existential zone $Z =$

$(m, \bar{P}, D)$, and a timed Petri net $N = (P, T, In, Out)$. From Lemma 2 we know that assuming Z to be normal does not affect the generality of our results.

For an interval $\mathcal{I} = [a : b]$, and $i \in m^+$, we define the *conjunction* $Z \otimes (\mathcal{I}, i)$ of Z with $\mathcal{I}$ at i to be the existential zone $Z' = (m, \bar{P}, D')$, where

- $D'(i, 0) = \min(b, D(i, 0))$.
- $D'(0, i) = \min(-a, D(0, i))$.
- $D'(k, j) = D(k, j)$, for each $j, k \in m^+$ with $k \neq j$, $(k, j) \neq (i, 0)$, and $(k, j) \neq (0, i)$.

Intuitively, the operation adds an additional constraint on the age of token i, namely that its age should be in the interval $\mathcal{I}$. For example, for a zone

$$Z = \left(2, \bar{P}, \begin{array}{c|ccc} & 0 & 1 & 2 \\ \hline 0 & - & 0 & 0 \\ 1 & 8 & - & 8 \\ 2 & 8 & 4 & - \end{array} \right)$$

the conjunction $Z \otimes ([1 : 6], 1)$ is the zone

$$\left(2, \bar{P}, \begin{array}{c|ccc} & 0 & 1 & 2 \\ \hline 0 & - & -1 & 0 \\ 1 & 6 & - & 8 \\ 2 & 8 & 4 & - \end{array} \right)$$

while the conjunction $Z \otimes ([0 : 10], 1) = Z$

For a place p and an interval $\mathcal{I} = [a : b]$, we define the *addition* $Z \oplus (p, \mathcal{I})$ of $(p, \mathcal{I})$ to Z to be the existential zone $Z' = (m + 1, \bar{P}', D')$, and

- $D'(m + 1, 0) = b$, and $D'(0, m + 1) = -a$.
- $D'(m + 1, j) = \infty$, and $D'(j, m + 1) = \infty$, for each $j \in m^+$.
- $\bar{P}'(m + 1) = p$.
- $D'(k, j) = D(k, j)$, for each $j, k \in m^*$, and $\bar{P}'(j) = \bar{P}(j)$, for each $j \in m^+$.

Intuitively, the new existential zone Z' requires one additional token to be present in place p such that the age of the token is in the interval $\mathcal{I}$. For example, for a zone

$$Z = \left(2, \begin{array}{l} \bar{P}(1) = B \\ \bar{P}(2) = C \end{array}, \begin{array}{c|ccc} & 0 & 1 & 2 \\ \hline 0 & - & 0 & 0 \\ 1 & 8 & - & 8 \\ 2 & 8 & 4 & - \end{array} \right)$$

the addition $Z \oplus (A, [1 : 2])$ is the zone

$$\left(3, \begin{array}{l} \bar{P}(1) = B \\ \bar{P}(2) = C \\ \bar{P}(3) = A \end{array}, \begin{array}{c|cccc} & 0 & 1 & 2 & 3 \\ \hline 0 & - & 0 & 0 & -1 \\ 1 & 8 & - & 8 & \infty \\ 2 & 8 & 4 & - & \infty \\ 3 & 2 & \infty & \infty & - \end{array} \right)$$

For $i \in m^+$, we define the *abstraction* $Z \backslash i$ of i in Z to be the zone $Z' = (m - 1, \bar{P}', D')$, where

- $D'(j, k) = D(j, k)$, for each $j, k \in (i - 1)^*$.
- $D'(j, k) = D(j, k + 1)$ and $D'(k, j) = D(k + 1, j)$, for each $j \in (i - 1)^*$ and $k \in \{i, \ldots, m - 1\}$.
- $D'(j, k) = D(j + 1, k + 1)$, for each $j, k \in \{i, \ldots, m - 1\}$.
- $\bar{P}'(j) = \bar{P}(j)$, for each $j \in (i - 1)^*$, and $\bar{P}'(j) = \bar{P}(j + 1)$, for $j \in \{i, \ldots, m - 1\}$.

Intuitively, the operation removes all constraints related to token i from Z, so the number of required tokens is reduced by 1 and the restrictions related to the age and place of the token disappear. For example, for a zone

$$
Z = \left(3, \begin{array}{c} \bar{P}(1) = B \\ \bar{P}(2) = C \\ \bar{P}(3) = A \end{array}, \begin{array}{c|cccc} & 0 & 1 & 2 & 3 \\ \hline 0 & - & 0 & 0 & -1 \\ 1 & 8 & - & 6 & 7 \\ 2 & 8 & 4 & - & 7 \\ 3 & 2 & 2 & 2 & - \end{array} \right)
$$

the abstraction $Z \backslash 2$ is the zone

$$
\left(2, \begin{array}{c} \bar{P}(1) = B \\ \bar{P}(2) = A \end{array}', \begin{array}{c|ccc} & 0 & 1 & 2 \\ \hline 0 & - & 0 & -1 \\ 1 & 8 & - & 7 \\ 2 & 2 & 2 & - \end{array} \right)
$$

Notice that the existential zones we obtain as a result of performing the three operations above need not be normal.

Now, we are ready to define *Pre*.

Lemma 4. *Consider a TPN* $N = (P, T, In, Out)$, *a transition* $t \in T$, *and an existential zone* $Z = (m, \bar{P}, D)$. *Let* $In(t) = ((p_1, \mathcal{I}_1), \ldots, (p_k, \mathcal{I}_k))$, *and* $Out(t) = ((q_1, \mathcal{J}_1), \ldots, (q_\ell, \mathcal{J}_\ell))$. *Then* $Pre_t(Z)$ *is the smallest set containing each existential zone* Z' *such that there is a partial injection* $h : m^+ \longrightarrow \ell^+$ *with a domain* $\{i_1, \ldots, i_n\}$, *and an existential zone* Z_1 *satisfying the following conditions.*

- $\bar{P}(i_j) = q_{h(i_j)}$, *for each* $j \in n^+$
- $Z \otimes (\mathcal{J}_{h(i_1)}, i_1) \otimes \cdots \otimes (\mathcal{J}_{h(i_n)}, i_n)$ *is consistent.*
- $Z_1 = Z \backslash i_1 \backslash \cdots \backslash i_n$.
- $Z' = Z_1 \oplus (p_1, \mathcal{I}_1) \oplus \cdots \oplus (p_k, \mathcal{I}_k)$.

Lemma 5. *For an existential zone* $Z = (m, \bar{P}, D)$, *the set* $Pre_\delta(Z)$ *is the existential zone* $Z' = (m, \bar{P}, D')$, *where* $D'(0, i) = 0$ *and* $D'(j, i) = D(j, i)$ *if* $j \neq 0$, *for each* $i, j \in m^*$, *with* $i \neq j$.

From Lemma 4 and Lemma 5 we get the following.

Lemma 6. *For an existential zone* Z, *the set* $Pre(Z)$ *is computable.*

7   Termination

In this section we show some results from the theories of well quasi-orderings and better quasi-orderings and explain their relation to termination of the reachability algorithm presented in Section 3. A *quasi-ordering* or a *qo* for short, is a pair $(A, \preceq)$ where $\preceq$ is a reflexive and transitive (binary) relation on a set A. We use $a_1 \equiv a_2$ to denote that $a_1 \preceq a_2$ and $a_2 \preceq a_1$. An infinite sequence $a_1, a_2, a_3, \ldots$ of elements of A is called a *bad sequence* iff $\forall i, j : i < j \Rightarrow a_i \npreceq a_j$. A qo $(A, \preceq)$ is a *well quasi-ordering* or a *wqo* for short, if there is no bad sequence of elements of A. Given a qo $(A, \preceq)$, we define a qo $(A^*, \preceq^*)$ on the set A^* of finite strings over A such that $x_1 \bullet \cdots \bullet x_m \preceq^* y_1 \bullet \cdots \bullet y_n$ if and only if there is a strictly monotone injection $h : \{1, \ldots, m\} \to \{1, \ldots, n\}$ where $x_i \preceq y_{h(i)}$ for $i : 1 \leq i \leq m$. A qo $(A^B, \preceq^B)$ on the set A^B of bags over A can be defined in a similar manner. We define the relation $\sqsubseteq$ on the set $\mathcal{P}(A)$ of subsets of A, so that $A_1 \sqsubseteq A_2$ if and only if $\forall b \in A_2 : \exists a \in A_1 : a \preceq b$.

In [AČJYK96a,AJ98a] we showed that the reachability algorithm is guaranteed to terminate if the constraint system is *well quasi-ordered (wqo)*. To prove well quasi-ordering of existential zones we apply a methodology presented in [AN00]. We use a tool which is more powerful than wqo, namely that of *better quasi-ordering (bqo)*. In the following theorem we state some properties of bqos.

Theorem 1.
1. *Each bqo is wqo.*
2. *If A is finite, then $(A, =)$ is bqo.*
3. *If $(A, \preceq)$ is bqo, then $(A^*, \preceq^*)$ is bqo.*
4. *If $(A, \preceq)$ is bqo, then $(A^B, \preceq^B)$ is bqo.*
5. *If $(A, \preceq)$ is bqo, then $(\mathcal{P}(A), \sqsubseteq)$ is bqo.*

A direct consequence of the last property is that bqo is closed under the operation of taking unions. Since bqo is a stronger relation than wqo it is sufficient to prove bqo of zones under entailment, to prove termination of the reachability algorithm.

In order to prove that existential zones are bqo we recall a constraint system related to existential zones, namely that of *existential regions* introduced in [AJ98b]. An existential region is a list of bags $(B_0, B_1, \ldots, B_n, B_{n+1})$ where $n \geq 0$ and B_i is a bag over $P \times \mathcal{N}$. In a similar manner to existential zones, an existential region R defines a set of conditions which should be satisfied by a configuration γ in order for γ to satisfy R. Intuitively B_0 represents tokens with ages which have fractional parts equal to 0. The bags $B_1, \ldots, B_n$ represent tokens whose ages have increasing fractional parts where ages of tokens belonging to the same bag have the same fractional part and ages of tokens belonging to B_i have a fractional part that is strictly less than the fractional part of the ages of those in B_{i+1}. Finally the bag B_{n+1} represents tokens with ages greater than the maximum natural number occuring in the enabling conditions of a given TPN (regardless of their fractional parts).

Lemma 7. *Existential zones are bqo (and hence wqo).*

1. *Existential regions are built starting from finite domains, and repeatedly building finite strings, bags, and sets. From the properties mentioned above, it follows that existential regions are bqo.*
2. *For each existential zone Z, there is a finite set Regions of existential regions such that $Z \equiv \bigcup Regions$. Since bqo is closed under union, it follows that existential zones are bqo.*

8 Existential CDDs and DDDs

CDDs [LPWY99] and DDDs [MLAH99] are constraint systems invented recently to give representations of real-time systems which are more compact than zones. In a similar manner to existential zones, we modify the definitions of CDDs (DDDs) into *existential CDDs (DDDs)*, in order to make them suitable for verifying systems with an unbounded number of clocks. Below we give the definition of existential DDDs. The definition of existential CDDs can be stated in a similar manner.

An *existential DDD* is a tuple $Y = (m, \bar{P}, \mathsf{V}, \mathsf{E})$, where m and $\bar{P}$ are defined as for existential zones (Section 4), and (V, E) is a finite directed acyclic graph where V is the set of vertices and E is the set of edges. We assume that V contains two special elements v^0 and v^1. The outdegrees of v^0 and v^1 are zero, while the outdegrees of the rest of vertices are two. Each vertex $\mathsf{v} \in \mathsf{V} - \{\mathsf{v}^0, \mathsf{v}^1\}$ has the following attributes: $pos(\mathsf{v}), neg(\mathsf{v}) \in m^*$, $op(\mathsf{v}) \in \{<, \le\}$, $const(\mathsf{v}) \in \mathcal{Z}$, and $high(\mathsf{v}), low(\mathsf{v}) \in \mathsf{V}$. The set E contains the edges $(\mathsf{v}, low(\mathsf{v}))$ and $(\mathsf{v}, high(\mathsf{v}))$, where $\mathsf{v} \in \mathsf{V} - \{\mathsf{v}^0, \mathsf{v}^1\}$. In a similar manner to BDDs, the internal nodes of Y correspond to the if-then-else operator $\phi \to \phi_1, \phi_2$, defined as $(\phi \wedge \phi_1) \vee (\neg \phi \wedge \phi_2)$. Intuitively, the attributes of the node represent the DBL-formula $\phi = x_{pos(\mathsf{v})} - x_{neg(\mathsf{v})} \, op(\mathsf{v}) \, const(\mathsf{v})$, and $high(\mathsf{v})$ and $low(\mathsf{v})$ are children of v corresponding to ϕ_1 and ϕ_2 respectively. The special vertices v^0 and v^1 correspond to *false* and *true*.

Consider an existential DDD $Y = (m, \bar{P}, \mathsf{V}, \mathsf{E})$, a vertex $\mathsf{v} \in \mathsf{V}$, a marking $M = ((p_1, c_1), \ldots, (p_k, c_k))$, and an injection $h : m^+ \to k^+$. We say that M satisfies Y at v *with respect to* h, written $M, h \models (Y, \mathsf{v})$, if $\bar{P}(i) = p_{h(i)}$, for each $i \in m^+$, and either

- $\mathsf{v} = \mathsf{v}^1$; or

$$- \left(\left(\begin{array}{c} x_{h(pos(\mathsf{v}))} \\ - \\ x_{h(neg(\mathsf{v}))} \end{array} \right) \sim const(\mathsf{v}) \right) \to \left(\begin{array}{c} M, h \models (Y, high(\mathsf{v})) \\ , \\ M, h \models (Y, low(\mathsf{v})) \end{array} \right),$$

  where $\sim = op(\mathsf{v})$.

In a similar manner to existential zones, we can modify the operations defined in [MLAH99] to compute predecessors of existential DDDs with respect to transitions of a TPN. To check entailment we must, as we did for existential zones, take into consideration all variable permutations.

For each existential DDD Y there is a finite set Z of existential zones such that $[\![Y]\!] = [\![Z]\!]$. Intuitively this means that an existential DDD can replace several existential zones, and hence existential DDDs give a more compact (efficient) representation of sets of states. Note that each existential DDD is a union of existential zones. This together with Lemma 7 and Theorem 1 (Property 5) gives us the following result.

Lemma 8. *Existential DDDs are bqo (and hence also wqo).*

9   Experimental Results

We have implemented a prototype to perform reachability analysis for TPNs. The constraints are represented by existential DDDs. The implementation is based on a DDD package developed at Technical University of Denmark [ML98]. We used the tool to verify a parameterized version of Fischer's protocol. The purpose of the protocol is to guarantee mutual exclusion in a concurrent system consisting of an arbitrary number of processes. The example was suggested by Schneider et al. [SBK92]. The protocol analysed here is in fact a weakened version of Fischer's protocol but since the set of reachable states of the weakened version is a superset of the reachable states of the original protocol, the results of our analysis are still valid.

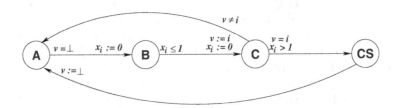

Fig. 3. Fischer's Protocol for Mutual Exclusion

The protocol consists of each process running the code which is graphically described in Figure 3. Each process i has a local clock, x_i, and a control state, which assumes values in the set $\{A, B, C, CS\}$ where A is the initial state and CS is the critical section. The processes read from and write to a shared variable v, whose value is either $\perp$ or the index of one of the processes.

All processes start in state A. If the value of the shared variable is $\perp$, a process wishing to enter the critical section can proceed to state B and reset its local clock. From state B, the process can proceed to state C within one time unit or get stuck in B forever. When making the transition from B to C, the process resets its local clock and sets the value of the shared variable to its own index. The process now has to wait in state C for more than one time unit, a period of time which is strictly greater than the one used in the timeout of state

B. If the value of the shared variable is still the index of the process, the process may enter the critical section, otherwise it may return to state A and start over again. When exiting the critical section, the process resets the shared variable to $\perp$.

We will now make a model of the protocol in our TPN formalism. The processes running the protocol are modeled by tokens in the places A, B, C, CS, $A^\dagger$, $B^\dagger$, $C^\dagger$ and $CS^\dagger$. The places marked with $\dagger$ represent that the value of the shared variable is the index of the process modeled by the token in that place. We use a place *udf* to represent that the value of the shared variable is $\perp$. A straightforward translation of the description in Figure 3 yields the Petri net model in Figure 4. q is used to denote an arbitrary process state.

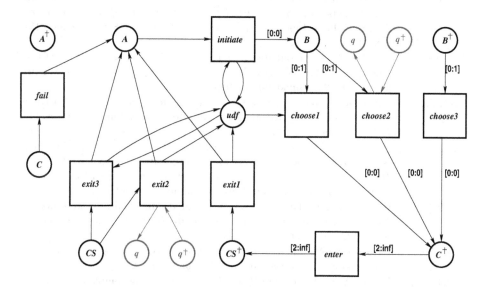

Fig. 4. TPN model of Fischer's Protocol for Mutual Exclusion

In order to prove mutual exclusion we examine the reachability of the existential zones stating that at least two processes are in the critical section, i.e., the following zones:

- $Z_1 = \left(2, \bar{P}_1, D\right)$ where $\bar{P}_1(1) = \bar{P}_1(2) = CS$
- $Z_2 = \left(2, \bar{P}_2, D\right)$ where $\bar{P}_2(1) = CS$ and $\bar{P}_2(2) = CS^\dagger$
- $Z_3 = \left(2, \bar{P}_3, D\right)$ where $\bar{P}_3(1) = \bar{P}_3(2) = CS^\dagger$

For all three zones $D(0,i) = 0$, $D(i,j) = \infty$ for $i \neq j$.

The reachable state space, represented by 45 existential DDDs, takes 3.5 seconds to compute on a Sun Ultra 60 with 512 MB memory and a 360 MHz UltraSPARC-II processor. In the process, pre was computed for 51 existential DDDs.

Acknowledgments. We are grateful to Alberto Marcone and Petr Jančar for discussions on the theory of better quasi-ordering. We would also like to thank Marc Boyer for making us aware of Ruiz, Escrig and Alonsos work on timed arc Petri nets, which is closely related to our work. Special thanks to Jesper Møller and Henrik Reif Andersen for letting us use their DDD implementation and for helping us solve the problems that arose in the process. Thanks also to the anonymous referees for many valuable comments.

References

[AČ98] Parosh Aziz Abdulla and Karlis Čerāns. Simulation is decidable for one-counter nets. In *Proc. CONCUR '98, 9th Int. Conf. on Concurrency Theory*, volume 1466 of *Lecture Notes in Computer Science*, pages 253–268, 1998.

[ACD90] R. Alur, C. Courcoubetis, and D. Dill. Model-checking for real-time systems. In *Proc. 5th IEEE Int. Symp. on Logic in Computer Science*, pages 414–425, Philadelphia, 1990.

[AČJYK96a] Parosh Aziz Abdulla, Karlis Čerāns, Bengt Jonsson, and Tsay Yih-Kuen. General decidability theorems for infinite-state systems. In *Proc. 11th IEEE Int. Symp. on Logic in Computer Science*, pages 313–321, 1996.

[AČJYK96b] Parosh Aziz Abdulla, Karlis Čerāns, Bengt Jonsson, and Tsay Yih-Kuen. General decidability theorems for infinite-state systems. In *Proc. 11th IEEE Int. Symp. on Logic in Computer Science*, pages 313–321, 1996. To appear in the journal of Information and Computation.

[AD90] R. Alur and D. Dill. Automata for modeeling real-time systems. In *Proc. ICALP '90*, volume 443 of *Lecture Notes in Computer Science*, pages 322–335, 1990.

[AJ96] Parosh Aziz Abdulla and Bengt Jonsson. Verifying programs with unreliable channels. *Information and Computation*, 127(2):91–101, 1996.

[AJ98a] Parosh Aziz Abdulla and Bengt Jonsson. Ensuring completeness of symbolic verification methods for infinite-state systems, 1998. To appear in the journal of Theoretical Computer Science.

[AJ98b] Parosh Aziz Abdulla and Bengt Jonsson. Verifying networks of timed processes. In Bernhard Steffen, editor, *Proc. TACAS '98, 4th Int. Conf. on Tools and Algorithms for the Construction and Analysis of Systems*, volume 1384 of *Lecture Notes in Computer Science*, pages 298–312, 1998.

[AN00] Parosh Aziz Abdulla and Aletta Nylén. Better is better than well: On efficient verification of infinite-state systems. In *Proc. 15th IEEE Int. Symp. on Logic in Computer Science*, pages 132–140, 2000.

[BD91] B. Berthomieu and M. Diaz. Modeling and verification of time dependent systems using time Petri nets. *IEEE Trans. on Software Engineering*, 17(3):259–273, 1991.

[Bow96] F. D. J. Bowden. Modelling time in Petri nets. In *Proc. Second Australian-Japan Workshop on Stochastic Models*, 1996.

[BS95] O. Burkart and B. Steffen. Composition, decomposition, and model checking of pushdown processes. *Nordic Journal of Computing*, 2(2):89–125, 1995.

[Čer94]   K. Čerāns. Deciding properties of integral relational automata. In Abite-
             boul and Shamir, editors, *Proc. ICALP '94*, volume 820 of *Lecture Notes
             in Computer Science*, pages 35–46. Springer Verlag, 1994.
[CES86]   E.M. Clarke, E.A. Emerson, and A.P. Sistla. Automatic verification of
             finite-state concurrent systems using temporal logic specification. *ACM
             Trans. on Programming Languages and Systems*, 8(2):244–263, April
             1986.
[dFERA00]   D. de Frutos Escrig, V. Valero Ruiz, and O. Marroquín Alonso. Deci-
             dability of properties of timed-arc Petri nets. In *ICATPN 2000*, number
             1825, pages 187–206, 2000.
[Dil89]   D.L. Dill. Timing assumptions and verification of finite-state concur-
             rent systems. In J. Sifakis, editor, *Automatic Verification Methods for
             Finite-State Systems*, volume 407 of *Lecture Notes in Computer Science*.
             Springer Verlag, 1989.
[Esp95]   J. Esparza. Petri nets, commutative context-free grammers, and basic
             parallel processes. In *Proc. Fundementals of Computation Theory*, vo-
             lume 965 of *Lecture Notes in Computer Science*, pages 221–232, 1995.
[Fin94]   A. Finkel. Decidability of the termination problem for completely speci-
             fied protocols. *Distributed Computing*, 7(3), 1994.
[GMMP91]   C. Ghezzi, D. Mandrioli, S. Morasca, and M. Pezzè. A unified high-level
             Petri net formalism for time-critical systems. *IEEE Trans. on Software
             Engineering*, 17(2):160–172, 1991.
[GS92]   S. M. German and A. P. Sistla. Reasoning about systems with many
             processes. *Journal of the ACM*, 39(3):675–735, 1992.
[GW93]   P. Godefroid and P. Wolper. Using partial orders for the efficient veri-
             fication of deadlock freedom and safety properties. *Formal Methods in
             System Design*, 2(2):149–164, 1993.
[Hen95]   T.A. Henzinger. Hybrid automata with finite bisimulations. In *Proc.
             ICALP '95*, 1995.
[Jan97]   P. Jančar. Bisimulation equivalence is decidable for one-counter proces-
             ses. In *Proc. ICALP '97*, pages 549–559, 1997.
[JLL77]   N. D. Jones, L. H. Landweber, and Y. E. Lyen. Complexity of some
             problems in Petri nets. *Theoretical Computer Science*, (4):277–299, 1977.
[JM95]   P. Jančar and F. Moller. Checking regular properties of Petri nets. In
             Proc. CONCUR '95, 6th Int. Conf. on Concurrency Theory, volume 962
             of *Lecture Notes in Computer Science*, pages 348–362. Springer Verlag,
             1995.
[JP93]   B. Jonsson and J. Parrow. Deciding bisimulation equivalences for a class
             of non-finite-state programs. *Information and Computation*, 107(2):272–
             302, Dec. 1993.
[KMM+97]   Y. Kesten, O. Maler, M. Marcus, A. Pnueli, and E. Shahar. Symbolic
             model checking with rich assertional languages. In O. Grumberg, editor,
             Proc. 9th Int. Conf. on Computer Aided Verification, volume 1254, pages
             424–435, Haifa, Israel, 1997. Springer Verlag.
[LPWY99]   K. G. Larsen, J. Pearson, C. Weise, and W. Yi. Efficient timed reacha-
             bility analysis using clock difference diagrams. In *Proc. 11th Int. Conf.
             on Computer Aided Verification*, 1999.
[LPY95]   Kim G. Larsen, Paul Pettersson, and Wang Yi. Model-checking for real-
             time systems. In Horst Reichel, editor, *Proceedings of 10th International
             Fundamentals of Computation Theory*, number 965 in LNCS, pages 62–
             88, Dresden, Germany, August 1995.

[LPY97]   K.G. Larsen, P. Pettersson, and W. Yi. Uppaal in a nutshell. *Software Tools for Technology Transfer*, 1(1-2), 1997.

[MF76]   P. Merlin and D.J. Farber. Recoverability of communication protocols - implications of a theoretical study. *IEEE Trans. on Computers*, COM-24:1036–1043, Sept. 1976.

[ML98]   Jesper Møller and Jakob Lichtenberg. Difference decision diagrams. Master's thesis, Department of Information Technology, Technical University of Denmark, Building 344, DK-2800 Lyngby, Denmark, August 1998.

[MLAH99]   Jesper Møller, Jakob Lichtenberg, Henrik R. Andersen, and Henrik Hulgaard. Difference decision diagrams. Technical Report IT-TR-1999-023, Department of Information Technology, Technical University of Denmark, February 1999.

[QS82]   J.P. Queille and J. Sifakis. Specification and verification of concurrent systems in cesar. In *5th International Symposium on Programming, Turin*, volume 137 of *Lecture Notes in Computer Science*, pages 337–352. Springer Verlag, 1982.

[RGdFE99]   V. Valero Ruiz, F. Cuartero Gomez, and D. de Frutos Escrig. On non-decidability of reachability for timed-arc Petri nets. In *Proceedings of the 8th Int. Workshop on Petri Net and Performance Models (PNPM'99)*, pages 188–196, 1999.

[RP85]   R. Razouk and C. Phelps. Performance analysis using timed Petri nets. In *Protocol Testing, Specification, and Verification*, pages 561–576, 1985.

[SBK92]   F. B. Schneider, Bloom B, and Marzullo K. Putting time into proof outlines. In de Bakker, Huizing, de Roever, and Rozenberg, editors, *Real-Time: Theory in Practice*, volume 600 of *Lecture Notes in Computer Science*, 1992.

[VW86]   M. Y. Vardi and P. Wolper. An automata-theoretic approach to automatic program verification. In *Proc. 1st IEEE Int. Symp. on Logic in Computer Science*, pages 332–344, June 1986.

[Wol86]   Pierre Wolper. Expressing interesting properties of programs in propositional temporal logic (extended abstract). In *Proc. 13th ACM Symp. on Principles of Programming Languages*, pages 184–193, Jan. 1986.

[Yov97]   S. Yovine. Kronos: A verification tool for real-time systems. *Journal of Software Tools for Technology Transfer*, 1(1-2), 1997.

CPN/Tools: A Post-WIMP Interface for Editing and Simulating Coloured Petri Nets

Michel Beaudouin-Lafon, Wendy E. Mackay, Peter Andersen, Paul Janecek,
Mads Jensen, Michael Lassen, Kasper Lund, Kjeld Mortensen, Stephanie Munck,
Anne Ratzer, Katrine Ravn, Søren Christensen, and Kurt Jensen

Department of Computer Science
University of Aarhus
IT-Parken, Aabogade 34
8200 Aarhus N - Denmark
cpn2000@daimi.au.dk

Abstract. CPN/Tools is a major redesign of the popular Design/CPN tool from
the University of Aarhus CPN group. The new interface is based on advanced,
post-WIMP interaction techniques, including bi-manual interaction, toolglasses
and marking menus and a new metaphor for managing the workspace. It chal-
lenges traditional ideas about user interfaces, getting rid of pull-down menus,
scrollbars, and even selection, while providing the same or greater functionality.
It also uses the new and much faster CPN simulator and features incremental
syntax checking of the nets. CPN/Tools requires an OpenGL graphics accelera-
tor and will run on all major platforms.

1 Introduction

Interaction techniques for desktop workstations have changed little since the creation
of the Xerox Star in the early eighties. The vast majority of today's interfaces are still
based on a single mouse and keyboard to manipulate windows, icons, menus, dialog
boxes, and to drag and drop objects on the screen. While these WIMP interfaces
(Windows, Icons, Menus, Pointing) are now ubiquitous, they are also reaching their
limits: as new applications become more powerful, the corresponding interfaces become more complex. Some users are at a breaking point and are less and less able to
cope with new software releases [11,12].

New interaction techniques, such as toolglasses [4] and marking menus [10], have
been proposed to reduce this trade-off between power and ease-of-use. Yet such post-
WIMP interaction techniques tend to be developed in isolation, as the focus of a particular research project. As a result, they have not made it into commercial tools even
though they have been shown to be significantly more efficient than traditional techniques. CPN/Tools is the first real-size application to combine such advanced interaction techniques into a consistent interface. The goal of this project is two-fold: first, it
will provide the CPN community with a new, cutting-edge interface to edit and simulate Coloured Petri Nets; second, it paves the way to a new generation of post-WIMP
applications that will take advantage of recent advances in graphical interfaces.

J.-M. Colom and M. Koutny (Eds.): ICATPN 2001, LNCS 2075, pp. 71–80, 2001.

The CPN2000 Project

CPN/Tools is a complete redesign of Design/CPN [9], a graphical editor and simulator of Coloured Petri Nets (CPNs) developed at Meta Software (USA) and the University of Aarhus (Denmark) over the past 10 years and a remote descendant of PeTriPote [1]. Design/CPN has a standard WIMP interface, based on direct manipulation, menus and dialog boxes. It is in use by over 600 organizations around the world, in both academia and industry. Production CPNs can have over a thousand places, transitions and arcs, structured into a hundred modules or more.

The CPN2000 project started in February 1999. We used a highly participatory design process, involving the users throughout the design process [14, 9]. Version 1 of CPN/Tools was released in April 2000 and is in use by a small group of CPN designers. Version 2 will be released to a selected set of users outside the project.

The CPN/Tools interface uses a combination of traditional, recent and novel interaction techniques, e.g. tool palettes, toolglasses, and magnetic guidelines. Integrating these interaction techniques together in a consistent way in a single tool proved quite challenging. To our knowledge, this had never been done before. We wanted to design a system that would strike a better balance between power and simplicity than current WIMP interfaces. This led us to define three design principles: reification, polymorphism and reuse [3]. Reification states that *any* entity in the interface should be accessible as a first-class object. Polymorphism states that commands should apply to as many different object types as possible. Reuse states that any output from the system and any input to the system should be reusable later, e.g. in the form of macros.

The resulting interface has no menu bars, no pull-down menus, no scrollbars, no dialog boxes and no notion of selection. Instead, it uses a unique combination of floating palettes, toolglasses and hierarchical marking menus, a novel windowing model based on pages and binders, and several new interaction techniques such as magnetic guidelines to align objects and bi-manual interaction to manipulate objects. This interface supports the same or higher level of functionality as the previous Design/CPN application, yet we have empirical evidence [14, 9, 3] that it is both simpler to use and more powerful.

The rest of this article presents the CPN/Tools interface and outlines the design process and implementation. The design process is further described in [14, 9, 3] and implementation details and performance data can be found in [2].

2 The CPN/Tools Interface

The CPN/Tools interface requires a traditional mouse and keyboard, plus a trackball (or other locator) for the non-dominant hand. For simplicity, we assume a right-handed user, but the mouse and trackball can be swapped for left-handed users. The keyboard is used only to input text and to navigate within and across text objects. The design of the bi-manual interaction follows Guiard's Kinematic Chain theory [7] in which the left hand manipulates the context (container objects such as windows and toolglasses) while the right hand manipulates objects within that context. The exception is direct interaction for zooming and resizing, which, according to Casalta et al.

[6], should give both hands symmetrical roles. CPN/Tools incorporates six primary interaction techniques: direct and bi-manual interaction, marking menus [11], keyboard input, floating palettes, and toolglasses [5].

Direct manipulation (i.e. clicking or dragging objects) is used for frequent operations such as moving objects, panning the content of a view and editing text. When a tool is held in the right hand, e.g. after having selected it in a floating palette, direct manipulation actions are still available via a long click, i.e. pressing the mouse button, waiting for a short delay (200ms) until the cursor changes, and then either dragging or releasing the mouse button. Because of the visual feedback, this multiplexing of tools in the right hand is easily understood by users.

Bi-manual manipulation is a variant of direct manipulation that involves using both hands for a single task. It is used to resize objects (windows, places, transitions, etc.) and to zoom the content of a page. The interaction is similar to holding an object with two hands and stretching or shrinking it. Bi-manual interaction could also be used to control the orientation and position of an object. This might be used in the future to control the orientation of our magnetic guidelines (see below).

Marking menus are radial, contextual menus that appear when clicking the right button of the mouse. Marking menus offer faster selection than traditional linear menus for two reasons. First, it is easier for the human hand to move the cursor in a given direction than to reach a target at a given distance. Second, the menu does not appear when the selection gesture is executed quickly, which supports a smooth transition between novice and expert use. Kurtenbach and Buxton [11] have shown that selection times can be more than three times faster than with traditional menus. Hierarchical marking menus involve more complex gestures but are still much more efficient than their linear counterparts.

Keyboard input is used only to edit text. Some navigation commands are available at the keyboard to make it easier to edit several inscriptions in a row without having to move the hands to the mouse and trackball. Keyboard modifiers and shortcuts are not necessary since most of the interaction is carried out with the two hands on the locator devices.

Floating palettes contain tools represented by buttons. Clicking a tool with the mouse activates this tool, i.e. the user conceptually holds the tool in his or her hand. Clicking on an object with the tool in hand applies the tool to that object. In many current interfaces, after a tool is used (especially a creation tool), the system automatically activates a "select" tool. This supports a frequent pattern of use in which the user wants to move or resize an object immediately after it has been created but causes problems when the user wants to create additional objects of the same type. CPN/Tools avoids this automatic changing of the current tool by getting rid of the notion of selection (see below) while ensuring that the user can always move an object, even when a tool is active, with a long click (200ms) of the mouse. This mimics the situation in which one continues holding a physical pen while moving an object out of the way. Floating palettes also support bi-manual interaction: the tool held in the right hand can be selected in the palette with the left hand, saving round trips to the palette. The floating palette can also be held in the left hand and moved next to the work area with the left hand, minimizing the time it takes to select a tool.

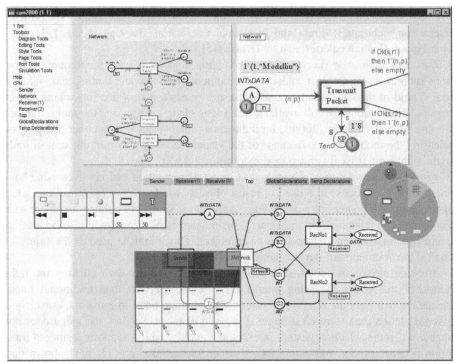

Fig. 1. The CPN/Tools interface. The index is in the left column. The upper-right binder contains a page with the simulation layer active. The upper-left binder contains a view of the same page, at a different scale. The lower binder contains six pages: the page on top shows several magnetic guideline (dashed lines). The VCR-like controls to the left belong to the simulation floating palette. The toolglass at the bottom is positioned over objects on the page and is ready to apply any of the attributes shown. To the right a circular, hierarchical marking menu has been popped up on the page and is ready to accept a gesture to invoke one of the commands displayed

Toolglasses, like floating palettes, contain a set of tools represented by buttons. Unlike floating palettes, they are semi-transparent and are moved with the left hand. A tool is applied to an object with a click-through action: The tool is positioned over the object of interest and the user clicks through the tool onto the object. The toolglass disappears when the tool requires a drag interaction, e.g., when creating an arc. This prevents the toolglass from getting in the way and makes it easier to pan the document with the left hand when the target position is not visible. This is a case where the two hands operate simultaneously but independently.

Since floating palettes and toolglasses both contain tools, it is possible to turn a floating palette into a toolglass and vice versa, using the right button of the trackball. Clicking this button when a toolglass is active drops it, turning it into a floating palette. Clicking this same button on a floating palette picks it up, turning it into a toolglass.

None of the above interaction techniques requires the concept of selection. All are contextual, i.e. the object of interest is specified as part of the interaction. For groups of objects, this requires some concept of group representation to specify which group to interact with. In Version 2 of CPN/Tools, we are incorporating features to create groups, including dynamic groups resulting from a search, and we are looking at different ways of addressing this issue. Also, some features of the interface, such as magnetic guidelines, described below, reduce the need to work with groups.

Preliminary results from our user studies [14, 9] make it clear that none of the above techniques is always better or worse. Rather, each emphasizes a different, but common, pattern of use by using a different syntax:

- *object-then-command*: point at the object of interest, then select the command from a contextual marking menu;
- *command-then-object*: select a command by clicking a tool in a floating palette, then apply the tool to one or more objects of interest;
- *command-and-object*: select the command and the object simultaneously by clicking through a toolglass or moving it directly.

As a result, marking menus work well when applying multiple commands to a single object. Floating palettes work well when applying the same command to different objects. Toolglasses work well when the work is driven by the structure of the application objects, such as working around a cycle in a Petri net.

The Workspace Manager

Coloured Petri Nets frequently contain a large number of modules. In the existing Design/CPN tool, each module is presented in a separate window and users spend time switching among them. In CPN/Tools we have designed a new window manager to improve this situation: the Workspace Manager.

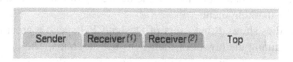

Fig. 2. Tabs for the pages in a binder

The workspace occupies the whole screen (figure 1) and contains window-like objects called *binders*. Binders contain *pages*, each equivalent to a window in a traditional environment. Each page has a tab similar to those found in tabbed dialogs (figure 2). Clicking the tab brings that page to the front of the binder. A page can be dragged to a different binder with either hand by dragging its tab. Dragging a page to the background creates a new binder for it. Dragging the last page out of a binder removes the binder from the screen. Binders reduce the number of windows on the screen and the time spent organizing them. Binders also help users organize their work by grouping related pages together and reducing the time spent looking for hidden windows.

CPN/Tools also supports multiple views, allowing several pages to contain a representation of the same data. For example, the upper-right page in figure 1 shows a module with simulation information, while the upper-left page shows the same module without simulation information and at a smaller scale.

The left part of the workspace is called the *index* (top left, figure 1) and contains a hierarchical list of objects that can be dragged into the workspace with either hand. Objects in the index include toolglasses, floating palettes and Petri net modules. Dragging an entry out of the index creates a view on its contents, i.e. a toolglass, a floating palette or a page holding a CPN module.

Pages and binders do not have scrollbars. If the content of a page is larger than its size, it can be panned with the left button of the trackball, even while the right hand is using the mouse to, for example, move an object or invoke a command from a marking menu. Getting rid of scrollbars saves valuable space but makes it harder to tell how much of the whole document is being displayed. A future version will use the borders of the page to show what portion of the document is viewed in a non-intrusive, space-saving way.

Resizing a binder and zooming the contents of a page involves direct bi-manual interaction (as described above). Unlike traditional window management techniques, using two hands makes it possible to simultaneously resize and move a binder, or pan and zoom the contents of a page at the same time. Clicking the right button of the mouse on the page tab or on the binder pops up a contextual marking menu with additional commands to close, collapse, expand the page or create a new page with the same content.

Creating and Laying out Objects

Creation tools are accessible via the following interaction techniques: The user may select the appropriate object from the floating palette, move to the desired position and click, or use the left hand to move the toolglass to the desired position and click-through with the right hand, or move to the desired location and make the appropriate gesture from the marking menu.

Our user studies showed that users of Design/CPN spend a great deal of time creating and maintaining the layout of their Petri net diagrams. The primary technique is a set of *align* commands, similar to those found in other drawing tools. The limitation is that they align the objects at the time the command is invoked, but do not remember that those objects have been aligned. We observed that most users use the same pattern to move an object: They manually select all objects aligned to the object of interest and move them as a group. This dramatically slows down the interaction.

In order to facilitate the alignment of objects, we have introduced horizontal and vertical *magnetic guidelines*. Guidelines are first-class objects that are created in the same way as the elements of the Petri net model, i.e. with tools found in a palette/toolglass or in a marking menu. Guidelines are displayed as dashed lines (figure 1) and are magnetic. Moving an object near a guideline causes the object to snap to the guideline. Objects can be removed from a guideline by clicking and dragging them away from the guideline. Moving the guideline moves all the objects that are snapped to it, thus maintaining the alignment. An object can be snapped simultaneously to a horizontal and a vertical guideline.

We have designed, and will implement, additional types of guidelines. For example, rectangular or elliptical guidelines would make it easier to layout the cycles commonly found in Petri nets. We also plan to support spreading or distributing objects over an interval within a line segment, since this is a common layout technique. Adding these new types of guidelines may create conflicts when an object is snapped to several guidelines. One solution is to assign weights to the guidelines and satisfy the alignment constraints of the guidelines with heaviest weight first. Such conflicts do not exist in the current system because only horizontal and vertical guidelines are available.

Editing Attributes

The tools to edit the graphical attributes of the CPN elements are grouped in a palette/toolglass that contains five rows (figure 3): two rows of color swatches, a row of lines with different thicknesses, a row of lines with different dash patterns and a row for user-defined styles. The first four rows are fairly standard and are not described further here.

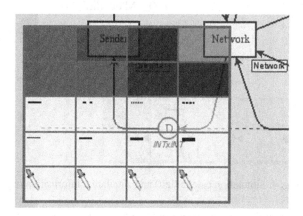

Fig. 3. Toolglass for editing attributes

Tools in the last row correspond to the reification of groups of graphical attributes into styles. Initially, each tool in this row is a *style picker*. Applying this tool to an object copies the object's color and thickness into the tool and transforms the tool into a *style dropper*. Applying a style dropper to an object assigns the tool's color and thickness to that object. Applying a style dropper to the background of the page empties it and turns it into a style picker. If this is done by mistake, the *undo* command restores its previous state. In practice, style pickers and style droppers make it very easy and efficient for users to define the styles they use most often and apply them to objects in the diagram.

In Version 2 objects will remember which style they belong to (like in, e.g., Microsoft Word) and it will be possible to edit the attributes of a style in the toolglass itself. This will affect all the objects that use this style, saving repetitive editing.

Simulation Tools

Once a CPN model has been created, the developer runs simulations to validate it. CPN/Tools uses the new simulator developed by the University of Aarhus CPN group [8], which is up to 1000 times faster than the previous one used by Design/CPN. The simulator runs as a separate process and communicates with the tool asynchronously over a TCP/IP network connection.

CPN/Tools displays simulation information in a *simulation layer* that can be added to any page via a tool in a palette, toolglass or marking menu. When the simulation layer is active, the background color of the page changes, the number of tokens are displayed as small green disks, the token colors are displayed as yellow text annotations, and enabled transitions are displayed with a green halo (figure 4). Each of these types of feedback can be toggled on or off using the tools in the simulation palette or toolglass (figure 4, top row of the palette).

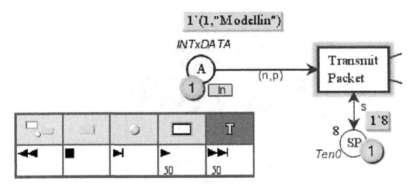

Fig. 4. Simulation palette (left) and simulation information (right)

Running the simulation involves compiling the net into ML code based on the structure of the net and the text inscriptions. The net is checked incrementally as the user edits it. This saves a lot of time compared with Design/CPN where switching to simulation mode could take several minutes. Checking and simulating the net may result in syntax errors and run-time errors. In both cases, error messages are displayed as red "bubbles" next to the location of the error. The object that caused the error has a red halo. Since the error may occur in a page that is not on top, the red halo also appears in the tab of any page that has an error.

CPN/Tools uses a video tape player metaphor to control the simulation (figure 4, bottom row of the palette). *Next frame* lets the user select a transition to fire. *Play* randomly fires enabled transitions until a deadlock is reached or the user hits the *stop* button. *Fast-forward* runs the simulation at full speed for a maximum number of steps set by the user, displaying only the final state. *Rewind* resets the net to its initial state. The *Next frame* command is polymorphic: If applied to an enabled transition, it fires that transition. If applied to a page, it fires a randomly selected transition within the page. If applied to a binder or to the workspace, it fires a randomly selected transition within the pages of the binder or the whole model, respectively.

3 Design Process and Implementation

The design process of CPN/Tools followed an incremental and user-centered approach. We studied both novice and expert users of Design/CPN and identified a number of areas for improvement, e.g. window management, layout, management of graphical attributes, alignment, interface with the simulator. We then worked with a group of users to design the new system together [9]. We showed them examples of novel interaction techniques such as toolglasses or pen input. We conducted brainstorming sessions to generate ideas for the new design. We used short scenarios such as creating editing an existing net or laying out a net created by someone else to inform the design of the new system. We organised workshops with users where we created paper prototypes of the new system and videotaped them [14]. These video prototypes became precise enough to be used by the developers to implement the system: they served as specification of what the interface should do. Finally, when the first version of the CPN/Tools was functional, we conducted user studies to evaluate the design. These studies showed that the new interface was easy to learn and confirmed that the preferred interaction technique changes according to the user and the context of use. This led us to integrate palettes, toolglasses and marking menus even further into a consistent interface rather than select a single technique [3].

The implementation of the system is based on a software architecture with three main components [2]: document management, input and rendering. This architecture is generic and could be reused for other applications. The simulator runs in a separate process and communicates with the editor using an asynchronous protocol. Documents represent top-level objects such as CPN diagrams, tool palettes and the index. Documents can be saved to disk in XML. This allows users to exchange CPN diagrams as well as tool palettes, supporting the customization of the system. Input management implements the interaction techniques. It manages a set of interaction instruments, which modify the document structure when activated by the user. Rendering is in charge of displaying the document structure after it has been modified. We decided to use OpenGL for rendering since hardware-accelerated graphics card are becoming cheaper and faster. OpenGL allows us to use advanced graphical effects, e.g. transparency. It also supports portability: CPN/Tools will run with the same code base on Windows, Unix/Linux and MacOs.

4 Conclusion and Future Work

We have described the interface of CPN/Tools and shown how it supports a combination of advanced interaction techniques in a post-WIMP interface. Version 1 is functional and already in use by a small group of users. We are currently working on the next version that will incorporate new features and improvements, based on the same design principles and overall approach.

Version 2 will support groups, specified either explicitly by designating the objects in the group or indirectly through a query, e.g. to find all places with a given color set. The interface will be customizable: users will be able to compose their own palettes/toolglasses and exchange them with other users. Styles and guidelines will be improved, and context-sensitive help will be available throughout the interface. We are looking forward to the release of CPN/Tools to the wider Petri Nets community to collect valuable feedback for the next iteration of the design.

Acknowledgments. We thank the members of the CPN group at the University of Aarhus for their participation in the design process. This work is supported by the University of Aarhus, the Danish Centre for IT Research (CIT), Hewlett-Packard and Microsoft Research.

References

1. Beaudouin-Lafon, M. PeTriPote: a graphic system for Petri-nets design and simulation. *Proc. 4th European Workshop on Applications & Theory of Petri-Nets*, 1983, p. 20-30.
2. Beaudouin-Lafon, M. & Lassen, M. The Architecture and Implementation of CPN2000, a Post-WIMP Graphical Application. In *Proc. ACM Symposium on User Interface Software and Technology*, UIST 2000, CHI Letters 2(2):181-190, ACM, 2000.
3. Beaudouin-Lafon, M. & Mackay, W. Reification, Polymorphism and Reuse: Three Principles for Designing Visual Interfaces. In *Proc. Conference on Advanced Visual Interfaces*, AVI 2000, Palermo, Italy, May 2000, p. 102-109.
4. Beaudouin-Lafon, M. Instrumental Interaction: An Interaction Model for Designing Post-WIMP User Interfaces. In *Proc. Human Factors in Computing Systems*, CHI'2000, CHI Letters 2(1):446:453, ACM Press, 2000.
5. Bier, E., Stone, M., Pier, K., Buxton, W., De Rose, T. Toolglass and Magic Lenses : the See-Through Interface. In *Proc. ACM SIGGRAPH*, ACM Press, 1993, p.73-80.
6. Casalta, D., Guiard, Y. and Beaudouin-Lafon, M. Evaluating Two-Handed Input Techniques: Rectangle Editing and Navigation. In *Proc. ACM Human Factors In Computing Systems*, CHI'99, Extended Abstracts, 1999, p. 236-237.
7. Guiard, Y. Asymmetric division of labor in human skilled bimanual action: The kinematic chain as a model. *Journal of Motor Behavior*, 19:486-517, 1987.
8. Haag, T.B. and Hansen, T.R. *Optimising a Coloured Petri Net Simulator*. Master's Thesis, University of Aarhus (Denmark), December 1994.
9. Janecek, P., Ratzer, A., and Mackay, W. Redesigning Design/CPN: Integrating Interaction and Petri-Nets-In-Use. In *Proc. International Workshop on Coloured Petri Nets*, Aarhus, Denmark, 1999.
10. Jensen, K. *Coloured Petri Nets*: Basic Concepts (Vol. 1, 1992), Analysis Methods (Vol. 2, 1994), Practical Use (Vol. 3, 1997). Monographs in Theoretical Computer Science. Springer-Verlag, 1992-97.
11. Kurtenbach, G. & Buxton, W. User Learning and Performance with Marking Menus. In *Proc. Human Factors in Computing Systems*, CHI'94, ACM, 1994, p. 258-264.
12. Mackay, W.E. *Users and Customizable Software: A Co-Adaptive Phenomenon*. Ph.D. Dissertation, Massachusetts Institute of Technology, 1990.
13. Mackay, W.E. Triggers and barriers to customizing software. In *Proc. ACM Human Factors in Computing Systems*, CHI'91, ACM Press, 1991, p. 153-160.
14. Mackay, W., Ratzer, A. & Janecek, P. Video Artifacts for Design: Bridging the Gap between Abstraction and Detail. In *Proc. ACM Conference on Designing Interactive Systems*, DIS 2000, New York, August 2000, p. 72-82.

Petri Net Based Design and Implementation Methodology for Discrete Event Control Systems

Slavek Bulach[1], Anton Brauchle*, Hans-Jörg Pfleiderer[1], and Zdenek Kucerovsky[2]

[1] Department of Microelectronics
University of Ulm, Ulm, D-89069, Germany
sbulach@mic1.e-technik.uni-ulm.de
[2] Department of Electrical and Computer Engineering
Faculty of Engineering Science
University of Western Ontario
London, Ontario, N6A 5B9, Canada

Abstract. This paper presents an embedded Discrete Event System (DES) design and realization methodology combining the advantages of system modeling based on the Petri Net (PN) formalism and implementational efficiency of a proposed dedicated programmable event-driven controller. A DES is initially modeled as communicating plant and controller nets which concisely capture concurrent behavior of the system and yield themselves to formal analysis techniques. The control specifications are subsequently compiled into the compact executable binary code according to the lean net encoding scheme and stored in a commercially available programmable parallel read only memory (PROM). The controller executes the PN control code in an event-driven manner responding to the external events and concurrently tracking multiple execution threads. The 8-bit prototype of the controller has been fabricated in 0.35 μm CMOS technology. Operating at 80 MHz it delivers fast response times, power efficiency and transition firing rates of up to 4 million transitions per second.

1   Introduction

Numerous industrial and consumer oriented electronic systems operate in environments where interaction with a user or another system is an important issue. Many of them perform specific control functions, exhibit interactive or reactive behavior and are implemented as *embedded systems*. Embedded systems are treated here as a special implementation case of a broader class of systems known as *Discrete Event Systems (DES)*. One of the firmly established formalisms used to design and analyze such systems is a family of formalisms known as *Petri Nets (PN)*. Though modeling of systems using PN has been pursued extensively [13,15], their implementations in hardware have received

---

* A. Brauchle has been with the Department of Microelectronics, University of Ulm. He is currently with SOREP Electronic Engineering GmbH, Munich.

J.-M. Colom and M. Koutny (Eds.): ICATPN 2001, LNCS 2075, pp. 81–100, 2001.

much less attention. This paper addresses modeling of embedded discrete event control systems and presents a *Petri Net Decision Unit (PNDU)* which is a dedicated programmable controller designed solely to process control tasks specified in terms of Petri Nets. The PNDU is a single chip device fabricated in the year 2000 and according to the literature survey appears to be the first integrated circuit of its kind. This paper is the first publication presenting the measured performance results of the PNDU.

The paper first gives an overview of different kinds of hardware implementations of Petri Net based controllers. Section 3 defines embedded DES and relates them to Petri Nets. Section 4 gives a formal definition of the executable Petri Net control specifications and presents the proposed PN-based system design methodology. Section 5 covers the compilation of the control specifications and outlines architectural features of the PNDU.

2   Hardware Implementations of Petri Nets: An Overview

Just within ten years after their inception by Carl Adam Petri in 1962 [14], Petri Nets were already applied to the modeling of asynchronous event-driven hardware structures [11]. Since that time, PN have been used in the design of systems at different levels of abstraction. Hardware structures manifest themselves in many different types of implementations. It is worthwhile to distinguish *hardwired implementations* from *programmable* ones, since only programmable hardware implementations are relevant to the subject of this paper.

2.1   Hardwired Implementations Based on Petri Nets

Hardwired implementations are realizations of a particular function or algorithm in a digital electronic system such that the system may only be used for one specified application. The executing hardware resources are fixed and the algorithm is also fixed as it is mapped directly into hardware. The very first attempt at mapping the executable Petri Net specifications directly in hardware was based on speed-independent switching circuits which operate without a global clock signal [11]. This initial work was subsequently applied to the design of an asynchronous logic array, which was already a programmable implementation [12].

The pioneering work on speed-independent circuits [11] was later taken up by numerous research groups. Today, there is a maturing research field known as *design of asynchronous circuits and systems* where special classes of Petri Nets are used to model asynchronous logic or system behavior. Remarkably, there exist research software tools which synthesize *self-timed* digital circuits specified in terms of labeled Petri Nets. A comprehensive overview of the hardwired digital circuit design based on Petri Nets, latest research achievements in this field and a list of software tools is given in [17].

2.2   Programmable Implementations Based on Petri Nets

Programmable implementations are realizations of an algorithm in a digital system such that the system itself may be used for more than one application. The computing resources are fixed; however, the algorithm is stored in a semiconductor memory and may be altered upon requirements. All system realizations based on processors, microprocessors or processor-like controllers are programmable implementations. Two kinds of programmable implementations are realizations using *commercially available* devices and *custom designed* or *dedicated* ones.

Here is a brief overview of programmable implementations based on commercially available computing platforms. The list is by no means exhaustive as it is only used to emphasize the variety of choices. For example, Programmable Logic Controllers (PLC) able to execute Petri Net control specifications were proposed in [16]. The PLC used processor cards with ZILOG 80A microprocessors. A controller for a machining workstation employing a VAX 11/780 was described in [5]. A low cost programmable controller suitable for real-time applications was presented in [10]. It was a single-chip design using the INTEL 8031 microcontroller. All these implementations were ultimately restricted by a general-purpose computing architecture of commercial devices which do not inherently behave in an event-driven manner. Since the computing hardware was provided, the remaining task was to properly configure devices. The main challenge in such implementations was to design software that would meet performance specifications.

Custom Programmable Controllers. Substantially more efficient hardware implementations are achieved with dedicated Petri Net based programmable controllers. This is due to special attention given simultaneously to both hardware and software aspects of the implementation. Since at the starting point of the design larger degrees of freedom are provided for hardware and software the final solution is expected to be more efficient compared to implementation solutions deploying off-the-shelf devices.

For example, a programmable controller based on a custom ASIC (Application Specific Integrated Circuit) memory was estimated to perform two orders of magnitude faster as compared to microprocessor-based implementations [8]. The executable specification based on Petri Nets was mapped to memory in a tabular form. The table had to have a fixed number of places and transitions corresponding to the desired net size. A special *fire unit* performed transition fire checks and actual firing in response to the incoming input events. However, at the heart of this implementation and also its subsequent proposals was the requirement to use ASIC memory with a very wide word. Furthermore, the encoding of the net demanded a substantial amount of memory per place-transition combination. Manipulating large amounts of data on wide buses consumes much power and custom chips with a large pin count are expensive. The cost effectiveness additionally suffers from the fact that the memory must be custom manufactured.

An independently developed concept proposing to utilize commercial semi-conductor memory in conjunction with a dedicated programmable controller was first mentioned in [1]. Here, an executable PN control specification was compiled into machine code employing a very lean net encoding scheme. The code could be programmed into a conventional EPROM (Electrically Programmable Read Only Memory) or EEPROM (Electrically Erasable PROM). The controller, known as a Petri Net Decision Unit (PNDU), was proposed to perform a *token player* or *Petri Net interpreter* function directly in hardware (Fig. 1c). Its responsibilities were to respond to the external stimuli in an event-driven manner. First, the appropriate place-transition block of information was loaded from the PROM. Subsequently, the decision which transitions (if any) should fire was made based on the validity of the input signals. Fired transitions produced new output signals. Interfacing the controller with conventional memory offered a high degree of flexibility because the external PROM could be easily re-programmed and the controller provided the computing resources necessary for executing the Petri Net control specifications.

The major limitation of this first PNDU version was the fact that the Petri Nets were restricted to the *state machine* class. This implied, first of all, that the hardware of the PNDU was relatively simple. However, complex nets had to be transformed before compilation. Concurrent execution traces or processes had to undergo parallel composition of their individual state spaces. Therefore, software size grew exponentially with respect to the number of concurrent processes.

These limitations were addressed in the second version of the PNDU which was designed to process *free choice* nets [2]. This required more intelligent hardware, namely, a larger circuit. However, now the PNDU could directly process both *marked graphs* and *state machines*. The amount of pre-compilation transformations was reduced and the concurrency was explicitly exploited in the net encoding. As a result, software size grew linearly with respect to the number of concurrent processes. The PNDU locally tracked movement of concurrent tokens in the net and fired appropriate transitions in response to the incoming external events. At any given time, it processed only *marked* or *active* places thus avoiding the need to consider the whole net at once.

The functionality of the PNDU was extended even further in its third version proposed in [3]. In the sequel, any reference to the PNDU will be understood to refer to the third version unless otherwise explicitly stated. The improved hardware architecture of the PNDU now allows direct processing of all possible Petri Net constructs belonging to *state machines*, *marked graphs*, *free choice*, *extended free choice*, *asymmetric choice* classes, and beyond. The degree of pre-compilation transformations is minimal and the explicit support of concurrency ensures that the implementation of complex interactive control behavior is as simple as possible. A lean net encoding scheme allocates 4 to 8 memory lines (16/32/64 bits/line) per *transition block*. Consequently, the software is very compact as its size is directly proportional to the number of transitions in the net. Section 5 describes hardware and software architectural features of the PNDU.

3   Embedded Discrete Event Systems

Discrete Event Systems are dynamic time-invariant non-linear systems [4]. Their
state space is a discrete set. State transitions are event-driven and are assumed
to be instantaneous. They are triggered in response to asynchronously occurring
discrete events. The state remains unchanged until new events arrive. A discrete
event *control* system is typically modeled as a *plant* (or *environment*) and a
controller as shown in Fig. 1a). This configuration will be referred to as an
aggregate system. An event produced by the plant is an action which takes place
within the plant and is referred to as a *plant event*. An event occurring within a
controller is called a *controller event*.

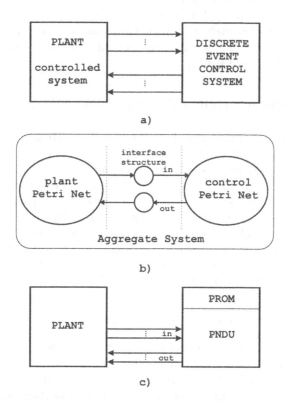

Fig. 1. A Discrete Event System from conception to implementation

In a *reactive* DES, the controller has an ongoing interaction with the plant [6].
It continuously receives plant events through *sensors*, processes them, keeps track
of the aggregate system's state, and produces controller events which are communicated back to the plant through *actuators*. It is assumed that the system
is always able to react to plant events and it responds fast enough such that
the plant is still receptive to controller events. An *implementation* of a system

is a realization of an abstraction in terms of hardware, software, or both. Reactive applications requiring strict timing constraints to be met are referred to as *real-time systems*. When an implementation is restricted by the number of input/output channels, performance, size, cost, and power consumption the system is said to be an *embedded system*.

This paper considers embedded implementations of reactive *untimed* discrete event control systems. They will be assumed to fulfill *soft real-time* requirements, such that the correct operation of the system is sustained even if there is an occasional failure to meet the deadline. Untimed DES concentrate on logical behavior of the system disregarding at which point in time a transition to a given state is triggered, how long a transition itself lasts or how long the system remains in that state. The sequence of input events arriving from the plant is denoted as $\{e_{p1}, e_{p2}, e_{p3}, ..., e_{pn}\}$ without precise reference to the exact arrival time. Such systems are adequately represented by state automation and Petri Nets. However, Petri Nets are inherently superior for problems involving interactive concurrent behavior (see Fig. 2).

4   Petri Nets for Embedded Systems

One of the popular definitions of Petri Nets as Place-Transition nets (PT-nets) is: $PN = (P, T, A, W, M_0)$, where P is a finite set of places, T is a finite set of transitions, A is a finite set of arcs determining a flow relation, W is a weight function, M_0 is the initial marking [9,13]. A set of input/output places to/from transition t_j are denoted by $I(t_j)$ and $O(t_j)$, respectively. A set of input/output transitions to/from place p_i are given as $I(p_i)$ and $O(p_i)$, respectively. A PN is called *ordinary* if all of its arc weights are 1's. A PN is *safe* if for all firing sequences a place may contain no more than one token. A PN where for all possible markings the total number of tokens in the net never exceeds some upper limit K will be referred to as *capacity-K conservative*.

4.1   Modeling Power versus Decision Power

Modeling power of a PN is its ability to properly and efficiently represent real systems or processes. It is inversely proportional to the *model size*. *Decision power* of a PN is inversely related to the complexity of computation and the amount of *computing resources* required to determine various properties and to *execute* the net. To *execute* a PN means to perform a control function described in terms of the PN. Structural restrictions produce new *subclasses* of PN, reduce modeling power of the net, lose important nuances of a real system and increase the model size. However, they increase decision power, ease the analysis and execution of the net. *Extended* Petri Nets, such as *High-level Petri Nets* (HLPN), better capture fine details of a real system and decrease the model size. At the same time, more intelligent computing resources are required to analyze and execute them. The relationships between modeling and decision powers, computing resources and model size are intuitively depicted in Fig. 2. The conflicting objectives in the design of embedded systems are to maximize the modeling and decision powers, and to minimize the model size and the amount of computing hardware.

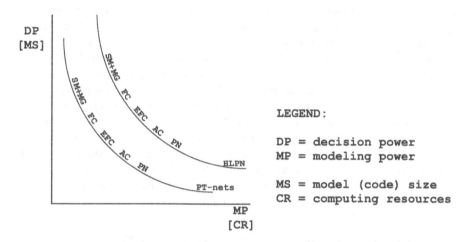

Fig. 2. Modeling power versus decision power of Petri Nets

Petri Net Classes and Extensions. Based on their structural features five subclasses of *general* Petri Nets are distinguished. Arranged in ascending order they are: *state machines* (SM) and *marked graphs* (MG) → *free choice* (FC) → *extended free choice* (EFC) → *asymmetric choice* (AC) → *general* Petri Nets [9,13]. State machines model *choice* situations through their *conflict* ($|O(p_i)| > 1$) and *merge* ($|I(p_i)| > 1$) constructs. They can equivalently represent state automation. Marked graphs model *synchronization* and *production* through their *fork* ($|O(t_j)| > 1$) and *join* ($|I(t_j)| > 1$) constructs. Free choice nets comprise both state machines and marked graphs but no transition may simultaneously be part of *choice* and *join* constructs. Extended Free Choice relax this restriction. A PN construct where *conflict* and *synchronization* are mixed together is known as a *confusion*. FC nets admit no confusion, AC nets allow *asymmetric confusion* and general PN permit *symmetric confusions*. Figure 2 illustrates that transformations from a higher to the lower PN class increase the net size. Using state automation to model concurrent behavior ultimately results in larger nets. In order to minimize the model size the PN executing hardware must be able to process all PN classes.

Extensions which increase PN modeling power (Fig. 2), and simplify modeling and implementation of DES control algorithms are: *token colour, transition guards, priorities* of conflicting transitions and *time delay expressions* [7].

4.2   Modeling Plant and Controller as Petri Nets

The complete logical behavior of an *aggregate system* consisting of a *plant* and a *controller* is captured by an *aggregate Petri Net* consisting of a *plant Petri Net, interface structure* and a *control Petri Net* (Fig. 1b). It is convenient to consider their behavior separately, but to allow an on-going interaction between them through the *input* and *output* communication channels. In each net, events are represented by transitions and places are associated with transition enabling conditions. For an event to occur certain *preconditions* which depend on the state of the aggregate system must be *true* or *valid*. When an event occurs, current preconditions become invalid and a set of *postconditions* is produced.

Interface arcs represent collections of signal wires or *buses* between the plant and the controller, while *interface places* represent *registers*. The two nets exchange tokens which are boolean logic valued vectors. An interface place can hold at most one *interface input* or *output token* since at any given instant a bus can only hold one value. Hence, interface places must be *safe* and interface arcs must be *ordinary*. *Plant tokens* and *controller tokens* model *execution threads* within the plant and the controller, respectively. Thus, an aggregate PN must be *ordinary* and *safe* in order to have a real meaning.

Control Executable Petri Net. An *Aggregate Petri Net* (APN) is formally defined as a tuple:

$$APN = (cePN, IS, pPN), \text{ where} \tag{1}$$

pPN is a plant Petri Net,
IS is an interface structure,
$cePN$ is a control executable Petri Net.

A *control executable Petri Net* (cePN) is defined at the level where it can be directly compiled into the binary executable format. It is a tuple:

$$cePN = (C, P, T, A, \kappa, \pi, Pre, Post, S, M_0), \text{ where} \tag{2}$$

C is a set of *control colours*, $C = \{control, iin, iout, state\}$, where
    $control \in B = \{0, 1\}$, is of boolean type;
    $iin = [i_1 i_2 i_3 ... i_i]$ is a boolean vector corresponding to the product of interface input variables, $|iin| = $ number of input channels; each $i_j \in B = \{0, 1\}$, is of boolean type;
    $iout = [o_1 o_2 o_3 ... o_o]$ is a boolean vector corresponding to the product of interface output variables, $|iout| = $ number of output channels; each $o_j \in B = \{0, 1\}$, is of boolean type;
    $state = [\sigma_1 \sigma_2 \sigma_3 ... \sigma_\sigma]$ is a boolean vector corresponding to the product of internal state variables, $|state| = $ number of bits in the internal state registers; each $\sigma_j \in B = \{0, 1\}$, is of boolean type;
$P = \{p_1, p_2, p_3, ..., p_p\}$ is a finite set of *control places*;
    p_{start} is a special starting source place, $I(p_{start}) = \emptyset$;

$T = \{t_1, t_2, t_3, ..., t_t\}$ is a finite set of *control transitions*;

$T_{dummy} \subset T$ is an optional set of *dummy* transitions required for correct resolution of asymmetric confusion constructs;

$A \subseteq (P \times T) \cup (T \times P)$ is a finite set of *control arcs* determining a flow relation; it is assumed that arcs transfer tokens only of the colour associated with their corresponding input or output places;

$\kappa : P \to C$ is a *colour function* defined from P into C, mapping each place to a specific type;

π is a *priority function* which must be specified for each output arc of a place:

$$\forall A \in O(p_j),\ \pi : A \to \{1, 2, 3, ..., m\}\ \text{if}\ |O(p_j)| = m,\ \text{and}$$
$$\pi(A(O(p_m))) = \pi(A(O(p_n))) = a\ \text{for each}\ t_a \in (O(p_m) \cap O(p_n))\ ; \tag{3}$$

$Pre = [g_1 g_2 g_3 ... g_g]$ is a *guard* or *precondition* expressed as a product of variables, such that $g_j \in B_x = \{0, 1, x\}$, where B_x is a boolean type extended with a *don't care* value x; note that $|Pre| = |iin| + |state|$;

$Post = [\varsigma_1 \varsigma_2 \varsigma_3 ... \varsigma_\varsigma]$ is a *postcondition* expressed as a product of variables, such that $\varsigma_j \in B_x = \{0, 1, x\}$, where B_x is a boolean type extended with a *don't care* value x; note that $|Post| = |iout| + |state|$;

$S = [s_1 s_2 s_3 ..., s_o]$ is an optional *subroutine address* expressed as a boolean vector, $s_j \in B = \{0, 1\}$;

M_0 is the initial control marking, such that $M(p_{start}) = 1$, $M(p_{iin}) = M(p_{iout}) = M(p_{state}) = [000...0]$, and the remaining places are unoccupied.

Token colour allows to distinguish between tokens and places which can be of the following types: *control, interface input, interface output*, and *state* (Fig. 3). Control and plant tokens are denoted as "•" and "*", respectively. They flow only through their respective nets. Tokens of the colour *iin* are boolean vectors of plant events received through the interface input place p_{iin}. Tokens of the *iout* colour are boolean vectors of control events that are sent out through the place p_{iout}. A *state* place p_{state} with its associated *state* colour is used only to simplify hardware resolution of synchronization constructs. To avoid visual cluttering, arcs to/from p_{iin}, p_{iout} and p_{state} should be omitted.

Place p_{start} is used as a starting source place for the execution of the control algorithm. Upon initialization, it contains a "•" token, places p_{iin}, p_{iout} and p_{state} contain boolean zero vectors $[000...0]$, while the remaining places are unoccupied. Hence, the resetting of a controller is correctly represented by the initialization of the control net. *Dummy* transitions are used for correct resolution of *asymmetric choice* constructs. Each regular control transition has a *precondition* or *guard*, *postcondition* and an optional *subroutine address* associated with it. The guard consists of two components: a *input precondition* denoted as $(Pre_{iin}(t_j))$ and a *state precondition* denoted as $(Pre_{state}(t_j))$. The *input precondition* is related to the input token coming from the place p_{iin} and is denoted by the prefix 'I' in front of the guard. The *state precondition* is related to the tokens coming from the *state place*, p_{state}, and is denoted by a prefix 'S'. Both components of the guard are located in the upper left corner of a transition (Fig. 3). The postcondition consists of a *state postcondition* denoted as $(Post_{iout}(t_j))$ and an *output postcondition* denoted as $(Post_{state}(t_j))$. They are denoted by the prefixes 'S' and 'O', respectively, and are located in the upper right corner of the transition box. The *subroutine address* is denoted by the prefix 'Sr' and is graphically located below the postcondition elements. It points to the subroutine which may be executed by an optionally interfaced computing device.

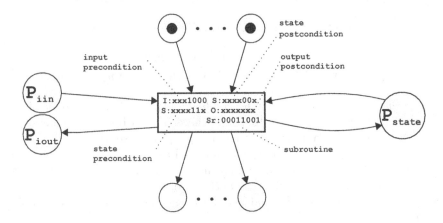

Fig. 3. Control transition of a *cePN*

Conflicting transitions belonging to the same input place must have a unique priority assigned in ascending order by the function $\pi : A \rightarrow \{1, 2, 3, ..., j\}$. Priority determines the sequence in which enabled conflicting transitions are tested for firing, with the highest priority denoted by '1'. This prevents nondeterministic behavior of an execution thread, increases the modeling power of the *cePN* and produces compact constructs even for complex enabling preconditions. Priority also simplifies hardware execution of *choice* and other constructs because the order in which conflicting transitions are tested is fixed by their position within a *place block*. Arc belonging to the same *join* construct must have identical priority to ensure correct processing in hardware.

A transition t_j is enabled when an *enabling condition*

$$E_{t_j} = \forall \, p_l \in \{I(t_j) : \kappa(I(t_j)) = control\}, \ M(p_k) = 1$$
$$\land \, \forall \, p_m \in \{I(t_j) : \kappa(I(t_j)) = iin\}, \ (Pre_{iin}(t_j)) = var(p_{iin}) \qquad (4)$$
$$\land \, \forall \, p_n \in \{I(t_j) : \kappa(I(t_j)) = state\}, \ (Pre_{state}(t_j)) = var(p_{state}) \ ,$$

is satisfied. $(Pre_{iin}(t_j)) = var(p_{iin})$ states that the value of the interface input token in place p_{iin} (denoted as $var(p_{iin}))$ must match the *input precondition*. Enabled transitions fire and remove control tokens from the control input places. However, the state token of colour B is never really removed from the p_{state} place but is overwritten with the new *state postcondition* value of colour B_x such that the bits denoted by x retain their previous values. Also the interface output token of colour B is overwritten with the *output postcondition* value of colour B_x.

In addition to being *safe* and *ordinary*, the *cePN* must also have a *max-K capacity* limit since an infinite number of control tokens cannot be supported in hardware. The value of K is dictated by the number of *slots* in the PNDU.

Interface Structure. An *interface structure* (IS) has all elements defined identically to the *cePN* such that $IS = (iC, iP, iT, iA, i\kappa, iM_0)$. It may be a simple register module as shown in Fig. 4, or a *First In First Out* (FIFO) module. In either case, it contains an *input bus* and an *output bus*, each transferring tokens of either *iin* or *iout* colour. Care should be taken when modeling the interface structure of a real physical system where new events overwrite old ones.

Plant Petri Net. A *plant Petri Net* (pPN) defines all elements except *pAD* identically to the *cePN*: $pPN = (pC, pP, pT, pA, p\kappa, pPre, pPost, pAD, pM_0)$, where *pAD* is an *output arc delay expression*, given as a positive real number.

The structure and behavior of the plant Petri Net is less restricted than the *cePN* since it is not an executable specification but a testbench. It does not have to be *safe* and *ordinary* and must only deposit tokens corresponding to the appropriate events into the interface places at desired time instances through the *plant postconditions*. For interactive applications where behavior of the plant depends on control stimuli, plant transitions use *preconditions* (Fig. 4).

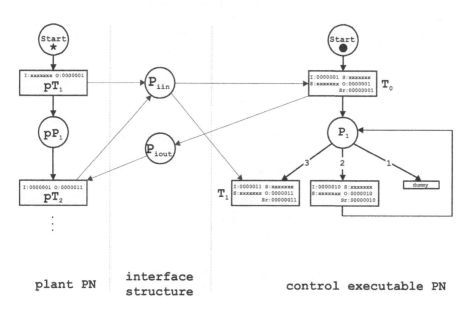

Fig. 4. Example of an Aggregate Petri Net

4.3 Petri Net Based System Design Methodology

The overall system design methodology based on Petri Net formalism is shown in Figure 5. The starting point may be as abstract as desired, such that a system may be initially specified using High-Level Petri Nets. However, it is advisable to separate the aggregate system into the plant, interface and control nets already at this level. Simulations and formal analysis techniques should be applied to verify the correctness of the modeled system. If it does not meet the specifications the system should be redesigned.

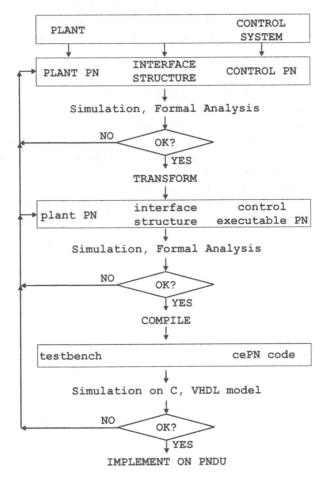

Fig. 5. PN based system design methodology

If the aggregate system meets the specification at the high level, it is transformed into the *control executable, interface structure* and *plant* Petri Nets such that they conform to the definitions described Sect. 4. At this level, the control executable Petri Net takes into account limitations imposed by the hardware architecture of the PNDU. The *cePN* must be *ordinary, safe* and have a fixed net *capacity*. If the control net at this level contains more tokens than allowed by the capacity, some execution threads must undergo parallel composition. This should be done until the maximum number of tokens does not exceed the net capacity. *Join* and *fork* constructs may also need to be transformed. First of all, all synchronizations become explicit through the use of the internal state register represented by the place p_{state}. Places that need to be synchronized must have appropriate state variables set through the postconditions of their respective input transitions. Hence, the number of input places per transition is limited by the width of the state vector of the control net. Therefore, if a *join* construct synchronizes more input places than is permitted by the width of the state vector, the synchronization should be cascaded into several steps. Due to the fixed net capacity, the *fork* construct should avoid producing more tokens than can be accommodated in hardware. Moreover, there is a limit on the number of output places per transition. Therefore, some execution threads may need to be combined through the parallel composition or their invocation should be postponed until some of the execution threads are terminated. Note that at this level all conflicting transitions belonging to the same enabling marking will be assigned a unique priority number. Additional transformations may involve converting abstract *input* and *output* tokens into binary vectors of the type *iin* and *iout*.

Again, simulations and formal analysis should be applied at this level to verify the correct behavior of the aggregate system. If it is satisfactory, the control executable Petri Net is ready for compilation. If not, redesign either at this or at the high level is required. The compiled object code may be first simulated on a C or VHDL executable model of the PNDU which is especially useful for verification of timing constraints. If the simulated code meets the specifications, it is then programmed into a programmable memory (EPROM or EEPROM). At this stage the control executable Petri Net is implemented in software and the controller is ready to take over the control of the plant.

5 Architecture of the Petri Net Decision Unit

5.1 Memory Format

The control executable Petri Net is compiled into machine code according to the memory format shown in Figures 6 and 7. The resulting code is a list of *place blocks* with their respective *transition blocks*. Place and transition blocks are unambiguously accessed since they have unique addresses in the compiled machine code. Figure 6 shows the encoding scheme for the *base* format with an 8-bit address bus and a 16-bit data bus. The ratio 1:2 between the address and data bus provides reasonable trade-offs between the bus power dissipation, latency of transition block access times and memory capacity measured in number of transition blocks. The memory format is scalable and Table 1 illustrates the effect of scaling.

transition block

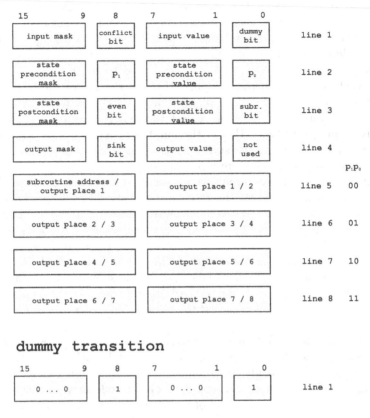

dummy transition

Fig. 6. Memory format of the net encoding

The first two lines of the transition block contain information associated with the precondition, and the following two lines encode postconditions. Each bit of the precondition or the postcondition may be masked by a *mask vector*. Only when the mask bit is '1', the actual values of the corresponding precondition bit and the input event bit are compared with each other. Bits 0 and 8 of the first 4 lines are used to encode additional indicators. The number of lines containing addresses of output places of a transition is encoded in line two with bits p_0 and p_1. They allow the transition block to be 4 to 8 bits long. If a place has more than one output transition, bit 8 of line 1 indicates this by setting the *conflict bit* set to '1'. The *subroutine* bit indicates whether the transition block has an optional subroutine address. This address is written to the SUB output port upon transition firing. It may be used by another processor which may be optionally interfaced with the PNDU. A *dummy* transition is indicated by a *dummy bit* set to '1' in which case the transition block consists of one line.

Table 1. Scalable system parameters of the PNDU

Address Bus (bits)	Data Bus (bits)	Inputs (bits)	Outputs (bits)	States (bits)	Memory Capacity (transitions)
8	16	7	7	7	$32 - 64$
16	32	15	15	15	$(8.2 - 16.4) \times 10^3$
32	64	31	31	31	$(0.54 - 1.1) \times 10^9$

This encoding scheme has no limit on the number of output transitions per place (*conflict* construct) since conflicting transitions with their *conflict* bits set are simply listed within the place block. The priority of conflicting transitions is derived by their relative position within the block. There is also no limitation on a *merge* construct since it is not explicitly encoded. The number of output places per transition in a *fork* construct is restricted to 8. Since *join* constructs explicitly use state register values, the number of input places per transition is limited to $n - 1$, where n is the width of the address bus.

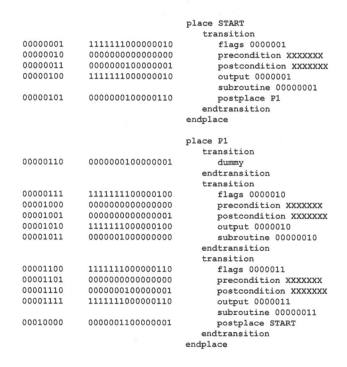

```
                                  place START
                                      transition
00000001    1111111000000010            flags 0000001
00000010    0000000000000000            precondition XXXXXXX
00000011    0000000100000001            postcondition XXXXXXX
00000100    1111111000000010            output 0000001
                                        subroutine 00000001
00000101    0000000100000110            postplace P1
                                      endtransition
                                  endplace

                                  place P1
                                      transition
00000110    0000000100000001            dummy
                                      endtransition
                                      transition
00000111    1111111100000100            flags 0000010
00001000    0000000000000000            precondition XXXXXXX
00001001    0000000000000001            postcondition XXXXXXX
00001010    1111111000000100            output 0000010
00001011    0000001000000000            subroutine 00000010
                                      endtransition
                                      transition
00001100    1111111000000110            flags 0000011
00001101    0000000000000000            precondition XXXXXXX
00001110    0000000100000001            postcondition XXXXXXX
00001111    1111111000000110            output 0000011
                                        subroutine 00000011
00010000    0000001100000001            postplace START
                                      endtransition
                                  endplace
```

Fig. 7. Compiled machine code of the *cePN* of Fig. 4

A Petri Net compiler written in *Perl* language compiles a textual description of the net into the binary code. The description is intuitively derived directly from the *cePN*. Figure 7 shows the correspondence between the description of an example Petri Net of Fig. 4 to the compiled code.

5.2   Hardware Architecture of the Petri Net Decision Unit

The architecture of the Petri Net Decision Unit is summarized in Figure 8. The four main modules are: *First In First Out* (FIFO) buffer, *Active Place Buffer* (APB), *Cache* and *Control*. The interface to the plant is done through the parallel IN and OUT ports. The IN port is fed into the FIFO buffer which is responsible for capturing new plant events. A new event is defined as an input vector whose value differs from the current event value. If a new event is detected, it is latched into the FIFO and the Control module receives a signal that the processing can begin. If none of the currently enabled transitions fires, the event is invalid since its value does not match any of the preconditions. Such an event is overwritten by a subsequently arriving event and in the absence of a valid event the execution of the PNDU suspends. Once a new event is detected, the new processing cycle is promptly initiated. This type of processing reflects the event-driven nature of the PNDU hardware architecture. The OUT port is a register whose value is updated with control event outputs of firing transitions. The SUB parallel port may be used to interface the PNDU to other computing devices.

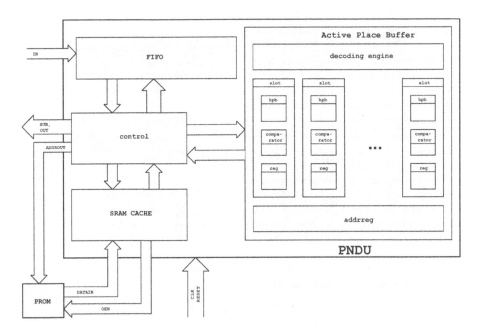

Fig. 8. Block diagram of the PNDU

The *Active Place Buffer* contains *slots* which store and process preconditions of currently enabled transition. One slot is allocated for each marked or *active* place. Preconditions of enabled transitions are first stored in *reg* registers. A firing check is performed on all slots in parallel using the *comparator* module. If preconditions are *true*, the slot is marked as *hot* in the corresponding *hpb* register indicating that the transition contained in this slot will fire. The *address register* module keeps track of the addresses of active places, while the *decoding engine* is in charge of correct operation of such tasks as resolution of conflicting transitions.

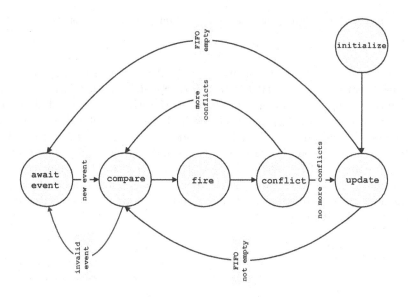

Fig. 9. PNDU execution algorithm

The overall execution of the PNDU is managed by the *Control* module according to the algorithm summarized in Fig. 9. The *Init* phase is the initialization stage triggered by the Reset signal. At any given time the PNDU is either in waiting mode *Flagevent* or in execution mode consisting of the *Compare, Fire, Conflict* and *Update* phases. Once a new event is detected, processing is initiated by testing preconditions stored in slots (*Compare* phase). All transitions marked as *hot* fire in sequence (*Fire* phase). First, postconditions and a subroutine address of a firing transition are read from PROM and are written to the OUT and SUB registers. Then, the addresses of output places are written into the APB module. If a place has conflicting transitions, their preconditions are read from PROM and tested according to the assigned priority (*Conflict* phase). At the end of the execution cycle, depending whether any transition has fired, some registers are overwritten with new information (*Update* phase).

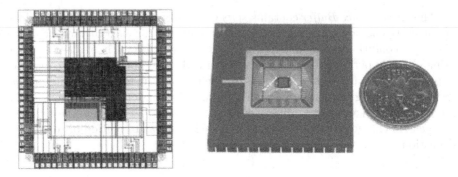

Fig. 10. Circuit layout and the fabricated integrated circuit of the PNDU

The PNDU was developed using a typical ASIC design methodology, starting with a parameterized VHDL (*Very High Speed Integrated Circuit* Hardware Description Language) model. The depth of the FIFO, number of slots in the APB, and other system parameters listed in Table 1 are defined by a user prior to the circuit synthesis. The base 8-bit model was synthesized into the 0.35 μm 5 metal Alcatel digital CMOS technology standard cell circuit. The prototype has 8 slots in the APB and the depth of the FIFO is 8. The circuit containing approximately 7500 standard cells features a 64-word cache block which speeds up memory accesses by an order of magnitude. The final 8.4 mm^2 circuit layout is shown in Fig. 10.

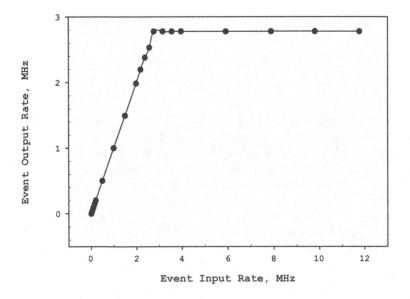

Fig. 11. Event response rate of the PNDU at $f_{clk,max}$

The prototype was fabricated at the IMEC, Belgium, through the Europractice Academic Program. Its Pin Grid Array (PGA-100, 33.5×33.5 mm) ceramic package is shown in Figure 10. When interfaced with two 8-bit MX271000L UV erasable EPROM the PNDU runs at 80 MHz (limited by the EPROM access times) and dissipates between 100 and 120 mW of power depending on the event input rate. At this speed the PNDU is capable of firing transitions within roughly 300 ns. Figure 11 shows a typical event response characteristic of the PNDU operating at the maximum clock frequency.

6 Conclusions

Control executable Petri Nets presented in this paper offer a convenient modeling paradigm for the design of discrete event control systems. The system design methodology offers elegant handling of concurrency complemented by the possibility of formal verification. It is oriented toward embedded implementations based on the proposed programmable event-driven controller which is optimized to process control specifications formulated in terms of *control executable Petri Nets*. The specifications are compiled into binary code according to an efficient net encoding scheme with 4 to 8 memory lines per transition. This is critical in embedded systems design where the code size is an important factor. The resulting compact code is programmed into conventional parallel EPROM/EEPROM.

The fabricated integrated circuit of the PNDU prototype is capable of tracking eight execution threads. It processes all classes of PN directly in hardware and fires transitions in an event-driven fashion. Its architecture is complemented with an on-chip cache. This ensures that the transition firing rate improves with increasing concurrency. Running at 80 MHz and executing 8 concurrent processes it fires up to 4 million transitions per second dissipating only 120 mW of power.

Continued work on precise characterization of the timing response of the controller will enable future incorporation of timing into the system design methodology. This will make it feasible to apply the controller to discrete event embedded systems with stringent real-time requirements.

Acknowledgments. The authors wish to thank Professors K. Jensen and S. Christensen of Computer Science Department, University of Aarhus, Denmark, L. Lavagno of Politechnico di Torino, and J. Cortadella with his colleagues of Universitat Politecnica de Catalunya, Barcelona, for discussions regarding this project. In addition, numerous constructive comments of the anonymous referees are gratefully acknowledged.

References

1. Bulach, S., Baur, H., Pfleiderer, H.-J., Kucerovsky, Z.: ALPiNe: A Hardware Computing Platform for High-Level Petri Nets. In: Jensen, K. (ed.): *Proceedings of the Workshop on Practical Use of Coloured Petri Nets and Design/CPN*, Computer Science Department, Aarhus University, Denmark (1998) 31-45

2. Bulach, S., Baur, H., Pfleiderer, H.-J., Kucerovsky, Z.: Design of Discrete Event Systems Using Petri Nets and a Dedicated Controller. *Proceedings of the IFAC Conference on Control System Design*. Elsevier Science Ltd. (2000) 317-322

3. Bulach, S., Brauchle, A., Pfleiderer, H.-J., Kucerovsky, Z.: An Architecture of a Petri Net Based Event-Driven Controller. In: Boel, R., Stremersch, G. (eds.): *Discrete Event Systems: Analysis and Control*. Kluwer Academic Publishers. (2000) 383-390

4. Cassandras, C.G.: *Discrete Event Systems: Modeling and Performance Analysis*. Richard D. Irwin, Inc., and Aksen Associates, Inc. (1993)

5. Crockett, D., Desrochers, A., DiCesare, F., Ward, T.: Implementation of a Petri Net Controller for a Machining Workstation. *Proceedings of the IEEE International Conference on Robotics and Automation*. (1987) 1861-1867

6. Harel, D., Pnueli, A.: On the Development of Reactive Systems. In: Apt, K.R. (ed.): *NATO ASI Series: Logics and Models of Concurrent Systems*. Springer-Verlag **13** (1985) 447-498

7. Jensen, K.: *Coloured Petri Nets: Basic Concepts, Analysis Methods and Practical Use*. Springer-Verlag **1** (1997)

8. Murakoshi, H., Dohi, J.: Petri Net Based High Speed Programmable Controller by ASIC Memory. *Proceeding of 29th Annual SICE Conference* **II** (1990) 697-700

9. Murata, T.: Petri Nets: Properties, Analysis and Applications. *Proceedings of the IEEE* **77(4)** (1989) 541-580

10. Nketsa, A., Courviosier, M.: A Petri Net Based Single Chip Programmable Controller for Distributed Local Controls. *Proceedings of Signal Processing and System Control Factory Automation IECON* **1** (1990) 542-547

11. Patil, S.S.: Circuit Implementation of Petri Nets. *Computation Structures Group Memo 73, MIT Project MAC*, Cambridge, Massachusetts (1972)

12. Patil, S.S.: An Asynchronous Logic Array. *Computation Structures Group Memo 111, MIT Project MAC*, Cambridge, Massachusetts (1975)

13. Peterson, J.L.: *Petri Net Theory and the Modeling of Systems*. Prentice-Hall, Englewood Cliffs, N.J. (1981)

14. Petri, C.A.: *Kommunikation mit Automaten*. Ph.D. Dissertation, University of Bonn, Schrift Nr.2. (1962)

15. Silva, M., Teruel, E., Valette, R., Pingaud, H.: Petri Nets and Production Systems In: Reisig, W., Rozenberg, G. (eds.): *Lectures on Petri Nets II: Applications*. Lecture Notes in Computer Science, Vol. 1492. Springer-Verlag (1998) 85-124

16. Valette, R., Courviosier, M., Bigou, JM., Albukerque, J.: A Petri Net Based Programmable Logic Controller. In: Warman, E.A., (ed.): Proceedings of the 1st International IFIP Conference on Computer Applications in Production and Engineering, North-Holland. (1983) 103-116

17. Yakovlev, A., Koelmans, A.: Petri Nets and Digital Hardware Design. In: Reisig, W., Rozenberg, G. (eds.): Lectures on Petri Nets II: Applications. Lecture Notes in Computer Science, Vol. 1492. Springer-Verlag (1998) 154-236

Condensed State Spaces for Timed Petri Nets

Søren Christensen[1], Lars Michael Kristensen[1,2], and Thomas Mailund[1]

[1] Department of Computer Science, University of Aarhus
IT-parken, Aabogade 34, DK-8200 Aarhus N., Denmark
{schristensen,lmkristensen,mailund}@daimi.au.dk
[2] School of Electrical and Information Engineering, University of South Australia
Mawson Lakes Campus, SA 5095, Australia
lars.kristensen@unisa.edu.au

Abstract. We present a state space method for Petri nets having a time concept based on a global clock and associating time stamps to tokens. The method is based on equivalence on states and makes it possible to condense the usually infinite state space of such timed Petri nets into a finite state space without loosing analysis power. The practical application of the method is demonstrated on a large example of an audio/video protocol by means of a computer tool implementing the method.

1 Introduction

It is generally recognised that time plays an important role in many concurrent and distributed systems. This has motivated the development and extension of several modelling languages and analysis methods to support validation of timed systems. In the area of Petri nets [16] different time concepts and extensions to the basic formalism have been introduced, making it possible to reason about timed systems. Some time concepts for Petri nets focus on time aspects when investigating logical correctness [17, 18], whereas others focus on performance analysis [15].

In this paper the focus is on time when investigating the logical correctness of systems by means of *state spaces*. State spaces analysis is one of the main analysis methods of Petri nets. The basic idea behind state spaces is to compute a directed graph with nodes representing the reachable states of the system and arcs representing the possible state changes. We consider the time concept of timed Coloured Petri Nets [12, 13, 14] (CP-nets or CPNs) as introduced in [13]. The time concept of CP-nets is inspired from [17, 18] and is based on the introduction of a *global clock* used to represent *model time*. In addition, tokens in a timed CP-net carry *time stamps*. Intuitively, the time stamp of a token describes the earliest model time at which the token can be consumed, i.e., be removed by the occurrence of a transition. The execution of a timed CP-net is time driven. The system remains at a given model time as long as there are enabled transitions. When no more transitions are enabled at the current model time, the global clock is incremented to the earliest next model time at

J.-M. Colom and M. Koutny (Eds.): ICATPN 2001, LNCS 2075, pp. 101–120, 2001.

which transitions are enabled. Despite its simplicity, this time concept has been successfully applied for simulation-based performance analysis in a number of case studies, e.g., in the areas of high-speed interconnects [6] and ATM networks [8]. For investigating logical correctness of timed systems, the main shortcoming of this time concept is that it does not work well with state space methods. The main problem with state spaces for timed CP-nets as defined in [13] is that for reactive/cyclic systems the state space often becomes infinite. This is because the absolute notion of time is carried over into the state space. As a consequence, state space analysis of timed CP-nets so far has had to rely on *partial state spaces*, i.e., finite subsets of the full state space. An example of this can be found in [5].

The contribution of this paper is a state space method which reconciles state spaces and a time concept as in timed CP-nets. The idea behind the method is to use equivalence on the states to factor out the absolute notion of time, and to ignore the time stamps of tokens which are in a certain sense not important. In this way the usually infinite state space can be condensed into a finite state space. This *condensed state space* can be computed using a variant of the standard algorithm for state space construction, but without constructing the full state space. We show that the quotient structure obtained can be used for model checking using discrete time temporal logics [1, 10] such as RTCTL [11].

Using equivalences for condensing infinite state spaces of timed CP-nets into finite state spaces has also been investigated in [17] and [2]. It was also suggested as a possible solution in [13]. We give a further discussion of the relationship between the methods in this paper and the results in [17] and [2] at the end of this paper. The state space method presented is not tied to timed CP-nets and works equally well for timed Place/Transition Nets (PT-nets). Because of this, and in order to keep the presentation simple, we present our results in the context of timed PT-nets.

The paper is organised as follows. Section 2 informally introduces timed PT-nets. Section 3 informally introduces our notion of equivalence and condensed state spaces of timed PT-nets. Section 4 formally defines timed PT-nets. Section 5 formally defines our notion of equivalence and condensed state spaces. Section 6 proves that our notion of equivalence is sound. Section 7 describes the properties preserved by our equivalence notion. Section 8 gives some numerical data on the performance of the method on a large case study. Finally, Sect. 9 contains the conclusions and a further discussion of related work. The reader is assumed to be familiar with untimed PT-nets [9].

2   Timed PT-Nets

In this section we informally introduce timed PT-nets. We formally define timed PT-nets in Sect. 4. A timed PT-net is a PT-net extended with a global clock, and with *time stamps* associated to tokens. The left-hand side of Fig. 1 shows a small timed PT-net modelling a mutex between two processes A and B. While markings of untimed PT-nets consist solely of a distribution of tokens on places, markings of timed PT-nets have time stamps associated to each token. In Fig. 1

each token (indicated by a black dot) has an associated dashed box giving the time stamps of the token. On the left-hand side of Fig. 1, the initial marking is indicated. Initially, there is one token on each of the places AIdle, Mutex, and BIdle. Initially, all other places contain no tokens. This models that initially both process A and B are idle and the mutex is unlocked. All tokens initially have time stamp 0.

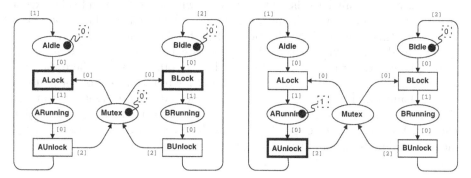

Fig. 1. An timed PT-net modelling a mutex between two processes.

A *state* of a timed PT-net consists of a marking (i.e., a distribution of tokens on the places including their time stamps), and the model time (the global clock). We will use 0 as the initial global clock value of the mutex in Fig. 1.

The basic idea behind the dynamic behaviour of timed PT-nets is that a transition will not be enabled at a given model time unless the tokens to be removed have time stamps which are smaller than the model time. The PT-net remains at a given model time as long as there are enabled transitions. When no more transitions are enabled at the current model time, the model time is incremented to the smallest time at which a transition becomes enabled. In addition to the weights associated to arcs in untimed PT-nets, arcs in timed PT-nets have associated time lists, i.e., arc labels on the form: $[r_1, r_2, \ldots, r_n]$. The meaning of the inscription on output arcs is that an occurrence of the transition will produce n tokens with time stamps which are respectively $r_1, r_2, \ldots, r_n$ time units larger than the current model time. To model that an event takes Δr time units, one can let the corresponding transition create time stamps for its output tokens which are Δr time units larger than the model time. These tokens will then be unavailable for Δr time units. For input arcs the meaning of the arc inscription is that the transition will only be enabled if there exists n tokens on the input place such that by adding the time stamps $r_1, r_2, \ldots, r_n$ to the existing time stamps of these tokens they obtain time stamps which are greater than or equal to the current model time. In the example of Fig. 1 the time stamps of all input arcs are equal to 0, i.e., the current model time must at least be equal to the time stamp of tokens before they can be consumed. All examples and informal explanations given in the rest of this paper will assume that all time stamps on input arcs are 0.

As an example, the transition ALock in the initial state shown in Fig. 1 is enabled, indicated by the thicker border. The transition is enabled since the tokens on places AIdle and Mutex both have a time stamp which is less than or equal to the initial model time which is 0. In a similar way, the transition BLock is enabled. If the transition ALock occurs in the initial state, we obtain the marking shown on the right-hand side of Fig. 1. In this state no transition is enabled and hence the model time will be incremented to 1 at which model time the transition AUnlock becomes enabled. When this transition occurs we reach a state which is similar to the initial state shown on the left-hand side, except that the place AIdle now contains a token with time stamp 2, Mutex contains a token with time stamp 3, and the global clock has been increased to 1.

The model time when a state is created is called the *creation time* of the state. The creation time of both states in Fig. 1 is 0. The time where the next transition can occur is called the *termination time* of the state. For the state on the left-hand side in Fig. 1 the termination time is 0. The termination time for the state on the right-hand side is 1, since the earlist time at which a transition (AUnlock) becomes enabled is 1. This means that each state exists inside a closed interval of time between creation- and termination time.

3   Condensed State Spaces for Timed PT-Nets

We now informally introduce the method for constructing condensed state spaces for timed PT-nets. We formally define the method in Sect. 5. The basic idea is to use an equivalence relation on the states to factor out the absolute notion of time, and to ignore the time stamps of tokens which are not important in a certain sense.

Figure 2 shows the initial part of the full state space for the mutex example from Fig. 1. The dashed box positioned next to a node specifies the marking, the creation time and the termination time of the state. Node 0 corresponds to the initial state of the PT-net. Node 1 corresponds to the state shown on the right-hand side of Fig. 1. The marking corresponding to node 1 has been written as ARunning[1] + BIdle[0] (see node 1). This should be read as place ARunning containing one token with time stamp 1 and place BIdle containing one token with time stamp 0. The state corresponding to node 1 has creation time 0 and termination time 1. This is written as Time : 0-1. Each edge has an associated label specifying the name of the occurring transition. Since time progresses in each loop executed by the two processes, both the model time and the time stamps of tokens will get larger and larger values and thus create new states. Thus the full state space of the mutex example is infinite as a result of increasing time stamps.

It can be noticed that time is only used to model *delays*. Only the relative differences between time stamps and model time, not the absolute values, affect the future behaviour. If we compare state 1 to state 7 in Fig. 2 they have identical token distributions and the only difference is that creation-, termination- and all time stamps of tokens have been increased by 3. This means that we will have the same possibility of future behaviour for state 1 and state 7. We generalise this to observing that when we reach a new state equal to a state already seen, except that the global clock and all time stamps of the tokens have been increased by

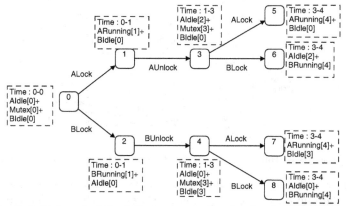

Fig. 2. Initial part of the full state space for the timed mutex PT-net.

the same amount, then we will have exactly the same future behaviors from the new state as from the state we have already seen. This means that we can consider two states as being similar/equivalent, if we can obtain one from the other by moving both creation time and token time stamps, by the same amount, thereby keeping the relative distance between time stamps and the global clock. Moreover, we can observe that time stamps less than or equal to creation time are interchangeable with respect to enabling of transitions. When the time stamp is lower than the model time, the token can be consumed. How much lower than the global clock is not important. We can ignore all time stamps lower than the creation time of the state without affecting the future behavior. We will call states which are similar in the above sense for *creation time equivalent*. Consider for instance Fig. 2. The states $\{1, 5, 7\}$ are creation time equivalent. As we have already discussed state 1 and state 7 are equivalent. State 5 and state 1 are also equivalent since, if we add 3 to the creation time and to all time stamps in state 1 we get a state equal to state 5, except for the time stamp on the token on BIdle. Here we notice that all time stamps which are less than the creation time of a state can be considered as equivalent. In a similar way, states $\{2, 6, 8\}$ are creation time equivalent. The states 0, 3 and 4 are only creation time equivalent to themselves.

The basic idea behind condensed state spaces for timed PT-nets is to group such equivalent states into equivalence classes. If we group creation time equivalent states of the initial part of the full state space in Fig. 2 into such equivalence classes, we obtain the condensed state space shown in Fig. 3. Each node now represents an equivalence class of states, e.g., node 1 represents states $\{1, 5, 7\}$ from Fig. 2. The labels associated with the nodes now indicate a *representative* of the corresponding equivalence class. In fact, the infinite full state space of the mutex example is represented by the condensed state space shown in Fig. 3.

There is yet another observation to make: there is not necessarily any transitions enabled at the creation time of a state. With creation time equivalence we distinguish time stamps between the creation time and the termination time of a state. For example, the states 3 and 4 in Fig. 3 are not creation time equivalent because they differ on the time stamps of the tokens on places AIdle and BIdle

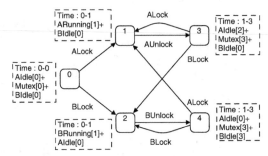

Fig. 3. Condensed state space for mutex PT-net using creation time equivalence.

and are thus different states. However, in both cases the only significant time stamp is the one on the token on place **Mutex**. At the termination time, the tokens on **AIdle** and **BIdle** will in both cases have time stamps less than or equal to the termination time. Hence, the difference between the two states does not affect the future behaviour. All time stamps less than or equal to the termination time might as well be equal to the termination time. States which are similar in this sense we call *termination time equivalent*. If we apply the termination time equivalence on the mutex PT-net, we obtain the condensed state space shown in Fig. 4. Compared to the condensed state space in Fig. 3, also state 3 and state 4 are now considered equivalent.

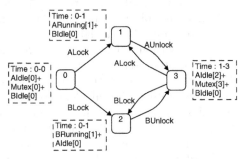

Fig. 4. Condensed state space for mutex PT-net using termination time equivalence.

The above shows that termination time equivalence is weaker than creation time equivalence, i.e., considers more states equivalent. Hence, termination time equivalence gives better reduction of the state space than creation time equivalence. Furthermore, the information lost when going from creation time to termination time equivalence seems to be of little practical importance.

4   Formal Definition of Timed PT-Nets

In this section we formally define timed PT-nets. Except for minor notational differences, this section is essentially a formulation of the time concept of timed CP-nets from [13] at the level of PT-nets. We include references to corresponding definitions even though they are not identical. Formally, the extension of untimed PT-nets to timed PT-nets is captured through the concept of *timed multi-sets*.

Definition 1. (Def. 5.1 [13]) *A **timed multi-set** tm over a set S, is a function* tm $: S \times \mathbb{R} \to \mathbb{N}$ *such that the formal sum* tm$(s) = \sum_{r \in \mathbb{R}}$ tm(s, r) *is finite for all* $s \in S$. S_{TMS} *denotes the set of all timed multi-set over S.*   □

The non-negative integer tm(s) is the number of appearances of the element s in the timed multi-set tm. The *time list* of s, tm$[s] = [r_1, r_2, \ldots, r_{\text{tm}(s)}]$ is defined to contain the time values $r \in \mathbb{R}$ for which tm$(s, r) \neq 0$. Each r appears tm(s, r) times in the list, which is sorted such that $r_i \leq r_{i+1}$ for $1 \leq i < $ tm(s). We usually represent a timed multi-set tm by a formal sum $\sum_{s \in S} s\,\text{tm}[s]$, e.g., ARunning$[1]$ + BIdle$[0]$ is the timed multi-set that maps (ARunning, 1) and (BIdle, 0) to 1, and everything else to 0. For a time list tm$[s] = [r_1, r_2, \ldots, r_n]$ and a time value $r \in \mathbb{R}$, tm$[s]^{+r}$ is the time list obtained from tm$[s]$ by adding r to each element, i.e., tm$[s]^{+r} = [r_1 + r, r_2 + r, \ldots, r_n + r]$. For a timed multi-set tm, tm^{+r} denotes the timed multi-set obtained by adding r to all time stamps.

It follows from Def. 1 that each timed multi-set over S is also an ordinary multi-set over $S \times \mathbb{R}$. This allows us to define $+$, $*$, and $=$ for timed multi-sets over S, to be identical to the corresponding operations for ordinary multi-sets over $S \times \mathbb{R}$. Comparison $(\leq)$ and subtraction $(-)$ could be defined in a similar fashion, but this is inadequate for our purposes, since then tm$_1 \leq$ tm$_2$ would require that each element in tm$_1$ appears in tm$_2$ with *exactly* the same time value. This is too strict since it is only required that the time stamp of a token is less than or equal to the current model time for the token to be available.

First we define comparison and subtraction of time lists. In the following we assume that $a = [a_1, a_2, \ldots, a_m]$ and $b = [b_1, b_2, \ldots, b_n]$ are two ascending time lists over $\mathbb{R}$. We define $a \leq b$ iff $m \leq n$ and $a_i \geq b_i$ for all $i = 1, \ldots, m$. This means that if we increase the time values in a time list we get a smaller time list. When $a \leq b$, $b - a$ is defined to be the time list of length $n - m$ obtained from b as follows: traverse the time list a from starting with a_1 and remove for each a_i the largest element still in b which is smaller than or equal to a_i. As an example, consider $a = [4, 4, 5]$ and $b = [2, 3, 4, 5, 8]$. Clearly, $a \leq b$ and $b - a = [2, 8]$.

Having defined comparison and subtraction of time lists we now define the corresponding operations on timed multi-sets. Intuitively, comparison is point-wise time list comparison, while subtraction is pointwise time list subtraction.

Definition 2. (Def. 5.2 [13]) *For all* tm$_1$, tm$_2 \in S_{\text{TMS}}$, ***comparison** is defined by:* tm$_1 \leq$ tm$_2 = \forall s \in S :$ tm$_1[s] \leq$ tm$_2[s]$. *When* tm$_1 \leq$ tm$_2$, ***subtraction** is defined by:* tm$_2 -$ tm$_1 = \sum_{s \in S} s\,($tm$_2[s] -$ tm$_1[s])$.   □

We define a timed PT-net as a PT-net together with a set of time values and a start time. Usually arcs of PT-nets are defined by a weight function $W : P \times T \cup T \times P \to \mathbb{N}$, with **pre** and **post** mappings given implicitly by stating that **pre**(t) consists of those places p for which $W(p, t) \neq 0$ while **post**(t) consists of those places p for which $W(t, p) \neq 0$ [9]. We choose to define timed PT-nets directly by the **pre** and **post** mappings since the introduction of time lists associated with arcs would complicate the definition of the weight function.

Definition 3. (Def. 5.3 [13]) *A **timed PT-net** is a tuple* (PTN, r_0) *where* PTN $= (P, T, \textbf{pre}, \textbf{post}, m_0)$ *satisfies that* P *is a set of **places**, T is a set* *of **transitions** such that $P \cap T = \emptyset$, **pre, post** : $T \to P_{\text{TMS}}$ are mappings from transitions to timed multi-sets over places, and $m_0 \in P_{\text{TMS}}$ is the **initial marking**. $r_0 \in \mathbb{R}$ is the **start time**.* □

Throughout this paper we assume, without loss of generality, that the start time is 0, i.e., $r_0 = 0$. Moreover, we will assume that all time stamps in timed multi-sets specified by **pre** and **post** are greater than or equal to 0. This assumption simplifies the presentation and is of little practical importance since negative time stamps are seldom used in practice for modelling of systems. The results in the paper can however be generalised to handle also negative time stamps. The relationship between the **pre** and **post** mappings and the graphical representation of timed PT-nets is as follows. There is an arc leading from a place p to a transition t labelled with the time list $[r_1, \ldots, r_n]$ iff $\mathbf{pre}(t)[p] = [r_1, r_2, \ldots, r_n]$. Similarly, there is an arc leading from a transition t to a place p labelled with the time list $[r_1, r_2, \ldots, r_n]$ iff $\mathbf{post}(t)[p] = [r_1, r_2, \ldots, r_n]$. We formally define markings and states as follows.

Definition 4. (Def. 5.4 [13]) *A **marking** of a timed PT-net is a timed multi-set over places. A **state** is a pair (m, r) where $m \in P_{\text{TMS}}$ is a marking and $r \in \mathbb{R}$ is the **creation time**. The **initial state** s_0 is the pair $\left(m_0^{+r_0}, r_0\right)$. $\mathbb{M}$ denotes the set of all markings. $\mathbb{S}$ denotes the set of all states.* □

We extend our notation m^{+r} from markings to states in a straightforward manner, i.e., $(m, r_0)^{+r} = (m^{+r}, r_0 + r)$. On the left-hand side of Fig. 1, the initial marking is indicated corresponding to the timed multi-set $m_0 = \mathsf{Aidle}\,[0] + \mathsf{Mutex}\,[0] + \mathsf{Bidle}\,[0]$. The informal explanation of the semantics of timed PT-nets given earlier is formalised in the following definition.

Definition 5. (Defs. 5.5 and 5.6 [13]) *A transition t is **enabled** in a state (m_1, r_1) at time $r_2 = r_1 + d$ iff $\mathbf{pre}(t)^{+r_2} \leq m_1$ and $d \geq 0$ and there exists no other pair $(t', d') \in T \times \mathbb{R}$ with $d' < d$ satisfying this. This is written $(m_1, r_1)[t, d\rangle$. When a transition t is enabled in state (m_1, r_1) at time r_2 it may **occur**, changing the state (m_1, r_1) to the state (m_2, r_2), where $m_2 = (m_1 - \mathbf{pre}(t)^{+r_2}) + \mathbf{post}(t)^{+r_2}$. This is written $(m_1, r_1)[t, d\rangle(m_2, r_2)$. We say that r_2 is the **termination time** of state (m_1, r_1).* □

To illustrate the definition above, consider the initial state $(m_0, 0)$ shown on the left-hand side of Fig. 1. The transition ALock is enabled at time 0, since $\mathbf{pre}(\mathsf{ALock})^{+0} = \mathsf{Aidle}\,[0] + \mathsf{Mutex}\,[0] \leq m_0 = \mathsf{Aidle}\,[0] + \mathsf{Mutex}\,[0] + \mathsf{Bidle}\,[0]$, and 0 is the smallest model time satisfying this. If ALock occurs in $(m_0, 0)$ it leads to the marking:

$$\left(m_0 - \mathbf{pre}(\mathsf{ALock})^{+0}\right) + \mathbf{post}(\mathsf{ALock})^{+0} = \mathsf{ARunning}\,[1] + \mathsf{Bidle}\,[0]$$

at time 0. At model time 0, no more transitions are enabled and the model time is increased to 1 at which model time the transition AUnlock is enabled since:

$$\mathbf{pre}(\mathsf{AUnlock})^{+1} = \mathsf{ARunning}\,[1] \leq \mathsf{ARunning}\,[1] + \mathsf{Bidle}\,[0].$$

An occurrence of transition AUnlock leads to the marking: AIdle [2] + Mutex [3] + BIdle [0] at time 1.

In contrast to the usual definition of timed CP-nets in [13], we associate a delay with enabling and occurrence rather than an absolute time value, i.e., we write $(m_1, r_1)[t, d\rangle$ and $(m_1, r_1)[t, d\rangle(m_2, r_2)$ rather than $(m_1, r_1)[t, r_2\rangle$ and $(m_1, r_1)[t, r_2\rangle(m_2, r_2)$. The d in the definition above is redundant, since it can be calculated from the time values r_1 and r_2. We have chosen to associate the delay between creation times to the occurring transitions since this simplifies the presentation of the results in Sect. 7.

An *occurrence sequence* of a timed PT-net is a sequence consisting of states $s_i = (m_i, r_i)$, and pairs consisting of a transition and a time value (t_i, d_i) denoted $s_1[t_1, d_1\rangle s_2 \ldots s_{n-1}[t_{n-1}, d_{n-1}\rangle s_n$ and satisfying $s_i[t_i, d_i\rangle s_{i+1}$ for $i = 1 \ldots, n-1$. A state s' is reachable from a state s iff there exists an occurrence sequence leading from s to s'. $[s\rangle$ denotes the set of all states reachable from a state s. $[s_0\rangle$ denotes the set of reachable states.

5   Formal Definition of Condensed State Spaces

In this section we formally define condensed state spaces and termination time equivalence for timed PT-nets. Full state spaces for timed PT-nets can be defined as for ordinary untimed PT-nets, except that the nodes now represent states instead of markings.

Definition 6. *The **full state space** of a timed PT-net is the directed graph* (V, E) *where* $V = [s_0\rangle$ *and* $E = \{(s, (t, d), s') \in V \times (T \times \mathbb{R}) \times V \mid s[t, d\rangle s'\}$. $\square$

Each node in the full state space corresponds to a reachable state of the PT-net and each edge corresponds to an occurring transition. An edge $(s, (t, d), s')$ between two nodes s and s' means that the transition t is enabled in state s at a model time which is d time units higher than the creation time of s, i.e., d is the difference between the creation time and the termination time of s. The occurrence of the transition at this model time leads to state s'.

The equivalence relations which will serve as a basis for obtaining the condensed state space will constitute a so-called equivalence specification.

Definition 7. *An **equivalence specification** for a timed PT-net is an equivalence relation* $\approx$ *on* $\mathbb{S}$. $\mathbb{S}_\approx$ *denotes the set of all equivalence classes for* $\approx$. *For a state* $s \in \mathbb{S}$, $[s]_\approx \in \mathbb{S}_\approx$ *denotes the equivalence class of* $\approx$ *containing* s. $\square$

When the relation $\approx$ is known from the context, we write $[s]$ instead of $[s]_\approx$. The definition of equivalence specifications can be generalised by including an equivalence relation on pairs of transitions and time values, i.e., the edges in the state space. This more general definition is the usual definition of equivalence specifications as given in [13]. However, to specify termination time equivalence we do not need this more general form.

In order to do verification based on condensed state spaces, we require that the equivalence specification has the property that equivalent states are known to have similar behavioural properties. This is captured by the concept of *consistency* of equivalence specifications.

Definition 8. *An equivalence relation $\approx$ is* **consistent** *iff the following condition holds for all states $s_1, s_1' \in \mathbb{S}$, $(t,d) \in T \times \mathbb{R}$, and $s_2 \in [s_1]_{\approx}$:*

$$s_1[t,d\rangle s_1' \Rightarrow \exists s_2' \in [s_1']_{\approx} : s_2[t,d\rangle s_2' \qquad \square$$

The consistency requirement ensures that equivalent states have identical sets of pairs of enabled transitions and delays, and equivalents sets of immediate successor states. Condensed state spaces for timed PT-nets is defined as follows.

Definition 9. *Let $\approx$ be a consistent equivalence specification. The* **condensed state space** *is the directed graph (V, E), where $V = \{ C \in \mathbb{S}_{\approx} \mid C \cap [s_0\rangle \neq \emptyset \}$ and $E = \{ (C_1, (t,d), C_2) \in \mathbb{S}_{\approx} \times (T \times \mathbb{R}) \times \mathbb{S}_{\approx} \mid \exists (s,s') \in C_1 \times C_2 : s[t,d\rangle s' \}$.* $\qquad \square$

The condensed state space has a node for each equivalence class containing a reachable state. The condensed state space has an edge between two nodes iff there is a state in the equivalence class of the source node in which a transition is enabled, and whose occurrence leads to a state in the equivalence class of the destination node.

We now formalise termination time equivalence. We define the operation $\lceil - \rceil_r$ on timed multi-sets to be the operation that sets all time stamps less than or equal to r to r. For any non-dead state $s \in \mathbb{S}$, let $m(s)$ be the time value $d \in \mathbb{R}$ for which there exists $t \in T$ such that $s[t,d\rangle$. For a dead state $s \in \mathbb{S}$ (i.e., a state in which creation time cannot be increased to enable some transition) we define $m(s) = 0$, i.e., for a dead state, termination time will be equal to creation time. We extend our notation from timed multi sets $\lceil - \rceil_r$ to states such that for a state $s = (m, r)$ we write $\lceil (m, r) \rceil_{r'}$ as a shorthand for $(\lceil m \rceil_{r'}, r)$ and we will use the notation $\lceil s \rceil$ to denote $\lceil s \rceil_{r+m(s)}$. Termination time equivalence is formally defined as follows.

Definition 10. *Two states s_1 and $s_2 \in \mathbb{S}$ are* **termination time equivalent**, *written $s_1 \approx_{TT} s_2$, iff there exists $r \in \mathbb{R}$ such that: $\lceil s_1 \rceil = \lceil s_2 \rceil^{+r}$.* $\qquad \square$

Informally, the definition above states that two states are termination time equivalent, if we are able to move all time values by r in one state and obtain the other when considering all time values less than or equal to termination time to be equal. It is rather straightforward to check that termination time equivalence is indeed an equivalence relation.

The condensed state space method is usually implemented by representing each equivalence class by a representative state for the class. Construction of the condensed state space then follows the same procedure as the construction of the full state space with one exception. Whenever a new state is to be inserted into the condensed state space it is checked whether an *equivalent* state is already included in the condensed state space. To be able to implement the equivalence check efficiently, we will select a canonical representative, i.e., a unique representative for each equivalence class. The check then amounts to transforming the new state into this unique representative for the equivalence class and then check (using ordinary equality) whether the resulting state has already been included in the condensed state space. The canonical representative for an equivalence class under termination time equivalence will be the unique state with creation time zero and with all time stamps of tokens between creation time and termination time set to termination time, i.e., for a state $s = (m, r)$ the canonical representative $\bar{s}$ is given by $\bar{s} = \lceil s \rceil^{-r}$. It is rather straightforward to check that this does indeed yield canonical representatives, i.e., for two states $s_1 = (m_1, r_1)$, $s_2 = (m_2, r_2)$: $\lceil s_1 \rceil^{-r_1} = \lceil s_2 \rceil^{-r_2}$ iff $s_1 \approx_{TT} s_2$. The overhead incurred from the above computation of a canonical representative is rather small. The algorithm for construction of condensed state spaces has to find the termination time as part of computing the enabled transitions in a given state. Subtraction of creation time and ceiling of time stamps to termination time can be implemented in linear time in the number of tokens of m.

We conclude this section by a result concerning finiteness of condensed state spaces. For a finite PT-net we denote by $TV = \{tv_1, tv_2, \ldots, tv_n\}$ the set of time values appearing as time stamps on tokens in the initial marking, in time lists on input and output arcs of transitions, and as start time. A timed PT-net is *bounded* iff there exists a $k \geq 0$ such that for all places p and reachable states $s = (m, r)$ we have that $|m(p)| \leq k$. We will prove that the condensed state space of a finite bounded timed PT-net is finite if $TV \subseteq \mathbb{Q}$. Before we prove this results we will need two propositions. Proposition 1 states that only finitely many rational numbers in a bounded interval can be expressed as linear combinations of a finite set of rational numbers.

Proposition 1. (Lemma 4 [3]) *Let $q_1, q_2, \ldots q_k \in \mathbb{Q}$ and $A, B \in \mathbb{R}$. The set $Q_{AB} = \{q = \sum_{i=1}^{k} n_i * q_i \mid A \leq q \leq B \wedge n_1, n_2, \ldots, n_k \in \mathbb{Z}\}$ is finite.* □

For a finite timed PT-nets, the largest time value is defined by $\max_{TV} = \max\{tv \in TV\}$. For a state s we denote by $TS(s)$ the set of time stamps of tokens in s, and $\max_{TS}(s) = \max\{ts \in TS(s)\}$ denotes the largest time stamp in s. Proposition 2 states that the creation time and the time stamps of tokens in any reachable state can be written as a linear combination of the time values of the PT-net, and it gives a bound on the time stamps of tokens and termination time relative to the creation time of a reachable state. The proposition follows by induction and from the enabling and occurrence rule of finite timed PT-nets.

Proposition 2. *Let $s = (m, r) \in [s_0\rangle$ then the following holds:*

1. *There exists $n_1, n_2, \ldots n_n \in \mathbb{N}_0 : r = \sum_{i=1}^{n} n_i * tv_i$*
2. *$\forall ts \in TS(s)\ \exists n_1, n_2, \ldots n_n \in \mathbb{N}_0 : ts = \sum_{i=1}^{n} n_i * tv_i$*
3. *$0 \leq \max_{TS}(s) \leq r + \max_{TV}$*
4. *If $s[(t, d)\rangle$ then $0 \leq r + d \leq r + \max_{TV}$* □

Theorem 1. *The condensed state space obtained using termination time equivalence is finite for all finite and bounded timed PT-nets with $TV \subseteq \mathbb{Q}$.* □

Proof. First we prove that the number of equivalence classes of states containing a reachable state is finite. This is done by proving that only finitely many canonical representatives for such equivalence classes exists. Let $s = (m, r) \in [s_0\rangle$. It follows from Prop. 2 (3) that $\max_{TS}(s) \leq r + \max_{TV}$. Since the PT-net is bounded and finite, there is only a finite number of ways to distribute tokens on the places. We complete the proof by showing that the tokens can only have a finite number of time stamps. The canonical representative for s is given by $\bar{s} = \lceil s \rceil^{-r}$, i.e., by setting time stamps less than or equal to termination time to termination time and subtract r from all resulting time stamps. If s is a dead state, then termination time is equal to creation time in which case $\max_{TS}(\bar{s}) = \max_{TS}(s) - r \leq \max_{TV}$. If s is non-dead state then by Prop. 2 (4) we have $r + d \leq r + \max_{TV}$. Hence, $\max_{TS}(\bar{s}) = \max(r + d, \max_{TS}(s)) - r \leq r + \max_{TV} - r = \max_{TV}$. Hence, in both cases $0 \leq \max_{TS}(\bar{s}) \leq \max_{TV}$. It follows from Prop. 2 (1)-(2) that all time stamps in $\bar{s}$ can be written on the form $\sum_{i=1}^{n} n_i * tv_i$ where $n_i \in \mathbb{Z}$. The tokens in $\bar{s}$ can therefore only have a finite number of time stamps by Prop. 1. The number of arcs of the condensed state space is bounded since the number of nodes is bounded and since only a finite number of transitions is enabled in a state.   □

6 Consistency of Termination Time Equivalence

In this section we prove the consistency of termination time equivalence. Consistency is the basic result which ensures that condensed state spaces obtained using termination time equivalence can be used for verification. To prove the consistency of termination time equivalence, we first give a proposition stating a number of basic properties of the timed multi-set operations $(-)^{+r}$ (adding r to all time stamps) and $\lceil - \rceil_r$ (setting all time stamps less that r to r).

Proposition 3. *The operations* $(-)^{+r}$ *and* $\lceil - \rceil_r$ *are distributive in the sense that for* $\mathrm{tm}_1, \mathrm{tm}_2 \in S_{\mathrm{TMS}}$ *the following holds:*

1. $(\mathrm{tm}_1 + \mathrm{tm}_2)^{+r} = \mathrm{tm}_1^{+r} + \mathrm{tm}_2^{+r}$ *and* $(\mathrm{tm}_1 - \mathrm{tm}_2)^{+r} = \mathrm{tm}_1^{+r} - \mathrm{tm}_2^{+r}$
2. $\lceil \mathrm{tm}_1 + \mathrm{tm}_2 \rceil_r = \lceil \mathrm{tm}_1 \rceil_r + \lceil \mathrm{tm}_2 \rceil_r$ *and* $\lceil \mathrm{tm}_1 - \mathrm{tm}_2 \rceil_r = \lceil \mathrm{tm}_1 \rceil_r - \lceil \mathrm{tm}_2 \rceil_r.$ □

Below we give two lemmas which states two fundamental properties about the relationship between time stamps, enabling, and occurrence. Lemma 1 below states that the absolute time values (but not the relative time values) can be moved without affecting enabling. Hence, enabling and occurrence does not depend on the absolute value of creation time and time stamps.

Lemma 1. *Let* $s_1, s_2 \in \mathbb{S}$ *be two states. The following holds for all* $r \in \mathbb{R}$ *:* $s_1[t, d\rangle s_2 \Leftrightarrow (s_1{}^{+r})[t, d\rangle(s_2{}^{+r}).$ □

Proof. We first prove the $\Rightarrow$ direction. Let $s_1 = (m_1, r_1)$, $r \in \mathbb{R}$, and assume $s_1[t, d\rangle$. By the definition of enabling, $d \geq 0$ is the minimal time value such that $\mathbf{pre}(t)^{+(r_1+d)} \leq m_1$. By adding r on both sides of the inequality, d is still the minimal time value such that $\mathbf{pre}(t)^{+(r_1+r+d)} \leq m_1{}^{+r}$. Hence, $s_1{}^{+r}[t, d\rangle$. Now, let $s_2 = (m_2, r_2)$ be the result of the occurrence of (t, d) in (m_1, r_1), i.e.,

$m_2 = (m_1 - \mathbf{pre}(t)^{+(r_1+d)}) + \mathbf{post}(t)^{+(r_1+d)}$ and $r_2 = r_1 + d$. From Prop. 3 (1) it now follows that: $m_2{}^{+r} = (m_1{}^{+r} - \mathbf{pre}(t)^{+(r_1+d+r)}) + \mathbf{post}(t)^{+(r_1+d+r)}$ which is exactly the result of the occurrence of (t, d) in $(m_1{}^{+r}, r_1 + r) = s_1{}^{+r}$. Hence, $(s_1{}^{+r})[t, d\rangle(s_2{}^{+r})$. The $\Leftarrow$ direction follows from the above by setting $s_1 = s_1{}^{+r}$, $s_2 = s_2{}^{+r}$, $r = -r$, and exploiting that for all states s: $(s^{+r})^{-r} = s$. $\qquad\Box$

Lemma 2 below states that time values less that or equal to the termination time do not affect enabling. Hence, enabling and occurrence are independent of tokens with time stamps less than or equal to termination time.

Lemma 2. *Let $s_1 = (m_1, r_1)$, $s_2 \in \mathbb{S}$ be two states. The following holds:*
$$s_1[t, d\rangle s_2 \Leftrightarrow (\lceil s_1 \rceil_{r_1+d})[t, d\rangle(\lceil s_2 \rceil_{r_1+d}).$$
$\qquad\Box$

Proof. We first prove the $\Rightarrow$ direction. Assume that $s_1[t, d\rangle$. By the definition of enabling, $d \geq 0$ is the minimal time value such that $\mathbf{pre}(t)^{+(r_1+d)} \leq m_1$. By applying the $\lceil - \rceil_{r_1+d}$ operation on the right hand side of the inequality, d is still the minimal time value such that $\mathbf{pre}(t)^{+(r_1+d)} \leq \lceil m_1 \rceil_{r_1+d}$. Hence, $(\lceil s_1 \rceil_{r_1+d})[t, d\rangle$. Now, let $s_2 = (m_2, r_2)$ be the result of the occurrence of (t, d) in (m_1, r_1), i.e., $m_2 = (m_1 - \mathbf{pre}(t)^{+(r_1+d)}) + \mathbf{post}(t)^{+(r_1+d)}$ and $r_2 = r_1 + d$. From Prop. 3 (2) it now follows that: $\lceil m_2 \rceil_{r_1+d} = \lceil m_1 \rceil_{r_1+d} - \lceil \mathbf{pre}(t)^{+(r_1+d)} \rceil_{r_1+d} + \lceil \mathbf{post}(t)^{+(r_1+d)} \rceil_{r_1+d}$. Since all time stamps in $\mathbf{pre}$ and $\mathbf{post}$ are positive by assumption, then $\lceil \mathbf{pre}(t)^{+(r_1+d)} \rceil_{r_1+d} = \mathbf{pre}(t)^{+(r_1+d)}$ and $\lceil \mathbf{post}(t)^{+(r_1+d)} \rceil_{r_1+d} = \mathbf{post}(t)^{+(r_1+d)}$. Hence $\lceil m_2 \rceil_{r_1+d}$ is exactly the result of the occurrence of (t, d) in $(\lceil m_1 \rceil_{r_1+d}, r_1) = \lceil s_1 \rceil_{r_1+d}$ and therefore $(\lceil s_1 \rceil_{r_1+d})[t, d\rangle(\lceil s_2 \rceil_{r_1+d})$. The $\Leftarrow$ direction is similar. $\qquad\Box$

Theorem 2. *The equivalence relation $\approx_{TT}$ is consistent for timed PT-nets.* $\qquad\Box$

Proof. Let $s_1 = (m_1, r_1)$, $s_2 = (m_2, r_2) \in \mathbb{S}$ be two states. Assume $s_1[t, d\rangle s_1'$ and that $s_1 \approx_{TT} s_2$, i.e., $\exists r \in \mathbb{R}$ such that: $(\lceil s_1 \rceil_{r_1+d})^{+r} = \lceil s_2 \rceil_{r_2+d}$ and $r_2 = r_1 + r$. Now, $s_1[t, d\rangle s_1'$

$\Rightarrow \quad (\lceil s_1 \rceil_{(r_1+d)})\, [t, d\rangle\, (\lceil s_1' \rceil_{(r_1+d)}) \qquad$ (Lemma 2)

$\Rightarrow \quad ((\lceil s_1 \rceil_{(r_1+d)})^{+r})[t, d\rangle(\lceil s_1' \rceil_{(r_1+d)})^{+r} \quad$ (Lemma 1)

$\Rightarrow \quad (\lceil s_2 \rceil_{(r_2+d)})\, [t, d\rangle\, \left((\lceil s_1' \rceil_{(r_1+d)})^{+r}\right) \quad$ (since $(\lceil s_1 \rceil_{(r_1+d)})^{+r} = \lceil s_2 \rceil_{(r_2+d)})$

$\Rightarrow \quad (\lceil s_2 \rceil_{(r_2+d)})\, [t, d\rangle\, \left(\lceil s_1'{}^{+r} \rceil_{r_2+d}\right) \quad$ (since $(\lceil s_1' \rceil_{(r_1+d)})^{+r} = \lceil s_1'{}^{+r} \rceil_{r_2+d})$

$\Rightarrow \quad s_2[t, d\rangle(s_1'{}^{+r}) \qquad\qquad$ (Lemma 2)

Since $s_1'{}^{+r} \approx_{TT} s_1'$ this proves that TT is consistent. $\qquad\Box$

7   Properties Preserved

We now turn to the properties preserved by termination time equivalence. To this end we consider quantitative temporal logics (see [1] for a survey), and introduce a discrete real-time temporal logic called RTCTL*. RTCTL* is obtained from CTL* [7] by the addition of time-bounded operators. This is done in the same way as the temporal logic RTCTL was obtained from CTL in [11]. It should be stressed that our aim here is not to suggest RTCTL* as a temporal logic for verification of timed PT-net. It is merely a tool to show which properties are preserved of the system when using condensed state spaces and termination time equivalence. One can then use a subset of RTCTL* such as RTCTL [11] which have efficient model checking algorithms, or use specially tailored algorithms for the concrete verification question at hand. What is important is that if the verification question can be expressed in RTCTL*, then the results in this section ensure that the answer to the verification question is preserved by the condensation, independently of which algorithm is then used to actually compute the answer. Since all standard dynamic properties of PT-nets can be expressed in RTCTL*, it immediately follows from this section that all these properties are preserved by termination time equivalence.

First we briefly define the syntax and semantics of RTCTL* in the context of timed PT-nets. There are two types of formulas in RTCTL*: *state formulas* (which are true/false in a specific state) and *path formulas* (which are true/false along a specific path/occurrence sequence). Let Π be a set of atomic state propositions, i.e., functions from $\mathbb{S}$ to $\{\mathsf{True}, \mathsf{False}\}$. State formulas are defined as:

- π if $\pi \in \Pi$;
- if ϕ, ϕ_1, and ϕ_2 are state formulas, then $\neg\phi$ and $\phi_1 \vee \phi_2$ are state formulas;
- if ψ is a path formula, then $\mathsf{E}\psi$ is a state formula.

Path formulas are defined as:

- if ϕ is a state formula, then $\mathsf{PF}(\phi)$ is a path formula;
- if ψ, ψ_1, and ψ_2 are path formulas, then $\neg\psi$, $\psi_1 \vee \psi_2$, $\mathsf{X}\psi$, and $\phi_1 \mathsf{U} \phi_2$ are path formulas;
- if ψ_1, and ψ_2 are path formulas and $r \in \mathbb{R}$, then $\psi_1 \mathsf{U}^{\leq r} \psi_2$ is a path formula.

RTCTL* is the set of state formulas generated by the above rules. The difference compared to CTL* is the addition of the time-bounded until operator $\mathsf{U}^{\leq r}$. For technical reasons related to the later proofs, we define explicitly the PF operator for converting state formulas into path formulas.

The semantics of RTCTL* is defined with respect to structures of the form $\mathcal{M} = (V, E, \mathcal{E}_\Pi)$, where V is a set of states, $E \subseteq V \times (T \times \mathbb{R}) \times V$ is the transition relation, and $\mathcal{E}_\Pi : V \to 2^\Pi$ is the proposition labelling function, which assigns to each $s \in V$ the set $\mathcal{E}_\Pi(s) \subseteq \Pi$ of atomic state propositions that hold in s. The transition relation E is required to be total, i.e., each state has at least one successor state. Except for the requirement on the transition relation being total, it is straighforward to see that the full state space (from now on denoted $\mathcal{S}$) and the condensed state space (from now on denoted $\bar{\mathcal{S}}$) of a timed PT-net each determine a structure on the form $\mathcal{M} = (V, E, \mathcal{E}_\Pi)$. To make the transition relation total, we add for each dead state s the triple $(s, (t_d, 1), s)$ to E for some transition $t_d \notin T$. This can be interpreted as when the system enters a dead state it remains in this state forever. This is well-defined also for the condensed state space, since either none or all states in an equivalence class are dead states.

A *path* in $\mathcal{M}$ is a sequence of states and pairs of transitions and time values: $\sigma = s_1(t_1, d_1)s_2(t_2, d_2)s_3, \ldots$ such that, for all $i \geq 1$: $(s_i, (t_i, d_i), s_{i+1}) \in E$. σ^i

denotes the *suffix* of σ starting in s_i. We use the standard notation $\mathcal{M}, s \models \phi$ to mean that the state formula ϕ holds at s in the structure $\mathcal{M}$. Similarly, $\mathcal{M}, \sigma \models \phi$ means that the path formula ϕ holds along the path σ in $\mathcal{M}$. The $\models$ relation is inductively defined below. ϕ, ϕ_1, and ϕ_2 are state formulas whereas ψ, ψ_1, and ψ_2 are path formulas.

- $\mathcal{M}, s \models \pi \Leftrightarrow \pi \in \mathcal{E}_\Pi(s)$
- $\mathcal{M}, s \models \neg\phi \Leftrightarrow \mathcal{M}, s \not\models \phi$
- $\mathcal{M}, s \models \phi_1 \vee \phi_2 \Leftrightarrow \mathcal{M}, s \models \phi_1$ or $\mathcal{M}, s \models \phi_2$
- $\mathcal{M}, s \models \mathsf{E}\psi \Leftrightarrow$ there exists a path σ starting in s such that $\mathcal{M}, \sigma \models \psi$
- $\mathcal{M}, \sigma \models \mathsf{PF}(\phi) \Leftrightarrow$ for the first state s of σ : $\mathcal{M}, s \models \phi$.
- $\mathcal{M}, \sigma \models \neg\psi \Leftrightarrow \mathcal{M}, \sigma \not\models \psi$
- $\mathcal{M}, \sigma \models \psi_1 \vee \psi_2 \Leftrightarrow \mathcal{M}, \sigma \models \psi_1$ or $\mathcal{M}, \sigma \models \psi_2$
- $\mathcal{M}, \sigma \models \mathsf{X}\psi \Leftrightarrow \mathcal{M}, \sigma^2 \models \psi$
- $\mathcal{M}, \sigma \models \psi_1 \mathsf{U} \psi_2 \Leftrightarrow$ for $\sigma = s_1(t_1, d_1)s_2(t_2, d_2)s_3, \dots$, there exists $k \geq 1$ such that $\mathcal{M}, \sigma^k \models \psi_2$ and for all $1 \leq j < k : \mathcal{M}, \sigma^j \models \psi_1$
- $\mathcal{M}, \sigma \models \psi_1 \mathsf{U}^{\leq r} \psi_2 \Leftrightarrow$ for $\sigma = s_1(t_1, d_1)s_2(t_2, d_2)s_3, \dots$, there exists $k \geq 1$ such that $\mathcal{M}, \sigma^k \models \psi_2$, for all $1 \leq j < k : \mathcal{M}, \sigma^j \models \psi_1$, and $\sum_{i=1}^{k-1} d_i \leq r$

We will usually leave the model $\mathcal{M}$ out of the relation and simply write $s \models \pi$ or $\sigma \models \psi$ when the model is clear from the context. For a state $s \in \mathcal{S}$, we denote by $\bar{s}$ the representative chosen to represent $[s]$ in $\bar{\mathcal{S}}$.

Lemma 3. *There is a bidirectional correspondence between paths in the full state spaces $\mathcal{S}$ and paths in the condensed state space $\bar{\mathcal{S}}$.*

1. *If $\sigma = s_1(t_1, d_1)s_2(t_2, d_2)s_3, \dots$ is a path in $\mathcal{S}$, then we have that $\bar{\sigma} = \bar{s}_1(t_1, d_1)\bar{s}_2(t_2, d_2)\bar{s}_3, \dots$ is a path of $\bar{\mathcal{S}}$.*
2. *If $\bar{\sigma} = \bar{s}_1(t_1, d_1)\bar{s}_2(t_2, d_2)\bar{s}_3, \dots$, is a path of $\bar{\mathcal{S}}$, then for every state $s'_1 \approx_\mathbb{S} s_1$ in $\bar{\mathcal{S}}$ there exists a path $\sigma' = s'_1(t_1, d_1)s'_2(t_2, d_2)s'_3, \dots$, in $\mathcal{S}$ such that $s'_i \approx_\mathbb{S} \bar{s}_i$ for all $i \geq 1$.*

Proof. The lemma follows from successive application of the consistency requirement for $\approx_\mathbb{S}$ and is similar to the proofs of Prop. 2.4 and Prop. 3.6 in [13]. $\square$

From the above lemma it follows that for a path σ in $\mathcal{S}$, we can talk about *the* corresponding path $\bar{\sigma}$ in $\bar{\mathcal{S}}$, and for a path $\bar{\sigma}$ in $\bar{\mathcal{S}}$ we can talk about *a* corresponding path σ in $\mathcal{S}$. In order to be able to use the condensed state space for model checking, we require that all the atomic state propositions $\pi \in \Pi$ are such that for all equivalence classes $[s]$ we have: $s_1, s_2 \in [s] \Rightarrow \pi(s_1) = \pi(s_2)$, i.e., the truth value wrt. to atomic state propositions is the same for all states in an equivalence class. In practice this means that the atomic state propositions may refer to the number of tokens on places, but cannot refer to time stamps of tokens nor to the creation time of the state. From now on we assume that the set of atomic state propositions satisfies this property.

Theorem 3. *Let ϕ be a state formula and ψ a path formula of RTCTL*. Let $\sigma = s_1(t_1, d_1)s_2(t_2, d_2)s_3, \dots$ be a path in $\mathcal{S}$, and let $\bar{\sigma} = \bar{s}_1(t_1, d_1)\bar{s}_2(t_2, d_2)(\bar{s}_3), \dots$ be the corresponding path in $\bar{\mathcal{S}}$. Then the following holds:*

$$\mathcal{S}, s_1 \models \phi \Leftrightarrow \bar{\mathcal{S}}, \bar{s}_1 \models \phi \text{ and } \mathcal{S}, \sigma \models \psi \Leftrightarrow \bar{\mathcal{S}}, \bar{\sigma} \models \psi \qquad \square$$

Proof. We prove the theorem by structural induction on state and path formulas. The base case is $\phi \equiv \pi \in \Pi$. In this case the theorem follows from the assumption on Π which ensures that $s \models \pi \Leftrightarrow \bar{s} \models \pi$. For the induction step, the proof is split in a number of cases. Cases $\phi \equiv \neg\phi_1$ and $\phi \equiv \phi_1 \vee \phi_2$ are straightforward (both in the case of state formulas and path formulas). We consider the remaining cases below.

$\phi \equiv \mathsf{E}\psi$: Assume that $s \models \phi$. Then there is a path σ starting in s such that $\sigma \models \psi$. By Lemma 3, there is a corresponding path $\bar{\sigma} \in \bar{\mathcal{S}}$ starting in $\bar{s}$. By the induction hypothesis: $\sigma \models \psi \Leftrightarrow \bar{\sigma} \models \psi$. Hence, $\bar{s} \models \mathsf{E}\psi$. The reverse direction is similar.

$\psi \equiv \mathsf{PF}(\phi)$: Assume that $\sigma \models \psi$. Then, $s_1 \models \phi$. By the induction hypothesis $\bar{s}_1 \models \phi$, and hence $\bar{\sigma} \models PF(\phi)$. The reverse direction is similar.

$\psi \equiv \mathsf{X}\psi_1$: Assume that $\sigma \models \mathsf{X}\psi_1$. Then $\sigma^2 \models \psi_1$ and since σ and $\bar{\sigma}$ correspond so do σ^2 and $\bar{\sigma}^2$. By the induction hypothesis: $\bar{\sigma}^2 \models \psi_1$. Hence, $\bar{\sigma} \models \psi_1$.

$\psi \equiv \psi_1 \mathsf{U}^{\leq r}\psi_2$: Assume that $\sigma \models \psi$. Then $\exists k \geq 1$ such that $\sigma^k \models \psi_2$, for all $1 \leq j < k : \sigma^j \models \psi_1$, and $\sum_{i=1}^{k-1} d_i \leq r$. Since σ and $\bar{\sigma}$ correspond so do σ^j and $\bar{\sigma}^j$ for all j and therefore by the induction hypothesis: $\bar{\sigma}^k \models \psi_2$ and $\sigma^j \models \psi_1$ for all $1 \leq j < k$. Moreover, since delays are preserved along corresponding paths, then $\bar{\sigma} \models \psi$.

$\psi \equiv \psi_1 \mathsf{U}\psi_2$: This case is similar to $\psi \equiv \psi_1 \mathsf{U}^{\leq r}\psi_2$. □

We proved earlier in Thm. 2 that termination time equivalence is consistent. Hence, we have the following corollary of the above theorem.

Corollary 1. *Let PTN be a timed PT-net with a full state space $\mathcal{S}$ and let $\bar{\mathcal{S}}_{\mathsf{TT}}$ be the condensed state space for PTN obtained using termination time equivalence. For all state formulas ϕ of RTCTL* and all states $s : \mathcal{S}, s \models \phi \Leftrightarrow \bar{\mathcal{S}}_{\mathsf{TT}}, \bar{s} \models \phi$. For all paths formulas ψ of RTCTL*, paths σ in $\mathcal{S}$, and corresponding paths $\bar{\sigma}$ in $\bar{\mathcal{S}}_{TT} : \mathcal{S}, \sigma \models \psi \Leftrightarrow \bar{\mathcal{S}}_{\mathsf{TT}}, \bar{\sigma} \models \psi$.* □

8   A Case Study

We have implemented the state space method presented in the previous sections on top of the state space tool of Design/CPN [4]. The prototype implements termination time equivalence for CP-nets. In this section we apply this prototype implementation on a larger CPN model. The CPN model is taken from the industrial case study [5] in which timed CP-nets and the DESIGN/CPN tool were used to validate vital parts of the BANG&OLUFSEN BeoLink system.

The BeoLink system makes it possible to distribute audio and video throughout a home via a dedicated network. The state space analysis in [5] focused on the *lock management protocol* of the BeoLink system. This protocol is used to grant devices exclusive access to various services in the system. Timed CP-nets were applied in [5] since timing is crucial for the correctness of the lock management protocol. The exclusive access is implemented based on the notion of a *key*. A device is required to possess the key in order to access services. When the system boots no key exists, and the lock management protocol is (among other things) responsible for ensuring that a key is generated when the system starts. The lock management protocol is also responsible for ensuring that a key is generated in case it is lost during the operation of the system. It is the obligation of the so-called *video* and/or *audio master* device to ensure that new keys are generated when needed. The three main correctness criteria of the lock management protocol is listed below.

Eventual key generation. When the system is booted, a key is eventually generated. The key is to be generated within approximately 2.0 seconds. This property can be expressed as the RTCTL* formula: $AF^2\pi_{key}$, where π_{key} is an atomic state proposition which is true iff a key is present in the system, and $AF^r\phi \equiv \neg E(\neg(\text{ True } U^{\leq r}\phi))$

Mutual exclusion. At any time during the operation of the system at most one key exists. This property can be expressed as the RTCTL* formula $AG\ \pi_{01-key}$, where π_{01-key} is an atomic state proposition which is true iff there is zero or one key in the system, and $AG\phi \equiv \neg E(\text{ True } U\neg\phi)$

Persistent key access. Any given device always has the possibility of obtaining the key. This property can be expressed as the RTCTL* formula $\bigwedge_i AGEF\pi_{i-haskey}$, where $\pi_{i-haskey}$ is an atomic state proposition which is true iff the device i has the key, and $EF\phi \equiv E(\text{ True } U\phi)$.

Since all three properties can be expressed in RTCTL* it follows from Sect. 7 that they are preserved by termination time equivalence. In [5] the *eventual key generation* was verified using partial state spaces, where successors were not generated for those nodes in which a key existed. Using this partial state space it was possible to show that the lock management protocol ensures that a key is generated in the initialisation phase of the system, i.e., when the system boots. The two other properties were not verified in [5], since the full state space of the CPN model of the BeoLink system is infinite. Below we show that with the new state space method for timed CP-nets presented in this paper, we are now able to verify all three properties of the system. In addition to the termination time equivalence we also apply the symmetry method of CP-nets [13] as implemented in the DESIGN/CPN tool to alleviate the state explosion problem. The symmetry in the BeoLink is based on the observation that the devices which are neither the video nor the audio master are identical, i.e., their identity is interchangeable. All results in this section were obtained using a HP Unix Workstation with 1 Gbyte of memory.

Table 1 gives some statistics for the state spaces of the full BeoLink system for different configurations. The Config column specifies the configuration in question. Configurations with one video master is written on the form VM:n, where n is the total number of devices in the system. Configurations with one audio master is written on the form AM:n. The TT columns give the size of the condensed state space (nodes and arcs) when applying termination time equivalence, and the CPU time it took to generate the condensed state space. CPU time is written on the form $hh : mm : ss$ where hh is hours, mm is minutes and ss is seconds. The TT+SYMM column give the corresponding numbers for a combined use of termination time equivalence and symmetry. For VM:4 and

Table 1. Experimental results – Full BeoLink system.

Config	TT			TT+SYMM		
	Nodes	Arcs	Time	Nodes	Arcs	Time
VM : 2	274	310	0:00:02	274	310	0:00:02
AM : 2	346	399	0:00:03	346	399	0:00:03
VM : 3	10,713	14,917	0:01:34	5,420	7,562	0:00:47
AM : 3	27,246	37,625	0:04:10	13,647	18,872	0:02:10
VM : 4	3,557,441	7,351,877	-	593,209	866,085	7:10:21
AM : 4	12,422,637	25,059,384	-	2,070,796	3,064,778	25:51:57

AM:4 the state space was not actually generated, but the size (i.e., number of nodes and arcs) was calculated from the corresponding condensed state spaces by calculating the size of the equivalence classes represented by the nodes and arcs. The TT column shows that termination time equivalence results in a finite state space, but that the size of the state space grows rapidly with the number of devices. Hence, the main virtue of termination time equivalence is that it results in a finite state space. Column TT+SYMM shows that termination time equivalence and the symmetry method combine very well. For an example, for 3 devices the reduction in terms of states is 49%, and the generation time is reduced by 48% for the AM:3 configuration, and by 50% for the VM:3 configuration.

9   Conclusions

We have presented a state space method for timed Petri nets. The method uses an equivalence relation on states to obtain a condensed state space which satisfy exactly the same RTCTL* formulas as the full state space. We have developed tool support for this state space method and made some initial experiments on a CPN model taken from an industrial case study. The condensed state space makes it possible to analyse a class of systems which could previously only be partially analysed. The main benefit of the approach is that it makes it possible to condense an infinite state space into a finite state space. In order to alleviate the state explosion problem, termination time equivalence needs to be combined with other state space reduction methods, such as the symmetry method.

Condensed state space methods for timed Petri nets based on aggregating states into equivalence classes have also been investigate in [17] and [2]. The time concept of *Interval Timed Coloured Petri Nets* (ITCPNs) considered in [17] is similar to the time concept considered in this paper in that time stamps are associated with tokens. Moreover, time stamps of tokens in ITCPNs are increasing since the time concept of ITCPNs also introduces a global clock. The difference between the time concept of ITCPNs and the time concept considered in this paper is that time delays in [17, 2] can be continuous intervals. The *Modified Transition System Reduction Technique* (MTSRT) used in [17] to obtain a condensed state space, groups states into equivalence classes to recover from states of an ITCPN having an infinite number of successors states due to the infinite number of time values in the continuous interval specifying time delays.

The condensed state spaces of [17] may however still be infinite due to increasing time stamps. Hence, the condensed state spaces based on termination time equivalence in this paper and the condensed state spaces of [17] focus on two different sources for the state space to become infinite.

The results in [17] for ITCPNs were later extended in [2], developing the notion of equivalence presented in [17] to be both sound as well as complete. The original equivalence in [17] was sound but not complete, i.e., the condensed state space could contain the set of reachable states of the ITCPN as a proper subset. The notion of equivalence presented in [2] ensures that the condensed state space represent exactly the set of reachable states of the ITCPN. In addition to this, [2] introduced a notion of equivalence between states which can factor out the global clock of ITCPNs. This equivalence notion is similar to the creation time equivalence which was presented in Sect. 2 as an *intermediate* step towards termination time equivalence. Termination time equivalence is therefore a strictly weaker notion of equivalence than the equivalence presented in [2], and hence allows for better condensation of the state space. On the other hand, the condensed state space in our paper only considers discrete intervals, whereas the method presented in [2] can deal with continuous intervals. It remains to be investigated whether the idea of termination time equivalence presented in this paper can be applied to ITCPNs. Another difference between the results presented in [2] and the results presented in this paper is that for termination time equivalence a canonical form for states can be computed efficiently. In [2] no canonical form is given for the computation of canonical representatives of equivalence classes of states. Having a canonical form which can be computed efficiently is of great importance for the practical use and implementation of the method.

The time concept considered in [2] and [17] is on the other hand more advanced/powerful than the discrete time concept considered in this paper. However, there exists many interesting systems where the appropriate level of abstraction can be modelled (and meaningful analysis results obtained) using discrete time. The lock management protocol of the BANG&OLUFSEN BeoLink system used as a case study in this paper is a good example of this.

References

1. R. Alur and T. Henzinger. Logics and Models of Real Time: A Survey. In *Real-Time: Theory in Practice*, volume 600 of *Lecture Notes in Computer Science*, pages 74–106. Springer-Verlag, 1991.
2. G. Berthelot. Occurrence Graphs for Interval Timed Coloured Nets. In *Proceedings of ICATPN'94*, volume 815 of *Lecture Notes in Computer Science*, pages 79–98. Springer Verlag, 1994.
3. B. Berthomieu and M. Diaz. Modelling and Verification of Time Dependent Systems using Time Petri Nets. *IEEE Transactions on Software Engineering*, 17(3):259–273, March 1991.
4. S. Christensen, J. B. Jørgensen, and L. M. Kristensen. Design/CPN - A Computer Tool for Coloured Petri Nets. In E. Brinksma, editor, *Proceedings of TACAS'97*, volume 1217 of *Lecture Notes in Computer Science*, pages 209–223. Springer-Verlag, 1997.

5. S. Christensen and J.B. Jørgensen. Analysis of Bang and Olufsen's BeoLink Audio/Video System Using Coloured Petri Nets. In P. Azéma and G. Balbo, editors, *Proceedings of ICATPN'97*, volume 1248 of *Lecture Notes in Computer Science*, pages 387–406. Springer-Verlag, 1997.

6. G. Ciardo, L. Cherkasova, V. Kotov, and T. Rokicki. Modeling a Scaleable High-Speed Interconnect with Stochastic Petri Nets. In *Proceeding of PNPM'95*, pages 83–93. IEEE Computer Society Press, 1995.

7. E.M. Clarke, E.A. Emerson, and A.P. Sistla. Automatic Verification of Finite State Concurrent Systems using Temporal Logic. *ACM Transactions on Programming Languages and Systems*, 8(2):244–263, 1986.

8. H. Clausen and P. R. Jensen. Validation and Performance Analysis of Network Algorithms by Coloured Petri Nets. In *Proceedings of PNPM'93*, pages 280–289. IEEE Computer Society Press, 1993.

9. J. Desel and W. Reisig. Place/Transition Petri Nets. In *Lecture on Petri Nets I: Basic Models*, volume 1491 of *Lecture Notes in Computer Science*, pages 122–173. Springer-Verlag, 1998.

10. E. A. Emerson. *Temporal and Modal Logic*, volume B of *Handbook of Theoretical Computer Science*, chapter 16, pages 995–1072. Elsevier, 1990.

11. E. A. Emerson, A.K. Mok, A.P Sistla, and J. Srinivasan. Quantitative Temporal Reasoning. In *Proceedings of CAV'90*, volume 531 of *Lecture Notes in Computer Science*, pages 136–145. Springer-Verlag, 1990.

12. K. Jensen. *Coloured Petri Nets. Basic Concepts, Analysis Methods and Practical Use. Volume 1, Basic Concepts*. Monographs in Theoretical Computer Science. Springer-Verlag, 1992.

13. K. Jensen. *Coloured Petri Nets. Basic Concepts, Analysis Methods and Practical Use. Volume 2, Analysis Methods*. Monographs in Theoretical Computer Science. Springer-Verlag, 1994.

14. L. M. Kristensen, S. Christensen, and K. Jensen. The Practitioner's Guide to Coloured Petri Nets. *International Journal on Software Tools for Technology Transfer*, 2(2):98–132, December 1998.

15. M. A. Marsan, G. Balbo, G. Conte, S. Donatelli, and G. Franceschinis. *Modelling with Generalized Stochastic Petri Nets*. Series in Parallel Computing. Wiley, 1995.

16. T. Murata. Petri Nets: Properties, Analysis and Application. In *Proceedings of the IEEE, Vol. 77, No. 4*. IEEE Computer Society, 1989.

17. W. M. P. van der Aalst. Interval Timed Coloured Petri Nets and their Analysis. In *Proceedings of ICATPN'93*, volume 691 of *Lecture Notes in Computer Science*, pages 453–472. Springer Verlag, 1993.

18. K. M. van Hee, L. J. Somers, and M. Voorhoeve. Executeable Specifications for Distributed Information Systems. In *Proceedings of IFIP TC8/WG 8.1 Working Conference on Information Systsm Concepts*, pages 139–156. Elsevier Science Publishers, 1989.

Unfolding of Products of Symmetrical Petri Nets

Jean-Michel Couvreur[1], Sébastien Grivet[1], and Denis Poitrenaud[2]

[1] LaBRI, Université de Bordeaux I, Talence, France
{couvreur, grivet}@labri.u-bordeaux.fr
[2] LIP6, Université Pierre et Marie Curie, Paris, France
Denis.Poitrenaud@lip6.fr

Abstract. This paper presents a general technique for the modular construction of complete prefixes adapted to systems composed of Petri nets. This construction is based on a definition of a well-adapted order allowing combination. Moreover, the proposed technique takes into account the symmetries of the system to minimize the size of the produced complete prefixes. Finally, the technique has been instantiated in an efficient algorithm for systems combining finite state machines and k-bounded queues with k *a priori* known or not.

1 Introduction

Unfoldings of Petri nets have been originally studied from a theoretical point of view by Nielsen ([12]) and Engelfriet ([4]). Since a decade, they are intensively used in the context of verification. All these methods are based on a structure introduced in [11], which is a prefix of the maximal unfolding of a system, sufficiently large enough to cover all the reachable states of the system. Such a prefix is said to be complete. We can distinguish two kinds of works in this area. The first one consists in the definition of efficient algorithms for the construction of complete prefixes. The papers [8,7] are significant examples of such optimizations. The second type of research is the design of verification techniques based on complete prefixes ([5,3,2,6]).

In this paper, we focus our attention on the efficiency of the construction algorithm. Our starting point is the works presented in [10] and [7]. Langerak and Brinksma have proposed an order over the processes composing a system which allows to minimize the size of the complete prefix. Indeed, such an order is used to detect the points from which the construction can be stopped. The one presented in [10], is defined for systems consisting of synchronization of finite state machines. Esparza and Römer, in [7], have proposed a very efficient algorithm based on this order and taken benefit of the particular structure of the components of the system.

The present paper generalizes these works in two directions. First, we want to deal with systems communicating by rendez-vous but also by message passing. Is the order presented in [10] suitable for this kind of systems? Can we define some more efficient order specific to queues? Second, we want to generalize the method to components other than finite state machines and, if possible, to deal

J.-M. Colom and M. Koutny (Eds.): ICATPN 2001, LNCS 2075, pp. 121–143, 2001.
© Springer-Verlag Berlin Heidelberg 2001

with general Petri nets. The solution that we propose for the communication by message passing is based on considering the symmetries induced by the chosen model of queues. This allows us to minimize the size of the constructed complete prefixes. To deal with components defined by Petri nets, we show that the key point of an efficient implementation is the way in which the components are synchronized. We define a constraint on the synchronization under which a modular computation of complete prefixes is then possible.

The main results presented in this paper are the following:

– The definition of a new suitable order for the computation of complete prefixes allowing the combination of orders in new ones.
– The consideration of symmetries in the unfolding technique to minimize the size of the complete prefixes.
– The definition of a modular computation of prefixes.
– The design of an efficient implementation for systems composed of finite state machines and queues.

First, we recall the theoretical background of unfolding. In the corresponding section, the new order is presented over processes. Then, symmetries are introduced in the unfolding technique and are illustrated through different models of queues. In Section 4, the unfolding of systems composed by the synchronization of Petri nets is studied and the constraint concerning the synchronization and its effects is illustrated. It follows a section dedicated to the implementation for systems composed by finite state machines and queues. Some concluding remarks and perspectives close the paper.

2   Preliminaries

After recalling the basic definitions related to Petri nets, this section is dedicated to the theoretical context of the unfolding method. Many of the definitions are adapted from the works of Engelfriet [4] and of Esparza [8].

2.1   Petri Nets

Definition 1. *A (Petri) net is a tuple $\langle P, T, F \rangle$, where P is a set of places, T is a set of transitions, P and T are disjoint, $F \subseteq (P \times T) \cup (T \times P)$ and $\forall t \in T, \exists p \in P : \langle p, t \rangle \in F$. The preset of a node $n \in P \cup T$, denoted by $\bullet n$, is the set of nodes $\{n' \mid (n', n) \in F\}$. The postset of a node $n \in P \cup T$, denoted by $n^\bullet$, is the set of nodes $\{n' \mid (n, n') \in F\}$.*

Here, a net is possibly infinite. The main natural restriction is of finite synchronization, i.e. for every transition t, $\bullet t$ and $t^\bullet$ are finite sets, and moreover, we assume $\bullet t$ to be nonempty.

Definition 2. *A marking of a net $N = \langle P, T, F \rangle$ is a multiset on P. A marked net $\langle N, m_0 \rangle$ is a net with an associated initial marking m_0.*

Definition 3. *The firing rule of net $N = \langle P, T, F \rangle$ is defined as follows. Let m and m' be two markings of N and t be a transition. Then $m \xrightarrow{t} m'$ iff $^\bullet t \subseteq m$ and $m' = m - {}^\bullet t + t^\bullet$. A marking m is reachable from a marking m_0 if there exists a firing sequence $t_1 \cdots t_n$ such that $m_0 \xrightarrow{t_1} m_1 \cdots m_{n-1} \xrightarrow{t_n} m$. The set of markings reachable from m_0 is denoted by $Reach(N, m_0)$.*

In this paper, we deal with systems constructed from a set of components which are synchronised on events having a common label. Hence, we introduce the notion of labelled nets.

Definition 4. *A labelled net is a tuple $\langle P, T, F, A, \lambda \rangle$, where $\langle P, T, F \rangle$ is a net, A is an alphabet (of actions) and $\lambda : T \to A$ is a labelling.*

Unfoldings are defined by way of homomorphisms from nets to nets (see [4]). Intuitively, a homomorphism h from net N_1 to net N_2 formalizes the fact that N_1 can be obtained by partially unfold a part of N_2. Firing sequences of different nets can be related through homomorphisms.

Definition 5. *Let $N_1 = \langle P_1, T_1, F_1 \rangle$, $N_2 = \langle P_2, T_2, F_2 \rangle$ be two nets. A homomorphism from N_1 to N_2 is a mapping $h : P_1 \cup T_1 \to P_2 \cup T_2$ such that:*

- *$h(P_1) \subseteq P_2$ and $h(T_1) \subseteq T_2$,*
- *$\forall t_1 \in T_1 : h_l(^\bullet t_1) = {}^\bullet h(t_1)$ and $h_l(t_1^\bullet) = h(t_1)^\bullet$*

where $h_l : \mathbb{N}^{P_1 \cup T_1} \to \mathbb{N}^{P_2 \cup T_2}$ is the linear extension of h. If N_1 and N_2 have initial markings m_1 and m_2 then we require that $h_l(m_1) = m_2$. Moreover, if $N_1 = \langle P_1, T_1, F_1, A, \lambda_1 \rangle$ and $N_2 = \langle P_2, T_2, F_2, A, \lambda_2 \rangle$ are labelled nets on the same alphabet A then $\lambda_1(t_1) = \lambda_2(h(t_1))$ for all $t_1 \in T_1$.

Proposition 1. *Let $N_1 = \langle P_1, T_1, F_1 \rangle$, $N_2 = \langle P_2, T_2, F_2 \rangle$ be two nets. Let h be a homomorphism from N_1 to N_2. If $m_0 \xrightarrow{t_1} m_1 \cdots m_{n-1} \xrightarrow{t_n} m$ is a firing sequence in N_1 then $h_l(m_0) \xrightarrow{h(t_1)} h_l(m_1) \cdots h_l(m_{n-1}) \xrightarrow{h(t_n)} h_l(m)$ is a firing sequence in N_2.*

2.2 Branching Processes

In this subsection, the partial order semantics of a net is defined as a particular net called its maximal branching process. Intuitively, a branching process formalizes a set of behaviors of the net and its structure allows to determine the causality of the events as well as the conflicts and the concurrency.

Definition 6. *Let $N = \langle P, T, F \rangle$ be a net and let $u, v \in P \cup T$. The precedence relation $\leq$, conflict relation $\sharp$ and concurrent relation $\parallel$ are defined by:*

- *$u \leq v$ iff (u, v) belongs to the reflexive transitive closure of F,*
- *$u \sharp v$ iff $\exists t_1, t_2 \in T : t_1 \neq t_2$, $^\bullet t_1 \cap {}^\bullet t_2 \neq \emptyset$, and $t_1 \leq u \wedge t_2 \leq v$,*
- *$u \parallel v$ iff neither $u \leq v$ nor $v \leq u$ nor $u \sharp v$).*

Branching processes belong to the subclass of Petri nets called occurrence nets.

Definition 7. *An occurrence net is a net $ON = \langle B, E, F \rangle$ where B and E are called conditions and events, and*

- $\forall b \in B :| \, {}^\bullet b \, | \leq 1$,
- *F is acyclic (i.e. the relation $\leq$ is a partial order),*
- *ON is finitely preceded: $\forall u \in B \cup E :| \, \{v \in B \cup E \mid v \leq u\}$ is finite,*
- *no event is in self-conflict: $\forall e \in E : \neg(e \sharp e)$*

Moreover, if $|b^\bullet| \leq 1$ for all $b \in B$ then N is called a causal net. An occurrence net has an implicit associated initial marking $Min(ON) = \{b \in B \mid {}^\bullet b = \emptyset\}$.

A branching process is simply an occurrence net associated to a homomorphism.

Definition 8. *A branching process of net N is a pair $\beta = \langle ON, h \rangle$ such that*

- *$ON = \langle B, E, F \rangle$ is an occurrence net,*
- *$h : ON \rightarrow N$ is a homomorphism,*
- *$\forall e_1, e_2 \in E$, if ${}^\bullet e_1 = {}^\bullet e_2$ and $h(e_1) = h(e_2)$ then $e_1 = e_2$.*

A branching process β of a net N has a corresponding initial marking in N, $Min_N(\beta) = h_l(Min(ON))$. For a branching process of a marked net $\langle N, m_0 \rangle$, we require that $Min_N(\beta) = m_0$. Moreover, if ON is a causal net, then β is called a process of N.

Homomorphisms are used to characterize the prefixes of a branching process as well as isomorphism of branching processes.

Definition 9. *Let $\beta_1 = \langle ON_1, h_1 \rangle$, $\beta_2 = \langle ON_2, h_2 \rangle$ be two branching processes of a marked net $\langle N, m_0 \rangle$. A homomorphism g from β_1 to β_2 is a homomorphism from ON_1 to ON_2 which fulfils $h_1 = h_2 \circ g$. Moreover,*

- *β_1 is a prefix of β_2 (denoted by $\beta_1 \sqsubseteq \beta_2$) iff there exists an injective homomorphism g from β_1 to β_2 such that $g(Min(ON_1)) = Min(ON_2)$,*
- *β_1 and β_2 are isomorphic (denoted by $\beta_1 = \beta_2$) iff $\beta_1 \sqsubseteq \beta_2 \wedge \beta_2 \sqsubseteq \beta_1$.*

Based on the fact that $\sqsubseteq$ forms a partial order on the branching processes of a net (up to isomorphism) and is a complete lattice (see [4]), the existence of a unique maximal branching process is ensured.

Theorem 1. *Let $\langle N, m_0 \rangle$ be a marked net. There exists a unique maximal branching process $\beta(N)$ of $\langle N, m_0 \rangle$ (up to isomorphism).*

For convenience, we will used the following notations: $\beta = \langle ON(\beta), h(\beta) \rangle$ and $ON(\beta) = \langle B(\beta), E(\beta), F(\beta) \rangle$ to denote the different components of a branching process β of a net $N(\beta)$, $Min(\beta) = Min(ON(\beta))$ and $Min_N(\beta) = h(\beta)(Min(\beta))$ are the initial markings of β in β and in $N(\beta)$.

2.3   Processes

The most important notion regarding occurence nets is that of a configuration. A configuration represents a possible partial run of the net.

Definition 10. *A configuration C of a branching process β is a set of events satisfying the two following properties:*

- *C is causally left-closed: $\forall e_1, e_2 \in E(\beta) : e_1 \in C \wedge e_2 \leq e_1 \Rightarrow e_2 \in C$,*
- *C is conflict-free: $\forall e_1, e_2 \in C : \neg(e_1 \sharp e_2)$.*

If U is a conflict-free set of events and conditions of β, the set $[U] = \{e \in E(\beta) \mid \exists u \in U : e \leq u\}$ is a configuration. A finite configuration has an associated terminal marking in β and in $N(\beta)$: $Cut(C) = (Min(\beta) \cup C^\bullet) \setminus {}^\bullet C$ and $Cut_N(C) = h(\beta)(Cut(C))$. Moreover, if β is a process, then $E(\beta)$ is a configuration, $Cut(\beta) = Cut(E(\beta))$ and $Cut_N(\beta) = Cut_N(E(\beta))$ are its terminal markings in β and $N(\beta)$.

Configurations and processes that are prefixes of a same branching process are closely related.

Proposition 2. *Let π be a process, prefix of a branching process β. Let g be an injective homomorphism from π to β. Then $g(E(\pi))$ is a configuration of β. Moreover, if C is a configuration of a branching process β then there exists a unique process (up to isomorphism), prefix of β, denoted by $\Pi(C)$, such that there exists an injective homomorphism g from $\Pi(C)$ to β satisfying $C = g(E(\Pi(C)))$.*

An important operation for the construction and the proofs is the concatenation.

Proposition 3. *Let π_1 be a finite process, and π_2 be a processes of a net N such that $Cut_N(\pi_1) = Min_N(\pi_2)$. Then there exist at least one process π_3, two injective homomorphisms $g_1 : \pi_1 \rightarrow \pi_3$ and $g_2 : \pi_2 \rightarrow \pi_3$ such that*

- *$g_1(E(\pi_1) \cup B(\pi_1)) \cup g_2(E(\pi_2) \cup B(\pi_2)) = E(\pi_3) \cup B(\pi_3)$,*
- *$g_1(E(\pi_1)) \cap g_2(E(\pi_2)) = \emptyset$,*
- *$g_1(B(\pi_1)) \cap g_2(B(\pi_2)) = g_1(Cut(\pi_1)) = g_2(Min(\pi_2))$,*
- *$g_1(Min(\pi_1)) = Min(\pi_3)$ and $g_2(Cut(\pi_2)) = Cut(\pi_3)$.*

Such a process π_3 is called a concatenation of π_1 and π_2. The set of concatenations of π_1 and π_2 are denoted by $\pi_1 \cdot \pi_2$.

2.4   Finite Complete Prefixes

We are now in the position to define the characteristics of finite prefixes which are the basis of the verification technique.

Definition 11. *A branching process β of a marked net $\langle N, m_0 \rangle$ is a finite complete prefix iff β is finite (i.e. the set $B(\beta) \cup E(\beta)$ is finite) and for any reachable marking m of $\langle N, m_0 \rangle$, there exists a process π such that $\pi \sqsubseteq \beta$ and $Cut_N(\pi) = m$.*

For the construction of such prefixes, some events are identified as points from which it is not necessary to extend the branching process. These particular events are called cutoffs and have to ensure the stability (all the firings are represented) and the completeness (all the reachable markings are represented) of the construction.

Definition 12. *Let β be a branching process of a marked net $\langle N, m_0 \rangle$. Let Cutoff be a set of events of β. The couple $\langle \beta, Cutoff \rangle$ is a stable branching process iff β is finite and for every process π of $\langle N, m_0 \rangle$ one of the two following properties holds:*

- $\exists C \in Conf(\beta) : C \cap Cutoff = \emptyset \wedge \Pi(C) = \pi$,
- $\exists e \in Cutoff : \Pi([e]) \sqsubseteq \pi$.

where $Conf(\beta)$ denotes the set of configurations of β.

An internal configuration C is a configuration of β which contains no event of Cutoff $(C \cap Cutoff = \emptyset)$. A marking m of N is an internal marking of β if m is the cut of an internal configuration C $(m = Cut_N(C))$. The sets of internal configurations and internal markings are respectively denoted by $Conf_I(\beta)$ and $Reach_I(\beta)$.

Now, we make precise the characteristics of the cutoffs.

Definition 13. *An unfolding of a marked net $\langle N, m_0 \rangle$ is a tuple $\langle \beta, Cutoff, \Phi \rangle$ where $\langle \beta, Cutoff \rangle$ is a stable branching process and Φ is a mapping from Cutoff to $Conf_I(\beta)$ such that $\forall e \in Cutoff : Cut_N(\Phi(e)) = Cut_N([e])$.*

We define the notion of well-adapted order over processes. These orders allow us to build finite complete prefixes. The definition of well-adapted order reduces that of adequate order used by Esparza and al. [8] to allow the design a new well-adapted order from the combination of other ones. Notice that well-adapted orders are defined over processes and not over configurations of a given branching process as in [8].

Definition 14. *A partial order $\preceq$ over finite processes of a marked net $\langle N, m_0 \rangle$ is a well-adapted order if:*

- *$\preceq$ is well-founded and refines $\sqsubseteq$,*
- *$\preceq$ is a pre-total order: for all processes π_1, π_2 of $\langle N, m_0 \rangle$: $\pi_1 \prec \pi_2 \vee \pi_2 \prec \pi_1 \vee \pi_1 \equiv_\preceq \pi_2$ where $\pi_1 \equiv_\preceq \pi_2$ iff for all processes π: $\pi_1 \prec \pi \Leftrightarrow \pi_2 \prec \pi$ and $\pi \prec \pi_1 \Leftrightarrow \pi \prec \pi_2$,*
- *$\preceq$ is compatible with concatenation: for all processes π_1, π_2 of $\langle N, m_0 \rangle$ such that $Cut_N(\pi_1) = Cut_N(\pi_2)$, the following two properties hold:*
  1. *$\pi_1 \prec \pi_2 \Rightarrow$ for all process π_3 of $\langle N, Cut_N(\pi_1) \rangle, \forall \pi_{13} \in \pi_1 \cdot \pi_3, \forall \pi_{23} \in \pi_2 \cdot \pi_3 : \pi_{13} \prec \pi_{23}$.*
  2. *$\pi_1 \equiv_\preceq \pi_2 \Rightarrow$ for all process π_3 of $\langle N, Cut_N(\pi_1) \rangle, \forall \pi_{13} \in \pi_1 \cdot \pi_3, \forall \pi_{23} \in \pi_2 \cdot \pi_3 : \pi_{13} \equiv_\preceq \pi_{23}$.*

One can remark that the McMillan's order $\preceq_{Mc}$ [11] is well-adapted. Let us recall that for two processes π_1 and π_2 of a marked net $\langle N, m_0 \rangle$ such that $Cut_N(\pi_1) = Cut_N(\pi_2)$, $\pi_1 \prec_{Mc} \pi_2$ holds iff $|E(\pi_1)| < |E(\pi_2)|$. Moreover, the order proposed by Esparza and al. in [8] for safe Petri nets is also well-adapted. We are now in the position to complete our characterisation of the cutoffs composing an unfolding.

Definition 15. *An unfolding $\langle \beta, Cutoff, \Phi \rangle$ of a marked net $\langle N, m_0 \rangle$ is well-adapted iff there exists a well-adapted order $\preceq$ such that $\forall e \in Cutoff : \Pi(\Phi(e)) \prec \Pi([e])$.*

Proposition 4. *If $\langle \beta, Cutoff, \Phi \rangle$ is a well-adapted unfolding of a marked net $\langle N, m_0 \rangle$ then β is a finite complete prefix. Moreover $Reach_I(\beta) = Reach(N, m_0)$. If the set $Reach(N, m_0)$ is finite and if for each marking in $Reach(N, m_0)$, the number of firable transitions is finite then $\langle N, m_0 \rangle$ has at least one well-adapted unfolding.*

Proof. One can apply Esparza and al. in [8] for adequate order. Indeed well-adapted order is adequate order, and then $Reach_I(\beta) = Reach(N, m_0)$. Because well-adapted order refine inclusion, the existence of a well-adapted unfolding is ensured in the finite case. □

We have claimed that the particularity of well-adapted orders is their capability to be combined into new a well-adapted order. This proposition is formalized here.

Definition 16. *Let $\preceq_1, \preceq_2$ be two orders over a set. The order $lex(\preceq_1, \preceq_2)$ is defined by $x \; lex(\preceq_1, \preceq_2) \; y$ iff $x \preceq_1 y \vee (x \equiv_{\preceq_1} y \wedge x \preceq_2 y)$. Let $\preceq_1, \ldots, \preceq_n$ be orders over a set, $lex(\preceq_1, \ldots, \preceq_n)$ is defined inductively as $lex(\preceq_1, lex(\preceq_2, \ldots, \preceq_n))$.*

Proposition 5. *Let $\preceq_1, \preceq_2$ be two well-adapted orders over finite processes of a marked net $\langle N, m_0 \rangle$. The order $lex(\preceq_1, \preceq_2)$ is well-adapted.*

Proof. The proof is simple and tedious. One has just to check that $lex(\preceq_1, \preceq_2)$ is a pre-order and fullfils the well-adapted conditions. □

As an example of the construction of a well-adapted order from others, one may define $\preceq_t$ through the number $|\pi|_t$ of events labelled by a given transition t appearing in a process π: $\pi \prec_t \pi'$ iff $|\pi|_t < |\pi'|_t$.

Suppose that one fixes an arbitrary total order over the transitions of a net $(T = \{t_1, \ldots, t_n\})$. Hence, the order $lex(\preceq_{Mc}, \preceq_{t_1}, \ldots, \preceq_{t_n})$ is a well-adapted order. Applying this new order on the examples presented in [8], the combinatorial explosion problem appearing with $\preceq_{Mc}$ is corrected.

3   Unfoldings of Symmetrical Petri Nets

In this section, we introduce the symmetrical Petri nets and show how one can take benefit of the symmetries to reduce the complete prefix of this kind of net. As our main goal is to analyse systems communicating by message passing, we will discuss the different modelling of queues and their consequences for the constructed prefixes.

3.1   Symmetrical Petri Nets

Symmetries are defined by the way of a group of mappings on the elements of the net.

Definition 17. *A symmetrical net is defined by a couple* $\langle N, G \rangle$ *where*

- *N is a labelled net,*
- *G is a group of automorphisms of N (i.e. a group of bijective homomorphisms from N to N).*

A marked symmetrical net is a tuple $\langle N, G, m_0 \rangle$ *where m_0 is a marking of N.*

From this definition, the notion of equivalent markings is introduced. This notion has been already used for the construction of reduced reachability graphs which are the basis of verification methods for qualitative properties [9] as well as for quantitative ones [1].

Definition 18. *Let $\langle N, G \rangle$ be a symmetrical net. Two markings m and m' of N are equivalent (denoted by $m \equiv m'$) iff $\exists g \in G$ such that $g_l(m) = m'$. For a marking m of N, we denote by $\widehat{m}$ the set of markings equivalent to m. For a set of markings M, we denote by $\widehat{M}$ the set $\{\widehat{m} \mid \exists m' \in M : m \equiv m'\}$.*

The introduction of queues in Petri nets leads to a strict extension of the ordinary (i.e. finite) model. Indeed, one has to deal with the order and the value of the tokens entered in the queue and consequently with its size. When the maximal size of the queue is known a priori, it is possible to represent it by a finite Petri net. On the contrary, the modelling of a queue for which the maximal size is unknown leads to an infinite Petri net. We will study these two cases.

Fig. 1 gives a first attempt of modelling a finite queue. The queue can receive two types of messages (a and b) and its maximal size is n. The transitions sx correspond to the entering of a message $x \in \{a, b\}$ where the firing of a transition rx indicates that a message x leaves the queue. Each position $i \in [1, n]$ of the queue is modelled by a set of places $\{Ma_i, Mb_i, F_i\}$ where the two first places indicate the presence of a message and the last one the emptiness of the position. Obviously, for any i, these three places are in mutual exclusion. When a message is enqueued, it is stored in the first position. Then, the message has to cross all the positions before being dequeued. The main disadvantage of this model is that it does not allow to identify two states where the queue contains the same messages in the same order but in different positions, and the symmetries cannot

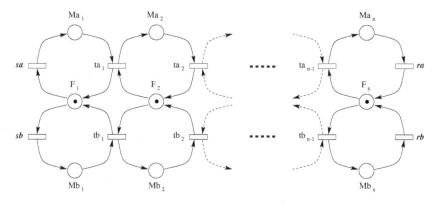

Fig. 1. A first modelling of a finite FIFO

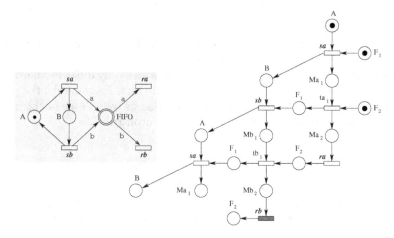

Fig. 2. A Petri net using a FIFO (size 2) of Fig. 1, and its complete prefix

help us in this task. Moreover, the internal transitions tx_i (with $x \in \{a, b\}$ and $i \in [1, n-1]$) lead to the production of numerous intermediary states.

A complete prefix of the net of Fig. 2 using this modelling of a queue and illustrating this situation is given in Fig. 2. The queue used in Fig. 2 is represented by a doubly bordered circle and the surrounding transitions sa, sb, ra and rb representing sending and receiving of messages have to be synchronized with the corresponding transitions of the queue model. One can remark that for a queue of size n, the complete prefix of the net is constituted of $\frac{1}{2}(n+1)(n+2)+1$ events where the number of reachable states of the system is $2(n+1)$.

A second modelling of a queue is presented in Fig. 3. It is based on a circular array and uses two counters In and Out. The value of In (resp. Out) indicates the position in the array where the following arriving (resp. departing) message must be stored (resp. taken). With this modelling, the intermediary states of the first solution are discarded. This is illustrated by the complete prefix of the net

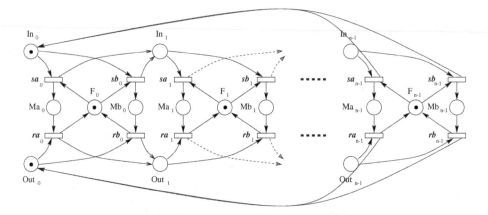

Fig. 3. A second modelling of a finite FIFO

of Fig. 2 presented in Fig. 4. We can remark that the number of events is then linear in the size of the queue. However, this modelling can also introduce some more states. Indeed, two states of the queue which only differ by the values of the counters In and Out will be considered as distinct. Fortunately, the symmetries of the net can be helpful to avoid this situation as we will see in the next section by considering equivalent markings for the cutoff rule.

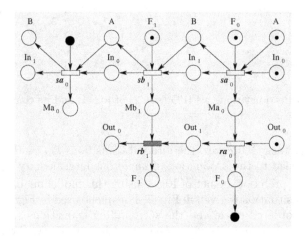

Fig. 4. A complete prefix of net of Fig. 2 using a FIFO (size 2) of Fig. 3

3.2   Finite Complete Prefixes of Symmetrical Petri Nets

A branching process of a symmetrical net is simply defined as a branching process of the underlying net.

Definition 19. *A branching process of a marked symmetrical net $\langle N, G, m_0 \rangle$ is a branching process of $\langle N, m_0 \rangle$.*

The symmetries allow us to relax the condition on the representation of all the markings for the completeness.

Definition 20. *A branching process β of a marked symmetrical net $\langle N, G, m_0 \rangle$ is a finite complete prefix iff β is finite (i.e. the set $B(\beta) \cup E(\beta)$ is finite) and for any reachable marking m, there exists a process π such that $\pi \sqsubseteq \beta$ and $Cut_N(\pi) \equiv m$.*

Similarly, the condition on the cutoffs can also be relaxed.

Definition 21. *An unfolding of a marked symmetrical net $\langle N, G, m_0 \rangle$ is a tuple $\langle \beta, Cutoff, \Phi \rangle$ where $\langle \beta, Cutoff \rangle$ is stable branching process and Φ is a mapping from Cutoff to $Conf_I(\beta)$ which fulfils $\forall e \in Cutoff : Cut_N(\Phi(e)) \equiv Cut_N([e])$.*

The partial orders used for the construction have to be reviewed under the angle of symmetries. We first define how a new process can be constructed from an automorphism.

Definition 22. *Let $\pi = \langle CN, h \rangle$ be a process of a marked symmetrical net $\langle N, G, m_0 \rangle$. Let g be an automorphism of G. We define $g(\pi)$ as the process $\langle CN, g \circ h \rangle$.*

Definition 23. *A partial order $\preceq$ on finite processes of a marked symmetrical net $\langle N, G, m_0 \rangle$ is a well-adapted order if:*

- *$\preceq$ is well-founded,*
- *$\preceq$ refines $\sqsubseteq$,*
- *$\preceq$ is a pre-total order,*
- *$\preceq$ is compatible with the concatenation: $\forall \pi_1, \pi_2$ processes of $\langle N, G, m_0 \rangle$ such that $\exists g \in G, g(Cut_N(\pi_1)) = Cut_N(\pi_2)$:*
    1. *$\pi_1 \prec \pi_2 \Rightarrow \forall \pi_3$ process of $\langle N, Cut_N(\pi_1) \rangle, \forall \pi_{23} \in \pi_2 \cdot g(\pi_3), \forall \pi_{13} \in \pi_1 \cdot \pi_3 : \pi_{13} \prec \pi_{23}.$*
    2. *$\pi_1 \equiv_{\prec} \pi_2 \Rightarrow \forall \pi_3$ process of $\langle N, Cut_N(\pi_1) \rangle, \forall \pi_{13} \in \pi_1 \cdot \pi_3, \forall \pi_{23} \in \pi_2 \cdot g(\pi_3) : \pi_{13} \equiv_{\prec} \pi_{23}.$*

We can remark that the order $\preceq_{Mc}$ is well-adapted for any symmetrical Petri net. Indeed, the identity of the event is not taken into account for this order. On the contrary, the order $\preceq_t$ (defined at the end of Sect. 2) is not necessary well-adapted. The number of events labelled by any transition equivalent to t has to be taken into account. For a given transition t of a symmetrical Petri net $\langle N, G, m_0 \rangle$, let us denote with $|\pi|_{\hat{t}}$ the number of events e of π such that $\exists g \in G, g(h(\pi)(e)) = t$. Thus, the order $\preceq_{\hat{t}}$ defined by $\pi \prec_{\hat{t}} \pi'$ iff $|\pi|_{\hat{t}} < |\pi'|_{\hat{t}}$ is well-adapted.

Definition 24. *An unfolding $\langle \beta, Cutoff, \Phi \rangle$ of a marked symmetrical net $\langle N, G, m_0 \rangle$ is well-adapted iff there exists a well-adapted order $\preceq$ such that $\forall e \in Cutoff : \Pi(\Phi(e)) \prec \Pi([e])$.*

Proposition 6. *If $\langle \beta, Cutoff, \Phi \rangle$ is a well-adapted unfolding of a marked symmetrical net $\langle N, G, m_0 \rangle$ then β is a finite complete prefix. Moreover, $\widehat{Reach_I}(\beta) = \widehat{Reach}(N, m_0)$. If the set $Reach(N, m_0)$ is finite and for any marking in $Reach(N, m_0)$, the number of firable transitions is finite then $\langle N, G, m_0 \rangle$ has at least one well-adapted unfolding.*

Proof. This proof is almost the same as the proof of Proposition 4. The proof of "β is a finite complete prefix" just used the well-founded property and the first assertion of the compatibility property of the well-adapted order. The existence of a well-adapted unfolding is also deduced from the fact that $\preceq$ refines $\sqsubseteq$. □

The following proposition claims that the combination of well-adapted orders remains possible for symmetrical Petri nets.

Proposition 7. *Let $\preceq_1, \preceq_2$ be two well-adapted orders on finite processes of a marked symmetrical net $\langle N, G, m_0 \rangle$. The order $lex(\preceq_1, \preceq_2)$ is well-adapted.*

Proof. This proof is almost the same as the proof of Proposition 5. □

The modelling of *a priori* unbounded queues is similar to that of finite ones. It is based on the use of an infinite array and two counters *In* and *Out*. This modelling is presented in Fig. 5 and allows to analyze systems for which the maximal size of the queue is unknown *a priori*.

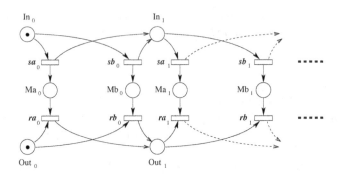

Fig. 5. A modelling of a finite unknown bound FIFO

Fig. 6 shows an example of a system using a finite unknown bound queue and its complete prefix. It is important to notice that the analysis of the complete prefix allows us to determine that the effective bound of the queue for this system is three.

4   Unfoldings of Products of Symmetrical Petri Nets

Considering a system as a set of inter-operating components has been the starting point of numerous optimisations of verification methods. Here, we present the

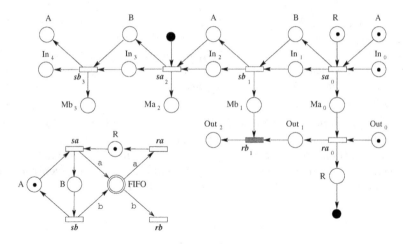

Fig. 6. A net using a finite unknown bound FIFO of Fig. 5 and its complete prefix

theoretical background which allows the design of a construction algorithm of the
complete prefix of a system by the analysis of the unfoldings of its components.
We first present the model that we consider, and then study its unfoldings.

4.1   Products of Petri Nets

We focus our attention on Petri nets synchronized on actions. Combined with
queues, the model can handle systems communicating by message passing as
well as rendez-vous.

Definition 25. *Let $N_1, \ldots, N_n$ be labelled nets, where $N_i = \langle P_i, T_i, F_i, A_i, \lambda_i \rangle$
(we assume for convenience that the sets P_i are pairwise disjoint as well as the
sets T_i). The product $N = \langle P, T, F, A, \lambda \rangle$ of the N_i is the labelled net defined as
follows:*

- $P = \bigcup_i P_i$
- $T = \{t \in \prod_i (T_i \cup \{\epsilon\}) \mid \exists a \in \bigcup_i A_i, \forall i : (a \in A_i \wedge t[i] \in T_i \wedge \lambda_i(t[i]) = a) \vee (a \notin A_i \wedge t[i] = \epsilon)\}$
- $\forall t \in T, \forall i, \forall p \in P_i : \langle p, t \rangle \in F$ *(resp.* $\langle t, p \rangle \in F$*) iff* $\langle p, t[i] \rangle \in F_i$ *(resp.*
  $\langle t[i], p \rangle \in F_i$*)*
- $A = \bigcup_i A_i$
- $\forall t \in T, \lambda(t) = \lambda_i(t[i])$ *where i is any $i \in [1, n] : t[i] \neq \epsilon$*

*Moreover if the N_i have m_i as initial markings then N has $m_0 = \sum_i m_i$ as
initial marking. We denote by $\bigotimes_i N_i$ (resp. $\bigotimes_i \langle N_i, m_i \rangle$) the product of the nets
N_i (resp. marked nets $\langle N_i, m_i \rangle$).*

We restrict the model to systems for which the rendez-vous mechanism is
not confusing. More precisely, we impose that when a service is asked for by a
component from another then there is no more than one candidate in the server.
A component having this property is said to be non-reentrant.

Definition 26. *Let* $N = \langle P, T, F, A, \lambda, m_0 \rangle$ *be a labelled marked net. A label* a *in* A *is non-reentrant in* N *iff* $\forall m \in Reach(N, m_0), \forall t, t' \in T, \lambda(t) = \lambda(t') = a \Rightarrow {}^{\bullet}t + {}^{\bullet}t' \not\subseteq m$. *Let* $N = \bigotimes_i \langle N_i, m_i \rangle$ *be a product. The product* N *has non-reentrant synchronization iff* $\forall a \in A(N), \forall i, j, (i \neq j \wedge a \in A_i \cap A_j) \Rightarrow a$ *is non-reentrant in* N_i. *For convenience such a product is said to be a non-reentrant product.*

Fig. 7 presents a set $\{N_1, N_2\}$ of labelled nets and their product (on the right). One can remark that this product has non-reentrant synchronization. On the contrary, if one does not distinguish x and y then the synchronisation is not non-reentrant any more. We will see in the following that the reentrant model causes problems for the modular view of the unfolding.

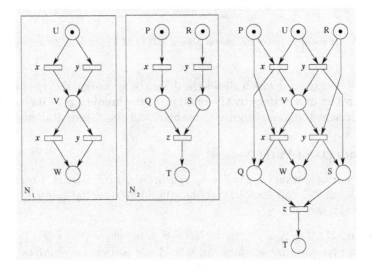

Fig. 7. A product having a non-reentrant synchronization

When the queues presented in Fig. 3 and 5 are considered as components of a product net, the transitions sx_i and rx_i are respectively labelled by sx and rx. These two nets are non-reentrant due to the counters In and Out.

4.2   Branching Processes of a Product of Petri Nets

From this point of the paper and in the following, we consider a set of marked labelled nets $\langle N_i, m_i \rangle, 1 \leq i \leq n$ and their product net $\langle N, m_0 \rangle = \bigotimes_i \langle N_i, m_i \rangle$. We denote for all i by $\beta_i = \langle \langle B_i, E_i, F_i \rangle, h_i \rangle$ the maximal branching process of $\langle N_i, m_i \rangle$ and by $\beta = \langle \langle B, E, F \rangle, h \rangle$ any branching process of $\langle N, m_0 \rangle$. One can consider a branching process β_i as a labelled net (the labelling function is $\lambda_i \circ h_i$). Thus, the product $\bigotimes_i \beta_i$ is well defined. Notice that this product is not necessarily an unfolding. However, it is used as an intermediary structure to define a labelling function which connects the elements of β to those of the β_i.

Proposition 8. *Let β be a branching process of the product $\langle N, m_0 \rangle = \bigotimes_i \langle N_i, m_i \rangle$. Then there exists a homomorphism Ψ from β to $\bigotimes_i \beta_i$ such that:*

- *$\forall b \in B, \forall i : \Psi(b) \in B_i \Rightarrow h(b) = h_i(\Psi(b))$*
- *$\forall e \in E, \forall i : if \Psi(e)[i] \in E_i$ then $h(e)[i] = h_i(\Psi(e)[i])$ else $h(e)[i] = \epsilon$*

Proof. We first set the values of the mapping Ψ for the initial conditions of β. Applying the definition of a branching process: $m_0 = h(Min(\beta)) = \sum_i h_i(Min(\beta_i))$. This proves that there exists a bijection θ between the sets $Min(\beta)$ and $\bigcup_i(Min(\beta_i))$ such that $\forall b \in Min(\beta), \forall i : \theta(b) \in B_i \Rightarrow h(b) = h_i(\theta(b))$. We set $\forall b \in Min(\beta) : \Psi(b) = \theta(b)$.

We define the value of the mapping Ψ for any event e and its output conditions inductively on the size of its local configuration $[e]$ such that the two requirements of the proposition and the homomorphism requirements hold for any event and any condition already defined. Let M_e be the marking in β obtained by firing all the events less than e. By induction the mapping Ψ is already defined for any condition in M_e, and hence $\Psi(M_e)$ is a reachable marking in $\bigotimes_i \beta_i$. If $h(e)[i] = \epsilon$, we set $\Psi(e)[i] = \epsilon$. If $h(e)[i] \neq \epsilon$, $\Psi(^\bullet e) \cap B_i$ is included in $\Psi(M_e)$, and hence is a nonempty set of concurrent conditions in β_i such that $h_i(\Psi(^\bullet e) \cap B_i) = {}^\bullet h(e)[i]$. We set $\Psi(e)[i] = e_i$ with $e_i \in E_i$ such that $h_i(e_i) = h(e)[i]$ and $^\bullet e_i = \Psi(^\bullet e) \cap B_i$. Because the partial markings $h(e^\bullet) \cap N_i$ and $h_i(e_i{}^\bullet)$ are equal, one can set the mapping Ψ for any condition b in $e^\bullet$ with $h(b) \in N_i$. This concludes the inductive construction of Ψ, and thus concludes the proof. $\square$

Fig. 8 illustrates the mapping Ψ on the product net of Fig. 7. On the left, the maximal branching processes of the components are shown. The maximal branching process of the product is presented on the right of the figure and for each element is given the image of Ψ. Moreover, the projection on each component of the local configuration $[e]$ is given as a tuple for each event e.

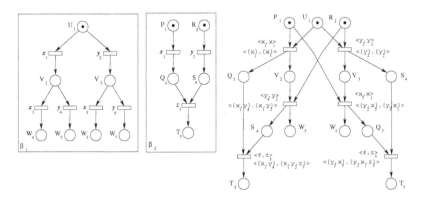

Fig. 8. Branching processes of a non-reentrant product and its components

The mapping Ψ is used to characterize the behavior of each component from a behavior of the product. Moreover, the property of non-reentrance ensures that this characterization is unique.

Proposition 9. *Let β be a branching process of the product $\langle N, m_0 \rangle$. Let C be a configuration of β. Let $\forall i, C[i] = \{\Psi(e)[i] \mid e \in C \wedge \Psi(e)[i] \neq \epsilon\}$. The set $C[i]$ is a configuration of β_i and $\Psi(Cut(C)) = \bigcup_i Cut(C[i])$. Moreover, if $\langle N, m_0 \rangle$ is non-reentrant then the mapping which associates to each configuration C the tuple $\langle C[1], \ldots, C[n] \rangle$ is injective.*

Proof. We first prove that $C[i]$ is a configuration of β_i. Let σ be a firing sequence in β containing all the events of C. Because Ψ is a homomorphism, $\Psi(\sigma)$ is a firing sequence in $\bigotimes_i \beta_i$. The projection $\Psi(\sigma)[i]$ of $\Psi(\sigma)$ is a firing sequence of β_i which contains exactly the events of $C[i]$. This proves that $C[i]$ is a configuration of β_i. The fact that $\Psi(Cut(C)) = \bigcup_i Cut(C[i])$ is immediate.

We prove the second part of the proposition by contradiction. We will use in advance Proposition 10. Let C and C' be two distinct configurations such that $C[i] = C'[i]$ for any branching process β_i. Thus, there exist a branching process β_i and two distinct events $e \in C$, $e' \in C'$ with $\Psi(e)[i] = \Psi(e')[i] \neq \epsilon$. Applying Proposition 10, $C \cup C'$ is also a configuration. If σ is a firing sequence associated to $C \cup C'$, the projection $\Psi(\sigma)[i]$ is a firing sequence on β_i which contains the event $\Psi(e)[i]$ at least twice. Because β_i is a branching process, this is impossible. □

Fig. 9 illustrates the confusing identification induced by a product having a reentrant synchronization. The two events $\langle \epsilon, z_3 \rangle$ have the same tuple of component configurations whereas their local configurations are obviously different.

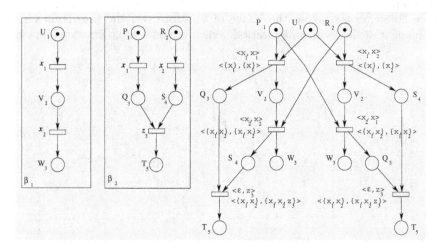

Fig. 9. Branching processes of a reentrant product and its components

One important property to construct a complete prefix is to check if two conditions b_1 and b_2 are concurrent. One way consists in determining if $C = [{}^\bullet b_1] \cup [{}^\bullet b_2]$ forms a configuration, and hence $b_1, b_2 \in Cut(C)$. The following proposition indicates how to determine in a modular way whether the union of two configurations is a configuration.

Proposition 10. *Let β be a branching process of the product $\langle N, m_0 \rangle$. Let C, C' be two configurations of β. If $\langle N, m_0 \rangle$ is non-reentrant, then $C \cup C'$ is a configuration of β iff $\forall i, C[i] \cup C'[i]$ is a configuration of β_i.*

Proof. The necessary condition part of the proposition is obvious. Indeed $(C \cup C')[i] = C[i] \cup C'[i]$ for any branching process β_i and is a configuration.

Let us prove that "$\forall i, C[i] \cup C'[i]$ is a configuration of β_i" is a sufficient condition for $C \cup C'$ to be a configuration. Let us prove it by contradiction. If $C \cup C'$ is not a configuration, $C \cup C'$ is not conflict-free: $\exists e \in C, e' \in C' : e \# e'$. Indeed, $C \cup C'$ is causally left-closed and if there is a conflict in $C \cup C'$, it must be between events in C and C'. Let select e, e' such that $[e]$ and $[e']$ are minimal: $[e] \cup [e'] \setminus \{e, e'\}$ is a configuration of β.

We show that the labellings of $[e]$ and $[e']$ are equal: $\lambda(h(e)) = \lambda(h(e'))$. Let b be an input condition of e and e'. If $\Psi(b_i) \in B_i$, then $\Psi(e)[i] = \Psi(e')[i]$. Otherwise $\Psi(e)[i]$ and $\Psi(e')[i]$ are conflicting events in $C[i] \cup C'[i]$. This proves that $\lambda(h(e)) = \lambda_i(h_i(\Psi(e)[i])) = \lambda(h(e'))$.

The events e and e' have two ways to be different: $h(e) \neq h(e')$ or ${}^\bullet e \neq {}^\bullet e'$. If $h(e) \neq h(e')$, there exists i: $h(e)[i] \neq h(e')[i]$. Because $C[i] \cup C'[i]$, $\Psi(e)[i]$ and $\Psi(e')[i]$ are concurrent events in β_i labelled by the same letter. This contradicts the non-reentrant property. If ${}^\bullet e \neq {}^\bullet e'$, there exists a condition b in ${}^\bullet e$ and not in ${}^\bullet e'$. Let β_i be the branching process containing the condition b ($\Psi(b) \in B_i$). If $\Psi(e)[i]$ and $\Psi(e')[i]$ are different events, then they are two concurrent events in β_i labelled by the same letter. This contradicts the non-reentrant property. Otherwise, there exists a condition $b' \in {}^\bullet e'$ such that $\Psi(b) = \Psi(b')$. The conditions b and b' are both marked in $M = Cut([e] \cup [e'] \setminus \{e, e'\})$, hence $\Psi(b)$ is not 1-bounded in $\Psi(M)$. This is impossible because β_i is a safe net as it is a branching process. □

4.3   Finite Complete Prefixes of a Product of Petri Nets

In this subsection, we study the processes of a product and show how to design a well-adapted order from well-adapted orders of its components.

Proposition 11. *Let $\pi = \langle \langle B, E, F \rangle, h \rangle$ a process of $\langle N, m_0 \rangle$. Let $\forall i, \pi[i] = \langle \langle B_i, E_i, F_i \rangle, h_i \rangle$ defined such that*

- $B_i = \{b \in B \mid h(b) \in P_i\}$
- $E_i = \{e \in E \mid h(e)[i] \neq \epsilon\}$
- $F_i = F \cap ((B_i \times E_i) \cup (E_i \times B_i))$
- $\forall b \in B_i, h_i(b) = h(b), \forall e \in E_i, h_i(e) = h(e)[i].$

Then $\pi[i]$ is a process of $\langle N_i, m_i \rangle$. Moreover, if $\langle N, m_0 \rangle$ is non-reentrant then the mapping which associates to each process π the tuple $\langle \pi[1], \ldots, \pi[n] \rangle$ is injective.

Proof. This proposition is a direct application of Proposition 10 when considering π as a branching process of N. □

A well-adapted order of a component defines a well-adapted order on the processes of the product. The lexicographical combination allows to design a well-adapted total order for the product when each component has a well-adapted total order.

Proposition 12. *Let $\forall i, \preceq_i$ be a well-adapted order over the finite processes of N_i. Let the relations $\preceq_{[i]}$ over finite processes of N be defined by $\pi \prec_{[i]} \pi'$ iff $\pi[i] \prec_i \pi'[i]$). These relations are well-adapted orders. Moreover, if N is non-reentrant and $\forall i, \preceq_i$ is a total order then $lex(\prec_{[1]}, \ldots, \prec_{[n]})$ is a total order.*

Proof. The fact that $\prec_{[i]}$ is well-adapted, is obvious. One has just to notice that $\pi \sqsubseteq \pi' \Rightarrow \pi[i] \sqsubseteq \pi'[i]$ and $\pi \in \pi_1 \cdot \pi_2 \Rightarrow \pi[i] \in \pi_1[i] \cdot \pi_2[i]$. The second part of the proposition is directly deduced from Proposition 11 which states that a process π is characterized by $\langle \pi[1], \ldots, \pi[n] \rangle$. $\square$

The previous proposition has an important consequence. It allows to select the best order depending on the type of components. For instance, in [10,7], each component is a state machine and the selected order is total and checked efficiently. In the following section, we will see how to extend this result to systems obtained by composition of state machines and queues.

4.4   Finite Complete Prefixes of a Product of Symmetrical Petri Nets

This part presents a natural adaptation of the previous results for products of symmetrical Petri nets. We first define a product of symmetrical Petri nets as a symmetrical net. As a consequence, a branching process of a product of symmetrical nets is well defined as a branching process of a product of nets. As a symmetrical net, the unfolding is also well defined. It remains to define the modular construction of a well-adapted order.

Definition 27. *Let $\langle N_1, G_1 \rangle, \ldots, \langle N_n, G_n \rangle$ be labelled symmetrical nets. The product $\langle N, G \rangle = \bigotimes_i \langle N_i, G_i \rangle$ is defined as $\langle \bigotimes_i N_i, \prod_i G_i \rangle$. A product of marked labelled symmetrical nets $\bigotimes_i \langle N_i, G_i, m_i \rangle$ is non-reentrant iff the product $\bigotimes_i \langle N_i, m_i \rangle$ is non-reentrant.*

The following proposition generalises Prop. 12 to products of symmetrical nets.

Proposition 13. *Let $\forall i, \preceq_i$ be a well-adapted order over finite processes of $\langle N_i, G_i \rangle$. Let the relations $\preceq_{[i]}$ over finite processes of $\langle N, G \rangle$ be defined by $\pi \prec_{[i]} \pi'$ iff $\pi[i] \prec_i \pi'[i]$). These relations are well-adapted orders. Moreover, if $\langle N, G \rangle$ is non-reentrant and $\forall i, \preceq_i$ is a total order then $lex(\prec_{[1]}, \ldots, \prec_{[n]})$ is a total order.*

Proof. This proof is as simple as the proof of Proposition 12. $\square$

5 Implementation of Unfoldings of Products of Symmetrical Petri Nets

In this section, we show how the construction of a complete prefix can benefit from the modular decomposition of the net. Moreover, the symmetries will be taken into account to limit the size of this prefix. Then, we present an efficient implementation concerning systems composed by state machines and queues.

5.1 Modular Construction

A generic algorithm for Petri nets is given in Alg. 5.1. The mainly required operations to build a complete prefix of a product of symmetrical nets are to manage a heap of events sorted on their local configurations, to detect cutoff and to compute the possible extensions (i.e. new events) of a prefix. To get a modular construction, all these operations have to be designed just using the following basic computations for each component. Let C_i and C_i' be two configurations of a component i and b_i a condition of i.

1. decide if $\Pi(C_i) \prec \Pi(C_i')$
2. compute $\widehat{Cut(C_i)}$
3. decide if $C_i \cup C_i'$ forms a configuration
4. decide if $b_i \in Cut(C_i)$

Propositions 8 and 9 indicate that any element $v \in B \cup E$ of a prefix can be encoded by $\Psi(v)$ together with the tuple of component configurations $\langle [v][1], \ldots, [v][n] \rangle$. For the computation of extensions, we introduce new data related to extensions local to a component. For a component i, a local extension l_i is a pair (e_i, Pre_i) with $e_i \in E_i$ and $Pre_i \subseteq \Psi^{-1}(B_i)$. We impose that ${}^\bullet e_i = \Psi(Pre_i)$ and that Pre_i is a set of concurrent conditions in the prefix already constructed. One can remark that the definition of Ψ ensures that such a local extension is valid (i.e. the homomorphism and the concurrency is respected in the component i). We denote by $\mathcal{L}_i$ the set of local extensions and by $\mathcal{L}_i(a)$ the ones labelled by $a \in A_i$ ($\mathcal{L}_i(a) = \{(e_i, Pre_i) \in \mathcal{L}_i \mid \lambda_i(h_i(e_i)) = a\}$. For convenience, we define $\mathcal{L}_i(a) = \{(\epsilon, \emptyset)\}$ when $a \notin A_i$. We are now in position to discuss the implementation for the computation of finite prefixes:

- *Sort configurations:* We use the lexicographic order defined in Prop. 13.
- *Check cutoff for an event e:* We compute the cut corresponding to the configuration $[e]$ using Prop. 9 as $Cut_N([e]) = \bigcup_i Cut_N([e][i])$. And then, we check if there exists an event e' with an equivalent cut ($\forall i, Cut_N([e][i]) \equiv Cut_N([e'][i])$) such that $[e'] \prec [e]$.
- *Compute extensions:* We assume that the sets of local extensions $\mathcal{L}_i(a)$ are coherent with respect to the already computed prefix. A global extension l is a tuple $\langle l_1, \ldots, l_n \rangle$ of local extensions $l_i \in \mathcal{L}_i(a)$ for a given label a. This definition ensures that the property of synchronization is satisfied. To construct a valid extension, we have to check that the set of all the conditions $\bigcup_k Pre_k$ are concurrent:

Algorithm 5.1 Unfold

```
Prefix func unfold(PetriNet ⟨N, m₀⟩) {
    Prefix prefix(N, m₀);
    SortedHeap heap;
    heap.put(prefix.InitEvent());
    while (not heap.isEmpty()) {
        Event event := heap.getMax();
        prefix.addIn(event);
        if (not prefix.isCutoff(event))
            foreach successor in extend(prefix, event)
                heap.put(successor);
    }
    return prefix;
}
```

- $C = \bigcup_k [Pre_k]$ is a configuration. This test can be done on each component $(\forall i, \bigcup_k [Pre_k][i]$ is a configuration of component $i)$.
- Each condition in $\bigcup_k [Pre_k]$ belongs to $Cut(C)$. This test can be done on each component $(\forall i, \forall b \in Pre_i, \Psi(b) \in Cut(C[i]))$.

5.2   Application to Finite State Machines and Queue Components

Applying the modular algorithm described previously to a given type of component induces the definition of a representation for the unfolding of components as well as the implementation of the basic operations. In the case of finite state machines and queues, we have explicit representations of maximal branching processes which allow to specialize the representations of a process, a configuration and a canonical form of equivalent states.

In the case of single-marked finite state machines, the maximal branching process has a tree structure and the paths from the root corresponds to sequences of the system. The following tables give a formal description of the branching process.

Element	Encoding
B	$\sigma : \sigma \in T^* \wedge m_0 \xrightarrow{\sigma}$
Min	$\{\epsilon\}$
E	$\sigma : \sigma \in T^+ \wedge m_0 \xrightarrow{\sigma}$

Event	${}^\bullet e$	$e^\bullet$
$\sigma \cdot t$	σ	$\sigma \cdot t$

Configurations and processes are encoded by firing sequences and states by places. We are now in the position to specify how the basically required operations are designed. Let σ and σ' be two sequences representing configurations and σ'' a sequence representing a condition. We use $\prec_{lex}$ for the lexicographical order, and $\sqsubset$ for the prefix order over words.

1. decide if $\Pi(\sigma) \prec \Pi(\sigma')$: $|\sigma| < |\sigma'|$ else $\sigma \prec_{lex} \sigma'$

2. compute $\widehat{Cut}(\sigma)$: $\{p \in P \mid m_0 \xrightarrow{\sigma} p\}$
3. decide if $\sigma \cup \sigma'$ forms a configuration: $\sigma \sqsubset \sigma'$ or $\sigma' \sqsubset \sigma$
4. decide if $\sigma'' \in Cut(\sigma)$: $\sigma = \sigma''$

For a finite unknown bound queue component, we consider two sets of transition labels *Send* and *Receive*, and a mapping $\mu : Send \cup Receive \to Mess$. The sets *Send* and *Receive* correspond to the enqueue and dequeue actions, and the mapping μ specifies the message for which the action is performed.

The maximal branching process has two types of events: *Send* and *Receive*. A *Send* event is specified by a sequence of enqueue actions of the system while a *Receive* event is specified by a sequence of dequeue actions of the system but also by a sequence of corresponding enqueue actions. We consider three types of conditions : *In*, *M* and *Out* associated to a place of the queue model. We characterize it by their input events (i.e. conditions *In* and *M* by their input *Send* event and conditions *Out* by their input *Receive* event). The following tables give a formal description of the branching process.

Element	Encoding
B	$In[\sigma_s] : \sigma_s \in Send^*$
	$M[\sigma_s] : \sigma_s \in Send^+$
	$Out[\sigma_s, \sigma_r] : \sigma_s \in Send^* \wedge \sigma_r \in Receive^* \wedge \mu(\sigma_s) = \mu(\sigma_r)$
Min	$\{In[\epsilon], Out[\epsilon, \epsilon]\}$
E	$Send[\sigma_s] : \sigma_s \in Send^+$
	$Receive[\sigma_s, \sigma_r] : \sigma_s \in Send^+ \wedge \sigma_r \in Receive^+ \wedge \mu(\sigma_s) = \mu(\sigma_r)$

Event	${}^\bullet e$	$e^\bullet$
$Send[\sigma_s \cdot t_s]$	$In[\sigma_s]$	$In[\sigma_s \cdot t_s], M[\sigma_s \cdot t_s]$
$Receive[\sigma_s \cdot t_s, \sigma_r \cdot t_r]$	$M[\sigma_s \cdot t_s], Out[\sigma_s, \sigma_r]$	$Out[\sigma_s \cdot t_s, \sigma_r \cdot t_r]$

Configurations and processes are encoded by pair of words $(\sigma_s, \sigma_r) \in Send^* \times Receive^*$ such that $\mu(\sigma_s) = \mu(\sigma_r)$ (i.e. the sequences of sending and receiving actions) and states by words on *Mess* (i.e. the message in the queue). We are now in the position to specify how the basically required operations are designed.

1. decide if $\Pi(\sigma_s, \sigma_r) \prec \Pi(\sigma'_s, \sigma'_r)$: $|\sigma_s| + |\sigma_r| < |\sigma'_s| + |\sigma'_r|$ else $\sigma_s \prec_{lex} \sigma'_s$ or else $\sigma_r \prec_{lex} \sigma'_r$
2. compute $\widehat{Cut}(\sigma_s, \sigma_r)$: $\{m \in Mess^* \mid \mu(\sigma_s) = \mu(\sigma_r) \cdot m\}$
3. decide if $(\sigma_s, \sigma_r) \cup (\sigma'_s, \sigma'_r)$ forms a configuration:
   - $\sigma_s \sqsubseteq \sigma'_s \wedge (\sigma_r \sqsubseteq \sigma'_r \vee (\sigma'_r \sqsubseteq \sigma_r \wedge \mu(\sigma_r) \sqsubseteq \mu(\sigma'_s)))$ or
   - $\sigma'_s \sqsubseteq \sigma_s \wedge (\sigma'_r \sqsubseteq \sigma_r \vee (\sigma_r \sqsubseteq \sigma'_r \wedge \mu(\sigma'_r) \sqsubseteq \mu(\sigma_s)))$
4. decide if $b \in Cut(\sigma_s, \sigma_r)$:
   - if $b = In[\sigma''_s]$ or $b = M[\sigma''_s]$: $\sigma''_s = \sigma_s$
   - if $b = Out[\sigma''_s, \sigma''_r]$: $\sigma''_s \sqsubseteq \sigma_s \wedge \sigma''_r = \sigma_r$

One can remark that all the basic operations are implemented using only simple computations on words. The implementation of a finite queue introduces a new type F of conditions representing the free slots of the queue. Its maximal branching process is then more intricate because a sending event is also depended on a sequence of *Receive* actions. However, the basic operations remain simple to implement.

Esparza and Römer in [7] compute the complete prefix of system for which the queues are bounded. Because they only manage finite state machines, they use the model of a queue of Fig. 1 viewed as a product of state machines. A first implementation of the unfolding algorithm presented in this paper which deals also with finite unknown bound queues has been designed. This tool has been used to compare the two implementations on an example of producer/consumer communicating through a queue. The producer sends a finite sequence of messages. In this experiment, our implementation leads to the construction of branching processes for which the number of events is constant independently of the size of the queues (finite or not). On the contrary, considering queue as a product of finite state machines leads to unfoldings for which the number of events increases linearly with the size of the queue.

6   Concluding Remarks

In this paper, we have presented a general technique for the modular construction of complete prefixes adapted to systems composed of Petri nets. This construction is based on a definition of a well-adapted order allowing combination. Moreover, the technique has been instantiated in an efficient algorithm for systems combining finite state machines and known or unknown bound queues.

It is important to note that all the verification techniques based on the analysis of a complete prefix can be used in our context. Moreover, the modular construction and the identification of the component parts of the prefix must allow the design of new verification techniques. As an example, one can easily compute the bound of a queue directly on the resulting prefix. We claim that the presented method is sufficiently generic to be instantiated to other types of components and we presently work in this direction.

References

1. G. Chiola, C. Dutheillet, G. Franceschinis, and S. Haddad. Well-formed colored nets and symmetric modeling applications. *IEEE Transactions on Computers*, 42(11):1343–1360, 1993.
2. J.-M. Couvreur, S. Grivet, and D. Poitrenaud. Designing a LTL model-checker based on unfolding graphs. In *Proc. of ICATPN'2000*, volume 1825 of *Lecture Notes in Computer Science*, pages 364–383. Springer Verlag, 2000.
3. J.-M. Couvreur and D. Poitrenaud. Detection of illegal behaviours based on unfoldings. In *Proc. of ICATPN'99*, volume 1639 of *Lecture Notes in Computer Science*, pages 364–383. Springer Verlag, 1999.
4. J. Engelfriet. Branching processes of Petri nets. *Acta Informatica*, 28:575–591, 1991.
5. J. Esparza. Model checking using net unfoldings. In *Proc. of TAPSOFT'93*, volume 668 of *Lecture Notes in Computer Science*, pages 613–628. Springer Verlag, 1993.
6. J. Esparza and K. Heljanko. A new unfolding approach to LTL model checking. In *Proceedings of ICALP'2000*, number 1853 in LNCS, pages 475–486. Springer-Verlag, 2000.

7. J. Esparza and S. Römer. An unfolding algorithm for synchronous products of transition system. In *Proceedings of CONCUR'99*, number 1664 in LNCS, pages 2–20. Springer-Verlag, 1999.

8. J. Esparza, S. Römer, and W. Vogler. An improvement of McMillan's unfolding algorithm. In *Proc. of TACAS'96*, volume 1055 of *Lecture Notes in Computer Science*, pages 87–106. Springer Verlag, 1996.

9. S. Haddad, J.-M. Ilié, and K. Ajami. A model checking method for partially symmetric systems. In *Formal Methods for Distributed System Development, Proc. of FORTE/PSTV'2000*, pages 121–136, Pisa, Italy, October 2000. Kluwer Academic Publishers.

10. R. Langerak and E. Brinksma. A complete finite prefix for process algebra. In *Proceedings of the 11th International Conference on Computer Aided Verification, Italy*, number 1633 in LNCS, pages 184–195. Springer, 1999.

11. K.L. McMillan. Using unfoldings to avoid the state explosion problem in the verification of asynchronous circuits. In *Proc. of the 4th Conference on Computer Aided Verification*, volume 663 of *Lecture Notes in Computer Science*, pages 164–175. Springer Verlag, 1992.

12. M. Nielsen, G. Plotkin, and G. Winskel. Petri nets, events structures and domains, part I. *Theoretical Computer Science*, 13(1):85–108, 1981.

Partial Order Verification of Programmable Logic Controllers*

Peter Deussen

Brandenburg University of Technology at Cottbus
Computer Science Institute
Cottbus, Germany
PDeussen@gmx.de,
http://www.Informatik.TU-Cottbus.DE/~wwwdssz/

Abstract. We address the verification of programmable logic controllers (PLC). In our approach, a PLC program is translated into a special type of colored Petri net, a so-called register net (RN). We present analysis methods based on the partial order semantics of RN's, which allow the generation of partial order traces as counter examples in the presence of programming errors. To that purpose, the behavior description 'concurrent automaton', introduced in [3] for safe Petri nets, is upliftet to the dedicated RN's.

1 Introduction

In this paper, we address the verification of industrially applied controllers. We concentrate on software for *programmable logic controllers* (PLC). The international norm IEC 1131-3 [10] defines several languages for PLC programming: *sequential function charts, structured text, ladder diagrams, function block diagrams,* and—most elementary—an assembler-like language called *instruction list* (IL), on which we will focus here.

Let us outline the main ideas:

Organization of the verification process. An IL program is compiled into a dedicated type of colored Petri net, called register net (RN). RN's represent the control flow of an IL (i. e. the order of computation of program parts as determined by jumps and labels) by means of a Petri net (to be precise, an elementary net system (ENS)), and the data flow (memory and hardware addresses and accumulators) by registers containing non-negative integers. Transitions can read and modify data, and their occurrence can depend on data.

The verification process is done as follows: After compiling the IL program into a RN, a model of the environment of the PLC is added to the resulting net, i. e. a model of the controlled facility (also in form of a RN). Such an environment model is necessary because of the following reason: Sensor values tested by the

---

* This work is supported by the German Research Council under grant ME 1557/1-1.

J.-M. Colom and M. Koutny (Eds.): ICATPN 2001, LNCS 2075, pp. 144–163, 2001.

PLC change because of actuator actions influenced by the PLC. This behavior has to be captured in some way.

Additionally, we want also to be able to deal with (parts of) plants comprising several facilities and associated PLC's. PLC's of different machines communicate in an implicit manner by the changes of sensor values: If a transport belt is activated by a PLC, a light barrier associated with some other PLC will observe the passing of the transported blank.

The analysis will be done on the composed model consisting of the RN of the environment and the RN's of the several PLC's. In this paper, we focus on three types of system properties:

1. absence/presence of deadlocks (which is merely a side effect of our analysis technique),
2. run-time errors like overflows and division by zero, and
3. simple safety properties. A simple safety property is an assumption on data values describing system states which are not allowed to occur.

Mathematical Considerations. Since RN will be used for analysis purposes, a mathematically sound model of the behavior of a RN is needed. We prefer a partial order semantics instead of the more familiar interleaving semantics for the following reason: Interleaving semantics is based on the notion of sequences of transition occurrences acting on a global state space of the system. In terms of PLC's: A global state is the product of the states of all the PLC's in the plant. Because concurrency and causality are not visible in sequences, such semantics gives the somewhat misleading picture that anything in a plant has to do something with everything else. But in a real plant, many processing steps can be (and are) performed independently of other processing steps.

In using partial order semantics, independence and causality become visible. The idea is to model causal dependence and independence of system actions by the mathematical concept of a partial order. If e_1 and e_2 are system actions (elementary processing steps or the computation of a single command in an IL program), then we use the notation $e_1 < e_2$ to indicate that e_1 has to precede e_2 in time, or that e_1 is a necessary precondition of e_2. On the other hand, if neither $e_1 < e_2$ nor $e_2 < e_1$ holds, then e_1 and e_2 are concurrent or independent of each other. We write e_1 co e_2.

There are many types of partial order semantics of Petri nets, for instance Mazurkiewicz traces, (prime) event structures, pomsets (partial words), or (branching) processes. We use so-called semi-words [11,15], which are basically partial orders of system actions.

Analysis Technique. For colored Petri nets (and therefore, for RN's), several analysis techniques are available, e. g. place invariant analysis or reachability graph (state graph) generation. Additionally, methods of the classical Petri net theory can be applied to the unfolding of the RN.

We focus on an alternative approach, namely the description of the behavior of a RN by means of a *concurrent automaton* (CA). CA are basically state graphs, however, with transitions comprising partial order representations of parts of the system behavior.

Our choice is motivated by the following reason: Partial order based techniques have their strength if the system under consideration exhibits a high degree of concurrency, and (more importantly) very few nondeterministic choices. In the given application area nondeterminism can be used to model random events like system faults or human inference. But we expect that the considered systems are 'almost' deterministic. Therefore, it is likely that partial order techniques behave well in the analysis of PLC programs.

Concurrent automata were introduced by Ulrich [12]. Ulrich uses CA for test case generation. A generation algorithm which bases on the input of an *unfolding* of a Petri net [8] is given.

The notion of CA has some similarities to *step covering graphs* [14]. Step covering graphs can be viewed as CA where each transition consists of a semi-order with empty ordering relation, i. e. a step.

Another CA-like approach are *process automata* introduced by Burns and Hulgaard [7]. Process automata comprise global states and transitions labelled by *processes* of safe Petri nets. A *stubborn reduced* reachability graph [13] is used as the input of an generation algorithm.

The paper is organized as follows: In section 2 we examine a very small control problem to explain how IL's are translated into RN's. The example serves only as a motivation for the definition of RN's, not as a running example throughout this paper. To meet the page limit, other examples had to be omitted. Section 3 lists the notions and basic definitions used. In section 4 we introduce RN's formally and define their partial order semantics. Section 5 discusses analysis methods for RN's and introduces CA. In section 6 we present a generation algorithm for CA of a given RN. Simple safety properties and run-time errors are defined in section 7, and a corresponding CA-based analysis algorithm is given. Section 8 summarizes our paper and gives an outlook on further work.

2   Instruction Lists

In this section we discuss a simple controlling problem [9] to motivate our verification method. Figure 1 shows a hydraulic piston. Two valves are used to increase and decrease the pressure of the liquid in the left and right hand part of the piston case. Activation/deactivation of the actors Y_l and Y_r opens/closes the left and the right hand valve, respectively. The sensors X_l and X_r indicate whether the piston has reached its leftmost or rightmost position. Finally, there is a switch to start and stop the piston's movement by a human operator.

The piston is expected to behave as follows: If the operator hits the switch (and keeps pressing it), the piston starts moving until it reaches its leftmost or rightmost position, then it is moving back into the opposite direction. If the operator releases the switch, the piston has to stop.

Figure 2 shows an IL program to solve this simple controlling problem. IL is an assembler-like language. Commands act on variables and on an accumulator.

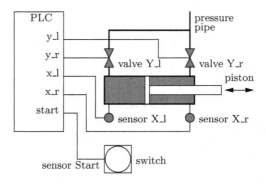

Fig. 1. Hydraulic Piston

For instance, the command LDN z reads: Load the negation of the value of the variable z into the accumulator. S and R are set/reset commands, AND is logical conjunction, and ST means store. Of course, conditional and unconditional jumps are supported by IL, and there are also subroutines in form of functions (procedures without memory) and function blocks (procedures with memory). These 'structured' programming constructs are not allowed to be recursive, i. e. IL's with functions or function blocks can be translated into a formalism with static control structure (e. g. a Petri net). Permitted data types are scalar types like integers, Booleans (the least significant bit (lsb) of a machine word determines its boolean value), or floats.

The processing cycle of a PLC is as follows: In a first step the sensor values of the controlled environment are read and mapped to input variables by the operating system of the PLC. (In our example, X_l is mapped to x_l, X_r to x_r, and so on.) The next step is the execution of the user program. Finally, the

```
VAR_INPUT              PROGRAM
    x_1: BOOL;             LDN z           R z
    x_2: BOOL;             AND x_r         LD z
    start: BOOL;           S z             AND start
END_VAR                    LD z            ST y_l
VAR_OUTPUT                 AND x_1         LDN z
    y_1: BOOL := FALSE;    R z             AND start
    y_r: BOOL := FALSE;    LDN start       ST y_r
END_VAR                                    END_PROGRAM
VAR
    z: BOOL := FALSE;
END_VAR
```

Fig. 2. An IL User Program

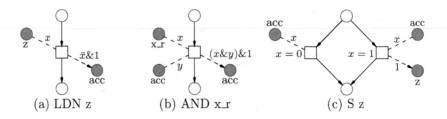

Fig. 3. Register Net Structures for Some IL Commands

values of output variables are mapped to the actuators of the plant. This process is repeated in a fixed time grid, the so-called cycle time.

RN's are Petri nets (ENS) augmented by registers containing non-negative integer values.[1] A transition of a RN is enabled if its pre-places are contained in the current case (marking), its post-places are not (safe firing rule), and if a predicate on register values associated with this transition yields not false (i. e. true or undefined, we will later discuss this point). Additionally, a function on tuples of integers is associated with each transition to determine the effect of the firing of this transition.

Consider fig. 3, 4, or 5 for some examples of RN's. White (unfilled) circles and boxes are places and transitions of the underlying ordinary Petri net, gray shaded circles are registers. A dashed line (without arrow) from a register to a transition means that the value of this register is read by the transition to determine its enabledness and its output values, a dashed arc from a transition to a register identifies this register as an output register. Dashed lines are labeled with variable symbols, dashed arcs are labelled with expressions on these variables—a mechanism adopted from colored Petri nets to define output functions. Predicates are given by Boolean expressions which appear as transition labels. Finally, to avoid edge crossings we fix the convention that the same register may have several graphical appearences.

Figure 3 shows some register net structures associated with several IL commands. Figures 3.(a) and 3.(b) build the RN-semantics of the LDN and AND commands. We use $\bar{x}$ to denote bitwise negation of x, and $x \& y$ for bitwise conjunction of x and y. (Recall that the Boolean value of an n-bit integer is determined by its lsb). The set command S (Fig. 3.(c)) and the reset command R are conditional commands: Their arguments get a new value only if the value of acc (the accumulator) is true. We therefore have to use two alternative transitions to describe the semantics for these commands, one which modifies the argument and one which does not: Modification of register clearly defines dependencies between transitions.

Since it is rather obvious, we refrain to give a complete list of RN structures for all IL commands. It should also be easy to imagine how the RN for the user program fig. 2 looks like. Figure 4 shows the RN for the 'operating system' of

---

[1] Clearly, all the scalar types mentioned above can be traced back to integers.

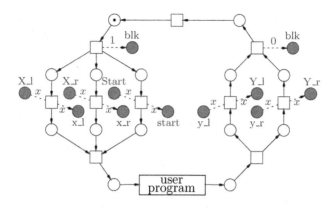

Fig. 4. System Program

a PLC: The system program which maps sensor values to input variables and output variables to actuators. Since we cannot make assumptions on the order in which both mappings occur, we decided to model the mappings as being concurrently executed.

Finally, we have to model the controlled plant (environment model) in order to describe the changes of sensor values in response to the control. In doing this, we are faced with an implicit assumption of any IL program: *The PLC is fast enough to observe any changes of sensor values.* We reflect this assumption by adding a so-called *blocking register* blk to our model: A transition is only allowed to modify a register associated with a sensor if the register blk contains the value 0. This transition modifies the blk register by writing the value 1 into it; therefore, no other transition modifying sensors is allowed to fire until blk is reset to 0. This is done by the RN model of the system program. It also makes sure that the environment model is blocked while the user program is executed.

A very analogous approach to block the environment is described by the authors of [6]: They use safe Place/Transition nets to model both the PLC program and its environment. A blocking place is added to prevent value changing while the PLC program is executed. However, in this approach the environment is never blocked if the PLC program is idle.

Figure 5 shows an environment model for the hydraulic piston. It comprises the basic states r (right hand position), m (middle position), and l (left hand position).

Finally, let us turn to another detail of our modelling approach. We mentioned already that registers contain non-negative integers, which actually means: values of PLC variables. These values cannot be arbitrarily large, but they are restricted to the range $[0..2^m - 1]$, where usually $m = 8$ or $m = 16$ is some constant. What happens if a user program tries to increase such a value beyond $2^m - 1$? Another question is: What happens in the case of dividing by zero? Such 'undefined behavior' (even if the programmer knows that $(2^8 - 1) + 1$ equals

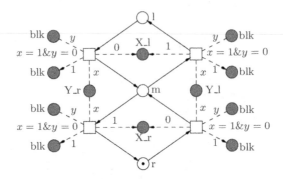

Fig. 5. Piston Environment Model

0 in 8-bit arithmetics) is in almost every case unintended, hence a programming fault.

To capture run-time errors of this type, we allow predicates and functions associated with transitions to return a special value $\perp$ (read 'undefined'), and we assume them to be *strict* (i.e. $f(\perp) = \perp$). $\perp$ models unknown behavior; we cannot define that a RN transition is not enabled if one of it's input registers contains $\perp$—especially, if this transition is located in the environment model, such a definition would model a rather unrealistic behavior. Therefore we decide to make a transition enabled even if the associated predicate yields $\perp$: In case of unknown behavior every behavior is possible.

3   Mathematical Background

This section summarizes the basic notations used throughout this paper.

To avoid tedious notions, we fix the following convention: If a structure $S = \langle A, B, \ldots \rangle$ is introduced, the components of S will always be denoted by $A_S, B_S, \ldots$ **N** denotes the set of non-negative integers, **B** = {false, true} is the set of Boolean values.

For some set A, $\mathcal{P}(A)$ is the set of all subsets of A. For $R \subseteq A \times B$ and $a \in A$, we denote the *image* of a under R by $R(a) =_{\mathrm{df}} \{b \in B : a \mathbin{R} b\}$. For $C \subseteq A$ we define $R(C) =_{\mathrm{df}} \bigcup_{a \in C} R(a)$. The *inverse* $R^{-1} \subseteq B \times A$ of R is defined by $b \mathbin{R^{-1}} a \Leftrightarrow_{\mathrm{df}} a \mathbin{R} b$. $R^{+} \subseteq A \times A$ denotes the least transitive relation containing $R \subseteq A \times A$. For every set A and $n \geq 0$, A^n is defined by $A^0 =_{\mathrm{df}} \emptyset$ and $A^n =_{\mathrm{df}} A^{n-1} \times A$. For sets $A_1, A_2, \ldots, A_n$ and some $i \leq n$, the *projection* $\mathrm{pr}_i : A_1 \times A_2 \times \cdots \times A_n \to A_i$ is defined to be $\mathrm{pr}_i(\langle a_1, a_2, \ldots, a_n \rangle) =_{\mathrm{df}} a_i$.

To deal with partial functions, we define $A_\perp =_{\mathrm{df}} A \cup \{\perp\}$ for any set A. If $f : A \rightharpoonup B$ is a partial function, then the total function $f^\perp : A_\perp \to B_\perp$ is defined as

$$f^\perp(x) =_{\mathrm{df}} \begin{cases} \perp, & \text{if } x = \perp \text{ or } f(x) \text{ is undefined;} \\ f(x), & \text{otherwise.} \end{cases}$$

This notion also applies to n-ary functions.

A (finite) *labeled partial order (lpo)* $a = \langle E, <, \lambda \rangle$ over some alphabet T consists of a finite set E of *events,* an (irreflexive) partial order $< \subseteq E \times E$, called the *precedence relation* of a, and a *labeling function* $\lambda : E \rightarrow T$. $\epsilon = \langle \emptyset, \emptyset, \emptyset \rangle$ is the *empty lpo*. If $t \in T$ is a symbol, then t is also be used to denote the *letter* $\langle \{0\}, \emptyset, \{\langle 0, t \rangle\} \rangle$.

The relation co_a is defined by $e_1 \text{ co}_a e_2 \Leftrightarrow_{\text{df}} \neg(e_1 \leq_a e_2) \ \& \ \neg(e_2 \leq_a e_1)$. A set $C \subseteq E$ is called a *co-set,* iff we have $e_1 \neq e_2 \Rightarrow e_1 \text{ co}_a e_2$ for all $e_1, e_2 \in C$. A *semi-order* is a lpo a where for all $e_1, e_2 \in E_a$, $e_1 \text{ co}_a e_2 \Rightarrow \lambda_a(e_1) \neq \lambda_a(e_2)$. $\mathbf{SO}(T)$ denotes the class of semi-orders over T.

We now introduce the prefix relation and the notion of sequentialization for lpo's. Since we want to abstract from the specific events of lpo's, both concepts are presented in terms of homomorphism between lpo's.

Let a and b be lpo's. A mapping $h : E_a \rightarrow E_b$ is called a *homomorphism,* iff $e_1 <_a e_2$ implies $h(e_1) <_b h(e_2)$ for all $e_1, e_2 \in E_a$ and furthermore, $\lambda_a = \lambda_b \circ h$. It is called an *embedding,* iff it is an injective homomorphism with the property $h(\leq_a^{-1}(e)) = \leq_b^{-1}(h(e))$. A bijective embedding is called an *isomorphism.* a is called a *prefix* of b, denoted by $a \leq b$, iff there is an embedding $h : E_a \rightarrow E_b$. We write $a \equiv b$, if $a \leq b$ and $b \leq a$ holds. a is called a *sequentialization* of b, denoted by $b \preceq a$, iff there is a bijective homomorphism $h : E_b \rightarrow E_a$.

In [3] we proved that if $a \leq b$ holds for semi-orders a, b over the same alphabet, then the embedding of a into b is unique. We denote it by $H_a^b : E_a \rightarrow E_b$.

Clearly, $\equiv$ is an equivalence relation. A *semi-word* is an equivalence class of semi-orders. We write $[a] = [E_a, <_a, \lambda_a]$ to denote the equivalence class of a lpo a. The same notion applies to semi-words. A *semi-language* is a set of semi-words. $\mathbf{SW}(T)$ denotes the class of semi-words over T.

We fix the following conventions: If a, b, c, $\ldots$ are semi-orders, then we use boldfaced lowercase letters $\boldsymbol{a}$, $\boldsymbol{b}$, $\boldsymbol{c}$, $\ldots$, to denote their equivalence classes $[a], [b], [c], \ldots$ Hence, for instance, E_a will always refer to the event set of a representative of $\boldsymbol{a} = [a]$. Especially, if t is a letter, then $\boldsymbol{t} = [t]$. The equivalence class of ϵ will also be denoted by ϵ.

Now it is easy to prove that both $<$ and $\prec$ are preorders on the class of lpo's. If we put $\boldsymbol{a} < \boldsymbol{b} \Leftrightarrow_{\text{df}} a < b$, and $\boldsymbol{a} \prec \boldsymbol{b} \Leftrightarrow_{\text{df}} a \prec b$ for all $a \in \boldsymbol{a}$, $b \in \boldsymbol{b}$, then $<$ and $\prec$ are partial orders on semi-words.

4   Register Nets and Their Partial Order Semantics

In this section we are going to introduce formally the notion of register nets (RN's). We start with the definition of a net, i.e the graph representation of the control flow part of a RN.

Definition 1 (Net). *A net* $\langle P, T, F \rangle$ *consists of non-empty, finite sets* P *and* T *such that* $P \cap T = \emptyset$, *where the elements of* P *and* T *are called* places *and* transitions, *respectively, and a flow relation* $F \subseteq (P \times T) \cup (T \times P)$. *We assume that* $F(t) \neq \emptyset$ *and* $F^{-1}(t) \neq \emptyset$ *for each* $t \in T$.

In the above definition, places are local system states, and transitions are used to model the changes of local system states according the flow relation.

Definition 2 (Register Net). *A register net* $V = \langle P, T, F, R, \rho, \boldsymbol{G}, \boldsymbol{P}, s \rangle$ *(a RN, for short) consists of the following components:*

1. *P, T, and F are such that $\langle P, T, F \rangle$ is a net.*
2. *R is a finite set of* registers. *We assume $R \cap (P \cup T) = \emptyset$.*
3. *A mapping $\rho : T_V \rightarrow R^* \times R^*$; $\rho(t)$ is called* signature *of t. To reduce the notational effort, we use the following shortcuts. Let $\rho(t) = \langle r_0 r_1 \ldots r_n, r_1' r_2' \ldots r_k' \rangle$:*
   a) $\text{in}(t) =_{\text{df}} n$ *(input arity of t), and $\text{out}(t) =_{\text{df}} k$ (output arity of t),*
   b) $\text{rd}(t, i) =_{\text{df}} r_i$ *for $1 \leq i \leq \text{in}(t)$ (input register selection of t), and $\text{wr}(t, j) =_{\text{df}} r_j'$ for $1 \leq j \leq \text{out}(t)$ (output register selection of t).*
   c) $\text{mod}(t) =_{\text{df}} \{\text{wr}(t, i) : 1 \leq j \leq \text{out}(t)\}$ *(the registers modified by t).*
   d) $\text{ac}(t) =_{\text{df}} \{\text{rd}(t, i) : 1 \leq i \leq \text{in}(t)\} \cup \text{mod}(t)$, *(access registers of t).*
   We assume $1 \leq i \neq j \leq \text{in}(t) \Rightarrow \text{rd}(t, i) \neq \text{rd}(t, j)$ and $1 \leq i \neq j \leq \text{out}(t) \Rightarrow \text{wr}(t, i) \neq \text{wr}(t, j)$
4. *A family $\boldsymbol{P} = \{P_t\}_{t \in T}$ of partial mappings $P_t : \boldsymbol{N}^{\text{in}(t)} \rightarrow \boldsymbol{B}$ (predicates).*
5. *A family $\boldsymbol{G} = \{G_t\}_{t \in T}$ of partial mappings $G_t : \boldsymbol{N}^{\text{in}(t)} \rightarrow \boldsymbol{N}^{\text{out}(t)}$.*
6. *$s \in \mathcal{P}(P) \times (R \rightarrow \boldsymbol{N}_\perp)$, an initial state.*

Let us discuss the above definition in more detail: The signature $\rho(t)$ of a transition gives us information of the arguments of the predicate and the function associated with this transition. If $\rho(t) = \langle r_1 r_2 \ldots r_n, r_1' r_2' \ldots r_k' \rangle$, then we will associate a function G_t with t, which obtains the values of the registers $r_1, r_2, \ldots, r_n$ and computes a new value for the register $r_1', r_2', \ldots r_k'$. The enabledness of t depends also on the registers $r_1, r_2, \ldots, r_n$: t becomes enabled only in the case where $P_t^\perp(r_1, r_2, \ldots, r_n)$ yields not false, i.e. either true or undefined.

Definition 3 describes system states of a RN V as pairs comprising a marking of V and an assignment of non-negative numbers to registers.

Definition 3 (State). *A* marking *of a RN V is a set $q \subseteq P_V$. A* register assignment *of V is a mapping $\sigma : R_V \rightarrow \boldsymbol{N}_\perp$. A* state *of V is a pair $s = \langle q, \sigma \rangle$, where q is a marking of V and σ is a register assignment of V. By $\Sigma(V) =_{\text{df}} \mathcal{P}(P_V) \times (R_V \rightarrow \boldsymbol{N}_\perp)$ we denote the set of possible states of V. A register assignment σ is K-restricted for some constant $K \geq 0$, if $\sigma(r) < K$ for each $r \in R$. $\langle q, \sigma \rangle \in \Sigma(V)$ is called K-restricted if σ is K-restricted.*

Data types permitted by the IEC 1131-3 are scalar types like integers, Booleans, floats, etc, but restricted to a representation by m bits. To simplify our definitions, each of these data types is mapped on the set $\boldsymbol{N}$ of non-negative integers; the specific interpretation of a value (e. g. the interpretation of the number 0 and 1 as the Boolean values false and true) is assumed to be done by the mappings G_t. However, to put the m-bit restriction into our model, we assume that there is an upper bound K of each possible register value.

Definition 4 (K-Restricted RN). V *is restricted to some constant $K \geq 0$ (K-restricted, for short), iff for each transition $t \in T_V$ and for all $k_i < K$ we have that $\mathrm{pr}_j(G_t^\perp(k_1, k_2, \ldots, k_n)) \neq \perp$ implies $\mathrm{pr}_j(G_t^\perp(k_1, k_2, \ldots, k_n)) < K$ ($1 \leq i \leq \mathrm{in}(t)$, $1 \leq j \leq \mathrm{out}(t)$). V is restricted if it is K-restricted and s_V is K-restricted.*

We are now going to define the partial order semantics of RN using the notion of semi-orders. The effect of the firing of a semi-order is described by the simultaneous effect of the order-respecting occurrence of each event of the semi-order to both the marking component and the register assignment component of a state.

To simplify the definition of a firing rule of semi-orders in a RN, we need some additional notations:

Definition 5. *Let $a \in \mathbf{SO}(T_V)$ for some RN V. We define $e^- =_{\mathrm{df}} F_V^{-1}(\lambda_a(e))$, and $e^+ =_{\mathrm{df}} F_V(\lambda_a(e))$ for each $e \in E_a$. If $C \subseteq E_a$, we put $C^- =_{\mathrm{df}} \bigcup_{e \in C} e^-$ and $C^+ =_{\mathrm{df}} \bigcup_{e \in C} e^+$. If no confusion is possible, we write $\mathrm{rd}(e, i)$ for $\mathrm{rd}(\lambda_a(e), i)$, $\mathrm{in}(e)$ for $\mathrm{in}(\lambda_a(e))$, G_e for $G_{\lambda_a(e)}$, and so on.*

Definition 6. *Let $a \in \mathbf{SO}(T_V)$ for some RN V and let $C \subseteq E_a$, q be a marking of V and σ be a register assignment of V. We define mappings $\delta_V^0 : \mathcal{P}(P_V) \times E_a \to \mathcal{P}(P_V)$, $\delta_V^1 : (R_V \to \mathbf{N}_\perp) \times E_a \to (R_V \to \mathbf{N}_\perp)$, and $\pi_V : (R_V \to \mathbf{N}_\perp) \times E_a \to \mathbf{B}_\perp$ by*

$$\delta_V^0(q, e) =_{\mathrm{df}} (q - e^-) \cup e^+$$

$$\delta_V^1(\sigma, e)(r) =_{\mathrm{df}} \begin{cases} \mathrm{pr}_j(G_e^\perp(\sigma(\mathrm{rd}(e, 1)), \ldots, \sigma(\mathrm{rd}(e, \mathrm{in}(e))))), \\ \qquad \text{if } r = \mathrm{wr}(t, j) \text{ and } 1 \leq j \leq \mathrm{out}(e); \\ \sigma(r), \text{ otherwise} \end{cases}$$

$$\pi_V(\sigma, e) =_{\mathrm{df}} P_e^\perp(\sigma(\mathrm{rd}(e, 1)), \ldots, \sigma(\mathrm{rd}(e, \mathrm{in}(e)))) \neq \text{false}$$

δ_V^1 and δ_V^0 are inductively lifted to subsets $C \subseteq E_a$ by putting $\Delta_V^0(q, C) \in \mathcal{P}(P_V)$ and $\Delta_V^1(\sigma, C) \in \mathcal{P}(R_V \to \mathbf{N}_\perp)$ to be the smallest sets such that $\delta_V^0(\Delta_V^0(q, C - \{e\}), e) \subseteq \Delta_V^0(q, C)$, and $\delta_V^1(\Delta_V^1(\sigma, C - \{e\}), e) \subseteq \Delta_V^1(\sigma, C)$, where $e \in \max_{<_a} C$ and $C \neq \emptyset$; moreover we put $\Delta_V^0(q, \emptyset) = \{q\}$, and $\Delta_V^1(\sigma, \emptyset) = \{\sigma\}$ to terminate this recursive computation rule. Finally, we define $\Delta_V(\langle q, \sigma \rangle, C) =_{\mathrm{df}} \langle \Delta_V^0(q, C), \Delta_V^1(\sigma, C) \rangle$.

The mappings δ_V^0 and δ_V^1 describe the effect of the occurrence of a single event to the marking component and the register assignment component of a state. π_V is used to determine whether a transition t with associated predicate P_t is able to fire or not. Δ_V^0 and Δ_V^1 describe the effect of a set of events C to a state. The idea is that the events of C occur in an order compatible with $<_a$ (i.e., if $e <_a e'$, then e occurs before e', independent events can occur in any order). But for now, we cannot say yet whether two events with e co_a e' are independent in V. Therefore, Δ_V^0 and Δ_V^1 cannot be defined as partial mapping $\Delta_V^0 : \mathcal{P}(P_V) \times \mathcal{P}(E_a) \to \mathcal{P}(P_V)$ and $\Delta_V^1 : (R_V \to \mathbf{N}_\perp) \times \mathcal{P}(E_a) \to (R_V \to \mathbf{N}_\perp)$. But as we will see below, if a as a certain property (V-consistence), both $\Delta^0_V(q, C)$ and $\Delta^1_V(\sigma, C)$ are singletons for each state $\langle q, \sigma \rangle$ and each subset $C \subseteq E_a$ of events, and therefore, they can be considered as partial functions.

Definition 7 (Dependence and Independence Relation). *If V is a RN, then the relation $D_V \subseteq T_V \times T_V$ is defined by*

$$t_1 \; D_V \; t_2 \Leftrightarrow_{\mathrm{df}} (\mathrm{mod}(t_1) \cap \mathrm{ac}(t_2) \neq \emptyset \vee \mathrm{mod}(t_2) \cap \mathrm{ac}(t_1) \neq \emptyset)$$
$$\vee \, (F_V(t_1) \cup F_V^{-1}(t_1)) \cap (F_V(t_2) \cup F_V^{-1}(t_2)) \neq \emptyset.$$

The complement of D_V, i. e the relation $(T_V \times T_V) - D_V$, is called the independence relation *of V and is denoted by I_V. A co-set C of a semi-order a is called* independent *in V iff for all $e, e' \in C$, $e \neq e'$ implies $\lambda_a(e) \; I_V \; \lambda_a(e')$. a is called V-consistent iff each co-set of a is independent in V. By $\mathbf{SO}_V(T_V)$ $(\mathbf{SW}_V(T_V))$ we denote the class of V-consistent semi-orders (semi-words) over T_V.*

Note: If a is a semi-order over T_V such that $<_a$ is a linear ordering, then a is V-consistent.

Lemma 8. *Let V be a RN and let $\langle q, \sigma \rangle \in \Sigma(V)$. For each V-consistent semi-order a and each $C \subseteq E_a$, the sets $\Delta^0_V(q, C)$ and $\Delta^1_V(\sigma, C)$ are singletons.*

Proof. We show the lemma only for Δ^1_V, as the other part follows the same line. If $C = \emptyset$, then we have $\Delta^1_V(\sigma, C) = \{\sigma\}$ by definition, and we are left with the case $C \neq \emptyset$. Let $e, e' \in \max_{<_a} C$. We have to prove that $\Delta^1_V(\sigma, C - \{e\}) = \Delta^1_V(\sigma, C - \{e'\})$ for all possible choices of e and e'. For induction let us assume that $\Delta^1_V(\sigma, C - \{e, e'\}) = \{\sigma'\}$ is a singleton. Now it is enough to show that $\delta^1_V(\delta^1_V(\sigma', e), e') = \delta^1_V(\delta^1_V(\sigma', e'), e)$. But this follows immediately from the fact that $\lambda_V(e) \; I_V \; \lambda_V(e')$ holds, i. e. $\lambda_V(e)$ and $\lambda_V(e')$ modify different registers (if any). $\qquad\square$

Because of this lemma, we will consider Δ^0_V and Δ^1_V as mappings, as they will be applied only to event sets of V-consistent semi-orders.

Definition 9 (Firing Rule). *Let a be a V-consistent semi-order over T_V for some RN V. Then a is* enabled *at a state $s = \langle q, \sigma \rangle \in \Sigma(V)$ iff for each co-set C of a the following conditions are satisfied:*

$$C^- \subseteq \Delta^0_V(q, <_a^{-1}(C)) \;\&\; (C^+ - C^-) \cap \Delta^0_V(q, <_a^{-1}(C)) = \emptyset, \; and \qquad (1)$$
$$\forall e \in C \left(\pi_V(\Delta^1_V(\sigma, <_a^{-1}(C)), e) \right). \qquad (2)$$

If a is enabled at s, we denote this by $s \overset{a}{\Longrightarrow}$. a fires *from s to $s' \in \Sigma(V)$ iff $s' = \Delta_V(s, E_a)$; this is denoted by $s \overset{a}{\Longrightarrow} s'$.*

Finally, we call a transition $t \in T_V$ enabled *at some state $s \in \Sigma(V)$ if t considered as a letter is enabled at s. A set $C \subseteq T_V$ of transitions is enabled at s if $t_1, t_2 \in C \;\&\; t_1 \neq t_2 \Rightarrow t_1 \; I_V \; t_2$ and $s \overset{t}{\Longrightarrow}$ for all $t \in C$ holds. If C is enabled at s, the we call C a* step.

Condition (1) is a generalization of the usual enabledness condition for ENS. It reads: C is enabled at q if the marking q' obtained by firing the history of C (the set $<_a^{-1}(C)$) subsumes the pre-conditions of C, and additionally, the places produced by C have not yet contained in q. Condition (2) says that if an event e of C has an associated predicate P_e, then P_e is not false at the register assignment obtained by firing the history of C.

Definition 10 (Reachable States and Semi-Language of a RN). *By* $\mathbf{R}_V(s) =_{\mathrm{df}} \{s' \in \Sigma(V) : \exists a \in \mathbf{SO}(T_V)(s \overset{a}{\Longrightarrow} s')\}$ *we denote the set of* states reachable *from some state s of V, and* $\mathbf{SL}_V(s) =_{\mathrm{df}} \{a \in \mathbf{SW}_V(T_V) : s \overset{a}{\Longrightarrow}\}$ *is the* semi-language *of V at s. Finally, we put* $\mathbf{SL}(V) =_{\mathrm{df}} \mathbf{SL}_V(s_V)$.

Lemma 11. *If s is a K-restricted state of a K-restricted RN V, then $\mathbf{R}_V(s)$ is finite, and moreover, each $s' \in \mathbf{R}_V(s)$ is also K-restricted.*

Proof. Simple induction on firing sequences. □

5 Analysis of Register Nets

It is not hard to see that RN's are of the computational power of Turing machines. The easiest way to prove this fact is to reduce *counter machines* to RN's. Counter machines are finite state machines equipped with a set of *counters* containing non-negative integer values. Each counter can be decremented or incremented by one. Additionally, a state change of a counter machine can depend on a test whether a counter contains the value zero. It is quite obvious how to translate a counter machine into a RN.

On the other hand, K-restricted RN's can be translated into an ENS and therefore, they are strictly less powerful than Turing machines. The translation is done by adding places for each pair $\langle r, k \rangle$, where r is a register and $0 \leq k < K$ is an integer. The marking of such a place represents the fact that the register r contains the value k. Transitions t which access registers are replaced by transitions which move tokens according to the possible values of the predicates P_t and functions G_t. This transformation is known as *unfolding* in the context of colored Petri nets. Clearly, the unfolding of RN tend to be very large, and therefore, many analysis tools are not applicable for those unfoldings because the size of the input net violates memory limitations.

There are better ways to unfold a RN . One way is to use a binary representation of the register values [5]: For each K-restricted register $N = 2 \times \log_2(K+1)$ places will be added (hence: $K \leq 2^{\frac{N}{2}}$). Let these places be called $r_0, r_1, \ldots, r_{N-1}$ and $\bar{r}_0, \bar{r}_1, \ldots, \bar{r}_{N-1}$, respectively, and let $k_0 k_1 \ldots k_{N-1}$ be the binary representation of a value k. Then the fact that r contains the value k is represented by the marking $\{r_i : k_i = 1\} \cup \{\bar{r}_i : k_i = 0\}$. It is not hard (although quite lengthy) to implement operations like integer addition or division in terms of a binary coding of values by means of ENS. Although binary value coding leads to smaller nets than unary value coding (ordinary unfolding), to our experience, the resulting nets are in many cases still too large for an analysis.

It should be noted that both transformations preserve the interleaving semantics of a RN (under a suitable notion of language homomorphism), but they *do not preserve* its partial order semantics. The reason is that transitions accessing a common set of registers can occur concurrently (see def. 7), but not those transitions which access a common set of places.

Another way to analyze RN nets is to find dedicated analysis methods for those nets. In this paper, we will describe the *partial order representation* of the behavior of a RN by means of *concurrent automata*. To do that, we need some more math.

Definition 12. *If V is a RN and $a \in \mathbf{SO}_V(T_N)$, then $\langle a \rangle_V$ is defined by $\langle a \rangle_V =_{df} [E_a, (<_a \cap D)^+, \lambda_a]$, where D is given by $e_1 \; D \; e_2 \Leftrightarrow_{df} \lambda_a(e_1) \; D_V \; \lambda_a(e_2)$. Moreover, $\odot_V : \mathbf{SW}(T_V) \times \mathbf{SW}(T_V) \to \mathbf{SW}(T_V)$ is the operator $a \odot_V b =_{df} [E_a \cup E_b, (<_a \cup <_b \cup D)^+, \lambda_a \cup \lambda_b]$, where $D \subseteq E_a \times E_b$ is as defined above (however, with a different domain), and E_a and E_b are assumed to be pairwise disjoint.*

The following lemma states that, if we consider a RN V, for each a member of the semi-language of V there is an uniquely defined least sequential semi-word $\langle a \rangle_V$. Moreover, if least sequential semi-words are concatenated using $\odot_V$, the result is also least sequential. The lemma resembles (the second part of) theorem 2.2.9 in [15].

Lemma 13. *For each restricted RN V we have*

1. $\langle a \rangle_V \odot_V \langle b \rangle_V = \langle a \odot_V b \rangle_V$
2. $a \in \mathbf{SL}_V(s)$ *implies* $\langle a \rangle_V \in \min_{\preceq}(\mathbf{SL}_V(s))$ *for each state s of V,*
3. $a \in \mathbf{SL}_V(s)$, $b \in \mathbf{SL}_V(s')$, *and* $s \xrightarrow{a} s'$ *imply* $a \odot_V b \in \mathbf{SL}_V(s)$

Proof. (1) holds by set theory. (2) Making a semiword less sequential than $\langle a \rangle_V$ would yield a V-inconsistent semi-word. (3) holds by definition. □

Definition 14. *The set $\mathbf{LSL}_V(s) =_{df} \{\langle a \rangle_V \in \mathbf{SW}_V(T_M) : a \in \mathbf{SL}_V(s)\}$ ist called the* least sequential semi-language *of a RN V at a state $s \in \Sigma(V)$. $\mathbf{LSL}(V) =_{df} \mathbf{LSL}_V(s_V)$ is the* least sequential semi-language *of V.*

Let us now turn to CA. Essentially, a CA of a net V is a finite automaton. Its state set consists of reachable markings of V. However, the transitions of a concurrent automaton are generally not labelled by single symbols, but by semi-orders.

The following questions arise: What is the language recognized by a CA? Under which circumstances does this language constitutes a complete and correct description of the behavior of a RN?

We choose the following answers to these questions: A CA is complete and correct if it recognizes exactly the least sequential semi-language of the associated RN. Recognition is defined by combining semi-orders obtained by traversing a CA via the $\odot$-operation defined above.

Definition 15 (Concurrent Automaton). *A concurrent automaton (CA) over an alphabet T is a structure $A = \langle S, X, \delta, s \rangle$ comprising a finite set S of states, a set $X \subseteq \mathbf{SO}(T)$ of semi-orders, partial transition function $\delta : S \times X \rightarrow S$, and an initial state $s \in S$. A CA of a RN V is a CA over T_V such that $S \subseteq \mathbf{R}_V(s_V)$ and $s = s_V$ holds.*

Examples for CA can be found in [3].

Definition 16 (Semi-Language of a Concurrent Automaton). *Let A be a CA of a RN V. A path through A is a finite sequence of semi-orders $\alpha = a_1 a_2 \ldots a_n$ ($a_i \in X_A$ for $1 \leq i \leq n$) such that there are states $s_0, s_1, \ldots, s_{n+1}$ with $s_0 = s_A$ and $\delta_A(s_i, a_i) = s_{i+1}$ is defined for $0 \leq i \leq n$. Let $P(A)$ denote the set of paths through A.*

If α is a path through A as given above, then the semi-word $\hat{\alpha}$ is defined by $\hat{\alpha} =_{\mathrm{df}} a_1 \odot_V a_2 \odot_V \cdots \odot_V a_n$. The semi-language is denoted by $\mathbf{SL}(A) =_{\mathrm{df}} \{a \in \mathbf{SW}(T_V) : \exists \alpha \in P(A) (a \leq \hat{\alpha})\}$.

Definition 17 (Correctness and Completeness). *A CA of a restricted RN V is called complete, iff $\mathbf{SL}(A) \supseteq \mathbf{LSL}_V(s_V)$ holds. It is called correct, iff we have $\mathbf{SL}(A) \subseteq \mathbf{LSL}_V(s_V)$.*

The following lemma is obvious:

Lemma 18 (Preservation of Dead States). *Let V be a RN and let A be a correct and complete CA of V. $s = \langle q, \sigma \rangle \in \mathbf{R}_V(s_V)$ is dead, iff $s \in S_A$ and $\delta_A(S, a)$ is undefined for all $a \in X_A$.*

6   Algorithm

In [2,3] we discussed an algorithm to generate a concurrent automaton A of a safe Petri net. In this section, we modify this algorithm to work with RN.

Basic Algorithm. The basic algorithm resembles the reachability graph construction algorithm. It works as follows: It starts by introducing the initial state $s_A = s_V$ of A into the set Q, which contains unprocessed states. If a state s is considered, a set of semi-orders enabled at s is generated and appropriate arcs are added to A. If a new state s' is encountered by the firing of a at s, s' is added to S_A and Q. The algorithm terminates if all states in Q have been completely processed.

We have to consider the following problems:

1. If s is a state of A already generated, how do we construct an appropriate set of semi-orders enabled at s, and
2. if a is such a semi-order under construction, do we add another event to a or do we stop extending a?

Let us discuss problem 1. Define the *forward conflict relation* $D_{\mathrm{f}} \subseteq T_V \times T_V$ by

$$t_1 \; D_{\mathrm{f}} \; t_2 \Leftrightarrow_{\mathrm{df}} \Big((\mathrm{mod}(t_1) \cap \mathrm{ac}(t_2) \neq \emptyset \vee \mathrm{mod}(t_2) \cap \mathrm{ac}(t_1) \neq \emptyset)$$
$$\vee \; F_V^{-1}(t_1) \cap F_V^{-1}(t_2) \neq \emptyset \Big) \; \& \; t_1 \neq t_2,$$

and the *forward independence relation* by $t_1 \; I \; t_1 \Leftrightarrow_{\mathrm{df}} \neg(t_1 \; D_{\mathrm{f}} \; t_2) \; \& \; t_1 \neq t_2$. At a state s under consideration, we generate the set C of all maximal steps in the set T of enabled transitions at s such that

$$\forall t_1, t_2 \in C(t_1 \neq t_2 \Rightarrow t_1 \; I \; t_2 \; \& \; D_{\mathrm{f}}(t_1) \subseteq T \; \& \; D_{\mathrm{f}}(t_2) \subseteq T). \qquad (3)$$

$t_1, t_2 \in C$ for different t_1, t_2 means that t_1 and t_2 are forward independent and each transition in static forward conflict to t_1 or t_2 is also enabled at s. Hence, for a transition $t \in T$ with $D_{\mathrm{f}}(t) \not\subseteq T$ a single step $\{t\}$ is generated. Using (3) enables us to deal with confusion situations; see [3] for a detailed discussion. Now each step C is turned into a semi-order a, i.e, if $C = \{t_1, t_2, \ldots, t_n\}$, then $a = \langle \{1, 2, \ldots, n\}, \emptyset, \{\langle i, t_i \rangle : 1 \leq i \leq n\}\rangle$. Events are added to a until some termination criterion holds (see below).

Problem 2 is solved in the following way: We suppose V to be extended by an *initialization part*, i.e. if V is a restricted RN, we construct a RN V^* from V and adding a transition t_I and a place p_I and the arcs $\langle p_I, t_I \rangle$ and $\langle t_I, p \rangle$ for all $p \in q_{s_V}$ to V. t_I has the signature $\rho(t_I) = \langle \epsilon, \epsilon \rangle$, and the predicate true. The initial state of V^* is $s_{V^*} = \langle \{p_I\}, \sigma_{s_V} \rangle$. Obviously, the extension of V to V^* does not change the behavior of the net significantly: We have $\mathbf{SL}_V(s_V) = \mathbf{SL}_{V^*}(s_V)$.

Define for some RN V the *backward conflict relation* $D_{\mathrm{b}} \subseteq T_V \times T_V$ by $t_1 \; D_{\mathrm{b}} \; t_2 \Leftrightarrow_{\mathrm{df}} F_N(t_1) \cap F_N(t_2) \neq \emptyset \; \& \; t_1 \neq t_2$. In [3] the following is shown for safe Place/Transition-nets:

Lemma 19. *Let V be a restricted RN. If there is a infinite sequence of semi-orders $a_0, a_1, a_2 \ldots$ such that $a_0 = \epsilon$ and for all $i \geq 0$, $a_i \in \mathbf{LSL}_{V^*}(s)$ and $a_i < a_{i+1}$, then there is some a_n with the following property: If $e \in E_{a_n}$ is an event of a_n, then there is another event $e' \in E_{a_n}$ with $e \leq_a e'$ such that*

1. $<_{a_k}(H_{a_n}^{a_k}(e')) = \emptyset$ for all $k \geq n$, or
2. $D_{\mathrm{b}}(\lambda_{a_n}(e')) \neq \emptyset$.

With other words, if we construct such an infinite sequence of semi-orders by adding successively transitions of V, we will finally end up by adding a transition with non-empty backward conflict relation. The proof of this lemma can be carried out exactly as in [3], because only the ENS part of a RN is used in the definition of D_{b}.

This solves problem 2. If a is a semiorder under consideration enabled at a state s of the concurrent automaton which we want to construct, a new event e, labelled with some transition t, is only added, if the following conditions hold:

```
procedure extend(a : in out SO_V(T_{V*}); s : in out Σ(V*)) is
    var T : set of T_{V*};
begin
    T ← addable(a, s);
    while T ≠ ∅ do
        select t ∈ T; a ← a ⊙_{V*} t; s ← Δ_{V*}(E_t, s);
        T ← addable(a, s)
    od
end extend;
```

Algorithm 1. Concurrent Automata Generation—Procedure *extend*.

T1. t is enabled after the firing of a at s;

T2. $D_f(t) = \emptyset$, i.e. events with non-empty forward conflict relation remain minimal in $a \odot_{V*} t$;

T3. $\forall e \in E_{a \odot_{V*} t}\left(D_b(\lambda_{a \odot_{V*} t}(e)) \neq \emptyset \Rightarrow e \in \max_{<_{a \odot_{V*} t}}(E_{a \odot_{V*} t})\right)$, i.e. if an event for t is added to a, then events of a with a non-empty backward conflict relation remain maximal events of $a \odot_{V*} t$.

Algorithm 1 shows the heart of the algorithm of [3], the procedure *extend*. It is called if a new state s is encountered. The input of this procedure is a semiorder a associated with a step C of enabled transitions at s, and the state $s' = \Delta_{V*}(s, E_a)$. It makes use of a function *addable(a, s)*, which returns a set T of transitions such that conditions (T1), (T2), and (T3) are satisfied for each $t \in T$.

We improve our basic algorithm in the following way. Let s be a state of $V*$ encountered by the generation of a concurrend automaton of $V*$, and let T be the set of enabled transitions at s. If we have a transition $t \in T$ such that $D_f(t) \not\subseteq T$, then a single step $C = \{t\}$ is generated at s. Let us further assume that $D_f(t) \cap T = \emptyset$ and let C' be another step in T generated at s. Let a be the semi-order associated with C'. The procedure *extend* adds only transitions with empty forward conflict relation to a, i.e. if a is extended to a', then t *remains enabled after the firing of a'*. Therefore, it is not necessary to fire t at s; we can postpone the consideration of t until the encountering of some other state where (hopefully) t belongs to a larger than a single step. This rule has two exceptions:

1. Each transition $t' \in T$ has the above property, i.e. $D_f(t') \not\subseteq T \Rightarrow D_f(t') \cap T = \emptyset$.
2. The firing of a semi-order a' at s leads to a cycle in the concurrent automaton, i.e the state $s' = \Delta_{V*}(s, E_{a'})$ is already generated, and moreover, s is reachable from s'. In this case, t would be postponed forever.

This idea leads to algorithm 2. The following data structures are used:

1. Q, a stack of states of $V*$, contains unprocessed states;
2. *num* is an array which assigns an unique number to each newly encountered state;

algorithm *generate* **is**
    input V, a RN;
    output A, a CA;
    local variables Q : **stack of** $\Sigma(V^*)$; R : **stack of N**;
                num : **array** $\Sigma(V^*)$ **of N**; i : $\mathbf{N} \leftarrow 0$; $loop$: **bool**;
                s, s' : $\Sigma(V^*)$; T, C : **set of** T_{V^*}; a : $\mathbf{SO}_{V^*}(T_{V^*})$;
                U, V : **set of set of** T_{V^*};
begin
(1) $s_A \leftarrow s_{V^*}$; $S_A \leftarrow \{s_A\}$; $R_A \leftarrow \emptyset$; $\delta_A \leftarrow \emptyset$; $push(Q, s_A)$; $num(s_A) \leftarrow i$;
(2) **while** $\neg empty(Q)$ **do**
(3) $s \leftarrow top(Q)$; $pop(Q)$; $push(R, num(s))$; $loop \leftarrow$ **false**;
(4) $T \leftarrow enabled(s)$; $V \leftarrow single\_steps(T)$; $T \leftarrow T - \bigcup_{C \in V} C$; $U \leftarrow steps(T)$;
(5)   **if** $U \neq \emptyset$ **then**
(6)   **foreach** $C \in U$ **do**
(7) $a \leftarrow so(C)$; $s' \leftarrow \Delta_{V^*}(s, E_a)$; $extend(a, s')$;
(8)   **if** $s' \notin S_A$ **then**
(9) $push(Q, s')$; $S_A \leftarrow S_A \cup \{s'\}$; $i \leftarrow i + 1$; $num(s') \leftarrow i$
(10)   **elsif** $member(R, num(s'))$ **then**
(11)   **while** $top(R) > num(top(Q))$ **do** $pop(R)$ **od**; $loop \leftarrow$ **true**
(12)   **fi**;
(13) $X_A \leftarrow X_A \cup \{a\}$; $\delta_A(s, a) \leftarrow s'$
(14)   **od**
(15)   **fi**;
(16)   **if** $loop \vee U = \emptyset$ **then**
(17)   **foreach** $C \in V$ **do**
(18) $a \leftarrow so(C)$; $s' \leftarrow \Delta_{V^*}(s, E_a)$; $extend(a, s')$;
(19)   **if** $s' \notin S_A$ **then**
(20) $push(Q, s')$; $S_A \leftarrow S_A \cup \{s'\}$; $i \leftarrow i + 1$; $num(s') \leftarrow i$
(21)   **elsif** $member(R, num(s'))$ **then**
(22)   **while** $top(R) > num(top(Q))$ **do** $pop(R)$ **od**;
(23)   **fi**;
(24) $X_A \leftarrow X_A \cup \{a\}$; $\delta_A(s, a) \leftarrow s'$
(25)   **od**
(26)   **fi**
(27) **od**
end *generate*;

Algorithm 2. Concurrent automata generation.

3. U and V are sets of transition sets. U contains those steps which can be fired at a state s, V contains single steps which probably can be postponed. The loop (6) – (14) deals with steps in the set U, single steps in V are considered in the loop (17) – (26).

4. R, a stack of state numbers, is used to detect cycles in the constructed concurrent automaton. For each pair of states s, s' considered in the outermost loop of algorithm 2, R contains the sequence of numbers of states from s_A to s which was computed to construct s'. Hence, if s' is already a member of

R, a cycle is detected. In this case, the cycle is removed from R (lines (11) and (22)) and every postponed transition is considered in the loop (17) – (26).

Furthermore, algorithm 2 uses the following subroutines:

1. *enabled(s)* returns the set of enabled transitions at a state s of V^*.
2. *single_steps(T)* returns for a transition set T a set of single steps of transitions which can be probably postponed.
3. *steps(T)* returns the set of all steps in T according to (3).
4. *so(C)* returns a semi-order a with empty ordering for the transition set C.

7 Simple Safety Properties and Run-Time Errors

By a simple safety property we mean a proposition φ on register values of a RN V, which is satisfied at all reachable states of V, i. e. φ characterizes those states of V which are 'good' states. Let $\varphi \equiv \varphi(r_1, r_2, \ldots, r_n)$ be denote a propositional formula containing $r_1, r_2, \ldots, r_n$ as 'parameters' where for $1 \leq i \leq n$, r_i is a register of a RN. Then φ is called a *simple safety property*. A RN satisfies a φ iff for each state $s = \langle q, \sigma \rangle \in \mathbf{R}_V(s_V)$ the formula $\varphi(\sigma(r_0), \sigma(r_1), \ldots, \sigma(r_n))$ is true.

Concerning our example from section 2, a simple safety property is Y_l·Y_r = 0; i. e. at every time point, at least one of the valves is closed.

The validation of simple safety properties in the state graph of a RN is obviously simple. If we want to use concurrent automata for those validations, we have to do some additional work. We use an idea which has deeply buried its origin in the history of Petri net theory: We will add *facts* for each simple safety property to be verified. A fact is a transition which is assumed to be dead at the initial marking of a Petri net, i. e. it is assumed to be never enabled.

A *fact* t_φ of a simple safety property $\varphi(r_1, r_2, \ldots, r_n)$ is a transition with the signature $\rho(t) = \langle r_1 r_2 \ldots r_n, \epsilon \rangle$, the associated predicate $P_t(k_1, k_2, \ldots, k_n) = \neg\varphi(k_1, k_2, \ldots, k_n)$, and the function $G_t = \emptyset$.

Let $\hat{V}$ be the RN obtained by adding a fact t_φ to V. To meet our definition 1, we assume a place p_φ such that $F_{\hat{V}}(t_\varphi) = \{p_\varphi\} = F_{\hat{V}}^{-1}(t_\varphi)$ and $F_{\hat{V}}(p_\varphi) = \{t_\varphi\} = F_{\hat{V}}^{-1}(p_\varphi)$. Clearly, $p_\varphi \in q_{s_{\hat{V}}}$, since otherwise t_φ would be dead regardless whether $\hat{V}$ fulfills φ or not.

Then it is clear that t_φ is not enabled at a reachable state s of V iff this state does not violate φ. Therefore, verification of simple safety properties can be performed on-the-fly while constructing a concurrent automaton of $\hat{V}$. If a semi-order a containing an event e with $\lambda_a(e) = t_\varphi$ is constructed, a violation of φ can be reported; additionally, it is possible to give a counter example for φ:

Lemma 20. *Let $s = \langle q, \sigma \rangle$ be a state of a RN V such that $s \in \mathbf{R}_V(s_0)$, which violates the simple safety property φ. Let a be a minimal semi-order in $\mathbf{LSL}_{\hat{V}}(s_0)$ such that $s_0 \xrightarrow{a}_V s$. Then there is an event $e \in E_a$ such that $\lambda_a(e) = t_\varphi$, and for all $e' \in E_a$ we have $e' \leq_a e$.*

The minimal semi-order a from the lemma above is called a *counter example* to φ.

An algorithm to determine counter examples is immediately at hand: We compute a shortest path $s_A = s_0, s_1, \ldots, s_n$ through a CA A such that $\delta_A(s_n, b)$ is defined and b contains this event e with $\lambda_b(e) = t_\varphi$; the shortest path algorithm by Dijkstra [1] can be used for this purpose. Next we select semi-orders $a_1, a_2, \ldots, a_n$ such that $\delta_A(s_{i-1}, a_i) = s_i$ is defined for $1 \le i \le n$, and build the semi-order $c = a_1 \odot_{\hat{V}} a_2 \odot_{\hat{V}} \cdots \odot_{\hat{V}} a_n \odot_{\hat{V}} b$. Now a is obtained by the restriction of c to the event set $\le_c^{-1}(e)$.

By a run-time error we mean the occurrence of the special value $\bot$ in some of the registers of a RN V. Run-time errors can be formulated as the simple safety property $\chi(r_1, r_2, \ldots, r_N) \equiv$ true, where $R_V = \{r_1, r_2, \ldots, r_N\}$. Note that the associated fact t_χ is enabled only if $\chi(\sigma(r_1), \sigma(r_2), \ldots, \sigma(r_N)) = \bot$ for some reachable state $S = \langle m, \sigma \rangle$ of V^*. Note also: If $\varphi(r_1, r_2, \ldots, r_n)$ is a simple safety property, and $\bot$ is assigned to one of the registers r_i at some reachable state s, then φ is violated at s, because t_φ is enabled at s.

8   Summary and Further Work

Starting with the special requirements to model adequately PLC programs given in IL, we defined (resticted) register nets (RN's), a variation of colored Petri net. RN's are tailored to a concise description of the operational semantics of IL. In order to get both efficient analysis methods and comprehensive behavior descriptions, RN's have been equipped with partial order semantics instead of the usual interleaving semantics. This gave us the chance to define the concurrent automaton (CA) as a semantic model for RN's. CA have been designed to combine the advantages of partial order semantics and state based models.

A generation algorithm for CA has been given. Finally, simple safety properties were defined and an analysis method for those properties based on CA has been given. If a violation of such a property is detected, a counter example is available to give the software developer information on the system behavior in which the error occurs. Due to our partial order semantics, the counter example is given by a concise semi-word instead of one of its arbitrary serializations. We concluded with the observation that run-time errors are expressible as a special type of a simple safety property.

However, the notion of simple safety properties is not powerful enough to capture every relevant analysis question. In industrial applications, it is sometimes important to determine whether a PLC program is able to react 'immediately', i. e. within the next processing cycle. Recalling our case study from section 2, an example of a property of this type is: If the operator releases the switch, does the piston stop moving right after the next processing cycle is completed.

Encouraging results of the available implementation of an algorithm for the construction of CA for safe Petri nets have been published in [2,4]. Our ongoing research focuses on an implementation of the approach presented in this paper to determine run-times and memory efforts in practice. The analysis of more challenging examples is under preparation.

References

[1] A. V. Aho, J. E. Hopcroft, and J. D. Ullman. *Data Structures and Algorithms.* Addison-Wesley, 1987.

[2] P. Deussen. Algorithmic aspects of concurrent automata. In H.-D. Burkhard, L. Czaja, and P. Starke, editors, *Workshop on Concurrency, Specification & Programming '98*, number 110 in Informatik-Berichte, pages 39–50, Berlin, 1998. Humboldt Univ. zu Berlin.

[3] P. Deussen. Concurrent automata. Technical Report 1-05/1998, Brandenburg Tech. Univ. Cottbus, 1998.

[4] P. Deussen. Improvements of concurrent automata generation. Technical Report I-08/1998, Brandenburg Tech. Univ. Cottbus, 1999.

[5] M. Heiner. Petri net based system analysis without state explosion. In *Proc. High Performance Computing '98, SCS Int. San Diego*, pages 394–403, 1998.

[6] M. Heiner and T. Menzel. Time-related modelling of PLC systems with time-less Petri nets. In R. Boel and G. Stremersch, editors, *Discrete Event Systems*, pages 275–282. Kluwer Academic Publishers, 2000.

[7] H. Hulgaard and S. M. Burns. Bounded delay timing analysis of a class of CSP programs. *Formal Methods in System Design*, 11:265–294, 1997.

[8] K. L. McMillan. Using unfoldings to avoid the state explosion problem in the verification of asynchronous circuits. In *Proc. of the 4th Workshop on Computer Aided Verification*, pages 164–174, Montreal, 1992.

[9] T. Mertke. Hydraulic piston example, 2000. private communications.

[10] Programmable logic controllers — programming languages, IEC 1131-3. International Electronical Commission, Technical Commitee No. 65, second edition. Commitee draft, 1998.

[11] P. H. Starke. Processes in Petri nets. *J. Inf. Process. Cybern. EIK*, 17(8/9):389–416, 1981.

[12] A. Ulrich. *Testfallableitung und Testrealisierung in verteilten Systemen.* Shaker Verlaq, Aachen, 1998.

[13] A. Valmari. A stubborn attack on state explosion. *Formal Methods in System Design*, 1:297–322, 1992.

[14] F. Vernadat and F. Michel. Covering step graph preserving failure semantics. In P.Azema and G.Balbo, editors, *18th International Conference on Application and Theory of Petri Nets*, volume 1248 of *LNCS*, pages 253–270. Springer-Verlag, 1997.

[15] W. Vogler. *Modular construction and partial order semantics of Petri nets*, volume 625 of *LNCS*. Springer-Verlag, 1992.

Structural Characterization and Qualitative Properties of Product Form Stochastic Petri Nets[*]

Serge Haddad[1], Patrice Moreaux[1,2], Matteo Sereno[3], and Manuel Silva[4]

[1] LAMSADE, Université Paris Dauphine
{haddad,moreaux}@lamsade.dauphine.fr
[2] LERI-RESYCOM, Université de Reims Champagne-Ardenne
[3] Dipartimento di Informatica, Università di Torino
matteo@di.unito.it
[4] Dep. de Ingenieria Elec.a e Informatica, Universidad de Zaragoza
silva@posta.unizar.es

Abstract. The model of Stochastic Petri nets (SPN) with a product form solution (Π-net) is a class of nets for which there is an analytic expression of the steady state probabilities w.r.t. markings, as for product form queueing networks w.r.t. queue lengths. In this paper, we prove new important properties of this kind of nets. First we provide a polynomial time (w.r.t. the size of the net structure) algorithm to check whether a SPN is a Π-net. Then, we give a purely structural characterization of SPN for which a product form solution exists regardless the particular values of probabilistic parameters of the SPN. We call such nets $\overline{\Pi}$-nets. We also present untimed properties of Π-nets and $\overline{\Pi}$-nets such like liveness, reachability, deadlock freeness and characterization of reachable markings. The complexity of the reachability and the liveness problems is also addressed for Π-nets and $\overline{\Pi}$-nets. These results complement previous studies on these classes of nets and improve the applicability of Product Form solutions.

1   Introduction

Stochastic Petri nets (SPNs) are a powerful tool for modelling and evaluating the performance systems involving concurrency, non determinism, and synchronization, such as parallel and distributed systems, communication networks, etc. The stochastic semantics of SPN have been proven to be a Continuous Time Markov Chain and steady state analysis can thus be expressed as the solution of a system of equilibrium equations, one for each possible marking of their state space. The major problem in the computation of performance measures using SPNs is thus the size of the reachability set of these models that increases exponentially both with the number of tokens in the initial marking and with the number of places in the net. As a consequence, the dimension of this reachability set and the time complexity of the solution procedure preclude, in the general

---

[*] At time of writing, P. Moreaux was visiting professor at the Dipartimento di Informatica, Università di Torino

J.-M. Colom and M. Koutny (Eds.): ICATPN 2001, LNCS 2075, pp. 164–183, 2001.

case, the direct exact numerical evaluation of many interesting models. To cope with this problem, we can first accept non exact performance measures. The two main approaches developed in this area are discrete-event simulation and approximate methods. Bounds computation methods provide more reliable information about the the performance indices. However, if we wish to obtain the exact values of performance measures, then we may improve numerical methods solving the underlying mathematical problem (linear or differential systems of equations) and/or we may relate the structure of the model to the properties of these underlying mathematical objects.

One successful approach in this last direction is the product form analysis (PFA) for Queueing Networks (QN), that is the expression of basic performance indices of QN, such as steady state probabilities, mean throughputs, utilization, etc., as functions of the model parameters (service rates, routing probabilities, properties of the service stations, etc.). The first structural property involved in PFA is obviously the setting up of the model as a collection of service stations bounded with paths taken by "clients". From this structure, PF solutions may be proven for several classes of QN by examination of sets of some kind of "local balance equations", for instance equations established for each station. Second, specific descriptions of the state space of PF-QN lead to important relations. For instance, the convolution algorithms [21] and the Mean Value Analysis (MVA) method [4] are based upon recursive relations between models with state spaces with different number of clients. Unfortunately, (the standard version of) PF-QN offer limited possibilities for what concerns synchronization between clients activities. This situation was one of the main motivations in the study of Stochastic Petri Nets (SPN) with a Product form solution (PF-SPN). First results about PF-SPN were established in [15] based on the structure of the reachability graph of the net. Recently, several authors proposed structural sufficient conditions for a Petri net to be a PF-SPN. These results are summarized in Section 2. The present paper supplements previous results for PF-SPN regarding four important issues.

Membership Problem for SPN with PF solution. As we will see in Section 2, a straightforward verification procedure for deciding whether a given SPN has a PF solution requires the computation of all minimal T-semiflows of the marked net (T-semiflows are structural invariants of Petri nets (PN), see Section 2). It is however known that the number of minimal T-semiflows can be exponential in the number of transitions (e.g., [17]). In fact, we establish a polynomial time algorithm to decide whether a SPN has a PF solution.

Rate independent structural characterization of PF-SPN. Previous criteria for PF-SPN have two drawbacks: they are only sufficient conditions, and they involve properties of the rates of the transitions of the net. We present a necessary and sufficient structural condition on nets to admit a PF solution whatever the rates of its transitions. Hence we prove a rate-independent structural characterization of PF-SPN. Moreover, this criterion can be checked in polynomial time.

Untimed properties of PF-SPN. We investigate untimed properties for the class of PF-SPN. Since many results (deadlock-freeness, liveness, etc.) have been established for several known classes of PN, it can be valuable to point out the relation between PF-SPN and these classes.

Reachability Set properties. Efficient numerical solutions for PF-SPN require to characterize subsets of reachable markings. It is hence important to have a structural criterion for reachable markings (e.g., a method based on the minimal P-semiflows, a method based on the net state equation, etc.). We present new results about these possible criteria.

The organization of the paper is the following: in Section 2, we review SPN and previous results about Π-nets. Section 3 presents the verification procedure for PF-SPN and a series of results about the class of PF-SPN in relation to other classes of Petri nets. In Section 3.3 we define the new class, $\overline{\Pi}$-nets, of PF-SPN corresponding to rate independent criteria for a PF solution together with globally dependent rates. Untimed properties of Π-nets are studied in Section 4. The conclusion summarizes results presented in the paper.

2   Background and Notations

2.1   Stochastic Petri Nets

One may find introductory presentations of Petri net concepts for instance in [19,20,26]. We remind the reader only with definitions necessary to understand product form results for stochastic Petri nets.

A *marked stochastic Petri net* is a 5-tuple $SPN = (\mathcal{P}, \mathcal{T}, W, \mathcal{Q}, \mathbf{m_0})$, where $\mathcal{P}$ and $\mathcal{T}$ are disjoint sets of *places* and *transitions* (with $|\mathcal{P}| = np$ and $|\mathcal{T}| = nt$), $W := (\mathcal{P} \times \mathcal{T}) \cup (\mathcal{T} \times \mathcal{P}) \to \mathbb{N}$ defines the *weighted flow relation*: if $W(j, i) > 0$ (resp. $W(i, j) > 0$) then we say that there is an *arc* from t_j to p_i, with *weight* or *multiplicity* $W(j, i)$ (resp. there is an arc from p_i to t_j with weight $W(i, j)$), $\mathcal{Q}$ is the set of transition firing rates drawn from exponential distributions, and $\mathbf{m_0}$ is the initial marking.

For a given transition $t_j \in \mathcal{T}$, its *preset* and *postset* are given by $^\bullet t_j = \{p_i \mid W(i, j) > 0\}$ and $t_j{}^\bullet = \{p_i \mid W(j, i) > 0\}$, respectively. In the same manner we can define the *preset* and *postset* of a given place.

For any transition t_j, from the weighted flow relation we can the define the *input vector* $\mathbf{i}(t_j) = [W(1, j), W(2, j), \ldots, W(|\mathcal{P}|, j)]$ and the *output vector* $\mathbf{o}(t_j) = [W(j, 1), W(j, 2), \ldots, W(j, |\mathcal{P}|)]$. From the weighted flow relation we can also define the *incidence matrix* $\mathbf{C}$ with entries $\mathbf{C}[i, j] = W(j, i) - W(i, j)$.

A transition t_j is *enabled* in a marking $\mathbf{m}$ iff $\mathbf{m} \geq \mathbf{i}(t_j)$. Being enabled, t_j may *occur* (or *fire*) yielding a new marking $\mathbf{m}' = \mathbf{m} + \mathbf{C}[., j]$ ($\mathbf{C}[., j]$ is the jth column of $\mathbf{C}$), and this is denoted by $\mathbf{m} \xrightarrow{t_j} \mathbf{m}'$. The set of all the markings reachable from $\mathbf{m_0}$ is called *reachability set*, and is denoted by $\mathrm{RS}(\mathbf{m_0})$.

Semiflows are non-null natural annullers of $\mathbf{C}$. Right and left annullers are called T- and P-(semi)flows respectively. A semiflow is called *minimal* when its support (i.e., the set $\|\mathbf{s}\|$ of the non-zero components of vector $\mathbf{s}$) is not a proper superset of the support of any other, and the g.c.d. of its elements is one.

2.2 Previous Product Form Solution Results for Stochastic Petri Nets

A class of SPNs characterized by the fact that the stationary probability distribution of any net in this class can be factored into a product of terms has been introduced [11,13]. Nets possessing this property are called *Product-Form Stochastic Petri Nets* (PF-SPNs) and are easily identified by the criteria proposed in [2,7,11,13].

Let $\mathbf{x_1}, \mathbf{x_2}, \ldots, \mathbf{x_h}$ denote the minimal T-semiflows found from the incidence matrix. The following definitions are essential to the analysis of the SPNs that have Product Form Solution.

Definition 1. *A subset of transitions $\mathcal{T}'$ ($\mathcal{T}' \subseteq \mathcal{T}$) is said to be* closed *if $\bigcup_{t_j \in \mathcal{T}'} \mathbf{i}(t_j) = \bigcup_{t_j \in \mathcal{T}'} \mathbf{o}(t_j)$. An alternative definition of a closed set of transitions is the following: let $\mathcal{R}(\mathcal{T}') = \bigcup_{t_j \in \mathcal{T}'} \{\mathbf{i}(t_j) \cup \mathbf{o}(t_j)\}$ be the set of input and output bags for transitions in $\mathcal{T}'$. The subset of transitions $\mathcal{T}'$ is said to be closed if for any $\mathbf{l} \in \mathcal{R}(\mathcal{T}')$ there exists $t_i, t_j \in \mathcal{T}'$ such that $\mathbf{l} = \mathbf{i}(t_i)$ and $\mathbf{l} = \mathbf{o}(t_j)$; that is, each output bag is also an input bag for some transition in $\mathcal{T}'$, and vice-versa each input bag is also an output bag.*

Definition 2. *$\mathcal{N}$ is a Π-net if $\forall t_j \in \mathcal{T}$ there exists a minimal T-semiflow $\mathbf{x}$ such that $t_j \in ||\mathbf{x}||$, and $||\mathbf{x}||$ is a closed set.*

In other words, $\mathcal{N}$ is a Π-net if all transitions are covered by closed support minimal T-semiflows.

Example of Π-net. Figure 1(a) shows a net satisfying Definition 2. We can see that there are two minimal T-semiflows $\mathbf{x_1} = [1, 0, 1, 0]$ and $\mathbf{x_2} = [0, 1, 0, 1]$, with $||\mathbf{x_1}|| = \{t_1, t_3\}$ and $||\mathbf{x_2}|| = \{t_2, t_4\}$. We can observe that $\bigcup_{t_j \in ||\mathbf{x_1}||} \mathbf{i}(t_j) = \{[1, 0, 0, 0], [0, 0, 1, 0]\} = \bigcup_{t_j \in ||\mathbf{x_1}||} \mathbf{o}(t_j)$ and $\bigcup_{t_j \in ||\mathbf{x_2}||} \mathbf{i}(t_j) = \{[1, 1, 0, 0], [0, 0, 0, 1]\} = \bigcup_{t_j \in ||\mathbf{x_2}||} \mathbf{o}(t_j)$. Both T-semiflows have closed support set. Since any transition belongs to a closed support minimal T-semiflow, this net is a Π-net.

The definition of Π-nets was originally motivated while studying the problem of finding product form solution for SPNs [2,7,11,13]. More precisely, for the SPNs having the Π property, there exists a positive solution for the traffic equations (see below). In a Π-net we denote by $\mathcal{X}_c = \{\mathbf{x_1}, \mathbf{x_2}, \ldots, \mathbf{x_l}\}$ the set of closed support minimal T-semiflows. Among the minimal closed support T-semiflows, we can identify a relation that can be used to derive the PFS. Two different minimal closed support T-semiflows $\mathbf{x}'$ and $\mathbf{x}''$ are said to be *freely related*, denoted as $(\mathbf{x}', \mathbf{x}'') \in FR$, if there exist $t_j \in ||\mathbf{x}'||$ and $t_h \in ||\mathbf{x}''||$ such that $\mathbf{i}(t_j) = \mathbf{i}(t_h)$. The relation FR^* is the transitive closure of FR. It is easy to see that the relation FR^* yields a partitioning of the set of minimal closed support T-semiflows. Because any t_j can belong to only one FR-class, the partition of

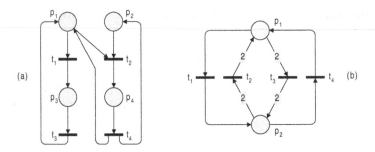

Fig. 1. Examples of Π-net s

T-semiflows leads to a partition of transitions. In the following we denote by $\mathcal{C}(t_j)$ the set of the partition to which transition t belongs.

As for Queueing Networks, PF solutions for SPN are based on the analysis of underlying Markov chains (MC). Instead of reasoning in terms of the MC with states as markings, it is more convenient to study an auxiliary MC with states being the input (or output) vectors $\mathbf{i}(t)$, called the *routing process* [11] of the SPN. The infinitesimal generator $\mathbf{Q}$ of this MC is defined by: $q(\mathbf{i}(t_j), \mathbf{o}(t_j)) = \mu(\mathbf{i}(t_j))\mathbf{P}[\mathbf{i}(t_j), \mathbf{o}(t_j)]$ with $\mu(\mathbf{i}(t_j)) = \sum_{\mathbf{i}(t_h)=\mathbf{i}(t_j)} \mu_h$. $\mathbf{P}[a, b]$ is the routing probability from $a = \mathbf{i}(t_j)$ to b: it can be computed by examining the various transitions enabled after the firing of t_j and μ_h is the usual rate of SPN transition t_h. For the sake of simplicity, we present all the results by assuming that the transition rates are marking independent. In [10] results are presented with several kinds of marking dependent transition firing rates.

The traffic equations of the routing process are the global balance equations of this MC. Denoting with $v(\mathbf{i}(t_j))$ the so-called visit-ratio to node $\mathbf{i}(t_j)$, these equations can be stated as:

$$\forall t_j \in \mathcal{T}, \quad v(\mathbf{i}(t_j)) = \sum_{t_h \in \mathcal{T}} v(\mathbf{i}(t_h))\mathbf{P}[\mathbf{i}(t_h), \mathbf{i}(t_j)] \tag{1}$$

Boucherie and Sereno [2] showed that traffic equations and structural properties of a net are closely related.

Theorem 1 (from [2]). *Let* $\mathcal{N} = (\mathcal{P}, \mathcal{T}, W, \mathcal{Q}, \mathbf{m_0})$ *be a SPN. There is a non null positive solution for the Traffic Equations (1) iff* $\mathcal{N}$ *is a* Π-*net.*

The existence of a positive solution for the Traffic Equations (1) is not a sufficient condition to assert a Product-Form Solution for the SPN. The following result from Coleman *et al.* [7] and [11], states that the equilibrium distribution has a product-form over the places of the SPN whenever one additional condition holds. Let us denote $f = v/\mu$ with v a solution for the traffic equations, and define the vector $\mathbf{w}_f = [w_1, \dots, w_n]$ as

$$\mathbf{w}_f = \left[\log\left(\frac{f(\mathbf{i}(t_1))}{f(\mathbf{o}(t_1))}\right), \log\left(\frac{f(\mathbf{i}(t_2))}{f(\mathbf{o}(t_2))}\right), \dots, \log\left(\frac{f(\mathbf{i}(t_{nt}))}{f(\mathbf{o}(t_{nt}))}\right) \right] \tag{2}$$

There may be many functions f that derive from solutions for the traffic equations. However each one is unique up to a multiplicative constant in each FR* class. This implies that the ratio $f(\mathbf{i}(t_i))/f(\mathbf{o}(t_i))$ is invariant.

Theorem 2 (Product-Form for Equilibrium Distribution of SPN, (from [7,11])). *Let $f = v/\mu$ with v a solution for the traffic equations. The equilibrium distribution for the SPN has the form*

$$\pi(\mathbf{m}) = \frac{1}{G} \prod_{i=1}^{np} y_i^{m_i} \qquad \forall\ \mathbf{m} \in RS(\mathbf{m_0}) \tag{3}$$

if and only if $Rank(\mathbf{C}) = Rank([\mathbf{C} \mid \mathbf{w}_f])$ where $[\mathbf{C} \mid \mathbf{w}_f]$ is the matrix $\mathbf{C}$ augmented with the row $\mathbf{w}_f$ and G a normalization constant. In this case, the np-component vector $\mathbf{r} = [\log(y_1), \ldots, \log(y_{np})]$, satisfies the matrix equation $-\mathbf{r}.\mathbf{C} = \mathbf{w}_f$.

It must be noted that, generally, the condition $Rank(\mathbf{C}) = Rank([\mathbf{C} \mid \mathbf{w}_f])$ depends on the *rates* of the transitions of the net and not only on the structure of the net.

2.3   Examples of Π-Nets

Let us present two detailed examples of Π-nets. The first one complements the study of the net of Figure 1(a) and the second one shows a more complex situation about the rank condition of Theorem 2. The reader will also find an example of an unbounded Π-net in Section 4.3.

Example 1. In this example we briefly review the procedure used to obtain the equilibrium distribution for the Π-net depicted in Figure 1(a). For additional details the reader is referred to [2,7,11,12,13]. Since we know that the SPN is a Π-net, there is a solution for the Traffic equations (1):

$$\begin{aligned} v(\mathbf{i}(t_1)) &= v(\mathbf{i}(t_3)) & v(\mathbf{i}(t_3)) &= v(\mathbf{i}(t_1)) \\ v(\mathbf{i}(t_2)) &= v(\mathbf{i}(t_4)) & v(\mathbf{i}(t_4)) &= v(\mathbf{i}(t_2)) \end{aligned}$$

One solution is $v(\mathbf{i}(t_1)) = v(\mathbf{i}(t_3)) = v(\mathbf{i}(t_2)) = v(\mathbf{i}(t_4)) = 1$, from which we obtain $f(\mathbf{i}(t_1)) = 1/\mu_1$, $f(\mathbf{i}(t_3)) = 1/\mu_3$, $f(\mathbf{i}(t_2)) = 1/\mu_2$, and $f(\mathbf{i}(t_4)) = 1/\mu_4$. The row vector $\mathbf{w}_f$ is:

$$\begin{aligned} \mathbf{w}_f = [&\log(f(\mathbf{i}(t_1))/f(\mathbf{i}(t_3))), \log(f(\mathbf{i}(t_2))/f(\mathbf{i}(t_4))), \log(f(\mathbf{i}(t_3))/f(\mathbf{i}(t_1))), \\ &\log(f(\mathbf{i}(t_4))/f(\mathbf{i}(t_2)))] = [\log(\mu_3/\mu_1), \log(\mu_4/\mu_2), \log(\mu_1/\mu_3), \log(\mu_2/\mu_4)] \end{aligned}$$

The rank condition $Rank(\mathbf{C}) = Rank([\mathbf{C} \mid \mathbf{w}_f])$ gives us:

$$Rank \begin{pmatrix} -1 & -1 & 1 & 1 \\ 0 & -1 & 0 & 1 \\ 1 & 0 & -1 & 0 \\ 0 & 1 & 0 & -1 \end{pmatrix} = Rank \begin{pmatrix} -1 & -1 & 1 & 1 \\ 0 & -1 & 0 & 1 \\ 1 & 0 & -1 & 0 \\ 0 & 1 & 0 & -1 \\ w_1 & w_2 & w_3 & w_4 \end{pmatrix}$$

The rank condition holds independently of the rate values because we can easily verify that $w_1 + w_3 = 0$ and $w_2 + w_4 = 0$ since $\log\left(\frac{\mu_3}{\mu_1}\right) + \log\left(\frac{\mu_1}{\mu_3}\right) = \log\left(\frac{\mu_3}{\mu_1}\frac{\mu_1}{\mu_3}\right) = \log(1) = 0$ and similarly for $w_2 + w_4 = 0$.

Since theorem 2 applies, we can obtain the expression of $\pi(\mathbf{m})$. To this end, we first solve the matrix equation $\mathbf{r}.\mathbf{C} + \mathbf{w}_f = \mathbf{0}$, that is to say:

$$-r_1 + r_3 + w_1 = 0 \quad r_1 - r_3 + w_3 = 0$$
$$-r_1 - r_2 + r_4 + w_2 = 0 \quad r_1 + r_2 - r_4 + w_4 = 0.$$

Then, setting $r_1 = r_2 = 0$, we obtain $r_3 = w_3$ and $r_4 = w_4$ from which we derive ($r_i = \log(y_i)$), $y_1 = y_2 = 1$, $y_3 = \mu_1/\mu_3$, and $y_4 = \mu_2/\mu_4$. Hence the equilibrium distribution of the SPN of Figure 1(a) is $\pi(\mathbf{m}) = \frac{1}{G}\left(\frac{\mu_1}{\mu_3}\right)^{m_3}\left(\frac{\mu_2}{\mu_4}\right)^{m_4}$.

Example 2. The SPN shown in Figure 1(b), taken form [7], represents an SPN in which the rank condition is not satisfied independently of the rate values. The incidence matrix $\mathbf{C}$ is given by $\mathbf{C} = \begin{pmatrix} -1 & 2 & -2 & 1 \\ 1 & -2 & 2 & -1 \end{pmatrix}$. This SPN is covered by four minimal T-semiflows whose support sets are $\|\mathbf{x_1}\| = \{t_1, t_4\}$, $\|\mathbf{x_2}\| = \{t_2, t_3\}$, $\|\mathbf{x_3}\| = \{2t_1, t_2\}$, and $\|\mathbf{x_4}\| = \{t_3, 2t_4\}$. Only $\mathbf{x_1}$ and $\mathbf{x_2}$ are closed, but they cover $\mathcal{T}$ so that the SPN satisfies Definition 2. Then the SPN is a Π-net and hence there exists a positive solution for the traffic equations. In particular we obtain $f(\mathbf{i}(t_i)) = \frac{1}{\mu_i}$ for $i = 1, \ldots, 4$. The vector $\mathbf{w}_f$ is given by

$$\mathbf{w}_f = \left[\log\left(\frac{f(\mathbf{i}(t_1))}{f(\mathbf{i}(t_4))}\right), \log\left(\frac{f(\mathbf{i}(t_2))}{f(\mathbf{i}(t_3))}\right), \log\left(\frac{f(\mathbf{i}(t_3))}{f(\mathbf{i}(t_2))}\right), \log\left(\frac{f(\mathbf{i}(t_4))}{f(\mathbf{i}(t_1))}\right)\right]$$
$$= \left[\log\left(\frac{\mu_4}{\mu_1}\right), \log\left(\frac{\mu_3}{\mu_2}\right), \log\left(\frac{\mu_2}{\mu_3}\right), \log\left(\frac{\mu_1}{\mu_4}\right)\right]$$

The augmented matrix $[\mathbf{C} \mid \mathbf{w}_f]$ is row equivalent to the fully row reduced matrix $\begin{pmatrix} 1 & 0 & 0 & 0 \\ -1 & 0 & 0 & 0 \\ w_1 & w_2 + 2w_1 & w_3 - 2w_1 & w_1 + w_4 \end{pmatrix}$. The rank conditions are $w_2 + 2w_1 = 0$, $w_3 - 2w_1 = 0$, and $w_1 + w_4 = 0$, which implies, $\frac{f(\mathbf{i}(t_2))}{f(\mathbf{i}(t_3))}\left(\frac{f(\mathbf{i}(t_1))}{f(\mathbf{i}(t_4))}\right)^2 = 1$, $\frac{f(\mathbf{i}(t_3))}{f(\mathbf{i}(t_2))}\left(\frac{f(\mathbf{i}(t_4))}{f(\mathbf{i}(t_1))}\right)^2 = 1$, and $1 = 1$ respectively. The first and second conditions are the same and arise because there is more than one way to produce the same change of marking. Substituting for the function f, the rank condition becomes $\frac{\mu_2}{\mu_3} = \left(\frac{\mu_4}{\mu_1}\right)^2$. If this condition is met, theorem 2 applies, and, letting $y_2 = 1$ gives $y_1 = \frac{f(\mathbf{i}(t_1))}{f(\mathbf{i}(t_4))}$. Finally, $\pi_f(\mathbf{m}) = \left[\frac{\mu_4}{\mu_1}\right]^{m_1}$.

3   The Class of Π-Nets

In this section we are interested in structural properties of Π-nets. We present first an important result which allows one to check, in polynomial time[1] , whether a given SPN is or not a Π-net. Then, trying to position the class of Π-nets with respect to classical structural classes of PN, we show that there is no simple relation between these classes and Π-nets.

3.1   Membership Problem

Algorithm Verify Π-net
$\mathcal{L} \leftarrow \mathcal{T}$
fail $\leftarrow$ *false*
repeat
    let $t \in \mathcal{L}$
    $\mathcal{A} \leftarrow \{t\}$
    $In \leftarrow \{\mathbf{i}(t)\}$
    $Out \leftarrow \{\mathbf{o}(t)\}$
    while $\exists t' \in \mathcal{L}$ s.t. $\mathbf{i}(t') \in Out$ **do**
        $\mathcal{A} \leftarrow \mathcal{A} \bigcup \{t'\}$
        $\mathcal{L} \leftarrow \mathcal{L} \setminus \{t'\}$
        $In \leftarrow In \bigcup \{\mathbf{i}(t')\}$
        $Out \leftarrow Out \bigcup \{\mathbf{o}(t')\}$
    endwhile
    fail $\leftarrow$ ($In \neq Out$)
    /* if not*fail* then $\mathcal{A}$ is a FR^* class */
until $\mathcal{L} = \emptyset$ or *fail*
/* *fail* is true iff the net is not a Π-net */

From the definition of Π-nets we can decide if a given net falls in this class. The problem that arises is the complexity of a straightforward application of Definition 2 because the number of minimal T-semiflows can be exponential in the number of transitions (e.g., [17]). We present now an algorithm that allows to recognize whether a net is a Π-net in polynomial time. The soundness of the algorithm is based on the following lemma (see [2] for the proof).

Lemma 1. *If* $\mathbf{x}$ *is a closed support minimal T-semiflow then (i) for each transition* $t_i \in \|\mathbf{x}\|$, $\mathbf{x}[i] = 1$ ($\mathbf{x}[i]$ *is the i-th component of* $\mathbf{x}$). *(ii)* $\|\mathbf{x}\|$ *may be ordered as* $\{t_{j_0}, t_{j_1}, \ldots, t_{j_{h-1}}\}$ *such that* $\mathbf{o}(t_{j_i}) = \mathbf{i}(t_{j_{i+1 \bmod h}})$ *(for* $i = 0, 1, \ldots, h-1$*), and* $l \neq l' \Rightarrow \mathbf{i}(t_{j_l}) \neq \mathbf{i}(t_{j_{l'}})$.

Algorithm for Π-net membership. The previous lemma states that a closed support minimal T-semiflow can be seen as a cycle of transitions $t_{j_0}, t_{j_1}, \ldots, t_{j_{h-1}}$ such that $\mathbf{o}(t_{j_i}) = \mathbf{i}(t_{j_{i+1 \bmod h}})$ (for $i = 0, 1, \ldots, h-1$). The algorithm **Verify** Π-net exploits this feature for checking if a net is a Π-net.

---

[1] Unless explicitly mentioned, all complexity results in the paper are w.r.t. the size of the net, i.e. the number of places, transitions, arcs and the binary representation of valuations.

We point out that the algorithm yields a covering set of closed support minimal T-semiflows (if the SPN is a Π-net). From then we can derive the routing probabilities and the partitions of the set of transitions $\mathcal{T}$ into FR*-classes.

From simple considerations we can see that both the inner and the outer cycles require $O(|\mathcal{T}|)$ steps. Hence the complexity of the algorithm that allows to recognize if a given net satisfy Definition 2 requires $O(|\mathcal{T}|^2)$ steps.

3.2 Π-Nets and Other Classes of PN

As usual for Petri net models, it is interesting to examine whether it is possible to structurally characterize behavioural properties of these nets and to deduce efficient checking of these properties. Since this is the case for some well known subclasses of nets, we first recall such subclasses and we compare Π-nets with them. For completeness, results include the class of $\overline{\Pi}$-nets which are introduced in section 3.3.

The following classes of Petri nets are particularly interesting for the analysis of behavioural properties:

- A state machine (SM) is a Petri net with binary valuations where any transition has exactly one input and one output place.
- A marked graph (MG) is a Petri net with binary valuations where any place has exactly one input and one output transition.
- A weighted transition system (WTS) is a Petri net where any place has exactly one input and one output transition (MG are special case of WTS).
- An extended free-choice net (EFC) is a Petri net with binary valuations where two transitions, sharing an input place, have the same set of input places.

Proposition 1. *Comparing Π-nets with some classical subclasses of Petri nets, we have:*

- *If $\mathcal{N}$ is a WTS and a Π-net, then it is behaviourally equivalent to a MG.*
- *Every SM is a Π-net (and even a $\overline{\Pi}$-net).*
- *There are MGs which are not Π-nets.*
- *There are Π-nets (and even $\overline{\Pi}$-nets) which are non EFC nets.*

Proof. Figure 2 explains the conversion from a WTS Π-net to a MG: in (a) we change the weights of arcs connecting isolated places ($k = w_1 - w_2$); in (b), we observe that any weighted Π-cycle is just equivalent to an ordinary cycle.

As a straightforward consequence of the definitions, every SM is a Π-net. In any SM, $\mathbf{r}$ vectors are $\mathbf{1_p}$: null components except on component p, input or output place of a transition t. Taking $\mathbf{a_r} = \mathbf{r}$ for each $\mathbf{r}$, we see that a SM is also a $\overline{\Pi}$-net (see below for definition of $\overline{\Pi}$-nets).

A net with an idle place followed by a parbegin-parend with intermediate action is a MG but not a Π-net. Note however, that any Π-net MG is a union of disjoint cycles, hence a $\overline{\Pi}$-net.

Finally, we will see that the net of Example 1 (Figure 1(a)) is a $\overline{\Pi}$-net, and it is clearly not an EFC.

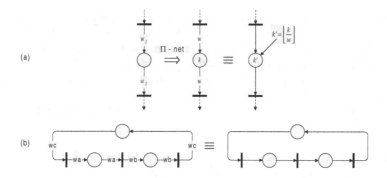

(a)

(b)

Fig. 2. Conversion of WTS Π-nets into MGs

3.3 $\overline{\Pi}$-Nets

In this section we define the class of $\overline{\Pi}$-nets which are exactly the set of Π-nets having a PF solution *for any stochastic specification* in contrast with previous results whose criteria are dependent on rates of transitions (see Example 2). Moreover, we introduce a more general dependency of the firing rates of transitions with respect to the global marking of the net system.

Definition of $\overline{\Pi}$-Nets. Criteria found by several authors since the late 80's for PF solution of SPN are only sufficient conditions, and moreover, they are made up of structural conditions and conditions on the stochastic parameters of the SPN. In search of a pure structural characterization of PF solution SPN, we were led to fully reconsider the concept of "virtual client state" of a Π-net system in the context of routing processes and to deeply analyze how to characterize these states. In previous works, T-semiflows identify concurrent "virtual clients" activities. These activities are "synchronized" by conflicting resources allocation, that is shared input places of transitions. For what concerns places, they are usually interpreted either as specific resources or as clients. But, indeed, this interpretation does not allows us to express the PF property at a structural level because virtual client states do not reduce to place markings, even in a Π-net. For instance, in the example net 1 (figure 1(a)), we may think of t_1, p_3, t_3 as batch jobs processing (activity 1), and of t_2, p_4, t_4, p_2 as interactive work of users (activity 2). The place p_1, modelling processor resources, cannot, alone, characterize the "idle" batch jobs state. This is the crucial point: in a Π-net, we have no information about the state of the virtual clients in the net system and this is the main reason which prevents us to state a necessary and sufficient condition for the existence of a PF solution. Actually, we have found that virtual client states are characterized by a relation $\mathbf{v}.\mathbf{C} = \mathbf{r}$, where $\mathbf{v}$ is a vector on places and $\mathbf{r}$ is a vector such that $\mathbf{r}[t] = 1$ if t adds a client to the "state", $\mathbf{r}[t] = -1$ if t removes a client and $\mathbf{r}[t] = 0$ otherwise. The $\overline{\Pi}$-net property expresses, by means of rational vectors $\mathbf{a_r}$, the relation which must hold between virtual clients states of a Π-net and input/output vectors of the net, to ensure that this Π-net has a PF solution,

Moreover, this explicit relation on states of virtual clients allows us to model the dependency of the firing rate of a transition t_j with respect to the global state of the system in parts (activities) of the net not related to the input/output vectors of transitions belonging to $\mathcal{C}(t_j)$. This kind of dependency, introduced by functions $\rho_{\mathcal{C}(t_j)}$ in the definition below, cannot be taken into account in the framework of Π-nets.

For the rest of this section, we set: $^\bullet\mathbf{r} = \{t_j \in \mathcal{T} \mid \mathbf{o}(t_j) = \mathbf{r}\}$ and $\mathbf{r}^\bullet = \{t_j \in \mathcal{T} \mid \mathbf{i}(t_j) = \mathbf{r}\}$ for every $\mathbf{r} \in \mathcal{R}(\mathcal{T})$.

Definition 3 ($\overline{\Pi}$-net). *A $\overline{\Pi}$-net (restricted Π-net) is a Π-net such that for every $\mathbf{r} \in \mathcal{R}(\mathcal{T})$, there exists $\mathbf{a_r} \in \mathbb{Q}^{|\mathcal{P}|}$ such that*

$$\mathbf{a_r}.\mathbf{C}[\mathcal{P}, j] = \begin{cases} 1 & \text{if } t_j \in {}^\bullet\mathbf{r} \\ -1 & \text{if } t_j \in \mathbf{r}^\bullet \\ 0 & \text{otherwise} \end{cases}$$

where $\mathbf{C}$ is the incidence matrix of the net (note that this excludes transitions t_h with $\mathbf{i}(t_h) = \mathbf{o}(t_h)$).
The firing rate of a transition t_j of a $\overline{\Pi}$-net system in the marking $\mathbf{m}$ is given by

$$\mu(t_j, \mathbf{m}) = \mu(\mathbf{i}(t_j)).\rho_{\mathcal{C}(t_j)}\left((\mathbf{a}_{\mathbf{r}''}.\mathbf{m})_{\mathbf{r}'' \notin \mathcal{C}(t_j)}\right).\mathbf{P}[\mathbf{i}(t_j), \mathbf{o}(t_j)] \qquad (4)$$

Positive, real valued functions $\rho_{\mathcal{C}(t_j)}\left((\mathbf{a}_{\mathbf{r}''}.\mathbf{m})_{\mathbf{r}'' \notin \mathcal{C}(t_j)}\right)$ make possible a homogeneous dependency of the transitions of the component $\mathcal{C}(t_j)$ w.r.t. the state of the virtual clients in the other components, given by the $\mathbf{a}_{\mathbf{r}''}.\mathbf{m}$ (see example below).

Note that the computation of the rational vectors $\mathbf{a_r}$ (or else the proof that there are no such $\mathbf{a_r}$), may be achieved in polynomial time with respect to the size of the net through a usual Gaussian elimination (but restricted to rational numbers).

The net of Example 2 is an example of a Π-net which is not a $\overline{\Pi}$-net(see its incidence matrix in Section 2.3). Let us set $\mathbf{r_1} = \{p_1\}$, so that $^\bullet\mathbf{r_1} = \{t_4\}$ and $\mathbf{r_1^\bullet} = \{t_1\}$. If we try to define the vector $\mathbf{a}_{\mathbf{r_1}} = [a, b]$, we get $a - b = 1$ (since $t_4 \in {}^\bullet\mathbf{r_1}$) and $a - b = 0$ (since $t_2 \notin {}^\bullet\mathbf{r_1} \bigcup \mathbf{r_1^\bullet}$). Hence, $\mathbf{a}_{\mathbf{r_1}}$ does not exist and this SPN is not a $\overline{\Pi}$-net. In fact, t_1 and t_3 have proportional input and output bags but belong to different T-semiflows and no distinction between these transitions is possible from $\mathbf{r_1} = [1, 0, 0, 0]$.

The Π-net of Example 1 (see Section 2.3) is a $\overline{\Pi}$-net. We have four input vectors $\mathbf{r}$, belonging to two classes: $\mathcal{C}_1 = \{\mathbf{r_1} = [1, 0, 0, 0], \mathbf{r_3} = [0, 0, 1, 0]\}$, $\mathcal{C}_2 = \{\mathbf{r_2} = [1, 1, 0, 0], \mathbf{r_4} = [0, 0, 0, 1]\}$. The $\mathbf{a_r}$ vectors are

$$\begin{aligned} \mathbf{a}_{\mathbf{r_1}} &= [0, 0, -1, 0] & \mathbf{a}_{\mathbf{r_3}} &= [0, 0, 1, 0] \\ \mathbf{a}_{\mathbf{r_2}} &= [0, 0, 0, -1] & \mathbf{a}_{\mathbf{r_4}} &= [0, 0, 0, 1] \end{aligned}$$

Let us assume that the rate of t_3 depends on the load of t_4 in such a way that if the marking of p_4 is greater than K_4, t_3 cannot fire (because no more resource is available for instance). Moreover, suppose that the rate of t_4 decreases linearly from μ_M to μ_m with the marking of p_4 varying from 0 to K_4. Then we can define

$$\rho_{\mathcal{C}(t_3)}\left((\mathbf{a_{r''}}.\mathbf{m})_{\mathbf{r''} \notin \mathcal{C}(t)}\right) = \begin{cases} 0 & \text{if } \mathbf{m}[p_4] \geq K_4 \\ \frac{\mu_m - \mu_M}{K_4}.\mathbf{m}[p_4] + \mu_M & \text{if } 0 \leq \mathbf{m}[p_4] < K_4 \end{cases}$$

since $\mathbf{a_{r_4}}.\mathbf{m} = \mathbf{m}[p_4]$ and we have still a PF steady state distribution.

Due to lack of space, we present in the rest of the paper, results without functions $\rho_{\mathcal{C}(t)}$. The reader will find full version of the results in the technical report [10].

Sufficient condition for PF-SPN. We first establish a sufficient condition for a Π-net to have a PF steady state distribution, *whatever* the parameters (i.e. rates of transitions) of the stochastic specification of the SPN.

Theorem 3. *Let* $(\mathcal{P}, \mathcal{T}, W, \mathcal{Q}, \mathbf{m_0})$ *be a* $\overline{\Pi}$*-net. Then, for any transition rates, the steady state distribution of the SPN has the product form*

$$\pi(\mathbf{m}) = \frac{1}{G} \cdot \prod_{\mathbf{a_r} \in \mathcal{R}(\mathcal{T})} \left(\frac{v(\mathbf{r})}{\mu(\mathbf{r})}\right)^{\mathbf{a_r}.\mathbf{m}} \qquad \forall \ \mathbf{m} \in RS(\mathbf{m_0}), \tag{5}$$

where G *is a normalization constant and* v *is a solution of Equations (1).*

Let us remark that this product form expression induces, of course, a product form with respect to $\mathbf{m}$, since:

$$\prod_{\mathbf{r} \in \mathcal{R}(\mathcal{T})} \left(\frac{v(\mathbf{r})}{\mu(\mathbf{r})}\right)^{\mathbf{a_r}.\mathbf{m}} = \prod_{\mathbf{r} \in \mathcal{R}(\mathcal{T})} \prod_{p_i \in \mathcal{P}} \left(\frac{v(\mathbf{r})}{\mu(\mathbf{r})}\right)^{\mathbf{a_r}[i].\mathbf{m}[i]} = \prod_{p_i \in \mathcal{P}} \left(\prod_{\mathbf{r} \in \mathcal{R}(\mathcal{T})} \left(\frac{v(\mathbf{r})}{\mu(\mathbf{r})}\right)^{\mathbf{a_r}[i]}\right)^{\mathbf{m}[i]}$$

Sketch of proof We give only a sketch of the proof (see [10] for a detailed proof). The starting point is the so-called the Group Local Balance Equation for a marking $\mathbf{m}$ with respect to a given vector $\mathbf{r}$ which is a splitting of the equilibrium (Chapman-Kolmogorov) equations of the Markov chain with markings as states:

$$\pi(\mathbf{m}) \sum_{t_j \in \mathbf{r}^\bullet} q(\mathbf{m}, \mathbf{m} - \mathbf{i}(t_j) + \mathbf{o}(t_j)) = \sum_{t_h \in {}^\bullet \mathbf{r}} \pi(\mathbf{m} + \mathbf{i}(t_h) - \mathbf{o}(t_h))q(\mathbf{m} + \mathbf{i}(t_h) - \mathbf{o}(t_h), \mathbf{m}]$$
$$\tag{6}$$

Then using the expression of the rates q, we introduce the proposed expression and after simplification, we get:

$$\mu(\mathbf{r}) = \sum_{t_h \in {}^\bullet \mathbf{r}} \prod_{\mathbf{r'} \in \mathcal{R}(\mathcal{T})} \left(\frac{v(\mathbf{r'})}{\mu(\mathbf{r'})}\right)^{\mathbf{a_{r'}}.(\mathbf{i}(t_h) - \mathbf{o}(t_h))} \mu(\mathbf{i}(t_h))\mathbf{P}[\mathbf{i}(t_h), \mathbf{r}]. \tag{7}$$

From $\mathbf{i}(t_h) - \mathbf{o}(t_h) = -\mathbf{C}[\mathcal{P}, h]$ and the definition of $\mathbf{a_r}$, (7) can be shown equivalent to the Traffic Equations (1).   ∎

Necessary condition for PF-SPN. The result of this section proves that the concept of $\overline{\Pi}$-net is the adapted one to capture the existence of a product form like the one of Theorem 3 *for any stochastic specification* of a Π-net. Combining Theorems 3 and 4, the "$\overline{\Pi}$-net property" appears as a necessary and sufficient structural condition for a net to have a product form steady state distribution *for any transition rates*.

Theorem 4. *Let* $(\mathcal{P}, \mathcal{T}, W, \mathcal{Q}, \mathbf{m_0})$ *be a* Π-net *and* v *a solution of the Traffic Equations. If there is a family* $(\mathbf{a_r})_{r \in \mathcal{R}(\mathcal{T})}$ *of rational vectors such that the distribution*

$$\pi(\mathbf{m}) = \frac{1}{G} \cdot \prod_{r \in \mathcal{R}(\mathcal{T})} \left(\frac{v(\mathbf{r})}{\mu(\mathbf{r})} \right)^{\mathbf{a_r} \cdot \mathbf{m}} \qquad \forall \, \mathbf{m} \in RS(\mathbf{m_0}),$$

satisfies the Group Local Balance Equations (6) for any $(\mu(\mathbf{r}))_{r \in \mathcal{R}(\mathcal{T})}$, *then we have*

$$\mathbf{a_r} \cdot \mathbf{C}[\mathcal{P}, j] = \begin{cases} 1 & \text{if } t_j \in {}^{\bullet}\mathbf{r} \\ -1 & \text{if } t_j \in \mathbf{r}^{\bullet} \\ 0 & \text{otherwise} \end{cases}$$

Sketch of proof (see [10] for a detailed proof). The Group Local Balance Equations for a given $\mathbf{m}$ with respect to a given $\mathbf{r}$ are (see (7))

$$\mu(\mathbf{r}) = \sum_{t_j \in {}^{\bullet}\mathbf{r}} \prod_{r' \in \mathcal{R}(\mathcal{T})} \left[\frac{v(\mathbf{r}')}{\mu(\mathbf{r}')} \right]^{-\mathbf{a_{r'}} \cdot \mathbf{C}[\mathcal{P}, j]} \mu(\mathbf{i}(t_j)) \mathbf{P}[\mathbf{i}(t_j), \mathbf{r}] \qquad (8)$$

since $a_{r'} \cdot (\mathbf{i}(t_j) - \mathbf{o}(t_j)) = -\mathbf{a_{r'}} \cdot \mathbf{C}[\mathcal{P}, j]$.
The idea is to express (8) as a multi-variables identically null "polynom" (i.e. extension of multi-variables polynom, with real valued exponents instead of integer) on $\mathbb{R}^+$ and to deduce the claimed properties of the $\mathbf{r}$ vectors from properties of the coefficients of this "polynom". To this end, we introduce the vectors with np components $\gamma(t_j)$ and γ_0 in the following way:

$$\gamma(t_j)[\mathbf{r}'] = \begin{cases} a_{r'} . \mathbf{C}[\mathcal{P}, j] & \text{if } \mathbf{r}' \neq \mathbf{i}(t_j) \\ a_{r'} . \mathbf{C}[\mathcal{P}, j] + 1 & \text{if } \mathbf{r}' = \mathbf{i}(t_j) \end{cases} \quad \text{and} \quad \gamma_0[\mathbf{r}'] = \begin{cases} 1 & \text{if } \mathbf{r}' = \mathbf{r} \\ 0 & \text{otherwise} \end{cases}$$

Using these vectors, transformation of Equation (8) provides a "polynom" with variables $\mu(\mathbf{r}')$. Via a technical result, it can then be shown that for all t_j, the set $\{t_j \in {}^{\bullet}\mathbf{r} \mid \gamma(t_j) = \gamma \neq \gamma_0\}$ is empty, so that $\forall t_j \in {}^{\bullet}\mathbf{r}$, $\gamma(t_j) = \gamma_0$. The result then follows from the evaluation of the numbers $a_{r'} . \mathbf{C}[\mathcal{P}, j]$. ∎

4   Functional Properties of PF-SPN

Although Π-nets and $\overline{\Pi}$-nets are not easily comparable to standard classes of PN, they nevertheless enjoy specific qualitative properties. This section first reviews liveness and deadlock freeness in Π-nets; second, some results about the complexity of the reachability and liveness in Π-nets and $\overline{\Pi}$-nets are presented.

Finally, we expose results about the characterization of reachable markings in Π-nets. Since we need to distinguish between structural and behavioural properties of (S)PN, in this section, we denote by $\mathcal{N} = (\mathcal{P}, \mathcal{T}, W, \mathcal{Q})$ a SPN and by $\Sigma = (\mathcal{N}, \mathbf{m_0})$ a marked SPN (also called SPN *system*) with initial marking $\mathbf{m_0}$.

4.1 Some Behavioural Properties of Π-Nets

Liveness is an important property of Petri net systems. Due to importance of T-semiflows in Π-nets, it is not surprising that liveness in Π-nets systems enjoys particular properties that we present below together with related results. The following lemma is a direct consequence of Proposition 1.

Lemma 2. *Let* $\Sigma = (\mathcal{N}, \mathbf{m_0})$ *be a* Π-*system. If* $t \in \mathcal{T}$ *is enabled at* $\mathbf{m} \in$ RS($\mathbf{m_0}$), *then,*
(1) all transitions of all minimal closed support T-semiflows to which t belongs can be fired.
(2) there is a firing sequence that fires all the remaining transitions in the FR*- class of t.

Proposition 2. *Let* $\Sigma = (\mathcal{N}, \mathbf{m_0})$ *be a* Π-*system.*
1. *If* $\exists\, t \in \mathcal{T}$, *enabled at* $\mathbf{m_0}$ *then* Σ *is deadlock-free (DF).*
2. Σ *is reversible.*
3. $\mathcal{N}$ *is structurally live (SL).*
4. *If there is an enabled transition in any FR*-class in the initial marking, then* Σ *is live. The converse is false.*
5. *If* Σ *is live then* $\Sigma' = (\mathcal{N}, \mathbf{m_0'})$ *with* $\mathbf{m_0} \leq \mathbf{m_0'}$ *is live too (i.e., liveness is monotonic w.r.t the initial marking in the net).*

Proof. We only give the detailed proof of (2).
If $\mathbf{m_0}$ is not a deadlock marking, for any $\mathbf{m} \in$ RS($\mathbf{m_0}$) there is a finite firing sequence $\sigma = t_{\delta_1}, t_{\delta_2}, \ldots, t_{\delta_l}$ such that $\mathbf{m_0}[t_{\delta_1}\rangle\mathbf{m_1} \ldots \mathbf{m_{l-1}}[t_{\delta_l}\rangle\mathbf{m}$. Now we prove that there is a finite firing sequence η such that $\mathbf{m}[\eta\rangle\mathbf{m_0}$. Let $\mathbf{x}$ be a closed support T-semiflow (not necessarily minimal) such that $\mathbf{x} \geq \boldsymbol{\sigma}$. Since $\mathbf{x}$ is a linear combination of minimal closed support T-semiflows, it follows from Lemma 2 that from $\mathbf{m}$, $\mathbf{x} - \boldsymbol{\sigma}$ must be firable and hence $\mathbf{m}[\mathbf{x} - \boldsymbol{\sigma}\rangle\mathbf{m_0}$.

Reverse of Π-net. Finally, the next proposition addresses properties of the reverse net of Π-nets. The reverse net of a Petri net $\mathcal{N} = (\mathcal{P}, \mathcal{T}, W)$ is $\mathcal{N}^{(-1)} = (\mathcal{P}, \mathcal{T}, W^{(-1)})$, that is, the net with same places and transitions, but reversed arcs $(W^{(-1)}(i, j) = W(j, i))$. Note that $(\mathcal{N}^{(-1)})^{(-1)} = \mathcal{N}$ and that the incidence matrix of $\mathcal{N}^{(-1)}$ is $-C$.

Proposition 3. *Let* $\mathcal{N}$ *be a* Π-*net,* $\Sigma = (\mathcal{N}, \mathbf{m_0})$, *and* $\Sigma^{(-1)} = (\mathcal{N}^{(-1)}, \mathbf{m_0})$.
1. *The reverse of a Π-net (resp. $\overline{\Pi}$-net) is a Π-net (resp. $\overline{\Pi}$-net).*
2. Σ *is deadlock free iff* $\Sigma^{(-1)}$ *is deadlock free.*
3. *The reachability graph of* $\Sigma^{(-1)}$ *is the reverse of the reachability graph of* Σ.
4. Σ *is live iff* $\Sigma^{(-1)}$ *is live.*

Proof. For space savings, we only develop proof of (3). If $\mathbf{m_0}$ is not a dead-lock marking, then from Proposition 2 (1) and (2), Σ is reversible. But in any reversible Petri net, the announced property holds. Indeed, we have first $RS(\Sigma^{(-1)}) = RS(\Sigma)$. Let $\mathbf{m} \in RS(\Sigma)$. Since Σ is reversible, there is a firing sequence τ such that $\mathbf{m}[\tau\rangle\mathbf{m_0}$. Therefore, $\mathbf{m_0}[\tau^{(-1)}\rangle\mathbf{m}$ in $\Sigma^{(-1)}$ where $\tau^{(-1)}$ is τ with "reversed" transitions. Now, let $\mathbf{m}[t\rangle\mathbf{m}'$ in Σ. We have $\mathbf{m}' \in RS(\Sigma^{(-1)})$ and, obviously, $\mathbf{m}'[t^{(-1)}\rangle\mathbf{m}$. We have proven that the reverse of the reachability graph Σ is a partial graph of $\Sigma^{(-1)}$. The result follows, applying the same proof to $\Sigma^{(-1)}$.

4.2 Complexity of Liveness and Reachability Problems for Π-Nets and $\overline{\Pi}$-Nets

Condition (4) in proposition 2 is only a sufficient condition. In fact, checking liveness seems no more easy for Π-nets, and even 1-safe[2] $\overline{\Pi}$-nets, than for many other classes of Petri nets. We have shown in Section 3.1 that the complexity of the computation of FR*-Classes is polynomial time. But checking liveness requires to verify that each FR*-class is live. If some FR*-class is not *initially* firable, this is still a very complex problem. Indeed the next lemma gives some insight into this point. We recall that for general Petri nets, Lipton's result [16] implies a $2^{O(\sqrt{n})}$ lower bound space complexity for the liveness problem (see [9,8] for recent surveys on decidability problems for Petri nets). In fact, we are able to give more precise results, although the exact complexity of the reachability/liveness for $\overline{\Pi}$-nets still remains an open problem.

It has been shown in [6] that the liveness problem for 1-safe nets is PSPACE complete. The next lemma gives a lower bound of the problem for 1-safe Π-nets.

Proposition 4. *The liveness problem for 1-safe $\overline{\Pi}$-nets is NP-hard.*

Proof. To prove it, we reduce in polynomial time the 3SAT problem to the liveness problem for $\overline{\Pi}$-nets, following the idea first presented in [14]. The 3SAT problem is a well known NP-complete problem. We have K logical formulae $C_1, \cdots, C_K$, each one being a disjunction of three boolean variables v_i or their negation $(-v_i)$, from a set of I variables: for instance, $C_k = v_1 \vee -v_3 \vee v_6$. The 3SAT problem is: is there a set of values for $v_1, \cdots, v_I$ such that $C_1 \wedge C_2 \wedge \cdots \wedge C_K$ is true? We explain the reduction through the example $C_1 = v_1 \wedge -v_2 \wedge v_3$, $C_2 = v_2 \wedge v_3 \wedge v_4$ ($K = 2$, $I = 4$) (Figure 3).

For each variable v_i, we have two places p_i and p_{-i} and two transitions t_i and t_{-i}. Arcs between places and transitions for v_i are as indicated in the figure. We have also K sets of places $p_{Ck,i}$ (the introduction of several places for each C formulae ensures 1-safeness). If v_i is in C_k (like v_2 and C_2) there is an arc from t_{-i} to $p_{Ck,i}$ and one arc from $p_{Ck,i}$ to t_i. In contrast if $-v_i$ is in C_k (like $-v_2$ in C_1), these arcs are reversed. Otherwise, there is no arc between t_i, t_{-i} and place $p_{Ck,i}$. Places detailed in the right dotted part ensure that the place p_{Ck} will contain at most one token (p_{C2x} is a mutex place). Finally, we have one

---

[2] A 1-safe marked Petri net is a (bounded) marked net with at most one token in every place of every reachable marking

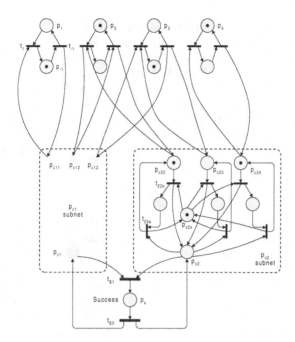

Fig. 3. Reduction of 3SAT to liveness in 1-safe $\overline{\Pi}$-nets

transition t_{s1} (for Success) and we added place p_s and transition t_{s2} to have a $\overline{\Pi}$-net and not only a Π-net.

We can easily verify that the net is a 1-safe $\overline{\Pi}$-net. The initial marking is chosen as follows: if v_i is true, there is one token in p_i and one token in place p_{C_ki} if v_i is in C_k; if v_i is false, there is one token in p_{-i} and one token in place p_{C_ki} if $-v_i$ is in C_k. In our example, we take $v_1 = v_3$ =false, $v_2 = v_4$ =true. Clearly the formula is true for a given set of boolean values of variables if the transition t_{s1} is live and the same for the reachability of a marking with one token in p_s.

Thus, there is still an open problem for $\overline{\Pi}$-nets since the upper bound of complexity for general Petri nets is in PSPACE. By contrast, the next proposition provides an exact characterization of the complexity of the problems for (1-safe) Π-nets. This distinctive result strengthens the specific character of the Π-nets class.

Proposition 5. *(1) The liveness and the reachability problems for 1-safe Π-nets are PSPACE complete.*
(2) The reachability problem for Π-nets is EXPSPACE-complete.

Proof. Due to lack of space, we only address the claim (2). For symmetric nets systems, we know [5,18] that the reachability problem is EXPSPACE complete.

A net is symmetric iff for every transition t, there is a "reverse" transition t' whose firing "undoes" the effect of the firing of t, i.e., the input places of t are the output places of t' and vice versa. Symmetric nets are clearly Π-nets. Thus, the reachability problem is EXPSPACE-hard for Π-nets. But any Π-net defines implicitly a symmetric net: for any transition t, we may add a reverse transition t' without changing the resulting reachability graph, because the closed T-semiflow (without t) of transitions to which t belongs acts exactly as t' when fired in a cyclic way. Thus, the reachability problem for Π-nets is reducible to the one for symmetric nets, hence in EXPSPACE, and finally EXPSPACE-complete.

4.3   Algebraic Properties of PF-SPN

The availability of a product form equilibrium distribution allows the development of computational algorithms that are analogous to those developed for product form solution queueing networks (e.g, [3,4,21]). For instance proposals for algorithms for the computation of performance measures throughout the normalization constant calculus can be found in [7,23]. In [1] a set of *Arrival Theorems*, similar to the analogous results developed for product form solution queueing networks [25] was proven, leading to a Mean Value Analysis (MVA) for the computation of performance measures for PF-SPNs. MVA for SPNs was also studied in [24].

This last section discusses reachability markings properties related to the solution of PF-SPN. For the development of computational algorithms for PF-SPN, the reachability set (RS) of the SPN must be partitioned according to certain criteria depending on the particular algorithm. For instance, the normalization constant computation algorithm requires a partitioning of the reachability set that groups together all the markings with a constant number of tokens in a given place. It is then important to know if reachable markings of a Π-net system may be characterized, among all markings, by some specific criterion based on their value and structural elements of the net. The most common such criteria are the so-called *state equation* and the one *based on the minimal P-semiflows* of the net. The difficulty then lies in the quality of those criteria, i.e. whether they allow to select all reachable markings and, *only* reachable markings.

Let us recall that the *state equation* $\mathbf{m} = \mathbf{m_0} + \mathbf{C}.\boldsymbol{\sigma}$ is an algebraic equation that gives a necessary condition for a marking to be reachable. The set of vectors $\mathbf{m} \in \mathbb{N}^{np}$ such that $\exists \, \boldsymbol{\sigma} \in \mathbb{N}^{nt} : \mathbf{m} = \mathbf{m_0} + \mathbf{C}.\boldsymbol{\sigma}$ is called the Potential Reachability Set (PRS) of the net. Obviously, $\mathrm{RS}(\mathbf{m_0}) \subseteq \mathrm{PRS}(\mathbf{m_0})$. In the literature, there are several proposals of computational algorithms for PF-SPN. They use a reachability characterization based on the minimal P-semiflows. Therefore, another set of "potential" markings has been defined. Let $\mathbf{B}$ be the matrix whose rows are the set of minimal P-semiflows of the net. The *Potential Reachability Set with respect to* $\mathbf{B}$ is the set $\mathrm{PRS}^{\mathbf{B}}(\mathbf{m_0}) = \{\mathbf{m} \mid \mathbf{B}.\mathbf{m} = \mathbf{B}.\mathbf{m_0}\}$. Clearly, $\mathrm{PRS}(\mathbf{m_0}) \subseteq \mathrm{PRS}^{\mathbf{B}}(\mathbf{m_0})$ since $\mathbf{B}.\mathbf{C} = 0$.

An unreachable marking belonging to one of these PRS is called a *spurious marking* (see [27] for a detailed study of several kinds of PRS). We show below that, unfortunately, none of these two characterizations is able to capture all

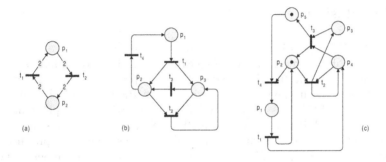

Fig. 4. Π-net and potential reachability: (a) $PRS(\mathbf{m_0}) \neq PRS^{\mathbf{B}}(\mathbf{m_0})$, unbounded (b) and bounded (c) Π-systems with spurious marking

the peculiarities of PF-SPN, that is to say that there are Π-net with spurious markings for PRS (thus for $PRS^{\mathbf{B}}$).

First we may have $PRS(\mathbf{m_0}) \neq PRS^{\mathbf{B}}(\mathbf{m_0})$ in Π-systems. This happens even in such simple case as the Π-cycle of Figure 4(a): the dead marking $\mathbf{m_1} = [1,1]^T$ has the same dot-product with the P-semiflow $\mathbf{Y} = [1,1]^T$ as the live one $\mathbf{m_0} = [2,0]$ although there is no $\boldsymbol{\sigma} \in \mathbb{N}^{nt}$ satisfying the state equation $\mathbf{m_1} = \mathbf{m_0} + \mathbf{C}.\boldsymbol{\sigma}$.

For what concerns the characterization of the reachability set of a Π-net in terms of potential reachability set, the proposition below (we omit the proof for sake of place) provides a rather positive result, but we give next, two examples which prove that properties of Π-nets are not strong enough to prevent the existence of spurious markings.

Proposition 6. *With respect to the state equation,*
(1) The potential reachability graph of $(\mathcal{N}, \mathbf{m_0})$ is equal to the reverse of the potential reachability graph of $(\mathcal{N}^{-1}, \mathbf{m_0})$.
(2) Spurious markings (if they exist) cannot be transient, i.e., if $\mathbf{m} \in PRS(\mathbf{m_0}) \setminus RS(\mathbf{m_0})$, then there is no firing sequence $\boldsymbol{\sigma}$ such that $\mathbf{m}[\boldsymbol{\sigma}\rangle\mathbf{m}'$ with $\mathbf{m}' \in RS(\mathbf{m_0})$.

The net of Figure 4(b) gives the first negative result. For the unbounded Π-net it is possible to see that $\mathbf{m} = [0,0,0]$ is a spurious marking. We can see that for any initial marking $\mathbf{m_0} = [k_1, k_2, k_3]$, $\mathbf{m_0}[t_1^{k_1}\rangle\mathbf{m_1} = [0, k_1 + k_2, k_1 + k_3][t_2^{k_1+k_2}\rangle\mathbf{m_2} = [0, 0, 2k_1 + k_2 + k_3]$. Setting $k = 2k_1 + k_2 + k_3$ we have $\mathbf{m_2}[(t_3t_2)^{k-1}\rangle\mathbf{m_3} = [0,0,1]$. Now "firing" t_2t_3 the null marking is spuriously reached.

The net of Figure 4(c) gives another and definitive negative result. This Π-net is bounded but it is possible to see that $\mathbf{m} = [0,0,1,0,1]$ is a spurious marking. Indeed, from the initial marking $\mathbf{m_0} = [0,1,0,0,1]$ and with the "firing" of t_2 we obtain the marking $[0,0,1,0,1]$ that it is not a reachable marking. Hence we have $\mathbf{m} \in PRS(\mathbf{m_0})$ but $\mathbf{m} \notin RS(\mathbf{m_0})$.

5   Conclusion

SPN with PF solution have been introduced some years ago as an extension of closed form solution methods of QN to SPN which allow to model systems with more complex synchronization schemes. In this paper we have presented four groups of new results giving a better insight in PF-SPN and allowing an efficient handling of this class of nets. We have first established a polynomial-time algorithm to check if a given SPN is a PF-SPN. This is an interesting result, in contrast with the general computation of T-semiflows which may produce an exponential number of T-semiflows (with respect to the size of the net). Then, we have proven a rate independent structural characterization of PF-SPN, which can also be checked in polynomial time. We call $\overline{\Pi}$-nets the subclass of Π-nets satisfying this criterion. Moreover, for $\overline{\Pi}$-nets, we are able to define transition rates globally dependent of components of the net "not related with" the considered transition, so that we can model complex dependency of activities on some other ones. Third, we have investigated untimed properties for the class of PF-SPN. We have shown that Π-nets, and even $\overline{\Pi}$-nets do not fit in any standard class of PN. Nevertheless, we have proved specific properties for deadlock-freeness, liveness and reverse nets for Π-nets. For what concerns liveness/reachability in Π-nets and $\overline{\Pi}$-nets, we were able to somewhat refine complexity bounds known for general PN. Finally, with examples and one proposition, we have given some answers, both positive and negative, to the problem of potential reachability, i.e. reachability based upon structural properties of the net. The interested reader will find detailed proofs and full versions of results in [10].

References

1. G. Balbo, S. C. Bruell, and M. Sereno. Arrival theorems for product-form stochastic Petri nets. In *Proc. 1994 ACM SIGMETRICS Conference*, pages 87–97, Nashville, Tennessee, USA, May 1994. ACM.
2. R. J. Boucherie and M. Sereno. On closed support t-invariants and traffic equations. *Journal of Applied Probability*, (35):473–481, 1998.
3. S. C. Bruell and G. Balbo. *Computational Algorithms for Closed Queueing Networks*. Elsevier North-Holland, New York, 1980.
4. J. P. Buzen. Computational algorithms for closed queueing networks with exponential servers. *Communications of the ACM*, 16(9):527–531, September 1973.
5. E. Cardoza, R.J. Lipton, and A.R. Meyer. Exponential space complete problems for Petri nets and commutative semigroups. In *Proc. of the 8th Annual Symposium on Theory of Computing*, pages 50–54, 1976.
6. A. Cheng, J. Esparza, and J. Palberg. Complexity results for 1-safe nets. In *Proc. of the 13th Conference on Foundations of Software Technology and Theoretical Computer Science*, Bombay, India, 1993.
7. J. L. Coleman, W. Henderson, and P. G. Taylor. Product form equilibrium distributions and an algorithm for classes of batch movement queueing networks and stochastic Petri nets. *Performance Evaluation*, 26(3):159–180, September 1996.
8. J. Esparza. *Decidability and Complexity of Petri nets problems – an introduction*, pages 374–428. 1998. in [22].

9. J. Esparza and M. Nielsen. Decidability issues for Petri nets – a survey. *Journal of Information Processing and Cybernetics*, 30(3):143–160, 1994. former version in Bulletin of the EATCS, volume 52, pages 245–262, 1994.

10. S. Haddad, P. Moreaux, M. Sereno, and M. Silva. Revisiting Product Form Stochastic Petri Nets. Technical report, LERI-*RESYCOM*, Université de Reims Champagne-Ardenne, Reims, France, June 2001.

11. W. Henderson, D. Lucic, and P.G. Taylor. A net level performance analysis of stochastic Petri nets. *Journal of Australian Mathematical Soc. Ser. B*, 31:176–187, 1989.

12. W. Henderson and P.G. Taylor. Aggregation methods in exact performance analysis of stochastic Petri nets. In *Proc. 3rd Intern. Workshop on Petri Nets and Performance Models*, pages 12–18, Kyoto, Japan, December 1989. IEEE-CS Press.

13. W. Henderson and P.G. Taylor. Embedded processes in stochastic Petri nets. *IEEE Transactions on Software Engineering*, 17:108–116, February 1991.

14. N.D. Jones, L.H. Landweber, and Y.E. Lien. Complexity of some problems in Petri nets. *Theoretical Computer Science*, 4:277–299, 1977.

15. A. A. Lazar and T. G. Robertazzi. Markovian Petri net protocols with product form solution. In *Proc. of INFOCOM '87*, pages 1054–1062, San Francisco, CA, USA, 1987.

16. R.J. Lipton. The reachability problem requires exponential space. Technical report, Dpt. of Computer Science, Yale University, 1976.

17. J. Martinez and M. Silva. A simple and fast algorithm to obtain all invariants of a generalized Petri net. In *Proc. 2nd European Workshop on Application and Theory of Petri Nets*, Bad Honnef, West Germany, September 1981. Springer Verlag.

18. E.W. Mayr and A.R. Meyer. The complexity of the word problems for commutative semigroups and polynomial ideals. *Advances in Mathematics*, (46):305–329, 1982.

19. T. Murata. Petri nets: properties, analysis, and applications. *Proceedings of the IEEE*, 77(4):541–580, April 1989.

20. J. L. Peterson. *Petri Net Theory and the Modeling of Systems*. Prentice-Hall, Englewood Cliffs, NJ, 1981.

21. M. Reiser and S. S. Lavenberg. Mean value analysis of closed multichain queueing networks. *Journal of the ACM*, 27(2):313–322, April 1980.

22. W. Reisig and G. Rozenberg, editors. *Lectures on Petri Nets I: Basic models.* Number 1491 in LNCS. Springer–Verlag, June 1998. Advances in Petri nets.

23. M. Sereno and G. Balbo. Computational algorithms for product form solution stochastic Petri nets. In *Proc. 5th Intern. Workshop on Petri Nets and Performance Models*, pages 98–107, Toulouse, France, October 1993. IEEE-CS Press.

24. M. Sereno and G. Balbo. Mean value analysis of stochastic Petri nets. *Performance Evaluation*, 29(1):35–62, 1997.

25. K. C. Sevcik and I. Mitrani. The distribution of queueing network states at input and output instants. *Journal of the ACM*, 28(2):358–371, April 1981.

26. M. Silva. *Las Redes de Petri en la Automatica y la Informatica*. Ed. AC, Madrid, Spain, 1985. In Spanish.

27. M. Silva, E. Teruel, and J.M Colom. *Linear algebraic and linear programming techniques for the analysis of Place/Transition net systems*, pages 309–372. 1998. in [22].

Generalized Conditions for Liveness Enforcement and Deadlock Prevention in Petri Nets

Marian V. Iordache and Panos J. Antsaklis*

Department of Electrical Engineering, University of Notre Dame,
Notre Dame, IN 46556, USA
{iordache.1, antsaklis.1}@nd.edu

Abstract. This paper presents new results concerned with liveness, liveness of a subset of transitions and deadlock in Petri nets. Liveness is seen as a particular case of what we call T-liveness: all transitions in the set T are live. The first results characterize the relation between supervisors enforcing liveness and T-liveness with supervisors preventing deadlock. Then we introduce a class of Petri net subnets allowing us to extend two well known results. Specifically we generalize the result relating deadlock to siphons and the extension to asymmetric choice Petri nets of the Commoner's Theorem. We conclude by considering how the theoretical results of this paper can be used for deadlock prevention, least restrictive deadlock prevention and least restrictive T-liveness enforcement.

Keywords liveness, deadlock, synthesis of liveness supervisors, structural properties of Petri nets.

1 Introduction

In this paper we consider three supervisory problems: deadlock prevention, liveness enforcement, and T-liveness enforcement, where the latter denotes enforcing that all transition in a transition subset T of a Petri net are live. Deadlock prevention corresponds to preventing the system from reaching a state of total deadlock. Liveness corresponds to the stronger requirement that no local deadlock occurs, or in other words, all transitions are live. T-liveness means that all transition in the set T are live. It is useful in problems where some transitions correspond to undesirable system events (such as faults).

A way to study the liveness properties of a Petri net uses the reachability graph. However this approach can only handle bounded Petri nets, needs the initial marking to be known, and due to the state explosion problem, requires reasonably small Petri nets. Unfolding has been proposed to reduce the computational burden [4], however the other two limitations remain. In this paper

* The authors gratefully acknowledge the partial support of the National Science Foundation (ECS-9912458) and of the Army Research Office (DAAG55-98-1-0199).

J.-M. Colom and M. Koutny (Eds.): ICATPN 2001, LNCS 2075, pp. 184–203, 2001.

we consider the structural approach to the liveness problem. The structural approach relies on the algebraic properties of the incidence matrix. Thus the initial marking is regarded as a parameter and unbounded Petri nets can be tackled. Our work has been inspired by the incidence matrix properties of repetitive Petri nets (e.g. [10]). Related work includes [1], presenting among others an extension of the relation between deadlocked Petri nets and siphons for generalized Petri nets, and a generalization of the extension to asymmetric choice Petri nets of the Commoner's Theorem. However, our supervisory perspective, our concern on T-liveness and our consideration of arbitrary Petri nets, including nonrepetitive Petri nets, differentiate this paper from previous works.

The contribution of this paper is described in sections 3, 4 and the appendix. To the authors' knowledge, all results presented in these sections and the appendix are new, except for part (b) of Proposition 3.

We begin in section 3.1 by characterizing the relation which exists among deadlock prevention, T-liveness enforcement and liveness enforcement. Thus we answer the following questions: (a) Which are the Petri nets in which deadlock prevention, or T-liveness enforcement, or liveness enforcement is possible? and (b) When deadlock prevention is equivalent to T-liveness enforcement or liveness enforcement? We answer question (a) in Proposition 3, and question (b) in Theorems 2 and 3. Theorem 2 considers the case of the deadlock prevention supervisors which are not more restrictive than liveness or T-liveness supervisors; Theorem 3 considers the general case. We conclude the first part of the paper with Theorem 4, which states that the transitions of a Petri net can be divided in two classes: transitions which can be made live under an appropriate supervisor for some initial markings, and transitions which cannot be made live under any circumstances. Theorem 4 is very important for the theoretical developments which follow in the remaining part of the paper.

The most important part of the paper is section 3.2. In this section we show how to characterize Petri nets for deadlock prevention and liveness enforcement based on a special type of subnets. Thus we begin by defining what we call the *active subnets* of a Petri net. Then we define a special class of siphons, which we call *active siphons*. Proposition 5 is a necessary condition for deadlock which generalizes the known result that a deadlocked ordinary Petri net contains an empty siphon. Proposition 6 is a further extension, as it gives a sufficient condition in terms of empty active siphons for deadlock to be unavoidable. Commoner's Theorem on free-choice Petri nets has been extended to asymmetric-choice Petri nets [2]; see also [1]. We further extend the result in Theorem 5: we show that each dead transition is in the postset of an uncontrolled siphon. Then in Theorem 6 we give a necessary and sufficient condition for T-liveness in an asymmetric choice Petri net. Polynomial complexity algorithms for the computation of the active subnets are included in the appendix.

We conclude our paper with section 4, which shows the significance of our results for deadlock prevention and liveness enforcement. Examples are included. In sections 4.1 and 4.3 we consider deadlock prevention and T-liveness enforcement. In section 4.2 we include Theorem 7, which shows how to do least restrictive deadlock prevention.

2   Preliminaries

We denote a Petri net by $\mathcal{N} = (P, T, F, W)$, where P is the set of places, T the set of transitions, F the set of transition arcs and W the transition arc weight function. We use the symbol μ to denote a marking and we write $(\mathcal{N}, \mu_0)$ when we consider the Petri net $\mathcal{N}$ with the initial marking μ_0. The incidence matrix of a Petri net is denoted by D, where the rows correspond to places and the columns to transitions. Also, by denoting a place by p_i or a transition by t_j, we assume that p_i corresponds to the i'th row of D and t_j to the j'th column of D. We use the notation $\mu \xrightarrow{\sigma} \mu'$ to express that the marking μ enables the firing sequence σ and μ' is reached by firing σ.

A Petri net $\mathcal{N} = (P, T, F, W)$ is **ordinary** if $\forall f \in F : W(f) = 1$. We will refer to slightly more general Petri nets in which only the arcs from places to transitions have weights equal to one. We are going to call such Petri nets *PT-ordinary*, because all arcs (p, t) from a place p to a transition t satisfy the requirement of an ordinary Petri net that $W(p, t) = 1$.

Definition 1. *Let $\mathcal{N} = (P, T, F, W)$ be a Petri net. We call $\mathcal{N}$ **PT-ordinary** if $\forall p \in P \, \forall t \in T$, if $(p, t) \in F$ then $W(p, t) = 1$.*

An **asymmetric choice** Petri net is defined by the property that $\forall p_1, p_2 \in P$ if $p_1 \bullet \cap p_2 \bullet \neq \emptyset$ then $p_1 \bullet \subseteq p_2 \bullet$ or $p_2 \bullet \subseteq p_1 \bullet$.

A **siphon** is a set of places $S \subseteq P$, $S \neq \emptyset$, such that $\bullet S \subseteq S \bullet$. A siphon S is **minimal** if there is no siphon $S' \subset S$. A siphon is **empty** at a marking μ if it contains no tokens. Given a Petri net $(\mathcal{N}, \mu_0)$, a **controlled siphon** is a siphon which is not empty at any reachable marking. A well known necessary condition for deadlock [11] is that a deadlocked ordinary Petri net contains at least one empty siphon. It can easily be seen that the proof of this result also is valid for PT-ordinary Petri nets.

Proposition 1. *A deadlocked PT-ordinary Petri net contains at least one empty siphon.*

In general we may not want all transitions to be live. For instance some transitions of a Petri net may model faults and we want to ensure that some other transitions are live. This is the motivation of the next definition.

Definition 2. *Let $(\mathcal{N}, \mu_0)$ be a Petri net and T a subset of the set of transitions. We say that the Petri net is **T-live** if all transitions $t \in T$ are live.*

A live transition is not the opposite of a dead transition. That is, a transition may be neither live nor dead. Indeed, a transition is live if there is no reachable marking for which it is dead. Note also that T-liveness corresponds to liveness when the set T equals the set of all Petri net transitions. In what follows we define what we mean by a supervisor.

Definition 3. *Let* $\mathcal{N} = (P, T, F, W)$ *be a Petri net,* $\mathcal{M}$ *the set of all markings of* $\mathcal{N}$, $\mathcal{M}_0 \subseteq \mathcal{M}$ *and*[1] $U \subseteq \mathcal{M} \times T^*$ *such that* $\forall \mu_0 \in \mathcal{M}_0 : (\mu_0, \varepsilon) \in U$.[2] *A* **supervisor** *is a map* $\Xi : U \rightarrow 2^T$ *such that* $\forall (\mu, \sigma) \in U$ $\forall t \in \Xi(\mu, \sigma)$, *if* $\mu \xrightarrow{t} \mu'$, *then* $(\mu', \sigma t) \in U$. *We say that* $\mathcal{M}_0$ *is the set of initial markings for which* Ξ *is defined. We also say that* Ξ *is a* **marking based supervisor** *if* $\Xi(\mu, \sigma)$ *depends only on* μ *and* $\forall (\mu, \sigma) \in U : \{\mu\} \times T^* \subseteq U$.

A Petri net $(\mathcal{N}, \mu_0)$ supervised by Ξ operates as follows: at every marking μ reached by firing some σ from μ_0 $(\mu_0 \xrightarrow{\sigma} \mu)$, only transitions in $\Xi(\mu, \sigma)$ may fire. We denote by $(\mathcal{N}, \mu_0, \Xi)$ the supervised Petri net and by $\mathcal{R}(\mathcal{N}, \mu_0, \Xi)$ its set of reachable markings. A marking based supervisor is *memoryless*, as it only depends on the marking. We say that Ξ_1 is **less restrictive** (or **more permissive**) than Ξ_2 w.r.t. $(\mathcal{N}, \mu_0)$ if the set of firing sequences firable from μ_0 in $(\mathcal{N}, \mu_0, \Xi_2)$ is a proper subset of the set of firing sequences firable from μ_0 in $(\mathcal{N}, \mu_0, \Xi_1)$. We say that **deadlock can be prevented** in a Petri net $\mathcal{N}$ if there is an initial marking μ_0 and a supervisor Ξ such that $(\mathcal{N}, \mu_0, \Xi)$ is deadlock-free. We say that **liveness (T-liveness) can be enforced** in $\mathcal{N}$ if there is an initial marking μ_0 and a supervisor Ξ such that $(\mathcal{N}, \mu_0, \Xi)$ is live (T-live). It is known that if $(\mathcal{N}, \mu_0)$ is live, then $(\mathcal{N}, \mu)$ with $\mu \geq \mu_0$ may not be live. The same is true for deadlock-freedom, as shown in Figure 1. The next result shows that if liveness (T-livenss) is enforcible at marking μ or if deadlock can be prevented at μ, then the same is true for all markings $\mu' \geq \mu$.

Proposition 2. *If a supervisor* $\Xi : U \rightarrow 2^T$ *which prevents deadlock (enforces (T-)liveness) in* $(\mathcal{N}, \mu_0)$ *exists, then for all* $\mu \geq \mu_0$ *there is a supervisor which prevents deadlock (enforces (T-)liveness) in* $(\mathcal{N}, \mu)$.

Proof. Let $\mu_1 \geq \mu_0$. A supervisor for $(\mathcal{N}, \mu_1)$ is Ξ_1 defined by

$$\Xi_1(\mu + \mu_1 - \mu_0, \sigma) = \begin{cases} \Xi(\mu, \sigma) & \text{for } (\mu, \sigma) \in U \\ \emptyset & \text{otherwise} \end{cases}$$

$\square$

As we prove in the next section, the Petri net structures in which liveness can be enforced (for some initial markings) are the *repetitive* Petri nets, and the Petri net structures in which deadlock can be prevented are the *partially repetitive* Petri nets. In what follows we formally define these two Petri net classes.

Definition 4. [10] *A Petri net is said to be* **(partially) repetitive** *if there is a marking* μ_0 *and a firing sequence* σ *from* μ_0 *such that every (some) transition occurs infinitely often in* σ.

A test allowing to check whether a Petri net is (partially) repetitive uses the incidence matrix D and is next presented. Linear programming techniques can be used to implement the test.

---

[1] T^* is the set of all firing sequences with transitions in T

[2] $\varepsilon \in T^*$ denotes the empty firing sequence

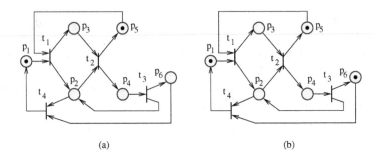

(a)   (b)

Fig. 1. A Petri net which is live for the initial marking μ_0 shown in (a) and not even deadlock-free for the initial marking $\mu \geq \mu_0$ shown in (b).

Theorem 1. [10] *A Petri net is (partially) repetitive iff a vector x of positive (nonnegative) integers exists, such that $Dx \geq 0$ and $x \neq 0$.*

3   Results

3.1   Conditions for Deadlock Prevention and Liveness Enforcement

In general it may not be possible to enforce liveness or to prevent deadlock in an arbitrary given Petri net. This may happen because the initial marking is inappropriate or because the structure of the Petri net is incompatible with such a supervision purpose. The next proposition characterizes the structure of Petri nets which allow supervision for deadlock prevention and liveness enforcement, respectively. It shows that Petri nets in which liveness is enforcible are repetitive, and Petri nets in which deadlock is avoidable are partially repetitive. Part (b) of the proposition also appears in [13].

Proposition 3. *Let $\mathcal{N} = (P, T, F, W)$ be a Petri net.*

(a) *Initial markings μ_0 exist such that deadlock can be prevented in $(\mathcal{N}, \mu_0)$ iff $\mathcal{N}$ is partially repetitive.*
(b) *Initial markings μ_0 exist such that liveness can be enforced in $(\mathcal{N}, \mu_0)$ iff $\mathcal{N}$ is repetitive.*
(c) *Initial markings μ_0 exist such that T-liveness can be enforced in $(\mathcal{N}, \mu_0)$ iff there is an initial marking μ_0 enabling an infinite firing sequence in which all transitions of T appear infinitely often.*

Proof.
(a) If deadlock can be avoided in $(\mathcal{N}, \mu_0)$ then μ_0 enables some infinite firing sequence σ, and by definition $\mathcal{N}$ is partially repetitive. If $\mathcal{N}$ is partially repetitive, then let μ_0 and σ be as in Definition 4; we define Ξ such that it only allows σ to fire from μ_0. Then Ξ prevents deadlock.
(b) and (c) The proof is similar to (a).   □

If $\mathcal{N}$ is partially repetitive, a constructive way to obtain an initial marking for which deadlock can be prevented or (T-)liveness can be enforced is implied by Theorem 1. Let x be as in Theorem 1 and $\sigma_x = t_{x,1} \dots t_{x,k}$ a firing sequence associated to a firing vector $q = x$. Let q_1 denote the firing vector after the first transition of σ_x fired, q_2 after the first two fired, and so on to $q_k = q$. If the rows of the D are $d_1^T, d_2^T, \dots, d_{|P|}^T$, $\delta_{i,j} = W(p_i, t_{x,j})$ if $t_{x,j} \in p_i\bullet$ and $\delta_{i,j} = 0$ otherwise, then a marking which enables σ_x is

$$\mu_0(p_i) = \max\{0, \delta_{i,1}, \max_{j=1\dots k-1}(\delta_{i,j+1} - d_i^T q_j)\} \quad i = 1 \dots |P| \tag{1}$$

At least one deadlock prevention strategy exists for μ_0: to allow only the firing sequence $\sigma_x, \sigma_x, \sigma_x, \dots$ to fire. This infinite firing sequence is enabled by μ_0 because $\mu_0 + Dx \geq \mu_0$ and μ_0 enables σ_x.

Note that if a deadlock prevention supervisor Ξ exists for $(\mathcal{N}, \mu_0)$, then a *marking based* deadlock prevention supervisor Ξ_m exists for $(\mathcal{N}, \mu_0)$ such that Ξ is at least as restrictive as Ξ_m. The same is true for liveness and T-liveness enforcing supervisors. Indeed, let $\sigma^{(j)} = t_1^{(j)} t_2^{(j)} t_3^{(j)} \dots$, for $j = 1, 2, \dots$, be the infinite firing sequences which can fire from μ_0 in $(\mathcal{N}, \mu_0, \Xi)$; for all $i, j = 1, 2 \dots$ let $\mu_i^{(j)}$ be the marking reached after firing $t_1^{(j)} \dots t_i^{(j)}$ from μ_0 and $\sigma_{i,\infty}^{(j)} = t_i^{(j)} t_{i+1}^{(j)} \dots$. We take $\Xi_m(\mu) = \{t : \exists i, j \geq 1 \text{ such that } \mu = \mu_{i-1}^{(j)} \text{ and } t = t_i^{(j)}\}$. Hence $\forall \mu \in \mathcal{R}(\mathcal{N}, \mu_0, \Xi_m)$: $\exists i, j$ such that $\sigma_{i,\infty}^{(j)}$ is firable from μ in $(\mathcal{N}, \mu_0, \Xi_m)$.

From a marking based supervisory perspective, it is known that if a liveness enforcing supervisor exists, the least restrictive liveness enforcing supervisor also exists [13]. The same is true for deadlock prevention and T-liveness enforcing supervisors. This is true also for the more general supervisors of Definition 3. This follows easily from the fact that given Ξ_1 and Ξ_2, a supervisor at least as permissive as each of Ξ_1 and Ξ_2 is $\Xi = \Xi_1 \vee \Xi_2$ which allows a transition to fire if either of Ξ_1 or Ξ_2 allows it.

Next we introduce a technical result which is necessary in order to prove some of the main results of this paper.

Lemma 1. *Let $\mathcal{N} = (P, T, F, W)$ be a Petri net of incidence matrix D. Assume that there is an initial marking μ_I which enables an infinite firing sequence σ. Let $U \subseteq T$ be the set of transitions which appear infinitely often in σ.*

(a) There is a nonnegative integer vector x such that $Dx \geq 0$, $\forall t_i \in U$: $x(i) \neq 0$ and $\forall t_i \in T \setminus U$: $x(i) = 0$.

(b) There is a firing sequence σ_x containing only the transitions with $x(i) \neq 0$, such that $\exists \mu_1^*, \mu_2^* \in \mathcal{R}(\mathcal{N}, \mu_I)$: $\mu_1^* \xrightarrow{\sigma_x} \mu_2^*$, each transition t_i appears $x(i)$ times in σ_x, σ can be written as $\sigma = \sigma_a \sigma_x \sigma_b$, and $\mu_I \xrightarrow{\sigma_a} \mu_1^*$.

Proof. Note that σ can be written as $\sigma_0 \sigma'$, where σ_0 is finite and σ' contains only transitions in U. Let μ_0 be the marking such that $\mu_I \xrightarrow{\sigma_0} \mu_0$. We further decompose σ' in $\sigma_1 \sigma_2 \dots \sigma_k \dots$ such that each σ_k is finite and in each σ_k all transitions of U appear at least once. Let $\mu_1, \mu_2, \dots \mu_k, \dots$ be such that $\mu_{k-1} \xrightarrow{\sigma_k} \mu_k$ for $k = 1, 2, \dots$. By Dickson's Lemma (see Lemma 17 in [3]) $\exists j, k, j < k$, such that $\mu_j \leq \mu_k$. Let q_j and q_k be the firing count vectors: $\mu_j = \mu_0 + Dq_j$ and $\mu_k = \mu_0 + Dq_k$; let $x = q_k - q_j$. Then $\mu_k - \mu_j \geq 0 \Rightarrow Dx \geq 0$, and by construction $x \geq 0$, $x(i) > 0 \ \forall t_i \in U$ and $x(i) = 0 \ \forall t_i \in T \setminus U$. Also we take $\sigma_a = \sigma_0 \sigma_1 \ldots \sigma_j$, $\sigma_x = \sigma_{j+1} \ldots \sigma_k$, $\sigma_b = \sigma_{k+1} \sigma_{k+2} \cdots$, $\mu_1^* = \mu_j$, and $\mu_2^* = \mu_k$. $\quad\square$

In order to characterize the supervisors which prevent deadlock, or enforce liveness or T-liveness, we define the properties P_1, P_2 and P_3 below, in which $\mathcal{N} = (P, T, F, W)$ is a Petri net, $T_x \subseteq T$ and σ denotes a nonempty firing sequence.

(P_1) $(\exists \sigma \ \exists \mu_1', \mu_1 \in \mathcal{R}(\mathcal{N}, \mu): \mu_1 \xrightarrow{\sigma} \mu_1'$ and $\mu_1' \geq \mu_1)$

(P_2) $(\exists \sigma \ \exists \mu_1', \mu_1 \in \mathcal{R}(\mathcal{N}, \mu): \mu_1 \xrightarrow{\sigma} \mu_1'$, $\mu_1' \geq \mu_1$ and all transitions of T appear in σ)

(P_3) $(\exists \sigma \ \exists \mu_1', \mu_1 \in \mathcal{R}(\mathcal{N}, \mu): \mu_1 \xrightarrow{\sigma} \mu_1'$, $\mu_1' \geq \mu_1$ and all transitions of T_x appear in σ)

The following theorem characterizes the relations existing between supervisors preventing deadlock and supervisors enforcing (T-)liveness. In general we may expect deadlock prevention supervisors to be at least as permissive as supervisors enforcing a stronger requirement, such as liveness or T-liveness. Such deadlock prevention supervisors are considered in the parts (d) and (e) of the following theorem.

Theorem 2. *Let $\mathcal{N} = (P, T, F, W)$ be a Petri net and $T_x \subseteq T$.*

(a) *Deadlock can be prevented in $(\mathcal{N}, \mu)$ iff (P_1) is true.*

(b) *Liveness can be enforced in $(\mathcal{N}, \mu)$ iff (P_2) is true.*

(c) *T_x-liveness can be enforced in $(\mathcal{N}, \mu)$ iff (P_3) is true.*

(d) *Let μ_0 be an arbitrary marking for which liveness can be enforced, Ξ_L the least restrictive liveness enforcing supervisor of $(\mathcal{N}, \mu_0)$, and S the set of all deadlock prevention supervisors of $(\mathcal{N}, \mu_0)$ at least as permissive as Ξ_L. Then all $\Xi \in S$ enforce liveness in $(\mathcal{N}, \mu_0)$ iff $\forall \mu \in \mathcal{R}(\mathcal{N}, \mu_0): (P_1) \Rightarrow (P_2)$.*

(e) *Let μ_0 be an arbitrary marking for which T_x-liveness can be enforced, Ξ_L the least restrictive T_x-liveness enforcing supervisor of $(\mathcal{N}, \mu_0)$, and S the set of all deadlock prevention supervisors of $(\mathcal{N}, \mu_0)$ at least as permissive as Ξ_L. Then all $\Xi \in S$ enforce T_x-liveness in $(\mathcal{N}, \mu_0)$ iff $\forall \mu \in \mathcal{R}(\mathcal{N}, \mu_0): (P_1) \Rightarrow (P_3)$.*

Proof.

(a) If (P_1) is true, then a deadlock prevention strategy is to first allow only a firing sequence that leads from μ to μ_1, and then only the infinite firing sequence $\sigma, \sigma, \sigma, \ldots$. Furthermore, if deadlock can be prevented, there is an infinite firing sequence enabled by the initial marking. Then, by Lemma 1, it follows that (P_1) is true.

(b) This is a particular case of (c) for $T = T_x$.

(c) The first part of the proof is similar to (a). If T_x-liveness can be enforced, there is an infinite firing sequence σ enabled by the initial marking, and the transitions in T_x appear infinitely often in σ. Then, by Lemma 1, it follows that (P_3) is true.

(d) This is a particular case of (e) for $T = T_x$.

(e) ($\Rightarrow$) Assume the contrary: $\exists \mu \in \mathcal{R}(\mathcal{N}, \mu_0)$ such that (P_1) is true and (P_3) is not. Note that the least restrictive deadlock prevention supervisor of $(\mathcal{N}, \mu_0)$, Ξ_D, is in $\mathcal{S}$. By part (a), deadlock can be prevented at the marking μ, so $\mu \in \mathcal{R}(\mathcal{N}, \mu_0, \Xi_D)$. However, by part (c), $(\mathcal{N}, \mu)$ cannot be made T_x-live, so Ξ_D does not enforce T_x-liveness, which is a contradiction.

($\Leftarrow$) Since T_x-liveness can be enforced at μ_0, deadlock can be prevented at μ_0, so $\mathcal{S}$ is nonempty. Let $\Xi \in \mathcal{S}$. The proof checks that for all $\mu \in \mathcal{R}(\mathcal{N}, \mu_0, \Xi)$ there is a firing sequence enabled by μ, accepted by Ξ, and which includes all transitions in T_x. Let $\mu \in \mathcal{R}(\mathcal{N}, \mu_0, \Xi)$. Since deadlock is prevented, (P_3) is true as (P_1) is true. Let Ξ_x be the supervisor that enforces T_x-liveness in $(\mathcal{N}, \mu_0)$ by firing $\sigma_1 \sigma_2 \sigma \sigma \sigma \ldots$, where $\mu_0 \xrightarrow{\sigma_1} \mu \xrightarrow{\sigma_2} \mu_1$, and σ, μ and μ_1 are the variables from (P_3). Since Ξ is at least as permissive as Ξ_L, Ξ is at least as permissive as Ξ_x. Thus Ξ allows $\sigma_2 \sigma$ to fire from μ. Therefore all transitions of T_x appear in some firing sequence enabled by μ and allowed by Ξ. $\quad\square$

In practice it may be difficult to check $(P_1) \Rightarrow (P_2)$ or $(P_1) \Rightarrow (P_3)$ in order to see whether a deadlock prevention supervisor will also enforce liveness or T-liveness. In contrast, the conditions of the next theorem can be easily verified using linear programming.

Theorem 3. *Let $\mathcal{N} = (P, T, F, W)$ be a Petri net, D its incidence matrix, $T_x \subseteq T$, $n = |T|$ the number of transitions, $M = \{x \in \mathbb{Z}_+^n : x \neq 0, Dx \geq 0\}$, $N = \{x \in M : \forall i = 1 \ldots n : x(i) \neq 0\}$ and $P = \{x \in M : \forall t_i \in T_x : x(i) \neq 0\}$.*

(a) *The following statements are equivalent:*
   (i) *$M \neq \emptyset$ and $M = N$*
   (ii) *supervisors which prevent deadlock exist for some initial marking, and for all such initial markings μ_0 all supervisors preventing deadlock in $(\mathcal{N}, \mu_0)$ also enforce liveness in $(\mathcal{N}, \mu_0)$*

(b) *The following statements are equivalent:*
   (i) *$M \neq \emptyset$ and $M = P$*
   (ii) *supervisors which prevent deadlock exist for some initial marking, and for all such initial markings μ_0 all supervisors preventing deadlock in $(\mathcal{N}, \mu_0)$ also enforce T_x-liveness in $(\mathcal{N}, \mu_0)$*

(c) *The following statements are equivalent:*
   (i) *$N \neq \emptyset$ and $N = P$*
   (ii) *supervisors which enforce T_x-liveness exist for some initial marking, and for all such initial markings μ_0 all supervisors enforcing T_x-liveness in $(\mathcal{N}, \mu_0)$ also enforce liveness in $(\mathcal{N}, \mu_0)$*

Proof.

(a) This is a particular case of (b) for $T = T_x$.

(b) ($\Rightarrow$) Since $M \neq \emptyset$, a marking μ_0 for which a deadlock prevention supervisor exists can be found as in equation (1). Let μ_0 be an initial marking for

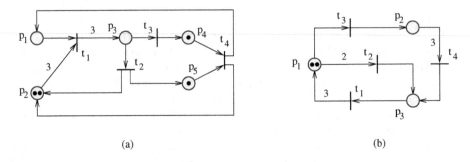

(a)   (b)

Fig. 2. Examples for Theorems 2 and 3

which deadlock prevention supervisors exist and Ξ a deadlock prevention supervisor of $(\mathcal{N}, \mu_0)$. We show that there is no reachable marking such that a transition in T_x is dead. Let $\mu \in \mathcal{R}(\mathcal{N}, \mu_0, \Xi)$. Since Ξ prevents deadlock, there is an infinite firing sequence σ enabled by μ and allowed by Ξ. Using Lemma 1 for $\mu_I = \mu$, we see that while firing σ a marking μ_1^* is reached such that μ_1^* enables σ_x corresponding to $x \in M$. But $M = P$, so all transitions in T_x appear in σ_x. Therefore no transition in T_x is dead at μ, so Ξ also enforces T_x-liveness.

($\Leftarrow$) Assume the contrary. Then there is a nonnegative integer vector x, $x \neq 0$, such that $Dx \geq 0$ and $x(i) = 0$ for some $t_i \in T_x$. Let Ξ be a deadlock prevention supervisor for $(\mathcal{N}, \mu_0)$, where μ_0 is such that it enables a firing sequence σ_x defined as follows: t_i appears in σ_x iff $x(i) \neq 0$, in which case it appears $x(i)$ times. If Ξ is defined to only allow firing $\sigma_x \sigma_x \sigma_x \ldots$, then deadlock is prevented but T_x-liveness is not enforced, as σ_x does not include all transitions of T_x. Contradiction.

(c) The proof is identical to (b) if we substitute in (b) T_x with T, and deadlock prevention with T_x-liveness enforcement.   □

Figure 2(a) shows an example for Theorem 3(a): all nonnegative vectors x such that $Dx \geq 0$ are a linear combination with nonnegative coefficients of $[1, 2, 1, 1]^T$ and $[2, 3, 3, 3]^T$. Figure 2(b) shows an example for Theorem 2(d). Indeed, all markings μ that enable any of t_1, t_2 or t_4 satisfy (P_2). Also, a marking that enables only t_3 either leads to deadlock or enables the sequence t_3, t_4 and hence satisfies (P_2). For instance, the deadlock prevention supervisor that repeatedly fires t_2, t_1 does not enforce liveness because it does not satisfy the requirement of Theorem 2(d) to be at least as permissive as any liveness enforcing supervisor.

With regard to Theorem 2(d-e), note that designing deadlock prevention supervisors less restrictive than liveness enforcing supervisors has been demonstrated for instance in [5,6,8,9].

Theorem 4. *Consider a Petri net $\mathcal{N} = (P, T, F, W)$ which is not repetitive. At least one transition exists such that for any initial marking it cannot fire infinitely often. Let T_D be the set of all such transitions. There are initial markings μ_0 and a supervisor Ξ such that $\forall \mu \in \mathcal{R}(\mathcal{N}, \mu_0, \Xi)$ no transition in $T \setminus T_D$ is dead.*

Proof. Let $\|x\|$ be the *support* of the vector x, that is $\|x\| = \{i : x(i) \neq 0\}$. There is an integer vector $x \geq 0$ with *maximum support* such that $Dx \geq 0$, that is, for all integer vectors $w \geq 0$ such that $Dw \geq 0$: $\|w\| \subseteq \|x\|$. Indeed if $y, z \geq 0$, are integer vectors and $Dy \geq 0$ and $Dz \geq 0$, then $D(z + y) \geq 0$, $y + z \geq 0$, and $\|y\|, \|z\| \subseteq \|y + z\|$.

If $t_j \in T$ can be made live, there is a marking that enables an infinite firing sequence σ such that t_j appears infinitely often in σ. Therefore by Lemma 1 $\exists y \geq 0$ such that $Dy \geq 0$ and $y(j) > 0$. Since x has maximum support, $\|y\| \subseteq \|x\|$ and so $t_j \in \|x\|$. This proves that all transitions that can be made live are in $\|x\|$. Moreover, only the transitions that can be made live are in $\|x\|$. Indeed, let σ_x be a firing sequence such that t_i appears in σ_x iff $x(i) \neq 0$, in which case it appears $x(i)$ times. Then there is a marking μ_0 given by equation (1) that enables the infinite firing sequence $\sigma_l = \sigma_x \sigma_x \sigma_x \ldots$. We may choose Ξ to only allow σ_l to fire from μ_0, and we note that all transitions in $\|x\|$ are live. However $T \not\subseteq \|x\|$, or else σ_l contains all transitions of T and so $\mathcal{N}$ is repetitive. Therefore $T \setminus \|x\| \neq \emptyset$. Since $\|x\|$ contains the transitions that can be made live, $T \setminus \|x\| \neq \emptyset$ contains the transitions that cannot be made live under any circumstances. So we have $T_D = T \setminus \|x\|$ and $T_D \neq \emptyset$. $\qquad\square$

A special case in Theorem 4 is $T \setminus T_D = \emptyset$, when the Petri net is not even partially repetitive, and so deadlock cannot be avoided for any marking. It was already shown that only repetitive Petri nets can be made live (Proposition 3). Theorem 4 shows that the set of transitions of a partially repetitive Petri net can be uniquely divided in transitions that can be made live and transitions that cannot be made live. So the liveness property of partially repetitive Petri nets is that all transitions that can be made live are live ($T \setminus T_D$-liveness). For an example, consider the Petri nets of Figure 4(a) and (b). For the first one $T_D = \{t_4, t_5\}$, and for the second one $T_D = \{t_1, t_2, t_3\}$.

3.2 Deadlock and (T-)Liveness Characterization Based on Active Subnets

We denote by the *active subnet* a part of a Petri net which can be made live for appropriate markings by supervision. In the following definition we use the notations from Theorem 4.

Definition 5. *Let $\mathcal{N} = (P, T, F, W)$ be a Petri net, D the incidence matrix and $T_D \subseteq T$ be the set of all transitions that cannot be made live for any initial marking. $\mathcal{N}^A = (P^A, T^A, F^A, W^A)$ is an **active subnet** of $\mathcal{N}$ if $P^A = T^A\bullet$, $F^A = F \cap \{(T^A \times P^A) \cup (P^A \times T^A)\}$, W^A is the restriction of W to F^A and T^A is the set of transitions with nonzero entry in some nonnegative vector $x \neq 0$ satisfying $Dx \geq 0$. $\mathcal{N}^A$ is the **maximal active subnet** of $\mathcal{N}$ if $T^A = T \setminus T_D$*

and $T \setminus T_D \neq \emptyset$. $\mathcal{N}^A$ is a **minimal active subnet** if there is no other active subnet $\mathcal{N}_1^A = (P_1^A, T_1^A, F_1^A, W_1^A)$ such that $T_1^A \subseteq T^A$.

Definition 6. *Given an active subnet $\mathcal{N}^A$ of a Petri net $\mathcal{N}$, a siphon of $\mathcal{N}$ is said to be an **active siphon** (with respect to $\mathcal{N}^A$) if it is or includes a siphon of $\mathcal{N}^A$. An active siphon is **minimal** if it does not include another active siphon (with respect to the same active subnet.)*

In Figure 3(a) and (c) two Petri nets are given. Figure 3(b) shows the minimal active subnets of the Petri net in Figure 3(a). The union of the two subnets is the maximal active subnet. Figure 3(d) shows the only active subnet of the Petri net of Figure 3(c). The minimal active siphons of the Petri net in Figure 3(a) with respect to the active subnet having $T^A = \{t_6, t_7, t_9\}$ are $\{p_1, p_5, p_6, p_7\}$ and $\{p_6, p_7, p_8\}$. The minimal active siphons of the Petri net of Figure 3(c) are $\{p_1, p_4, p_7\}$, $\{p_2, p_5, p_7\}$, $\{p_3, p_5, p_7\}$ and $\{p_6, p_7\}$.

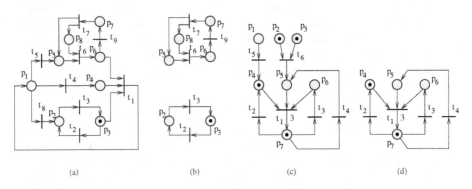

Fig. 3. Two Petri nets: (a) and (c), and their active subnets: (b) and (d), respectively.

Proposition 4. *A siphon which contains places from an active subnet is an active siphon with respect to that subnet.*

Proof. We use the notations from Definition 5. Let σ_x be a firing sequence such that a transition t_i appears in σ_x iff $x(i) \neq 0$, in which case it appears $x(i)$ times. Let S be a siphon such that $S \cap P^A \neq \emptyset$. We are to prove that there is a siphon s of $\mathcal{N}^A$ such that $s \subseteq S$. $\bullet S \subseteq S \bullet$ implies that $\bullet S \cap T^A \subseteq S \bullet \cap T^A$. Using the construction of equation (1) there is a marking enabling $\sigma_x \sigma_x \sigma_x \ldots$. Since $T^A = \|x\|$, this implies $\forall t \in T^A$: $\bullet t \subseteq P^A$. Hence $S \bullet \cap T^A \subseteq (S \cap P^A) \bullet$ and so $S \bullet \cap T^A = (S \cap P^A) \bullet \cap T^A$. Note also that $\bullet(S \cap P^A) \cap T^A \subseteq \bullet S \cap T^A$. Therefore $\bullet S \cap T^A \subseteq S \bullet \cap T^A$ implies $\bullet(S \cap P^A) \cap T^A \subseteq (S \cap P^A) \bullet \cap T^A$, which proves that $s = S \cap P^A$ is a siphon of $\mathcal{N}^A$. $\qquad\square$

The significance of the active subnets for deadlock prevention can be seen in the following propositions. First we prove a technical result.

Lemma 2. *Let $\mathcal{N}^A = (P^A, T^A, F^A, W^A)$ be an active subnet of $\mathcal{N}$. Given a marking μ of $\mathcal{N}$ and μ^A its restriction to $\mathcal{N}^A$, if $t \in T^A$ is enabled in $\mathcal{N}^A$, then t is enabled in $\mathcal{N}$.*

Proof. By definition, there is an nonnegative integer vector $x \geq 0$ such that $Dx \geq 0$ (D is the incidence matrix) and $x(i) > 0$ for $t_i \in T^A$ and $x(i) = 0$ for $t_i \in T \setminus T^A$. This implies that there are markings such that the transitions of T^A can fire infinitely often, without firing other transitions (see equation (1).) If t is not enabled in $\mathcal{N}$, there is $p \in \bullet t$ such that $p \notin P^A$ since t is enabled in $\mathcal{N}^A$. (The preset/postset operators $\bullet$ are taken with respect to $\mathcal{N}$, not $\mathcal{N}^A$.) Note that $p \notin P^A$ implies $\bullet p \cap T^A = \emptyset$. If $\bullet p = \emptyset$, t cannot fire infinitely often, which contradicts the definition of T^A (Definition 5), since $t \in T^A$. If $t_x \in \bullet p$, the transitions of T^A cannot fire infinitely often without firing t_x, which again contradicts the definition of T^A. Therefore t is also enabled in $\mathcal{N}$. $\square$

Note that in a repetitive Petri net all siphons are active with respect to the maximal active subnet. The next result is a generalization of the well known Proposition 1. It is a more powerful result since it not only states that deadlock implies an empty siphon, but also that for any active subnet $\mathcal{N}^A$ there is an empty active siphon with respect to $\mathcal{N}^A$.

Proposition 5. *Let $\mathcal{N}^A$ be an arbitrary active subnet of a PT-ordinary Petri net $\mathcal{N}$. If μ is a deadlock marking of $\mathcal{N}$, then there is at least one empty minimal active siphon with respect to $\mathcal{N}^A$.*

Proof. Since μ is a deadlock marking and $\mathcal{N} = (P, T, F, W)$ is PT-ordinary, $\forall t \in T \ \exists p \in \bullet t: \mu(p) = 0$. The active subnet is built in such a way that if the marking μ restricted to the active subnet enables a transition t, then μ enables t in the total net (Lemma 2.) Therefore, because the total net $(\mathcal{N}, \mu)$ is in deadlock, the active subnet is too. In view of Proposition 1, let s be an empty minimal siphon of the active subnet. Consider s in the total net. If s is a siphon of the total net, then s is also a minimal active siphon; therefore the net has a minimal active siphon which is empty. If s is not a siphon of the total net: $\bullet s \setminus T^A \neq \emptyset$. Let S be the set inductively constructed as follows: $S_0 = s$, $S_i = S_{i-1} \cup \{p \in \bullet(\bullet S_{i-1} \setminus S_{i-1} \bullet) : \mu(p) = 0\}$, where μ is the (deadlock) marking of the net. In other words S is a completion of s with places with null marking such that S is a siphon. By construction S is an active siphon and is empty for the marking μ. Hence an empty minimal active siphon exists. $\square$

The practical significance of Proposition 5 is that it can be used for deadlock prevention, since deadlock is not possible when all active siphons with respect to an active subnet cannot become empty.

Proposition 6. *Deadlock is unavoidable for the marking μ if for all minimal active subnets $\mathcal{N}^A$ there is an empty active siphon with respect to $\mathcal{N}^A$.*

Proof. All transitions in the postset of an empty siphon are dead. Therefore every minimal active subnet has some dead transitions. Assume that deadlock is avoidable. Then, in view of Lemma 1, after some transitions firings a marking can be reached which enables $\sigma\sigma\sigma\ldots\sigma\ldots$, where σ is a finite firing sequence. Let q be the firing count vector for σ. Then $Dq \geq 0$. If the active subnet for q is minimal, we let $x = q$, but if it is not, there is x such that $\|x\| \subset \|q\|$, $x \neq 0$, $x \geq 0$, $Dx \geq 0$ and the active subnet associated to x is minimal. But there must be an empty active siphon with regard to that active subnet, therefore not all of the transitions of $\|x\|$ can fire, which implies that not all of the transitions of σ can fire, which is a contradiction. $\square$

Propositions 5 and 6 generalize Proposition 1. Thus a Petri net will certainly enter deadlock if for all minimal active subnets $\mathcal{N}^A$ there is an empty active siphon with respect to $\mathcal{N}^A$. Conversely a deadlock state implies that for each active subnet there is an empty active siphon with regard to that subnet. Propositions 5 and 6 suggest an approach for least restrictive deadlock prevention, and we consider it in section 4.2.

While Commoner's Theorem is a necessary and sufficient condition, its extension to asymmetric choice Petri nets is usually presented as a sufficient condition (e.g. Theorem 10.4 in [2]). The reason for this is that the attention has been restricted to a particular class of controlled siphons, namely trap controlled siphons. In terms of the general notion of controlled siphons, the extension of Commoner's Theorem is a necessary and sufficient condition (see Corollary 27 in [1]). We go one step further: the next result not only states that a dead transition t implies an empty siphon for some reachable marking, but also that there is such an empty siphon S such that $t \in S\bullet$. This fact is important when we try to verify or ensure that t is live, since we only have to look at the siphons S such that $t \in S\bullet$.

Theorem 5. *Consider a PT-ordinary asymmetric choice Petri net $\mathcal{N}$ and a marking μ such that a transition t is dead. Then there is $\mu' \in \mathcal{R}(\mathcal{N}, \mu)$ such that S is an empty siphon for the marking μ' and $t \in S\bullet$.*

Proof. It is known that if a transition t of an ordinary Petri net with asymmetric choice is dead at a marking μ, then $\exists\mu_1 \in \mathcal{R}(\mathcal{N}, \mu)\ \exists p_1 \in \bullet t\ \forall\mu_x \in \mathcal{R}(\mathcal{N}, \mu_1)$: $\mu_x(p_1) = 0$. This is proved for instance in Lemma 10.2 of [2], and the proof applies without change to PT-ordinary asymmetric choice Petri nets. We inductively use this property to construct S. Note that all transitions in $\bullet p_1$ are dead at μ_1. Let $S_0 = \emptyset$ and $S_1 = \{p_1\}$. We inductively construct S by generating $S_2, \ldots S_{n+1}$ and the markings $\mu_2, \ldots \mu_{n+1}$. S_i for $i \geq 1$ is such that all transitions in $\bullet S_i$ are dead for some marking μ_i. The construction in a iteration is as follows. Let $T_i = \bullet(S_i \setminus S_{i-1})$ and $\mu_{i+1} \in \mathcal{R}(\mathcal{N}, \mu_i)$ such that $\forall t_x \in T_i\ \forall\mu_x \in \mathcal{R}(\mathcal{N}, \mu_{i+1})\ \exists p \in \bullet t$: $\mu_x(p) = 0$. Then we let $G_i = \bigcup_{t_x \in T_i} \{p \in \bullet t_x : \forall\mu_x \in \mathcal{R}(\mathcal{N}, \mu_{i+1}) : \mu_x(p) = 0\}$ and $S_{i+1} = S_i \cup G_i$. There is n such that $S_{n+1} = S_n$, for the Petri net has a finite number of places. We let $S = S_n$ and $\mu' = \mu_n$. By construction S is a siphon (note that $\bullet S_i \subseteq S_{i+1}\bullet$ for $i = 0 \ldots n$), S is empty at μ', $\mu' \in \mathcal{R}(\mathcal{N}, \mu)$, and $t \in S\bullet$ (since $p_1 \in S$). $\square$

Definition 7. *Let $\mathcal{N}$ be a Petri net, T a nonempty subset of the set of transitions and $\mathcal{N}^A = (P^A, T^A, F^A, W^A)$ an active subnet. We say that $\mathcal{N}^A$ is* **T-minimal** *if $T \subseteq T^A$ and $T_x^A \not\subseteq T^A$ for any other active subnet $\mathcal{N}_x^A = (P_x^A, T_x^A, F_x^A, W_x^A)$ such that $T \subseteq T_x^A$.*

In general a T-minimal active subnet may not be unique. However, as shown in the next theorem, any T-minimal active subnet can be used to characterize T-liveness. We also note that computing a T-minimal active subnet has polynomial complexity (it involves solving linear programs). The following new result may be seen as a correspondent for T-liveness of the Commoner's Theorem.

Theorem 6. *Given a PT-ordinary asymmetric choice Petri net $\mathcal{N}$, let T be a set of transitions and $\mathcal{N}^A$ a T-minimal active subnet which contains the transitions in T. If all the minimal siphons with respect to $\mathcal{N}^A$ are controlled, the Petri net is T-live (and T^A-live). If the Petri net is T-live, there is no reachable marking such that for each T-minimal active subnet $\mathcal{N}^A$ there is an empty minimal active siphon with respect to $\mathcal{N}^A$.*

Proof. For the first part, assume that there is a reachable marking such that a transition $t \in T^A$ is dead. Since $T \subseteq T^A$, by Theorem 5 there is a reachable marking such that a siphon S is empty and $t \in S\bullet$. However $t \in S\bullet$ implies $S \cap P^A \neq \emptyset$, and by Proposition 4 S is an active siphon. However S empty contradicts the fact that all active siphons are controlled.

For the second part, let $\mathcal{N}_i^A$ denote a T-minimal active subnet, $i = 1 \ldots k$, where k is the number of T-minimal active subnets. Let μ be a marking such that an active siphon S_i is empty. Let $T_i = S_i \bullet \cap T_i^A$, where T_i^A is the set of transitions of $\mathcal{N}_i^A$. Since S_i is active, T_i is nonempty; because S_i is empty, the transitions of T_i are dead. Assume that there is an infinite firing sequence σ_x such that all transitions of T appear infinitely often in σ_x and after a part of σ_x is fired, (let μ_x be the marking reached) all T-minimal active subnets $\mathcal{N}_i^A$ have an empty active siphon S_i. Let σ be the remaining part of σ_x which is enabled by μ. All transitions of T appear infinitely often in σ. Therefore, by Lemma 1, there is $x \geq 0$ such that $Dx \geq 0$ (D is the incidence matrix) and $T \subseteq \|x\|$. However, $\|x\|$ does not contain all transitions of any of the T-minimal subnets $\mathcal{N}_i^A$: $T_i \subseteq T_i^A \setminus \|x\|$, for $i = 1 \ldots k$. This implies that $\|x\|$ defines another T-minimal active subnet, which contradicts the fact that $\mathcal{N}_i^A$ $i = 1 \ldots k$ are all T-minimal active subnets. $\qquad\square$

In the particular case in which there is a single T-minimal active subnet, Theorem 6 shows that the net is T-live iff all siphons are controlled. When T equals the total set of transitions of the net and the Petri net is repetitive, the T-minimal active subnet exists, is unique, and equals the total net; in this case we obtain the extension of the Commoner's Theorem to asymmetric choice nets.

4   Implications and Discussion

In this section we discuss our results and show how they relate to the supervisory problems of deadlock prevention, liveness enforcement and T-liveness en-

forcement. Some of the theoretical results only consider particular classes of
Petri nets, specifically PT-ordinary and asymmetric choice nets. However, for
our supervisory problems this is a surmountable difficulty, since it is possible to
transform a Petri net to a PT-ordinary or PT-ordinary asymmetric choice Petri
net; then, it is possible to derive a deadlock prevention (or a $(T\text{-})$liveness enforcement) supervisor from a supervisor for deadlock prevention (or $(T\text{-})$liveness
enforcement) of the transformed net [6,7,12]. Briefly, a possible solution for our
supervisory problems is as follows: given a target net $\mathcal{N}_0$, generate a sequence
of increasingly enhanced nets $\mathcal{N}_1$, $\mathcal{N}_2$ ... until we reach a net $\mathcal{N}_k$, such that we
can use Proposition 5 or Theorem 6 on $\mathcal{N}_k$ to guarantee deadlock-freedom or
$(T\text{-})$liveness; then a supervisor for $\mathcal{N}_0$ is derived based on the construction of
$\mathcal{N}_k$. For more details the reader is referred to [6,7,8].

4.1   Deadlock Prevention

Proposition 1 implies that if the marking of any of the minimal siphons of a
Petri net can never become zero, the Petri net is deadlock-free. This is an useful
property for repetitive Petri nets, but not always for nonrepetitive Petri nets. For
partially repetitive Petri nets Proposition 5 is much more useful. For instance
consider the Petri net of Figure 4(a). The only active subnet has $T^A = \{t_1, t_2, t_3\}$.
After firing t_4, $\{p_4\}$ is an empty siphon. However, there is no empty active siphon
(the minimal active siphons are $\{p_1, p_3, p_4\}$, $\{p_2, p_3, p_5\}$ and $\{p_2, p_3, p_6\}$), and
thus we can see from Proposition 5 that the Petri net is not in deadlock, while
this cannot be ascertained from Proposition 1. The same is true for the Petri net
in Figure 4(b): $\{p_1, p_3\}$ is an empty siphon, but the only minimal active siphon,
$\{p_4, p_5, p_6, p_7\}$, is not empty, and therefore the Petri net is not in deadlock by
Proposition 5.

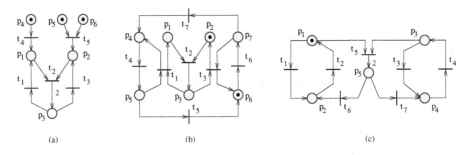

(a)   (b)   (c)

Fig. 4.

Proposition 5 is more useful than Proposition 1 even for repetitive Petri nets,
as seen in Figure 4(c). The Petri net of Figure 4(c) has several active subnets.
While with respect to some of them there are empty active siphons, if we take

the active subnet $\mathcal{N}^A$ defined by $T^A = \{t_1, t_2\}$, the only minimal active siphon with respect to $\mathcal{N}^A$ is $\{p_1, p_2, p_5\}$, which is not empty. Thus we are able to detect based on Proposition 5 that the Petri net is not in deadlock.

In the applications in which deadlock prevention is desired to approximate liveness enforcement, Proposition 5 can be used for the maximal active subnet. Thus it would be desirable that no active siphon with respect to the maximal active subnet ever becomes empty. Indeed, if an active siphon S with respect to the maximal active subnet is empty, all transitions in $S\bullet$ are dead, and some of them are in the set of $T \setminus T_D$ of Theorem 4.

For the applications in which least restrictive deadlock prevention is desired rather than a liveness approximation, see the next section.

Proposition 5 can be used for deadlock prevention by extending the target Petri net to a net in which all siphons are controlled. The usual technique for siphon control involves adding a new place to each siphon to be controlled, such that place invariants are created. Such additional places can be seen as implementing a (marking based) supervisor for deadlock prevention. We have designed a deadlock prevention methodology based on Proposition 5 in [6]. The methodology of [6] produces two sets of constraints: $L\mu \geq b$ and $L_0\mu \geq b_0$. Thus $L\mu \geq b$ defines the supervisor (the set of additional places ensuring that all active siphons are invariant controlled), defined for all initial markings μ_0 satisfying both $L\mu_0 \geq b$ and $L_0\mu_0 \geq b_0$. For an example, consider the Petri nets in Figure 5(a) and (b). They are supervised for deadlock prevention using the methodology of [6]. The additional places defining the supervisor are, in both cases, the places C_1, C_2 and C_3. It can be easily checked that all minimal active siphons are invariant controlled in both cases. In the case (a) the inequalities $L\mu \geq b$ are $\mu(p_1) + \mu(p_3) + \mu(p_4) \geq 1$ (so $\mu(C_1) = \mu(p_1) + \mu(p_3) + \mu(p_4) - 1$), $\mu(p_2) + \mu(p_3) + \mu(p_5) \geq 1$ ($\mu(C_2) = \mu(p_2) + \mu(p_3) + \mu(p_5) - 1$) and $\mu(p_2) + \mu(p_3) + \mu(p_6) \geq 1$ ($\mu(C_3) = \mu(p_2) + \mu(p_3) + \mu(p_6) - 1$); $L_0\mu_0 \geq b_0$ contains the inequalities $\mu_0(p_1) + \mu_0(p_2) + \mu_0(p_3) + \mu_0(p_4) + \mu_0(p_5) \geq 2$ and $\mu_0(p_1) + \mu_0(p_2) + \mu_0(p_3) + \mu_0(p_4) + \mu_0(p_6) \geq 2$. In the case (b), the inequalities $L\mu \geq b$ are $\mu(p_1) + \mu(p_2) \geq 1$ ($\mu(C_1) = \mu(p_1) + \mu(p_2) - 1$), $\mu(p_3) + \mu(p_4) \geq 1$ ($\mu(C_2) = \mu(p_3) + \mu(p_4) - 1$) and $\mu(p_1) + \mu(p_2) + \mu(p_3) + \mu(p_4) \geq 3$ ($\mu(C_3) = \mu(C_1) + \mu(C_2) - 1$); there are no constraints $L_0\mu_0 \geq b_0$. Moreover, by Theorem 3, the supervisors also enforce $\{t_1, t_2, t_3\}$-liveness in case (a), and liveness in case (b).

4.2   Least Restrictive Deadlock Prevention

Assume that we have u supervisors for deadlock prevention in $\mathcal{N}_0$: Ξ_1, Ξ_2, ... Ξ_u. Each supervisor can prevent deadlock if the initial marking is in the sets $\mathcal{M}_1$, $\mathcal{M}_2$, ... $\mathcal{M}_u$, respectively. Let Ξ be the supervisor defined on $\mathcal{M} = \bigcup\limits_{i=1...u} \mathcal{M}_i$, which allows a transition to fire only if at least one of the supervisors Ξ_i, defined for the current marking, allows that transition to fire. We denote the supervisor by $\Xi = \bigvee\limits_{i=1}^{u} \Xi_i$. Obviously, Ξ is a deadlock prevention supervisor, and Ξ is at least as permissive as any of Ξ_i.

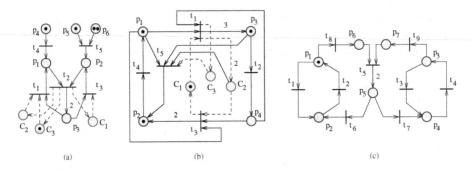

Fig. 5.

Theorem 7. *Let $\mathcal{N}_0$ be a Petri net and $\mathcal{N}_i^A$, for $i = 1 \ldots u$, the minimal active subnets of $\mathcal{N}_0$. Let T_i denote the set of transitions of $\mathcal{N}_i^A$ and let Ξ_i, for $i = 1 \ldots u$, be deadlock prevention supervisors. Assume that each Ξ_i is defined for all initial markings for which T_i-liveness can be enforced and that each Ξ_i is at least as permissive as any T_i-liveness enforcing supervisor. Then $\Xi = \bigvee_{i=1}^{u} \Xi_i$ is the least restrictive deadlock prevention supervisor of $\mathcal{N}_0$.*

Proof. The only thing which is to be proved is that a marking unacceptable to Ξ leads to deadlock. Consider such a marking μ. Let $x_1, x_2, \ldots x_u$ be the nonnegative integer vectors defining $\mathcal{N}_1^A, \mathcal{N}_2^A, \ldots \mathcal{N}_u^A$ in Definition 5. Thus $T_i = \|x_i\|$ for $i = 1 \ldots u$. Since μ is unacceptable to all of Ξ_i and each Ξ_i is at least as permissive as any T_i-liveness enforcing supervisors, for all $i = 1 \ldots u$ not all transitions of T_i can be made live given the marking μ. Assume that deadlock can be prevented at μ. Then there is an infinite firing sequence σ enabled by μ. Let T_x be the set of transitions which appear infinitely often in σ. By Lemma 1 there is a nonnegative integer vector x such that $T_x = \|x\|$ and $Dx \geq 0$, where D is the incidence matrix. Since $\mathcal{N}_1^A, \mathcal{N}_2^A, \ldots \mathcal{N}_u^A$ are all the minimal active subnets of $\mathcal{N}_0$, there is $j \in \{1, 2, \ldots u\}$ such that $\|x_j\| \subseteq \|x\|$. But this contradicts the fact that not all transitions of $\|x_j\|$ can be made live at μ. $\qquad\square$

Each of the supervisors Ξ_i satisfying the requirements of the theorem above can be found with the procedure for deadlock prevention that we present in [6], by starting it with an initial active subnet $\mathcal{N}_i^A$. As an example, consider the Petri net of Figure 5(c). There are three minimal active subnets $\mathcal{N}_1^A$, $\mathcal{N}_2^A$ and $\mathcal{N}_3^A$, defined by $T_1^A = \{t_1, t_2\}$, $T_2^A = \{t_3, t_4\}$ and $T_3^A = \{t_2, t_4, t_5, t_6, t_7, t_8, t_9\}$, respectively. Three deadlock prevention supervisors corresponding to $\mathcal{N}_1^A$, $\mathcal{N}_2^A$ and $\mathcal{N}_3^A$ are Ξ_1, Ξ_2 and Ξ_3, defined as follows. For simplicity of notation, we let $\mu_i = \mu(p_i)$. Ξ_1 requires $\mu_1 + \mu_2 + \mu_5 + \mu_6 \geq 1 \wedge \mu_1 + \mu_2 + \mu_3 + \mu_4 + \mu_5 + \mu_7 \geq 1$ (the inequalities correspond to the two minimal active siphons with respect to $\mathcal{N}_1^A$); Ξ_2 requires $\mu_3 + \mu_4 + \mu_5 + \mu_7 \geq 1 \wedge \mu_1 + \mu_2 + \mu_3 + \mu_4 + \mu_5 + \mu_6 \geq 1$; Ξ_3 requires $\mu_1 + \mu_2 + \mu_5 + \mu_6 \geq 1 \wedge \mu_3 + \mu_4 + \mu_5 + \mu_7 \geq 1$, and the initial marking μ_0

to satisfy in addition $\sum\limits_{i=1...7} \mu_{0,i} \geq 2$. It can be easily seen that $\Xi = \Xi_1 \vee \Xi_2 \vee \Xi_3$ is the least restrictive deadlock prevention supervisor. In this particular case $\Xi_1 \vee \Xi_2 \vee \Xi_3 = \Xi_1 \vee \Xi_2$.

4.3 *T*-Liveness Enforcement

We demonstrate a procedure for least restrictive *T*-liveness enforcement in [7]. The procedure is based on Theorem 6.

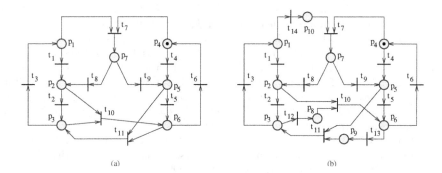

(a)   (b)

Fig. 6.

Consider the Petri net of Figure 6(a), in which it is desired to ensure *T*-liveness for $T = \{t_1, t_2, t_3\}$. For the displayed marking all of t_1, t_2 and t_3 are dead. However we cannot use Theorem 5, as the Petri net is not with asymmetric choice. Figure 6(b) shows the same Petri net transformed to be with asymmetric choice. Theorem 5 is verified, as the minimal active siphon $S = \{p_1, p_2, p_3, p_4, p_5, p_6, p_7\}$ (with respect to the active subnet with set of transitions T) is uncontrolled. Indeed, by firing t_4, t_5 and t_{13}, S becomes empty. The Petri net of Figure 6(a) is not *T*-live for most initial markings. By applying our *T*-liveness enforcement approach from [7], the least restrictive *T*-liveness supervisor of the Petri net of Figure 6(a) enforces $2\mu_1 + 2\mu_2 + 2\mu_3 + \mu_4 + \mu_5 + \mu_6 + 2\mu_7 \geq 2$.

5 Conclusion

We have introduced new theoretical results which are practical for deadlock prevention and (*T*-)liveness enforcement. The first part of the paper characterizes the relation between deadlock prevention and (*T*-)liveness enforcement. The second part extends literature results on deadlock and liveness. The extensions are based on the concept of active subnets and siphons. The usage of active subnets has allowed us to extend the well known sufficient condition for deadlock to necessary and sufficient conditions, and to derive T-liveness results generalizing Commoner's Theorem. We note that our results are effective not only on repetitive Petri nets but also on partially repetitive Petri nets.

References

1. K. Barkaoui and J. F. Pradat-Peyre. On liveness and controlled siphons in Petri nets. In *Lecture Notes in Computer Science: 17th International Conference in Application and Theory of Petri Nets (ICATPN'96), Osaka, Japan*, volume 1091, pages 57–72. Springer-Verlag, June 1996.
2. J. Desel and J. Esparza. *Free Choice Petri nets*. Number 40 in Cambridge Tracts in Theoretical Computer Science. Cambridge University Press, 1995.
3. J. Desel and W. Reisig. Place/transition petri nets. *Lecture Notes in Computer Science: Lectures on Petri Nets I: Basic Models*, 1491:122–173, 1998.
4. K. X. He and M. D. Lemmon. Liveness verification of discrete event systems modeled by n-safe ordinary Petri nets. In Nielsen, M. and Simpson, D., editors, *Lecture Notes in Computer Science: 21st International Conference on Application and Theory of Petri Nets (ICATPN 2000), Aarhus, Denmark, June 2000*, volume 1825, pages 227–243. Springer-Verlag, 2000.
5. M. V. Iordache, J. O. Moody, and P. J. Antsaklis. A method for deadlock prevention in discrete event systems using Petri nets. Technical report of the isis group, isis-99-006, University of Notre Dame, July 1999.
6. M. V. Iordache, J. O. Moody, and P. J. Antsaklis. Automated synthesis of deadlock prevention supervisors using Petri nets. Technical report of the isis group, isis-2000-003, University of Notre Dame, May 2000.
7. M. V. Iordache, J. O. Moody, and P. J. Antsaklis. Automated synthesis of liveness enforcement supervisors using Petri nets. Technical report of the isis group, isis-2000-004, University of Notre Dame, September 2000.
8. M. V. Iordache, J. O. Moody, and P. J. Antsaklis. A method for the synthesis of deadlock prevention controllers in systems modeled by Petri nets. In *Proceedings of the 2000 American Control Conference*, pages 3167–3171, June 2000.
9. K. Lautenbach and H. Ridder. The linear algebra of deadlock avoidance – a Petri net approach. Technical report, University of Koblenz, Institute for Computer Science, 1996.
10. T. Murata. Petri nets: Properties, analysis and applications. In *Proceedings of the IEEE*, pages 541–580, April 1989.
11. W. Reisig. *Petri Nets*, volume 4 of *EATCS Monographs on Theoretical Computer Science*. Springer-Verlag, 1985.
12. S. R. Sreenivas. On a free-choice equivalent of a Petri net. In *Proceedings of the 36th IEEE Conference on Decision and Control*, pages 4092–4097, San Diego, California, December 1997.
13. S. R. Sreenivas. On the existence of supervisory policies that enforce liveness in discrete event dynamic systems modeled by controlled Petri nets. *IEEE Transactions on Automatic Control*, 42(7):928–945, July 1997.

Appendix: The Computation of the Active Subnets

The active subnets of special significance are the minimal, T-minimal and maximal active subnets. Note that the minimal subnets of a Petri net are the t-minimal subnets, for each transition t of the Petri net. The following algorithm computes a T-minimal subnet or, if none exists, a T_x-minimal subnet such that $T_x \subset T$ and there is no $T_y \subset T$, $T_x \subset T_y$ such that a T_y-minimal subnet exists. A T-minimal subnet does not exist iff some of the transitions of T cannot be made live under any circumstances.

Input: The Petri net $\mathcal{N}_0 = (P_0, T_0, F_0, W_0)$ and its incidence matrix D; a nonempty set of transitions $T \subseteq T_0$; an optional set Z of transitions which are not desired to be made live; by default $Z = \emptyset$.

Output: The active subnet $\mathcal{N}^A = (P^A, T^A, F^A, W^A)$.

1. Check the feasibility of $Dx \geq 0$ s.t. $x \geq 0$, $x(i) \geq 1$ $\forall t_i \in T$ and $x(i) = 0$ $\forall t_i \in Z$.
      If feasible **then** let x_0 be a solution; $T^A = minactn(T_0, x_0, D, T)$
      else $T^A = maxactn(T_0, D, T, Z)$ (no T-minimal active subnet exists, and so an approximation is constructed)
2. The active subnet is $\mathcal{N}^A = (P^A, T^A, F^A, W^A)$, $P^A = T^A \bullet$, $F^A = F_0 \cap \{(T^A \times P^A) \cup (P^A \times T^A)\}$ and W^A is the restriction of W_0 to F^A.

minactn(T_0, x_0, D, T)

>   Let $M = \|x_0\|$ and $x_s = x_0$.
>   **For** $t_i \in M \setminus T$ **do**
>   Check feasibility of $Dx \geq 0$ subject to $x \geq 0$, $x(i) = 0$, $x(j) = 0$ $\forall t_j \in T_0 \setminus M$ and $x(j) \geq 1$ $\forall t_j \in T$.
>   **If** feasible **then** let x^* be a solution; $M = \|x^*\|$ and $x_s = x^*$.
>   **Return** $\|x_s\|$

maxactn(T_0, D, T, Z)

>   Let $M = T$ and $x_s = \mathbf{0}_{|T_0| \times 1}$
>   **While** $M \neq \emptyset$ **do**
>   Check feasibility of $Dx \geq 0$ subject to $x \geq 0$, $\sum\limits_{t_i \in M} x(i) \geq 1$ and $x(i) = 0$ $\forall t_i \in Z$.
>   **If** feasible **then** let x^* be a solution; $M = M \setminus \|x^*\|$ and $x_s = x^* + x_s$.
>   **Else** $M = \emptyset$.
> $N = minactn(T_0, x_s, D, T \cap \|x_s\|)$
>   **Return** N

A Concurrent Semantics of Static Exceptions in a Parallel Programming Language

Hanna Klaudel and Franck Pommereau

LACL, Université Paris 12
61, avenue du général de Gaulle
94010 Créteil, France
{klaudel,pommereau}@univ-paris12.fr

Abstract. This paper aims at introducing a mechanism of exceptions in a parallel programming language, giving them a formal concurrent semantics in terms of preemptible and composable high-level Petri nets. We show that, combined with concurrency, exceptions can be used as a basis for other preemption related constructs. We illustrate this idea by presenting a generalized timeout and a simple UNIX-like system of concurrent preemptible threads.

Keywords: Exceptions, Petri nets, semantics, parallel programming.

1 Introduction

The starting point of our approach is $B(PN)^2$ [3,9] (*Basic Petri Net Programming Notation*) which is a high-level programming language comprising in a simple syntax most traditional concepts of parallel programming. It includes nested parallel composition, iteration, guarded commands, procedures and communications *via* both handshake and buffered communication channels, as well as shared variables. One of the most interesting aspects of $B(PN)^2$ is its simplicity: it features most classical concepts in a simple syntax. So, it becomes possible to use it as a test language and then to extend or apply the results found for $B(PN)^2$ to "real-life" languages.

$B(PN)^2$ has an original formal semantics in terms of *boxes* [1], a class of labelled Petri nets provided with a set of composition operations, and *M-nets* [2], a high-level version of boxes. M-nets are strongly related to boxes by an *unfolding* of M-nets into boxes and allow to represent in a clear and compact way large (possibly infinite) systems. $B(PN)^2$, boxes and M-nets are implemented in PEP toolkit [7], allowing to simulate a modeled system and also to verify its properties *via* model checking.

Recent works [11] led to the definition of the model of P/M-nets which extends M-nets with preemption, introducing for this purpose a new operator, π. Given a net N, $\pi(N)$ is a net which can be aborted, *i.e.*, it's termination can be forced immediately. Despite this augmented capability, it is proved in [11] that P/M-nets stay strictly equivalent to M-nets in terms of computational power (both may be transformed into 1-safe Petri nets, but P/M-nets lead to much

J.-M. Colom and M. Koutny (Eds.): ICATPN 2001, LNCS 2075, pp. 204–223, 2001.

bigger nets) and have also a concurrent semantics. Having preemption naturally leads to consider enhancing $B(PN)^2$ with related constructs. This paper proposes a modeling of static exceptions in $B(PN)^2$, giving their semantics with P/M-nets.

The presented approach allows to propagate exceptions through a nested block structure. However, the *resolution* procedure proposed here is one of the simplest possible : the actually handled exception is choosen arbitrarily between exceptions occurring concurrently. On the top of this system, a more sophisticated resolution system could be introduced, as proposed for instance in [15,16, 17].

We also show that combining exceptions with parallelism allows to express other constructs like a generalized timeout and a simple multi-threaded system.

2   M-Nets, P/M-Nets, and Their Algebras

This section is devoted to introduce *P/M-nets*, high-level composable and preemptible Petri nets [11], which are used as semantic model for exceptions. P/M-nets are an extension of a high-level net model, called M-nets, which are introduced first. P/M-nets (as well as M-nets) may be considered as an efficient abreviation for safe place/transition Petri nets, into which they may be unfolded [2,11].

2.1   Basic Definitions

Let E be a set. A *multi-set* over E is a function $\mu : E \to \mathbb{N}$, generally denoted with an extended set notation, e.g., $\{a, a, b\}$ for $\mu(a) = 2$, $\mu(b) = 1$ and $\mu(e) = 0$ for all $e \in E \setminus \{a, b\}$. A multi-set μ is finite if so is its support set $E \setminus \mu^{-1}(0)$. We denote by $\mathcal{M}(E)$ (resp. $\mathcal{M}_f(E)$) the set of multi-sets (resp. finite multi-sets) over E, by $\oplus$ and $\ominus$ the sum and difference of multi-sets. We may also use the usual set notations, for example, if μ_1 and μ_2 are two multi-sets over E, $\mu_1 \subseteq \mu_2$ stands for $\forall x \in E : \mu_1(x) \le \mu_2(x)$.

2.2   M-Nets

M-nets (introduced in [2] and developed in [4,10]) form a class of high-level Petri nets provided with a set of operations giving them a structure of process algebra.

An M-net N is a triple (S, T, ι), where S is the set of places, T is the set of transitions, $(T \times S) \cup (S \times T)$ is the set of arcs, and ι is the annotation function on places, transitions and arcs. The annotation of a place has the form $\lambda.\tau$, where λ is a *label* (entry e, exit x or internal i) and τ is a *type* (a non-empty set of values from a fixed set *Val*). As usual, for each node (place or transition) $r \in S \cup T$, we denote by $^\bullet r$ the set of nodes $\{r' \in S \cup T \mid \iota(r', r) \ne \emptyset\}$ and, similarly, $r^\bullet = \{r' \in S \cup T \mid \iota(r, r') \ne \emptyset\}$.

Transitions annotations are of the form $\lambda.\gamma$ where λ is a *label* (which can be hierarchical or for communications) and γ is a *guard* (a finite set of predicates from a set Pr). Hierarchical labels are composed out of a single hierarchical action (*e.g.*, $\mathcal{X}$) indicating a future refinement (*i.e.*, a substitution) by an M-net. A transition may perform different kind of communications when it fires:

- *synchronous* ones, similar to CCS ones [13], *e.g.*, between transitions labelled by synchronous communication actions such as $A(a_1,\ldots,a_n)$ or $\widehat{A}(a_1',\ldots,a_n')$, where A is a *synchronous communication symbol*, $\widehat{A}$ is its *conjugate* and each a_i and a_i' is a value or a variable (belonging to a fixed set *Var*);
- *asynchronous* ones, *e.g.*, between transitions labelled by asynchronous links such as $b^+(a_1)$ or $b^-(a_2)$, where b is an *asynchronous communication symbol* and each a_i is a variable or a value ranging in $type(b) \subseteq Val$. The communication is done *via* a place s_b of type $\tau(s_b) = type(b)$ which plays the rôle of a heap buffer. Link $b^+(a_1)$ means that a_1 can be sent to s_b and $b^-(a_2)$ means that a_2 can be received from s_b;
- or possibly both types at the same time.

Communication labels are then of the form $\lambda = \alpha.\beta$ where α is a finite multi-set of synchronous communication actions and β is a finite multi-set of asynchronous links.

Arcs are incribed by annotations which encode the values consumed or produced in places by a firing of an adjacent transition. If no refinement is concerned, they are simply multi-sets of values or variables; otherwise they are constructed in a systematic way from the arc annotations coming from the refined and refining nets [4,5].

2.3   Dynamic Behavior and Concurrent Semantics of M-Nets

For each transition $t \in T$ we shall denote by $var(t)$ the set of all the variables occurring in the annotations of t and in the arcs coming to and from t. A *binding* for a transition t is a substitution $\sigma : var(t) \to Val$; it will be said *enabling* if it satisfies the guard, if it respects the types of the asynchronous links, and if the flow of tokens it implies respects the types of the places adjacent to t.

A *marking* of an M-net (S, T, ι) is a mapping $M : S \to \mathcal{M}(Val)$ which associates to each place $s \in S$ a multi-set of values from $\tau(s)$. In particular, we shall distinguish the *entry marking*, denoted M_e, where, for each $s \in S$, $M_e(s) = \tau(s)$ if $\lambda(s) = e$ and the empty multi-set otherwise; the *exit marking*, M_x, is defined similarly.

The transition rule specifies the circumstances under which a marking M' is reachable from a marking M. A transition t is *enabled* at a marking M, this is denoted $M[t\rangle$, if there is an enabling binding σ of t such that $\forall s \in S : \iota(s,t)[\sigma] \subseteq M(s)$, *i.e.*, there are enough tokens of each type to satisfy the required flow. The effect of an occurrence of t is to remove from its input places all the tokens used for the enabling binding σ and to add to its output places the tokens according to σ; this leads to a marking M' such that

$$\forall s \in S : \; M'(s) = M(s) \ominus \iota(s,t)[\sigma] \oplus \iota(t,s)[\sigma].$$

The above transition rule defines the *interleaving semantics* of an M-net which consists in a set of occurrence sequences. This semantics can be generalized by introducing the *step sequence semantics* [6], which allows any number of transitions to occur simultaneously.

2.4   Algebra of M-Nets

For compositionality, we are particularly interested in a sub-class of M-nets: we assume that each M-net has at least one entry and one exit place, that each transition has at least one input and one output place (*T-restrictness* property), and that there are neither arcs going to entry places nor from exit places. Such M-nets are said ex-*good*.

The algebra of ex-good M-nets comprises the operations listed below, where N_1, N_2 and N_3 are M-nets, $\mathcal{X}$ is a hierarchical symbol, A is a synchronous communication symbol, b is an asynchronous link symbol and f is a renaming function on synchronous and asynchronous symbols.

$N_1[\mathcal{X} \leftarrow N_2]$	refinement	$N_1[f]$	renaming
$N_1 \| N_2$	parallel composition	$N_1 \,\mathbf{sy}\, A$	synchronization
$N_1 ; N_2$	sequence	$N_1 \,\mathbf{rs}\, A$	restriction
$N_1 \,\Box\, N_2$	choice	$[A : N_1]$	scoping
$[N_1 * N_2 * N_3]$	iteration	$N_1 \,\mathbf{tie}\, b$	asynchronous links

The sequential composition "$N_1 ; N_2$" means that N_1 is executed first and then N_2. The parallel composition puts nets side by side without any link between them so they can execute in total concurrency. The choice composes nets in such a way that only one of them can be executed. The iteration composes three nets such that the first one is executed once as an initialization part, then the second one is executed an arbitrary number of times as a loop part, and finally the third one is executed once as an exit part. The synchronization w.r.t. a synchronous symbol A adds to a net new transitions anticipating all possible synchronous communications on A. The restriction w.r.t. A removes from the net all unsatisfied communication capabilities on A. The scoping w.r.t. A is defined as a synchronization w.r.t. A followed by a restriction w.r.t. A, it is used to setup all synchronous communications w.r.t. A, making them local to the net and no longer aviable for the other synchronizations. The asynchronous links operation w.r.t. b, applied to a net, adds a new buffer place s_b and arcs between transitions which export or import values, through b^+ or b^-, into or from the buffer, and removes all asynchronous link capabilities w.r.t. b from the inscriptions of the transitions. The refinement of the transitions labelled $\mathcal{X}$ in a net by another net is a kind of substitution which allows the refining net to be executed each time (for every enabling binding) the hierarchical transition in the refined net could fire. Renaming allows to change the names of synchronous or asynchronous communication symbols. Detailed explanations and some examples of these operations are given in [2,5,10].

2.5   Pairwise Priorities and Priority M-Nets

Let $N = (S, T, \iota)$ be an M-net. A *pairwise priority relation* over T is a binary relation $\rho \subseteq T \times T$. Intuitively, $(t_1, t_2) \in \rho$ means that during an execution of N, the firing of transition t_2 is always preferred to the firing of t_1 when both are possible; in other words, t_1 has a lower priority than t_2. We use standard mathematical notations, in particular, for $\rho \subseteq T \times T$, we denote:

$$dom(\rho) = \{t_1 \in T \mid \exists t_2 \in T \text{ such that } (t_1, t_2) \in \rho\},$$
$$cod(\rho) = \{t_2 \in T \mid \exists t_1 \in T \text{ such that } (t_1, t_2) \in \rho\}.$$

A *priority M-net* is a pair $P = (N, \rho)$ where $N = (S, T, \iota)$ is an M-net (possibly having some non T-restricted communication transitions) and ρ is a pairwise priority relation over T. We call N the *net part* of P.

Definition 1. *Let $P = (N, \rho)$ be a priority M-net, M a marking of $N = (S, T, \iota)$ and t a transition of N such that $M[t\rangle$; then t is ρ-enabled in P at M, denoted $M[t\rangle_\rho$ iff $\nexists t' \in T$ such that $M[t'\rangle$ and $(t, t') \in \rho$.*

Notice that ρ allows to disable a transition which would have been enabled with the usual M-nets transition rule, but not the contrary. In other words, we have $M[t\rangle_\rho \Rightarrow M[t\rangle$. As for M-nets, this transition rule can be generalized in order to define the step semantics of priority M-nets [11].

An algebra of priority M-nets can also be considered. The extension of the usual M-net operations to priority M-nets is immediate for most of them. In order to make the paper self-contained, we recall here the definition from [11] which is an important for introducing the preemption operation π and preemptible M-nets.

Definition 2. *Let $P_i = (N_i, \rho_i)$, for $i \in \{1, 2, 3\}$, be priority M-nets, where $N_i = (S_i, T_i, \iota_i)$, and let $\mathcal{X}$ be a hierarchical symbol, A a synchronous communication symbol, b an asynchronous link symbol, and f a renaming function on communication symbols. The usual M-net operations are extended as follows for priority M-nets:*

- $P_1[\mathcal{X} \leftarrow P_2] = (N_1[\mathcal{X} \leftarrow N_2], \rho)$ *where*
  $\rho = \{(t, t') \in \rho_1 \mid \lambda_1(t) \neq \mathcal{X} \neq \lambda_1(t')\}$
  $\qquad \uplus \{(t_\mathcal{X}.t, t_\mathcal{X}.t') \mid (t, t') \in \rho_2 \wedge t_\mathcal{X} \in T_1 \wedge \lambda_1(t_\mathcal{X}) = \mathcal{X}\}$
  $\qquad \uplus \{(t_\mathcal{X}.t, t') \mid t \notin cod(\rho_2) \wedge (t_\mathcal{X}, t') \in \rho_1 \wedge t_\mathcal{X} \in T_1 \wedge \lambda_1(t_\mathcal{X}) = \mathcal{X}\};$
- $P_1 \operatorname{\mathbf{tie}} b = (N_1 \operatorname{\mathbf{tie}} b, \rho_1);$
- $P_1[f] = (N_1[f], \rho_1);$
- $P_1 \operatorname{\mathbf{sy}} A = (N_1 \operatorname{\mathbf{sy}} A, \rho)$ *where $N_1 \operatorname{\mathbf{sy}} A = (S, T, \iota)$ and ρ is the smallest set including ρ_1 such that if $t' \in T$ results from a basic synchronization of t_1 with t_2, and*
  - *if $\exists t''$ such that $(t_1, t'') \in \rho$ or $(t_2, t'') \in \rho$, then $(t', t'') \in \rho$,*
  - *if $\exists t''$ such that $(t'', t_1) \in \rho$ or $(t'', t_2) \in \rho$, then $(t'', t') \in \rho$.*
- $P_1 \operatorname{\mathbf{rs}} A = (N_1 \operatorname{\mathbf{rs}} A, \rho)$, *where $N_1 \operatorname{\mathbf{rs}} A = (S, T, \iota)$ and $\rho = \rho_1 \cap (T \times T)$.*

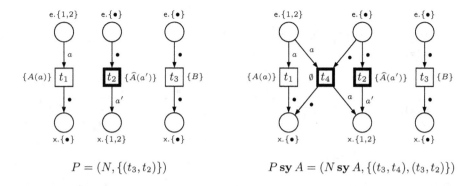

$$P = (N, \{(t_3, t_2)\}) \qquad\qquad P \operatorname{\mathbf{sy}} A = (N \operatorname{\mathbf{sy}} A, \{(t_3, t_4), (t_3, t_2)\})$$

Fig. 1. Example of synchronization of a priority M-net. (Only synchronous labels are represented.) Restricting on A would remove from the net transitions t_1 and t_2 (with their surrounding arcs) and pair (t_3, t_2) from the priority relation.

Control flow operators, as sequential composition, iteration, parallel composition and choice, are based on refinement and so they are defined canonically. Scoping, as for M-nets, is defined as a synchronization followed by a restriction: $[A : P] = (P \operatorname{\mathbf{sy}} A) \operatorname{\mathbf{rs}} A$.

Figure 1 shows an example of synchronization in a priority M-net. Transitions t_1 and t_2 are synchronized, leading to a new transition t_4 which "inherits" from t_2 its priority over t_3. In figures, transitions with thick borders are those belonging to $cod(\rho)$ and thus have the priority over some other ones. This notation is used in all the sequel.

The definition of operation π is very important in our context because it allows to make abortable an arbitrary priority M-net $P = (N, \rho)$. The definition of π presented here is slightly different from the original one from [11], but the modification only concerns some labels involved in the semantics of exceptions. We use for it the priority M-net $P_\pi = (N_\pi, \rho_\pi)$ where N_π is represented in figure 2, the priority relation being

$$\rho_\pi = \{(t_8, t_2), (t_8, t_5), (t_7, t_2), (t_7, t_5), (t_\chi, t_5), (t_3, t_4)\}.$$

In order to produce $\pi(P)$, P is embedded in P_π by refining transition t_χ and the resulting net is synchronized w.r.t. *throw*. This way, if P does not throw any exception, it completes and so does $\pi(P)$ by firing transition t_0 (and no other transition in N_π can fire). However, in the case where P throws an exception, by firing a transition labelled with *throw*, transition t_1 fires too and enables the abortion of $\pi(P)$. This abortion is performed by consuming tokens in P through the loop on t_2 which is synchronized with the "emptying transitions", $t_s \in T_s$, added to P when π is applied. Transitions t_3 to t_7 are used to transmit abortion to all π's nested in P. (More detailed explanations of this mechanism can be found in [11].) Abortion is not elementary but, thanks to priorities, it is atomic in the sense defined in [12]: when started, abortion cannot be interrupted. When

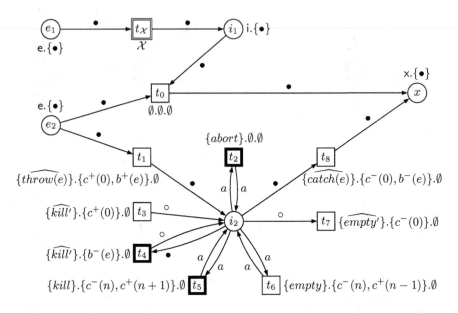

Fig. 2. N_π, net part of P_π where $type(b) = Val$, $type(c) = \mathbb{N}$ and $\iota(i_2) = \mathsf{i}.\{\bullet, \circ\}$.

abortion is terminated, transition t_8 fires, triggering the handler for the raised exception.

Definition 3. *Let P be a priority M-net. Then,*

$$\pi(P) = \left[\left[\{abort, throw, kill, empty\} : \ Ab\big(P_\pi[\mathcal{X} \leftarrow P]\big) \ \mathbf{tie}\{c, b\} \right] \right]$$
$$[\widehat{kill'} \mapsto \widehat{kill}, \widehat{empty'} \mapsto \widehat{empty}]$$

where P_π is the priority M-net defined above and Ab is an auxiliary operation which includes the additional emptying transitions; if $P_\pi[\mathcal{X} \leftarrow P] = P' = ((S', T', \iota'), \rho')$, then $Ab(P') = ((S'', T'', \iota''), \rho'')$ with:

- $S'' = S'$, and $\forall s \in S'' : \iota''(s) = \iota'(s)$;
- $T'' = T' \uplus T_s$ where $T_s = \{t_s \mid s \in S' \setminus \{x\} \wedge s^\bullet \cap cod(\rho') = \emptyset\}$

  and $\forall t \in T'' : \iota''(t) = \begin{cases} \iota'(t) & \text{if } t \in T', \\ \{\widehat{abort}\}.\emptyset.\emptyset & \text{if } t \in T_s; \end{cases}$

- $\forall (t, s) \in T'' \times S'' : \iota''(t, s) = \begin{cases} \iota'(t, s) & \text{if } t \in T', \\ \emptyset & \text{if } t \in T_s; \end{cases}$

- $\forall (s, t) \in S'' \times T'' : \iota''(s, t) = \begin{cases} \iota'(s, t) & \text{if } t \in T', \\ \{a\} \subset Var & \text{if } t = t_s \in T_s, \\ \emptyset & \text{if } t \in T_s \setminus \{t_s\}; \end{cases}$

- $\rho'' = \rho' \uplus \{(t, t_s) \mid t_s \in T_s \wedge t \in (^\bullet t_s)^\bullet\}$.

This mechanism is directly applied to the semantics of exceptions in a programming language. A net semantics P of (a part of) a concurrent program usually contains some $throw(e)$-labelled transitions. It means that if such a transition fires, P should not continue its normal behavior but should start an exceptional one. Operation π embeds P in such a way that the firing of such a transition in $\pi(P)$ is taken into account and brings about abortion of all the part corresponding to P in $\pi(P)$. This abortion is atomic (even if composed of several events) and when it is finished, a $catch(e)$-labelled transition can fire in $\pi(P)$. Also, $catch(e)$ is the only action related to exceptions and visible outside $\pi(P)$. It is used to trigger the handler for the thrown exception.

2.6   Preemptible M-Nets: P/M-Nets

Preemptible M-nets (*P/M-nets* for short) are defined as a sub-class of priority M-nets with some structural constraints. These constraints allow P/M-nets to have interesting properties, such as to be transformable into safe Petri nets. This sub-class is reasonably wide (it includes ex-good M-nets) and sound with respect to the semantics of preemption [11].

Definition 4. *Let* $P = (N, \rho)$ *be a priority M-net.* P *is a P/M-net iff either:*

- N *is an ex-good M-net and* $\rho = \emptyset$, *or*
- P *is defined as* $\pi(P_1)$, $P_1[\mathcal{X} \leftarrow P_2]$, $P_1 \| P_2$, $P_1; P_2$, $P_1 \,\square\, P_2$, $[P_1 * P_2 * P_3]$, $P_1 \,\mathbf{sy}\, A$, $P_1 \,\mathbf{rs}\, A$, $[A : P_1]$, $P_1 \,\mathbf{tie}\, b$ *or* $P_1[f]$, *where* P_i, *for* $i \in \{1, 2, 3\}$, *are P/M-nets,* $\mathcal{X}$ *is a hierarchical symbol,* A *is a synchronous communication symbol,* b *is an asynchronous links symbol, and* f *is a renaming function on communication symbols.*

In the following, we often use some basic P/M-nets as that shown in figure 3. Their net parts are denoted by the label of their unique transition (*i.e.*, $\alpha.\beta.\gamma$ or $\mathcal{X}.\gamma$).

Fig. 3. A basic M-net used in this paper. Transition t may be either a communication transition in which case $\iota(t) = \alpha.\beta.\gamma$ or a hierarchical one with $\iota(t) = \mathcal{X}.\gamma$.

3   Syntax and Semantics of B(PN)^2

B(PN)^2 is a parallel programming language comprising shared memory parallelism, channel (FIFO buffer) communication with arbitrary capacities, and allowing the nesting of parallel operators, blocks and procedures.

The following is a fragment of the syntax of B(PN)^2 (with keywords typeset in bold face, non-terminal in roman face and italic denoting values supplied by the program):

```
program ::= program block
block    ::= begin scope end
scope    ::= com | decl ; scope
com      ::= ⟨expr⟩ | proc-call
             | com ‖ com | com ; com | do alt-set od
             | block | (com)
decl     ::= var name: set
             | var name: chan k of set
             | procedure name(formal-parlist) block
             | decl, decl
proc-call ::= name(effective-parlist)
```

An atomic action is a $B(PN)^2$ expression "⟨expr⟩", *i.e.*, a term constructed over operators, constants (here again, *Val* is the set of all possible values) and identifiers of *program variables* and *channels*. A program variable v can appear in an expression as $'v$ (pre-value) or v' (post-value), denoting respectively its value just before and just after the evaluation of the expression during an execution of the program. A channel variable c can appear in an expression as $c!$ (sending) or $c?$ (receiving), denoting respectively the value sent or received in a communication on the channel c. An atomic action can execute if the expression evaluates to true. Thus, for example, $⟨'v > 0 \wedge v' = c?⟩$ corresponds to a guarded communication which requires v to be greater than zero and a communication to be available on channel c, in which case the value communicated on c is assigned to variable v.

A command "com" is either an atomic action, a procedure call ("proc-call"), one of a number of command compositions operator or a block comprising some declarations for a command. Parentheses allow to combine the various command compositions arbitrarily.

The domain of relevance of a variable, channel or procedure identifier is limited to the part of a $B(PN)^2$ program, called "scope", which follows its declaration. As usual, a declaration, in a new block, with an already used identifier results in the masking of the existing identifier by the new one. A procedure can be declared with or without parameters (in which case its "formal-parlist" is empty); each parameter can be passed by *value*, by *result* or by *reference*. A declaration of a program variable or a channel is made with the keyword "**var**" followed by an identifier with a type specification which can be "*set*", or "**chan** k **of** *set*" where *set* is a set of values. For a type "*set*", the identifier describes an ordinary program variable which may carry values within *set*. Clause "**chan** k **of** *set*" declares a channel of capacity k (which can be 0 for handshake communications, 1 or more for bounded capacities, or ∞ for an unbounded capacity) that may store values within *set*.

Besides traditional control flow constructs, sequence and parallel composition, there is a command "**do...od**" which allows to express all types of loops and conditional statements. The core of statement "**do...od**" is a set of clauses of two types: repeat commands, "com; **repeat**", and exit commands, "com; **exit**". During an execution, there can be zero or more iterations, each of them being an execution of one of the repeat commands. The loop is terminated by an execution of one of the exit commands. Each repeat and exit command is typically a sequence with an initial atomic action, the executability of which determining whether that repeat or exit command can start. If several are possible, there is a non-deterministic choice between them.

3.1 P/M-Net Based Semantics of B(PN)^2

The definition of the M-net semantics of B(PN)^2 programs (having no preemptible constructs) is given in [3] through a semantical function Mnet. A P/M-net semantics of such programs is easy to obtain through the canonical transformation from M-nets to P/M-nets (as defined in [11]).

In this paper, we introduce new constructs in B(PN)^2 in order to provide it with exceptions. The associated semantics is given through a semantical function PM which extends the canonical semantics obtained from Mnet: function PM maps directly the new B(PN)^2 constructs or overrides the semantics of some existing ones, in particular, of blocks, which may include now the treatment of exceptions.

The semantics of a program is defined *via* the semantics of its constituting parts. The main idea in describing a block is (i) to juxtapose the nets for its local resources declarations with the net for its command followed by a termination net for the declared variables, (ii) to synchronize all matching data/command transitions and (iii) to restrict these transitions in order to make local variables invisible outside of the block.

The access to a program variable v is represented by synchronous action $V(v^i, v^o)$ which describes the change of value of v from its current value v^i (i for *input*), to the new value v^o (*output*).

Each declared variable is described by some *data P/M-net* of the corresponding type, e.g., $N_{Var}(v, set)$ for a variable v of type *set* or $N_{Chan,k}(c, set)$ for a variable c being a channel of capacity k which may carry values of type *set*. The current value of the variable v is stored in a place and may be changed through a $\{\widehat{V}(v^i, v^o)\}$-labelled transition in the data net, while $\{\widehat{C!}(c^!)\}$- and $\{\widehat{C?}(c^?)\}$-labelled transitions are used for sending or receiving values to or from channel c.

Sequential and parallel compositions are directly translated into the corresponding net operations, e.g., $\mathsf{PM}(com_1; com_2) = \mathsf{PM}(com_1); \mathsf{PM}(com_2)$. The semantics of the "**do ... od**" construct involves the P/M-net iteration operator.

The semantics of an atomic action "$\langle$expr$\rangle$" is $(\alpha.\emptyset.\gamma, \emptyset)$ where α is a set of synchronous communication actions corresponding to program variables involved in "expr", and γ is the guard obtained from "expr" with program variables appropriately replaced by net variables, e.g., v^i for $'v$ and v^o for v'. For instance, we have:

$$\mathsf{PM}\Big(\langle 'v > 0 \wedge v' = c? \rangle\Big) = \Big(\{V(v^i, v^o), C?(c^?)\}.\emptyset.\{v^i > 0 \wedge v^o = c^?\}\,,\ \emptyset\Big).$$

The P/M-net above has one transition as shown in figure 3. Its synchronous label performs a communication with the resource net for variable v and for channel

c: it reads v^i and writes v^o with action $V(v^i, v^o)$, and it gets $c^?$ on the channel with action $C?(c^?)$. The guard ensures that $v^i > 0$ and that v^o is set to the value got on the channel.

4   Modeling Exceptions

In order to model exceptions we introduce in the syntax of B(PN)^2 a new command, **throw**, which takes one argument which may be either a constant in *Val* in which case it is denoted by w, or a program variable in which case it is denoted by v. It actually represents *the* exception to throw. Moreover, we change the syntax for the blocks as follows:

$$
\begin{aligned}
\text{block} \quad &::= \textbf{begin} \text{ scope } \textbf{end} \\
&\mid \textbf{begin} \text{ scope catch-list } \textbf{end} \\
\text{catch-list} \quad &::= \text{catch-clause} \\
&\mid [\text{catch-clause } \textbf{or}] \textbf{ catch-others } [v] \ [\textbf{then} \text{ com}] \\
\text{catch-clause} &::= \textbf{catch } w \ [\textbf{then} \text{ com}] \\
&\mid \text{catch-clause } \textbf{or} \text{ catch-clause}
\end{aligned}
$$

Each catch-clause specifies how to react to an exception w (a value in *Val*). The optional clause **catch-others** can be used to catch any exception uncaught by a previous catch-clause; in this case, it is possible to save the caught exception in a variable v whose type must be *Val*.

The semantics for a block "**begin** *scope* cc_1 **or** ... **or** cc_k **end**" where *scope* is the scope for the block and the cc_i's are the catch-clauses (cc_i handles exception w_i, cc_k may be a clause **catch-others**) is the following:

$$
\begin{aligned}
\text{PM}\big(\textbf{begin} \text{ scope } cc_1 \text{ } \textbf{or} \text{ } \ldots \text{ } \textbf{or} \text{ } cc_k \text{ } \textbf{end}\big) = \\
\Big[\!\!\Big[\{catch, noexcept\} : \quad \pi\Big(\text{PM}(scope) \, ; \, (\{noexcept\}.\emptyset.\emptyset \, , \, \emptyset)\Big) \\
\big\| \Big(\text{PM}(cc_1) \,\square\, \cdots \,\square\, \text{PM}(cc_k) \,\square\, P_{transmit} \,\square\, (\{\widehat{noexcept}\}.\emptyset.\emptyset \, , \, \emptyset)\Big)\Big]\!\!\Big]
\end{aligned}
$$

where $P_{transmit}$ and *noexcept* are explained below.

If the block finishes without throwing any exception, action $\{noexcept\}.\emptyset.\emptyset$ is reached in $\pi\big(\text{PM}(scope) \, ; \, (\{noexcept\}.\emptyset.\emptyset, \emptyset)\big)$ and the block can exit by firing the transition which results from the synchronization w.r.t. *noexcept*. If an exception e (e is a net variable) is thrown in the block, it is either caught by one of the catch-clauses cc_i in the block and then a corresponding $\text{PM}(cc_i)$ is executed, or there is no *specific* catch-clause for it and there are still two cases:

– a catch-clause has been specified explicitly in cc_k using **catch-others**, in which case the corresponding $\text{PM}(cc_k)$ is executed;
– there is no **catch-others** specified in the block and so, the uncaught exception e is simply re-thrown by P/M-net $P_{transmit}$.

P/M-net $P_{transmit}$ is defined has follows:

$$
P_{transmit} = \begin{cases}
(N_{stop}, \emptyset) \\
\quad \text{if } cc_k \text{ is a clause } \textbf{catch-others}; \\
(\{catch(e), throw(e)\}.\{e \neq w_1 \wedge \cdots \wedge e \neq w_k\}.\emptyset \, , \, \emptyset) \\
\quad \text{otherwise.}
\end{cases}
$$

where N_{stop} is shown in figure 4 and $w_1, \ldots, w_k$ are the exceptions caught in the catch-clauses of the block.

$$\text{e.}\{\bullet\} \bigcirc \qquad \bigcirc \text{x.}\{\bullet\}$$

Fig. 4. The net N_{stop} used in the semantics of blocks.

The new constructs added to the syntax have the following semantics:

$\mathsf{PM}(\textbf{throw}(w)) = (\{throw(w)\}.\emptyset.\emptyset \, , \, \emptyset)$

$\mathsf{PM}(\textbf{throw}(v)) = (\{V(v^i, v^o), throw(e)\}.\emptyset.\{e = v^i \wedge v^o = v^i\} \, , \, \emptyset)$
  where e is a net variable

$\mathsf{PM}(\textbf{catch } w) = (\{catch(w)\}.\emptyset.\emptyset \, , \, \emptyset)$

$\mathsf{PM}(\textbf{catch } w \textbf{ then } com) = \mathsf{PM}(\textbf{catch } w) \, ; \, \mathsf{PM}(com)$

$\mathsf{PM}(\textbf{catch-others}) = (\{catch(e)\}.\emptyset.\{e \neq w_1 \wedge \cdots \wedge e \neq w_{k-1}\} \, , \, \emptyset)$
  where $w_1, \ldots, w_{k-1}$ are the exceptions caught in the previous catch-clauses

$\mathsf{PM}(\textbf{catch-others } v) = (\{catch(e), V(v^i, v^o)\}.\emptyset$
$\qquad\qquad\qquad .\{e \neq w_1 \wedge \cdots \wedge e \neq w_{k-1} \wedge v^o = e\} \, , \, \emptyset)$

$\mathsf{PM}(\textbf{catch-others then } com) = \mathsf{PM}(\textbf{catch-others}) \, ; \, \mathsf{PM}(com)$

$\mathsf{PM}(\textbf{catch-others } v \textbf{ then } com) = \mathsf{PM}(\textbf{catch-others } v) \, ; \, \mathsf{PM}(com)$

The propagation of exceptions is ensured by alternating scoping w.r.t. *throw* and *catch*, as shown in figure 5. First a *throw*(e) is "emitted" somewhere in *scope*. Operation π aborts the scope and "converts" the *throw*(e) into a $\widehat{catch}(e)$ which synchronizes with an appropriate *catch*(w_i), then the associated com_i is executed.

In figure 5, we assume that cc_k is a clause **catch-others** which re-throws the exception outside the block. Otherwise, the semantics of the blocks would have ensured this behavior. For $1 \leq i < k$, we assume that cc_i is a clause "**catch** w_i **then** com_i".

Notice that several exceptions may be thrown concurrently (from different concurrent parts of the block); in such a case the choice operation in the semantics of the blocks ensures that only one of them may be caught and the others are ignored (this choice is non deterministic).

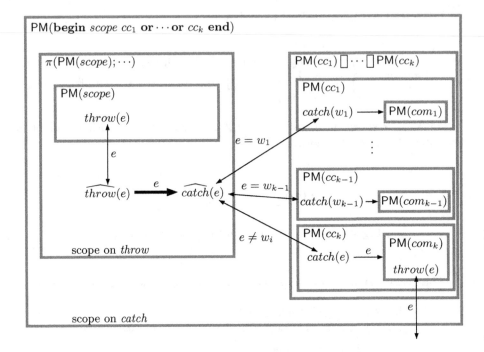

Fig. 5. The semantics of a block with exceptions. A simple arrow denotes a causal dependence between two actions, a double arrow links two synchronized actions. The thick arrow denotes that abortion is performed between the occurrence of the two linked actions.

4.1 Preprocessing

On the top of the semantics given above, we build a preprocessor which rewrites programs, before PM is applied, in order to enforce a more intuitive behavior. Until now:

1. The variables declared in a block are not visible from the commands given in the catch-clauses of this block. The reason is that the declarations for the block are made in a scope nested in π and so they are local to it.
2. Exceptions at the top level of the program are not handled.

The first rewriting rule fixes the first point. It applies when a block comprises some declaration followed by a command and some catch-clauses (all three at the same time):

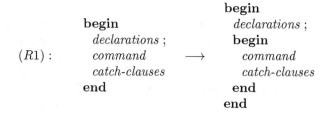

Then, we use a simple rule for the second point:

$$(R2): \quad \textbf{program } name\ block \quad \longrightarrow \quad \begin{array}{l} \textbf{program } name \\ \textbf{begin} \\ \quad block \\ \textbf{catch-others} \\ \textbf{end} \end{array}$$

In order to avoid a recursive application of this rule, the preprocessor has to jump to *block* after the first application of (*R2*).

Notice that this rules changes the behavior of the program since it now silently discards an unhandled exception. This behavior may be undesirable and one may prefer a more sophisticated rule which would warn about the problem. Our purpose is just to show that embedding the whole program into a generic environment is an easy solution of this problem.

4.2   Semantics of Procedures in the Context of Static Exceptions

It is well known that static exceptions may lead to an unexpected behavior of procedures when they raise an exception. Consider for example the following block and its sub-block (where w is a given value):

```
begin
  procedure P() begin throw(w) end ;
  begin
    P()
  catch w
  end
end
```

One could expect the clause "**catch** w" to catch the exception thrown by procedure call. But it is not the case with static exceptions: a **throw** occurs "physically" where it was declared, and so, not inside the sub-block.

In order to have the intuitively expected behavior, we extend the preprocessor in such a way that it encapsulates procedure declarations and procedure calls into some additional $B(PN)^2$ constructs. The usual way to solve this is to consider an exception coming out of a procedure call as a hidden return value. If this value is set when the procedure returns, then this value is re-thrown at the call point. This way, the thrown exception continues to be propagated from the point where the procedure was called and not from where it was declared. To do this, the preprocessor adds two additional parameters to all procedure call and declaration, one is used to know if an exception was thrown during the procedure call and the other carries the value of the exception when needed.

A call to a procedure P is encapsulated into a block which declares two additional variables, ex and v, which are assumed not to be already used as parameters for P nor as variables already visible from P (in such a case, we just need to choose other names). Variable ex is set to $\bot$ if no exception is thrown in P, otherwise it is set to $\top$ and, in this case, v stores the value of the thrown exception. So, for procedure calls, we have:

$$(R3): \quad P(\textit{effective-parlist}) \quad \longrightarrow \quad \begin{array}{l} \textbf{begin} \\ \quad \textbf{var } ex : \{\top, \bot\} \, , \ \textbf{var } v : \textit{Val} \, ; \\ \quad P(\textit{effective-parlist}, ex, v) \, ; \\ \quad \textbf{do} \\ \qquad \langle 'ex = \top \rangle \, ; \ \textbf{throw } v \, ; \ \textbf{exit} \\ \qquad \langle 'ex = \bot \rangle \, ; \ \textbf{exit} \\ \quad \textbf{od} \\ \textbf{end} \end{array}$$

where ex and v are fresh identifiers and $\textit{effective-parlist}$ is the list of effective parameters for the procedure call.

For a procedure declaration, we have:

$$(R4): \quad \begin{array}{l} \textbf{procedure } P(\textit{formal-parlist}) \\ \textit{block} \end{array} \quad \longrightarrow \quad \begin{array}{l} \textbf{procedure } P(\textit{formal-parlist}, \\ \qquad\qquad\qquad \textbf{ref } ex, \textbf{ref } v) \\ \textbf{begin} \\ \quad \textit{block} \, ; \ \langle ex' = \bot \rangle \\ \textbf{catch-others } v \\ \qquad \textbf{then } \langle ex' = \top \rangle \\ \textbf{end} \end{array}$$

where ex and v are fresh identifiers not already used in $\textit{formal-parlist}$ (the list of formal parameters). These two new parameters are passed by reference.

Since these rules could be applied recursively, the preprocessor uses the following additional directives: for $(R3)$, it jumps directly after the text produced since it does not match any other rule; for $(R4)$, the preprocessor just has to jump to $\textit{block}$ since no other rule matches the rest.

5   Applications

Combined with concurrency, exceptions allow to express some other preemption related constructs. As an illustration, we give in this section two applications of the exceptions introduced in the paper. First, we use the exceptions in order to introduce in the language a generalized timeout. Second, we show how to model systems composed of concurrently running blocks, called $\textit{threads}$, which can be killed from other parts of the program, in particular from the other threads.

5.1   Generalized Timeout

A timeout is usually expressed through a construct such as:

> **run** com_1 **then** com_1'
> **before** com_2 **then** com_2'

which intuitively means "start concurrently commands com_1 and com_2, if com_1 finishes before com_2, then run com_1' else run com_2'". Usually, com_2 just waits for a timeout event. This may be expressed using exceptions: the command, which finishes first throws an exception which is caught in order to run either com_1' or com_2'. So, the syntax given above may be rewritten as the following $B(PN)^2$ block:

> **begin**
> (com_1 ; **throw**(1)) $\|$ (com_2 ; **throw**(2))
> **catch** 1 **then** com_1'
> **or catch** 2 **then** com_2'
> **end**

This construct can be easily generalized to an arbitrary number of commands running concurrently, each one trying to finish first. The "winner" kills the others and is the only one allowed to execute its clause **then**. It would also be useful to allow one of the clauses to be a *timeout*. This may be made easily using, for instance, the *causal time model* introduced in [10]. Thus, the syntax would become:

> **run** com_1 **then** com_1'
> **and** com_2 **then** com_2'
>
> $\vdots$
>
> **and** com_n **then** com_n'
> [**timeout** d **then** com_0']
> **end run**

where d is the number of clock ticks to be counted before timeout occurs. This generalizes the **run/before** construct given above and its semantics is easy to obtain: all com_i and, optionally, a chronometer for at most d clock ticks run concurrently; the first which finishes stops the chronometer and throws an exception which is caught in order to execute the corresponding com_j'.

5.2   Simple Threads

As they are defined above, exceptions model what we could call "internal abortion": an exception is propagated through the nesting of blocks, from internal to external ones. In the following, we show that exception can be used in order to model "external abortion" where a block can be aborted by another (non nested) one. For this purpose, we model a simple multi-threaded system in which *processes* (or *threads*), identified by *process identifiers* (*pid* for short), are able to be killed from any part of the system. The execution of a command "$kill(s, p)$" somewhere in the program has the effect to send signal s to the thread identified by p. When it receives a signal, a thread is allowed to run a command and then it finishes. This behavior is a simplification of what happens in UNIX-like systems.

We use the following syntax for threads:

> **thread**
>   *declarations for the thread* ;
>   *command for the thread*
> **signal** sig_1 **then** com_1
>
> $\vdots$
>
> **signal** sig_n **then** com_n
> **end thread**

Constants sig_1 to sig_n are the signals captured by the thread; we assume that there exists a reserved constant $SIGKILL$ which cannot be used in a clause **signal** (in UNIX, it is the name of a signal which cannot be captured). This restriction can be checked syntactically and will be useful in the following.

The programmer is also provided with a new command "$\mathbf{kill}(s, p)$" which may be used to send a signal s (any constant in Val) to a thread identified by p. One could prefer to restrict signals to a predefined set but this is not necessary for our purpose. The semantics for this new command is simply

$$\mathsf{PM}\big(\mathbf{kill}(s, p)\big) = \big(\{kill(s, p)\}.\emptyset.\emptyset \, , \, \emptyset\big).$$

Inside each thread, a variable called *pid* is implicitly declared, it contains the pid allocated for the thread. This variable *must not be changed* by the program (this is easy to check syntactically).

In order to attribute pids and to transmit signals, we use a *pid server*, which is a kind of data P/M-net, its priority relation is empty and its net part is defined by the following expression:

$$\left[\begin{array}{l} \emptyset.\{b^+((p_1, \bot)), \dots, b^+((p_k, \bot))\}.\emptyset \\[4pt] \quad * \quad \{\widehat{kill(s, p)}, \widehat{transmit(s, p)}\}.\emptyset.\emptyset \\[4pt] \qquad \square \{\widehat{allocpid(p)}\}.\{b^-((p, \bot)), b^+((p, \top))\}.\emptyset \\[4pt] \qquad \square \{\widehat{freepid(p)}\}.\{b^-((p, \top)), b^+((p, \bot))\}.\emptyset \\[4pt] \quad * \{\widehat{PS_t}\}.\{b^-((p_1, x_1)), \dots, b^-((p_k, x_k))\}.\emptyset \end{array} \right] \mathbf{tie}\, b$$

This iteration is composed of three parts (separated by stars):

- $\emptyset.\{b^+((p_1, \bot)), \dots, b^+((p_k, \bot))\}.\emptyset$ is the initialization which sets up the server by filling the heap buffer represented by the asynchronous links on b. It is filled with pairs $(p_i, \bot)$, for $1 \leq i \leq k$, where the p_i's are the pids and $\bot$ mark them free;
- conversely, the termination, $\{\widehat{PS_t}\}.\{b^-((p_1, x_1)), \dots, b^-((p_k, x_k))\}.\emptyset$, clears the buffer; it can be triggered from the outside with a synchronization on action $\widehat{PS_t}$;

- the repeated part is the most complicated. It is a choice between three actions, this choice being proposed repeatedly as long as the iteration is not terminated. It offers the following services:
  - allocation of a pid when a thread starts: when the transition labelled by $\{\overwidehat{allocpid}(p)\}.\{b^-((p, \bot)), b^+((p, \top))\}.\emptyset$ fires, a token $(p, \bot)$ is chosen through asynchronous links on b and marked used (with $\top$ on its second component);
  - symmetrically, when a thread terminates, it frees its pid by synchronizing with $\{\overwidehat{freepid}(p)\}.\{b^-((p, \top)), b^+((p, \bot))\}.\emptyset$;
  - part $\{\widehat{kill}(s, p), \overwidehat{transmit}(s,p)\}.\emptyset.\emptyset$ is used to transmit a signal to a thread identified by p. It just "converts" a $\widehat{kill}$ into a $\overwidehat{transmit}$.

The iteration is under the scope of a **tie** b which sets up the asynchronous links.

Notice that because of the choice in the loop part of the iteration, only one thread action (starting, terminating or killing a thread) can be executed at one time, allowing in this way to avoid, for instance, mutual killings. However, a server with more concurrent behavior may be designed.

Provided this server, we define three internal commands (*i.e.*, not available for the programmer) with the following semantics:

- "**alloc-pid**(v)" asks the server to allocate a pid which is written in variable v, so we have
  PM(**alloc-pid**(v)) = $\left(\{allocpid(p), V(v^i, v^o)\}.\emptyset.\{v^o = p\} , \emptyset \right)$
- "**free-pid**(v)" frees an allocated pid, reading it in v and so
  PM(**free-pid**(v)) = $\left(\{freepid(p), V(v^i, v^o)\}.\emptyset.\{p = v^i\} , \emptyset \right)$
- "**capture-signals**" receives a signal relayed by the server and converts it into an exception:
  PM(**capture-signals**) =
  $$\Big(\{transmit(s, p), throw(s), PID(pid^i, pid^o)\}.\emptyset.\{p = pid^i \wedge pid^o = pid^i\}$$
  $$\square \, \{transmit(s, p), throw(SIGKILL), PID(pid^i, pid^o)\}.\emptyset$$
  $$.\{s \neq sig_1 \wedge \cdots \wedge s \neq sig_n \wedge p = pid^i \wedge pid^o = pid^i\} , \emptyset \Big)$$
  where $sig_1, \ldots, sig_n$ are the signal already captured by the thread.

Then, the semantics for the threads given above is a nested block structure as follows:

```
begin
    var pid : {p₁, ..., pₖ} ,
    var ex : Val ,
    declarations for the thread ;
    alloc-pid(pid) ;
    begin
    ( command for the thread ; throw(SIGKILL) ) || capture-signals
    catch sig₁ then com₁
        ⋮
    or catch sigₙ then comₙ
```

>   **or catch** *SIGKILL*
>   **end** ;
>   **free-pid**(pid) ;
> **catch-others** ex **then free-pid**(pid) ; **throw**(ex)
> **end**

First, we declare fresh variables pid (whose name is reserved for threads and must not be declared by the programmer) and ex (used to re-throw an exception thrown by the thread). Then we make the declarations for the thread in such a way that they are visible from clauses **signal**. The first instruction initializes pid with a call to the pid server. Then the commands (i) which forms the body of the thread and (ii) **capture-signals**, areput in parallel. Command **capture-signals** waits for any signal coming from outside. If this happens, the signal is converted into an exception which is caught accordingly to what is specified in the thread. If a signal which is not handled by the thread comes, it is converted into a *SIGKILL*. If the command for the body of the thread terminates, command **throw**(*SIGKILL*) is used to abort command **capture-signals** and to terminate the block. When the internal block is finished, the pid for the thread is freed and, if the termination comes from an unexpected exception, the first command **free-pid** is by-passed. A **catch-others** allows to free the pid in this case and to re-throw the unexpected exception so it is propagated to the block which declared the thread.

Finally, the semantics for the program just put the pid server in parallel to the most external block (as for a global variable declaration), with the scoping on actions *transmit*, *kill*, *allocpid*, *freepid* and PS_t.

6 Conclusion

Concurrent exceptions has been addressed in literature, for instance in the context of *Coordinated Atomic Actions* [14] or *Place Charts Nets* [8]. In this paper, we introduced static exceptions in a parallel programming language, $B(PN)^2$, which is provided with a concurrent semantics based on Petri nets and for which implemented tools can be used [7].

It turned out that combining these exceptions with concurrency allowed to express other preemption related constructs like a generalized timeout and a simple multi-threading system.

Future works may emphasize the links with real-time, for instance by introducing *causal time*, already defined in [10] for M-nets, at the level of $B(PN)^2$. This would allow one to express timed systems using statements like delays and deadlines, and thus would turn $B(PN)^2$ into a full featured real-time language. Another interesting work would be to apply this kind of semantics to other languages. We believe that, in the present state of the development, these ideas could be used to give a semantics for a reasonably rich (even if not fully general) part of the Ada programming language.

Acknowledgements. We are very grateful to Tristan Crolard: our discussions were fruitful and helped us to clarify our mind about exceptions. We also thank the anonymous referees who pointed out some mistakes and missing references.

References

[1] E. Best, R. Devillers, and J. G. Hall. The box calculus: a new causal algebra with multi-label communication. *LNCS 609:21–69*, 1992.

[2] E. Best, W. Fraczak, R. Hopkins, H. Klaudel, and E. Pelz. M-nets: An algebra of high-level Petri nets, with an application to the semantics of concurrent programming languages. *Acta Informatica*, 35, 1998.

[3] E. Best and R. P. Hopkins. B(PN)^2 — A basic Petri net programming notation. *PARLE'93, LNCS 694:379–390*, 1993.

[4] R. Devillers, H. Klaudel, and R.-C Riemann. General refinement for high-level Petri nets. FST&TCS'97, *LNCS 1346:297–311*, 1997.

[5] R. Devillers, H. Klaudel, and R.-C. Riemann. General parameterised refinement and recursion for the M-net calculus. *Theoretical Computer Science*, to appear (available at http://www.univ-paris12.fr/klaudel/tcs00.ps.gz).

[6] H. J. Genrich, K. Lautenbach, and P. S. Thiagarajan. Elements of General Net Theory. *Net Theory and Applications*, Proceedings of the Advanced Course on General Net Theory of Processes and Systems, *LNCS 84:21–163*, 1980.

[7] B. Grahlmann and E. Best. PEP — more than a Petri net tool. *LNCS 1055*, 1996.

[8] M. Kishinevsky, J. Cortadella, A. Kondratyev, L. Lavagno, A. Taubin and A. Yakovlev. Coupling asynchrony and interrupts: place chart nets and their synthesis. ICATPN'97, *LNCS 1248:328–347*, 1997.

[9] H. Klaudel. Compositional high-level Petri net semantics of a parallel programming language with procedures. *Science of Computer Programming*, to appear (available at http://univ-paris12.fr/lacl/klaudel/proc.ps.gz).

[10] H. Klaudel and F. Pommereau. Asynchronous links in the PBC and M-nets. *ASIAN'99, LNCS 1742:190–200*, 1999.

[11] H. Klaudel and F. Pommereau. A concurrent and compositional Petri net semantics of preemption. *IFM'2000, LNCS 1945:318–337*, 2000.

[12] P. A. Lee and T. Anderson. Fault tolerance: principle and practice. Springer, 1990.

[13] R. Milner. A calculus of communicating systems. *LNCS 92*, 1980.

[14] B. Randell, A. Romanovsky, R. J. Stroud, J. Xu and A. F. Zorzo. Coordinated Atomic Actions: from concept to implementation. Submitted to *IEEE TC Special issue*.

[15] A. Romanovsky. Practical exception handling and resolution in concurrent programs. *Computer Languages*, v. 23, N1, 1997, pp. 43-58.

[16] A. Romanovsky. Extending conventional languages by distributed/concurrent exception resolution. *Journal of systems architecture*, Elsevier science, 2000

[17] J. Xu, A. Romanovsky and B. Randell. Coordinated Exception Handling in Distributed Object-Oriented Systems: Improved Algorithm, Correctness and Implementation. Computing Dept., University of Newcastle upon Tyne, TR 596, 1997.

Modelling the Structure and Behaviour of Petri Net Agents

Michael Köhler, Daniel Moldt, and Heiko Rölke

University of Hamburg – Computer Science Department
Theoretical Foundations of Computer Science
Vogt-Kölln-Straße 30 – 22527 Hamburg
{koehler,moldt,roelke}@informatik.uni-hamburg.de

Abstract. This work proposes a way to model the structure and behaviour of agents in terms of executable coloured Petri net protocols. Structure and behaviour are not all aspects of agent based computing: agents need a world to live in (mostly divided into platforms), they need a general structure (e.g. including a standard interface for communication) and their own special behaviour. Our approach tackles all three parts in terms of Petri nets. This paper skips the topic of agent platforms and handles the agent structure briefly to introduce a key concept of our work: the graphical modelling of the behaviour of autonomous and adaptive agents.

A special kind of coloured Petri nets is being used throughout the work: reference nets. Complex agent behaviour is achieved via dynamic composition of simpler sub-protocols, a task that reference nets are especially well suited for. The inherent concurrency of Petri nets is another point that makes it easy to model agents: multiple threads of control are (nearly) automatically implied in Petri nets.

Keywords: agent, behaviour, concurrency, modelling, multi agent system, nets within nets, Petri net, reference net, structure

1 Motivation

To date agents are generally programmed using high-level languages such as Java (namely in agent frameworks as Jackal [8]) or they are defined by simple scripts. A graphical modelling technique that captures all parts of agents and their systems – as UML[1] in the context of object-orientation – is neither proposed nor in general use.[2]

The proper treatment of encapsulation, structuring, and flexibility in the scope of modelling is at the same time a major challenge in software engineering as well as in theoretical computer science.

---

[1] UML stands for Unified Modeling Language. See for example [16].

[2] The authors are aware of the upcoming proposals that base on UML i.e. from Odell et al. [2,30] (AUML). To our opinion these proposals capture only parts of the agent modelling tasks and leave out important areas such as agent mobility.

J.-M. Colom and M. Koutny (Eds.): ICATPN 2001, LNCS 2075, pp. 224–241, 2001.

Our department has expertise in examining to what extent Petri nets are suitable in the mastering of these questions. Starting from fundamental Petri net concepts especially Petri nets as active tokens (see [36]) and the basic construct of object oriented Petri nets (see [27]) have been investigated. The latter have led to a first approach to agent oriented Petri nets (see [29]). A complete redesign of agent oriented Petri nets (see [32] for the motivation and starting point) is now partly presented in this work.[3] Our approach aims at modelling a complete multi agent system in terms of Petri nets. This work decomposes into three parts: The system (e.g. the set of platforms) that hosts the agents, the agent itself, and its behaviour. A complete presentation of all three parts is out of the scope of this document, as the title implies, only the modelling of the structure and behaviour of a single agent is focused in this contribution.

Agent orientation is marked by intelligence, autonomy and mobility [4]. Whilst mobility requires an interplay of one agent and the agent system intelligence and autonomy are to be found in the overall architecture of agents (autonomy) and in their behaviour (intelligence). Mobility raises some hard questions e.g. concerning safety and is therefore not handled in this paper. Nevertheless our approach is easily expandable to allow forms of mobility as described in [28].

Our work uses the formalism of reference nets as presented by Kummer [23]. Reference nets are based on the "nets within nets"-paradigm that generalises tokens to arbitrary data types and even nets. The general idea behind our work is that an agent controls (sub-)nets as tokens which implement special kinds of behaviour. To (re-)act, the agent simply has to select (and instantiate) such a net. Additional concepts are dynamic resolution and binding of these nets.

The remainder of the document is organised as follows: in the following section 2, the formalism of reference nets is briefly introduced. Section 3 gives an overview of how the "nets within nets" paradigm can be used to model agents, their behaviour, and their environment. Section 4 describes the structure of the agents whose behaviour is determined by Petri net protocols. The net protocols are introduced in section 5 on the basis of an example. Section 6 cites some related material while the outlook in the closing section 7 names further points that are yet being discussed or are out of the scope of this paper.

2   Reference Nets

It is assumed throughout this text that the reader is familiar with Petri nets in general as well as coloured Petri nets. Reisig [31] gives a general introduction, Jensen [20] describes coloured Petri nets. Generally speaking coloured Petri nets permit a more compact representation while offering the same computational power compared to for example P/T-nets.

Reference nets [23] are so-called higher (coloured) Petri nets, a graphical notation that is especially well suited for the description and execution of complex,

---

[3] The redesign was necessary for several reasons. The most important one is that the original work did not use the nets within nets paradigm and therefore lacked of architectural clearness when introducing new layers in the multi agent system.

concurrent processes. As for other net formalisms there exist tools for the simulation of reference nets [24]. Reference nets show some expansions related to
"ordinary" coloured Petri nets: nets as token objects, different arc types, net
instances, and communication via synchronous channels. Beside this they are
very similar to coloured Petri nets as defined by Jensen. The differences now are
shortly introduced.

Nets as tokens. Reference nets implement the "nets within nets" paradigm of
Valk [36]. This paper follows his nomenclature and denominates the surrounding
net *system net* and the token net *object net*. Certainly hierarchies of net within
net relationships are permitted, so the denominators depend on the beholder's
viewpoint.

Arc types. In addition to the usual arc types reference nets offer *reservation
arcs*, that carry an arrow tip at both endings and reserve a token solely for one
occurrence of a transition, *test arcs*, and *inhibitor arcs*. Test arcs do not draw-off
a token from a place allowing a token to be tested multiple times simultaneously,
even by more than one transition (test on existence). Inhibitor arcs prevent
occurrences of transitions as long as the connected places are marked.

Net instances. Net instances are similar to the objects of an object oriented programming language. They are instantiated copies of a template net like objects
are instances of a class. Different instances of the same net can take different
states at the same time and are independent from each other in all respects.

Synchronous channels. Synchronous channels [6] permit a fusion of transitions
(two at a time) for the duration of one occurrence. In reference nets (see [23])
a channel is identified by its name and its arguments. Channels are directed,
i.e. exactly one of the two fused transitions indicates the net instance in which the
counterpart of the channel is located. The other transition can correspondingly
be addressed from any net instance. The flow of information via a synchronous
channel can take place bi-directional and is also possible within one net instance.

3   Multi Agent System

This section gives a short introduction to a multi agent system modelled in terms
of "nets within nets" (see figure 1). This survey is given to make the general ideas
visible that are prerequisite to the understanding of the concepts that follow in
later sections of this paper. It is neither an introduction to multi agent systems
nor the assets and drawbacks of dividing the system into platforms is discussed
here. For a broad introduction see for example [37], the special view taken in our
work is a standard proposal of the "Foundation for Intelligent Physical Agents"
(FIPA) [14]. The latest publications of the FIPA can be found in [13].

   Take a look at figure 1: The grey rounded boxes enclose nets (net instances)
of their own right. The ZOOM lines enlarge object nets that are tokens in the

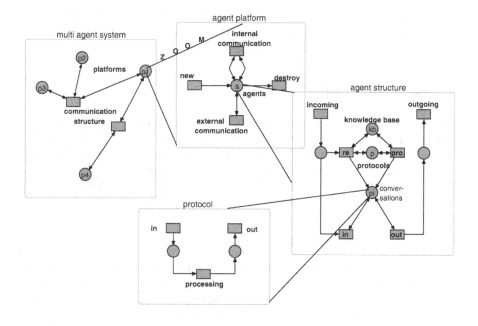

Fig. 1. MAS as nets within nets

respective system net.[4] The upper left net of the figure is an arbitrary agent system with places containing agent platforms and transitions modelling communication channels between the platforms. This is just an illustrating example, the number of places and the form of interconnection has no further meaning.

The first zoom leads to a closer view of a simplified agent platform. The central place agents contains all agents that are currently hosted on the platform. New agents can be generated (or moved from other platforms) by transition new, agents can be destroyed or migrate to another platform (transition destroy). Internal message passing differs from the external case so it is conceptually separated: The internal communication transition binds two agents (the sender and the receiver of a message) and allows them to hand over a message via call of synchronous channels. External communication involves only one agent of the platform. For external communication as well as for agent migration the communication transitions of the top level agent system net are needed. The interaction of the multi agent system and the agent platform is made possible by inscribing the transitions with synchronous channels connecting for example the transition external communication of an agent platform with that of another one via the communication structure transition of the multi agent system. These inscriptions are not visible in the figure.

---

[4] Beware not to confuse this net-to-token relationship with place refinement.

The remaining nets that show the structure of an agent and an example of its (dynamic) behaviour in form of protocols (protocol nets) are explained in more detail in section 4 and 5, respectively.

4   Agent Structure

An agent is a message processing entity, that is, it must be able to receive messages, possibly process them and generate messages of its own. In this context it is to be noted that a completely synchronous messages exchange mechanism as it is used in most object oriented programming systems, frequently violates the idea of autonomy among agents.[5] The fundamental concepts that characterise agents are (in our opinion) *autonomy*, *intelligence*, and *mobility*.

4.1   Abstract Agent Model

As a Petri net model, figure 2 shows the most abstract view on an agent. The input and output transitions are a speciality of the reference nets that are used in the figure. They can communicate with other (input and output) transitions in other net copies through the use of synchronous channels (Because they are agents this is done via message passing). The basic agent model takes advantage of the ability of a transition to occur concurrently with itself.[6] So the agent is able to receive, process, and send several messages at the same time, it does not block. The transition processing can be refined for concrete agents as desired. In this and all following net figures all not unconditionally necessary inscriptions have been omitted. This leads to simpler models but may sometimes bring in the danger of confusing the reference nets (special coloured Petri nets) with more basic net formalisms (e.g. P/T-nets). So the reader is kindly asked to keep in mind the power of the used net formalism.

The introduced basic agent model implies an encapsulation of the agents: regardless of their internal structure, access is only possible over a clearly defined communication interface. In figure 2, this interface is represented by the transitions incoming and outgoing. In the figure, the realisation of the interface (through connection of both transitions to a messages transmission network via synchronous channels) is not represented. Obviously several (then virtual) communication channels can be mapped to both transitions.

Providing a static interface is the key to interoperability amongst agents. The agents presented in this paper speak and understand FIPA messages [13]. Neither the content of the messages nor the way that they are used is limited, only their syntactical structure is fixed. Some advantages of the use of a standardised communication mechanism can be found in [9].

---

[5] To our understanding agents are not exclusively (artificial) intelligent agents, but rather a general software structuring paradigm on top of the ideas of object orientation [19].

[6] Please note that this is not a special feature of reference nets but of all proper net formalisms.

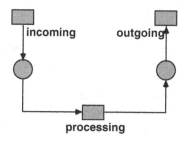

Fig. 2. Most abstract view on an agent

The presented agent model corresponds to the fundamental assumptions about agents: Because agents should show autonomy, they must be able to exercise an independent control over their actions. Autonomy implies the ability to monitor (and, if necessary, filter) incoming messages before an appropriate service (procedure, method...) is called. The agent must be able to handle messages of the same type (e.g. asking for the same service) differently just because of knowing about the message's sender. This is one of the major differences between objects and agents: A *public* object method can be executed by any other object, *protected* methods offer a static access control that is very often inconvenient to the programmer and user. Without regard to the fundamental autonomy, an agent can obviously be sketched (or take the obligation for itself) to appear like an object to the outside world, therefore to be perfect cooperative. Another reason to prefer messages over for example method calls is that methods are fixed both in respect to their arguments (number and type) and the number of methods that are offered by one object. Using method calls makes it tricky to adapt to new situations.

Autonomy is the major difference between agents and *active objects*: The latter may show some of the properties that characterise agents (an often used example is mobility) but they do not have the ability to control who is calling their (public) methods. Agents may have arbitrary fine grained access control.

The model does not affect nor restrict the intelligence nor the mobility of the agents: Intelligent behaviour is achieved through refinements of the transition processing, mobility requires interaction of the agent and the agent system.[7] Therefore, mobility is not a topic of this work, intelligence is raised again in the chapter on agent behaviour protocols (chapter 5).

---

[7] Note that the proper handling of mobility is a research area of its own rights. There is already some work done [28] in this direction, that will soon be brought into our agent definitions.

4.2   Refined Agent Model

Beside the fundamental agent concepts the agent model has to meet additional requirements concerning their ease of use: On the one hand it has to offer a high degree of flexibility in particular during execution and on the other hand the modelling process has to be manageable and adaptable. In addition for a broader acceptance, intuitive intelligibility of the processes within the agents is necessary. These considerations have played an important role in the development of the protocol-driven agents sketched here.

The abstract agent net of figure 2 is refined in the following manner (see figure 3): The agent net as shown in the figure is further used as the interface of the agent to the outside world. The transition processing is refined to a selection mechanism for specialised subnets, that implement the functionality of the agent, therefore (beside the selection process) its behaviour. These subnets are named *protocol nets* (or short *protocols*) in the following.

Each agent can control an arbitrary number of such protocols, possesses however only one net (in reference net nomenclature: one net page), that represents its interface to the agent system and therewith its identity. This main net (page) is the visible interface of an agent in the multi agent system. As mentioned before all messages that an agent sends or receives have to pass this net.

The main net of the (protocol-driven) agents introduced here is given in figure 3. It is a refinement of the abstract agent net given in figure 2.

The central point of activity of a protocol-driven agent is the selection of protocols and therewith the commencement of conversations [7,33]. The protocol selection can basically be performed pro-actively (the agent itself starts a conversation) or reactively (protocol selection based on a conversation activated by another agent).[8] This distinction corresponds to the bilateral access to the place holding the protocols (protocols). The only difference in enabling and occurrence of the transitions reactive and pro-active is the arc from the place input messages to the transition reactive. So the latter transition has an additional input place: the incoming messages buffer. It may only be enabled by incoming messages. Both the reaction to arriving messages and the kick-off of a (new) conversation is influenced by the knowledge of an agent. In the case of the pro-active protocol selection, the place knowledge base is the only proper enabling condition, the protocols are a side condition. In simple cases the knowledge base can be implemented for example as a subnet, advanced implementations as the connection to an inference engine are also possible (and have been put into practise). Unfortunately this topic can not be deepened here any further.

An agent has several possibilities to react to dynamically changing environments: It may not react at all (if it decides that no changing of its behaviour is needed or possible), it may alter its protocol selection strategy (choose different

---

[8] The fundamental difference between pro-active and reactive actions is of great importance when dealing with agents. An introduction to this topic is e.g. given by Wooldridge in [38] (in: [37]).

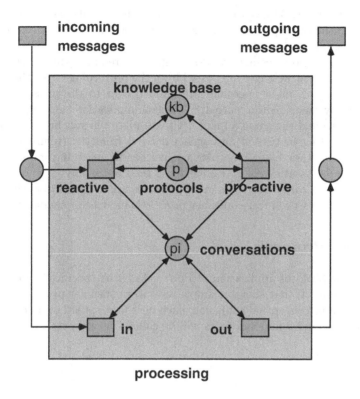

Fig. 3. Refined protocol-driven agent

protocols for the same message type), adapt one or more of its protocols[9] or ask other agents for protocols that suit to the new situation (protocols may be communicated as well).

A selected and activated protocol[10] is also called a conversation because it usually includes the exchange of messages with other agents. A conversation can however also run agent internal, therefore without message traffic. A freshly invoked conversation holds an unambiguous identification that is not visible in the figure. All messages belonging to a conversation carry this identification as a parameter to assign them properly. If an agent receives a messages carrying such a reference to an existing conversation, transition in is enabled instead of transition reactive. The net inscriptions that guarantee this enabling are not

---

[9] Protocol adaptation is done in a way similar to the "'reconfigurable nets"' formalism [1], i.e. restricted self-modifying nets [35]. The adaptation of protocols together with the agent's knowledge base unfortunately has to be skipped in this paper.

[10] Following the object oriented nomenclature one speaks of an instantiated net or protocol (that is represented in form of a net).

represented in figure 3 for reasons of simplicity. The transition in passes incoming messages to the corresponding conversation protocol in execution. Examples for this process follow in section 5.

If the sending of messages to other agents is required during the run of a conversation, these messages are passed from the protocol net over the transition out to the agent's main page and are handed over to the message transport mechanism by the transition output.[11] The communication between protocol net (conversation) and the agent's main net takes place via synchronous channels.

An interesting feature of any agents derived from the (template) agent in figure 3 is that they cannot be blocked, neither by incoming messages nor by their protocols[12] and therefore cannot loose their autonomy.

Examples for concrete conversation protocols are to be found in the following chapter 5 where a producer-consumer process is modelled exemplary.

5   Agent Protocols

An important field of application of Petri nets is the specification of processes as that in figure 4, that shows a simple producer-consumer process. In order to give no room to conceptual confusion, such nets that spread over several agents and/or distributed functional units will be called "survey nets".

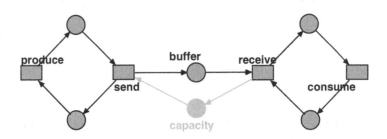

Fig. 4. Producer-consumer (survey net)

The place buffer in the middle of the figure represents an asynchronous coupling between the process of producing and that of consuming. This coupling is however to that extent independent that it for example blocks the consumer if it is empty or, given the case that it is inscribed with a capacity, blocks the producer when this maximal filling is reached. In the following, producer and consumer are introduced as autonomous agents and are modelled according to

---

[11] The message transport mechanism is part of the agent system (or platform) and is therefore only sketched in section 3.

[12] Unless it is strictly necessary for a protocol to block the entire agent and this is explicitly modelled.

figure 3 by means of a reference net. The buffer is not modelled as an independent agent, nevertheless this would both syntactically (this will be explained in the following) and semantically (in consideration of the level of autonomy the buffer owns) be no problem.

An interesting point is the re-usability of the protocols: Consider a refined model in which the buffer should play an active role and should therefore be modelled as an agent of its own. The protocols of the producer and the consumer remain structurally unchanged, only the addressees of their messages have to be adapted. But these should be modelled dynamically in any case.

Note that this is not the first work that uses the consumer-producer process as an example to illustrate new ideas of how to model and structure software systems by means of Petri nets. It is for example used by Reisig to introduce Petri nets in general [31] and by Valk to show different models of synchronisation [21].

The following example assumes that the buffer is restricted by a capacity of one item. This restriction is for simplification purposes only and may be lifted easily. The restriction is indicated in figure 4 by the grey place capacity under the buffer place.

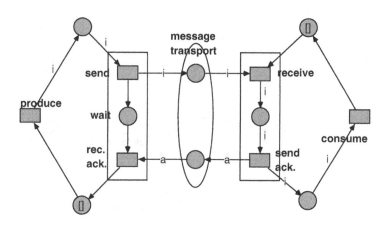

Fig. 5. Synchronous producer-consumer

The producer-consumer survey net is refined to the net in figure 5 which uses an explicitly modelled synchronous message exchange. The buffer and its capacity get carried away to form the message transport system. The message transport system is also the borderline of the two agents producer and consumer that will be introduced in the following subsections. The producer produces one item (denoted as i in the figure) and sends it to the consumer. The producer has to wait for an acknowledge (a) from the consumer to fire the transition rec.ack. in order to reach its initial state. When the consumer receives an item it sends the acknowledge (send ack.) and consumes the item received. After consuming it is ready to receive the next item.

The marking of the net in figure 5 indicates the starting points of the protocol nets telling the agents how to produce or consume: The production protocol has to be selected pro-actively and starts with the production of an item. The consume protocol is selected in a reactive manner to process an incoming message from the producer containing an item.

5.1   Producer

The protocol of the producer agent is represented in figure 6. The upper transitions with the channels :start, :out, :in, and :stop are typical for all types of protocol nets. The :start channel serves as a means to pass possibly necessary parameters to the protocol. It is called on the agent main page (see figure 3) either by transition reactive or pro-active. The channels :in and :out are responsible for the communication of an operating protocol with the environment. They connect to the transitions of the same denominators on the agent's main page. When a protocol has finished its task, the transition inscribed with channel :stop is enabled. By calling of this channel the agent may delete the protocol or, more correctly, the protocol instance.

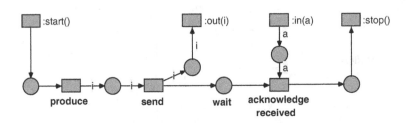

Fig. 6. Produce protocol

After the start of the protocol the transition produce produces a performative[13] (here i) containing an item, that is directed to the consumer. Note that in the example the performative is the only thing that is produced. The performative will be sent over the :out channel; subsequently the protocol is blocked waiting for an answer message. The blocking behaviour is necessary to simulate a synchronous communication between producer and buffer. Without waiting for

---

[13] Some of the ideas that led to the agent model introduced here are partially originated in the area of the KQML- ([12]) or FIPA-agents ([15]). Roughly speaking a performative is a message. KQML stands for "Knowledge Query and Manipulation Language", FIPA is the abbreviation of "Foundation for Intelligent Physical Agents".

an answer the producer would be able to "inundate" the buffer with messages, what requires an infinite buffer capacity. An arriving confirmation enables the transition acknowledge received. After occurrence of that transition the protocol is not blocked any further and terminates (by enabling the stop transition). The producer agent is now able to select and instantiate the produce protocol again.

5.2   Consumer

The protocol net that models the consume behaviour of the consumer agent (see figure 7) is selected (*reactively*) by the agent's main page to process an incoming performative from the producer agent. It is instantiated and the :start channel is used to pass the performative to the protocol. Beside others the performative is needed to send a acknowledge performative to the originator of the conversation (the producer). Note that the consumer agent does not know the producer or if there is one or several of them. The protocol works in either case.

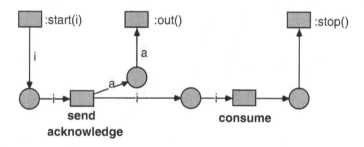

Fig. 7. Consume protocol

The consumer can block an arriving message as long as it wants, until it is ready to "consume" the carried item. In figure 7, this is represented by the transition send acknowledge. After acknowledging the receipt of the item the transition consume may occur. After that the protocol terminates and can be deleted.

Figures 6 and 7 show the protocols that model a conversation between producer and consumer. They are executed within agents of the type of figure 3. The figures form a simple example that illustrates how to model a producer-consumer process by means of agent oriented Petri nets. The proposed methodology to implement protocol nets in a top-down manner starting with so-called survey nets is not the only possibility to develop protocols. One can easily think of a bottom-up style or mixed cases especially for hierarchical protocol relationships as in the following subsection. Unfortunately this topic can not be deepened here.

5.3   Refined Producer

Consider a producer capable of producing several different items. The naive approach to model this more powerful producer is to enlarge the producer protocol of figure 6 and to use this new protocol afterwards. For several reasons this is not a good idea:[14]

- Redefinition of existing nets is tiresome and error-prone.
- Large nets tend to be complex and thus difficult to understand and maintain (see above).
- Further enhancements become more and more difficult.

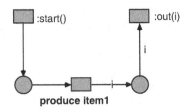

Fig. 8. Internal produce protocol

For these reasons the new produce process is split into pieces. First protocols for the different produce procedures like that in figure 8 have to be modelled. Now the production of the items can be driven from other protocols inside the producer agent. An example of such a higher-order protocol is a description of the following specification: imagine a consumer that does not care if it receives items of type 1 or type 2. In that case the producer can decide to produce items maybe on reasons of availability or price.

This decision is independent of the production process[15] and should therefore be carried out in an independent protocol. This protocol is shown in figure 9. The protocol is an extension to the producer protocol of figure 6. At first a decision is made about the type of item to produce (transitions i1 and i2). After that the protocol has to send a message to the selected production protocol, e.g. to that of figure 8. The protocol waits for the item to arrive, sends it to the consumer and so on. It may appear that the protocol net in figure 9 shows a conflict introduced through several transitions carrying the same synchronous channel. This possible conflict is resolved through additional inscriptions concerning the type of message that is the argument of the synchronous channels.

---

[14] This enumeration could easily be continued.
[15] Certainly only to that extent that an informed decision needs to know something about the needs of the production.

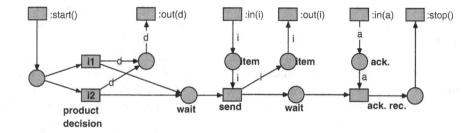

Fig. 9. Production alternatives

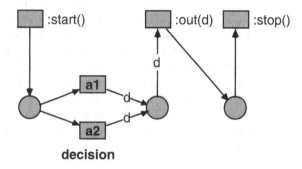

Fig. 10. Decision protocol

Note that this solution is not optimal considering the reasons for not enlarging existent nets. As a consequence the process of decision making is sourced out to a protocol of its own; it may be used by other protocols, too. To illustrate this, figure 10 shows the new decision protocol. The new produce protocol is quite similar to the original one, just the decision protocol is called instead of producing.

6 Related Work

Note that this is not the first approach to use Petri nets to model or implement agents. This section will give an idea of similarities and differences between our work and that of other researchers.

A major difference between related work of other authors and our approach is the use of reference nets and therefore the nets within nets idea. To our knowledge there are no other implementations of this net paradigm beside reference nets. Despite of this difference there are certainly similarities to the results of other researchers. This work aims at modelling multi agent systems as a whole.

Therefore it is not possible to relate it to work dealing with smaller parts of agents (e.g. the planning process or reasoning in uncertainty). This will be made up when discussing these parts of our work in future publications.

Sibertin-Blanc et.al. [5] implement agent behaviour in form of *Cooperative Nets* [34]. Cooperative nets model the behaviour of objects, this approach is enlarged to agents. While the basic idea to model the behaviour is somehow similar to our work there are also differences: The behaviour of a (cooperative) object is statically defined by one net, it can not be altered or adapted at runtime, which is a major drawback for agents. Furthermore, the work of Sibertin-Blanc et.al. contains no notion of an multi agent system.

Holvoet [18] proposes Petri net agents that shall communicate in a manner similar to synchronous channels. Like Sibertin-Blanc (see above) he does not explicitly handle multi agent systems. The proposed agents are not autonomous in the strict sense presented in section 4 of this paper, because their interaction (via transition synchronisation) is not filtered by an interface but directly concerns the transitions modelling the agent's behaviour. So a proper encapsulation of the agents can not be assured, the agents are rather active objects. The agents Holvoet proposes are "special agents": They are specialised and restricted to do tasks that are exchangeable protocols to the agents introduced in this paper.

Fernandes and Belo [10] introduce a modelling case study that uses coloured Petri nets to model multi agent systems activities. They do not model single agents (that are token in their approach) but the overall system behaviour for one example system.

Miyamoto and Kumagai [25,26] enhance the Cooperative nets (by Sibertin-Blanc, mentioned above) in several ways to multi agent nets. Their work is closer to our approach because it uses quite similar protocols to define the agents' behaviour. Their protocols offer public interfaces, therefore they are not autonomous in our strict sense.

Xu and Shatz [39] build agents on top of the G-Net formalism of Figueiredo and Perkusich [11]. Despite of the different formalisms their work has some similarities to ours, namely the planning process (how to react to incoming messages). The main difference between the approaches is that their multi agent system is completely unstructured. Therefore mobility is not taken into account. Furthermore their agents have a fixed set of methods and may only adapt by changing their goals, plans and knowledge, not by reconfiguring, adapting or exchanging their actions (methods).

7   Outlook

In consideration of the topic of this paper some interesting aspects of the agent introduced here could not be mentioned: the protocol nets an agent possesses are interchangeable at run time (so the agent allows this). A mobile agent (mobility is possible in the multi agent framework but not introduced here) that arrives at an agent platform can adopt the protocols valid there as his own.

The modelling process is a topic of its own rights. The transformation of a survey net that describes an entire conversation (as given in figure 4) to the protocol nets is exemplarily carried out in [17] (in German).

A further point concerns the protocols themselves: only very simple procedures were shown. More complicated protocols include a hierarchical nesting that is constructed at run time by mutual calls of protocols (in contrast to protocols forming a conversation this is done within one agent). In this way a dynamic adaptation of the agents via self modification becomes possible.

Besides intuitive and compact modelling of concurrent processes Petri nets are particularly well-known for their provability. It should not be concealed that there are no standard proving techniques nor frameworks for reference nets. This is a major drawback because exhaustive testing via simulation is not always wanted nor possible. There is an ongoing PhD. thesis in our department that deals with the question of property conserving composition in agent Petri nets, first results of this work can be found in [22]. This is a very promising approach because it fits to the compositionality of the protocols presented in this work: Once a property like liveness has been proven for a protocol net, it may be used in any conversation without loosing this property (subject to the condition that all other protocol offer this property, too).

At present the knowledge base is in an internal discussion. There exists a Prolog interpreter that is implemented in Java and therefore can be integrated into the reference net simulation tool Renew. It is used as an inference engine inside the agents in more elaborated examples. Desirable is a graphically modelled knowledge model.

References

1. Eric Badouel and Javier Oliver. Reconfigurable Nets, a Class of High Level Petri Nets Supporting Dynamic Changes. Research report pi 1163, INRIA, 1998.
2. Bernhard Bauer, James Odell, and H. van Dyke Parunak. Extending UML for Agents. In *Proceeding of Agent-Oriented Information Systems Workshop*, pages 3 – 17, 2000.
3. Jeffrey M. Bradshaw, editor. *Software Agents*. AAAI Press, 1997.
4. J.M. Bradshaw. *An Introduction to Software Agents*, chapter 1. in: [Bra97a], 1997.
5. Walid Chainbi, Chihab Hanachi, and Christophe Sibertin-Blanc. The Multi-agent Prey/Predator problem: A Petri net solution. In P. Borne, J.C. Gentina, E. Craye, and S. El Khattabi, editors, *Proceedings of the Symposium on Discrete Events and Manufacturing systems*, Lille, France, July 9-12 1996. CESA'96 IMACS Multiconference on Computational Engineering in System Applications.
6. S. Christensen and N.D. Hansen. Coloured Petri nets extended with channels for synchronous communication. In Rober Valette, editor, *Application and Theory of Petri Nets 1994, Proc. of 15th Intern. Conf. Zaragoza, Spain, June 1994*, LNCS, pages 159–178, June 1994.
7. R. Scott Cost, Ye Chen, T. Finin, Y. Labrou, and Y. Peng. Modeling agent conversation with colored Petri nets. In *Working notes on the workshop on specifying and implementing concersation policies (Autonomous agents '99)*, 1999.

8. R.S. Cost, T Finin, Y. Labrou, X. Luan, Y. Peng, L. Soboroff, J. Mayfield, and A. Voughannanm. Jackal: A Java-based Tool for Agent Development. In *Working Notes of the Workshop on Tools for Developing Agents, AAAI'98*, pages 73–82. AAAI Press, 1998.
9. Jonathan Dale and Ebrahim Mamdani. Open standards for interoperating agent-based systems. Software Focus, Wiley, 2001.
10. Joao M. Fernandes and Orlando Belo. Modeling Multi-Agent Systems Activities Through Colored Petri Nets. In *16th IASTED International Conference on Applied Infomatics (AI'98)*, pages 17–20, Garmisch-Partenkirchen, Germany, Feb. 1998.
11. Jorge C. A. Figueiredo and Angelo Perkusich. G-Nets: A Petri Net Based Approach for Logical and Timing Analysis of Complex Softwaree Systems. *Journal of Systems and Software*, 39 (1)(39-59), 1997.
12. Tim Finin and Yannis Labrou. A Proposal for a new KQML Specification. Technical report, University of Maryland, Februar 1997.
13. FIPA. Homepage. http://www.fipa.org.
14. FIPA. FIPA 97 Specification, Part 1 - Agent Management. Technical report, Foundation for Intelligent Physical Agents, http://www.fipa.org, Oktober 1998.
15. FIPA. FIPA 97 Specification, Part 2 - Agent Communication Language. Technical report, Foundation for Intelligent Physical Agents, http://www.fipa.org, Oktober 1998.
16. Martin Fowler. *UML Distilled*. Addison-Wesley Longman, Inc., 1. edition, 1997.
17. Daniela Hinck, Michael Köhler, Roman Langer, Daniel Moldt, and Heiko Rölke. Bourdieus Habitus-Konzept als prägendes Strukturelement für Multiagentensysteme. Mitteilung 298, Universität Hamburg, Fachbereich Informatik, 2000.
18. Tom Holvoet. Agents and Petri Nets. *Petri net Newsletter*, 49:3–8, 1995.
19. Nicholas R. Jennings. On agent-based software engineering. *Artificial Intelligence*, 177(2):277–296, 2000.
20. K. Jensen. *Coloured Petri nets, Basic Metods, Analysis Methods and Practical Use*, volume 1 of *EATCS monographs on theoretical computer science*. Springer-Verlag, 1992.
21. E. Jessen and R. Valk. *Rechensysteme – Grundlagen der Modellbildung*. Springer-Verlag, 1987.
22. Michael Köhler, Daniel Moldt, and Heiko Rölke. Liveness preserving composition of agent Petri nets. Technical report, Universität Hamburg, Fachbereich Informatik, 2001.
23. Olaf Kummer. Simulating synchronous channels and net instances. In J. Desel, P. Kemper, E. Kindler, and A. Oberweis, editors, *Forschungsbericht Nr. 694: 5. Workshop Algorithmen und Werkzeuge für Petrinetze*, pages 73–78. Universität Dortmund, Fachbereich Informatik, 1998.
24. Olaf Kummer, Frank Wienberg, and Michael Duvigneau. *Renew - User Guide*. University of Hamburg, Vogt-Kölln-Straße, Hamburg, 1.4 edition, November 2000.
25. Toshiyuki Miyamoto and Sadatoshi Kumagai. A Multi Agent Net Model of Autonomous Distributed Systems. In *Proceedings of CESA'96, Symposium on Discrete Events and Manufacturing Systems*, pages 619–623, 1996.
26. Toshiyuki Miyamoto and Sadatoshi Kumagai. A Multi Agent Net Model and the Realization of Software Environment. In *20th International Conference on Application and Theory of Petri Nets, Proceedings of the Workshop: Applications of Petri nets to intelligent system development*, pages 83–92, June 1999.
27. Daniel Moldt. *Höhere Petrinetze als Grundlage der Systemspezifikation*. Dissertation, Universität Hamburg, Fachbereich Informatik, Vogt-Kölln Str. 30, 22527 Hamburg, Deutschland, 1996.

28. Daniel Moldt and Ivana Tričkovič. The paradigm of nets in nets as a framework for mobility. Unpublished Technical Report, 2001.
29. Daniel Moldt and Frank Wienberg. Multi-agent-systems based on coloured Petri nets. In P. Azéma and G. Balbo, editors, *Lecture Notes in Computer Science: 18th International Conference on Application and Theory of Petri Nets, Toulouse, France*, volume 1248, pages 82–101, Berlin, Germany, June 1997. Springer-Verlag.
30. James Odell and H. van Dyke Parunak. Representing Social Structures in UML. In *Proceedings of Autonomous Agents'01*, Montreal, Canada, May/June 2001.
31. Wolfgang Reisig. *Petri nets: an introduction*. Springer, 1985.
32. Heiko Rölke. Modellierung und Implementation eines Multi-Agenten-Systems auf der Basis von Referenznetzen. Diplomarbeit, Universität Hamburg, 1999.
33. Heiko Rölke. Die Mulan Architektur. Technical report, Universität Hamburg, 2000.
34. Christophe Sibertin-Blanc. Cooperative Nets. In *Proceedings of the 15th International Conference on Application and Theory of Petri nets*, volume LNCS 815, Saragossa, June 1994.
35. Rüdiger Valk. Self-modifying nets. Technical Report Nr. 34, Universiät Hamburg, 1977. R 6033.
36. Rüdiger Valk. Petri nets as token objects: An introduction to elementary object nets. In Jörg Desel and Manuel Silva, editors, *Application and Theory of Petri Nets*, volume 1420 of *Lecture Notes in Computer Science*, pages 1–25, June 1998.
37. Gerhard Weiß, editor. *Multiagent Systems: A Modern Approach to Distributed Artificial Intelligence*. The MIT Press, 1999.
38. Michael Wooldridge. *Intelligent Agents*, chapter 1. The MIT Press, 1999.
39. Haiping Xu and Sol M. Shatz. A Framework for Modeling Agent-Oriented Software. In *Proc. of the 21th International Conference on Distributed Computing Systems (ICDCS-21)*, Phoenix, Arizona, April 2001.

Model Checking LTL Properties of High-Level Petri Nets with Fairness Constraints

Timo Latvala*

Laboratory for Theoretical Computer Science
Helsinki University of Technology
P.O. Box 9700, FIN-02015 HUT, Espoo, Finland
Timo.Latvala@hut.fi
http://www.tcs.hut.fi/Personnel/timo.html

Abstract. Latvala and Heljanko have presented how model checking of linear temporal logic properties of P/T nets with fairness constraints on the transitions can be done efficiently. In this work the procedure is extended to high-level Petri Nets, Coloured Petri Nets in particular. The model checking procedure has been implemented in the MARIA tool. As a case study, a liveness property of a sliding window protocol is model checked. The results indicate that the procedure can cope well with many fairness constraints, which could not have been handled by specifying the constraints as a part of the property to be verified.

Keywords: Model checking, fairness, LTL, high-level Petri Nets.

1 Introduction

Model checking [2,23] has established itself as one of the most useful methods for reasoning about the temporal behavior of Petri Nets. Currently there are several Petri Net tools which offer model checking of either linear or branching time properties of Petri Nets (see e.g. [29,37]).

Model checking liveness properties differs from model checking safety properties. In many cases certain unwanted behaviors of the model must be ignored, to facilitate model checking of liveness properties. This is usually done using fairness assumptions [7]. For Petri Nets this means that the behavior of a transition or several transitions is restricted according to the fairness assumption. The only support Petri Net model checkers so far have offered for this is by model checking properties with formulas of the form *"fairness ⇒ property"*. Using many fairness assumptions makes the formula long and therefore quickly makes the model checking intractable, since the model checking problem is PSPACE-complete [30] in the size of the formula.

---

* The financial support of TEKES, Nokia Research Center, Nokia Networks, Genera, EKE Electronics and Helsinki Graduate School in Computer Science and Engineering (HeCSE) is gratefully acknowledged.

J.-M. Colom and M. Koutny (Eds.): ICATPN 2001, LNCS 2075, pp. 242–262, 2001.

Latvala and Heljanko presented in [19] how P/T nets with fairness constraints on the transitions can be model checked efficiently by giving semantics to the P/T net with a fair Kripke structure (FKS) [14]. The FKS is then model checked with a procedure similar to the one presented in [14]. It can, however, be difficult to model large systems with P/T nets, and therefore there is a need to extend the procedure to high-level nets.

The main contributions and results of this work are the following. A procedure for model checking high-level Petri Nets with fairness constraints on the transitions is presented. The procedure is a generalization of the procedure in [19]. Included is also the proof of correctness of the construction, which were only briefly touched upon in [19]. The procedure has been implemented in the MARIA analyzer. Using a model of a sliding window protocol, the implementation is tested. A liveness property of the protocol is verified with a window size up to 11, without any reduction methods. This could not have been done with the fairness constraints as part of the property to be verified.

The rest of the paper is structured as follows. In Section 2 Coloured Petri Nets are defined and Fair CPNs are introduced. Section 3 covers the necessary automata theory and presents the model checking procedure. The implementation of the model checker is covered in Section 4. The sliding window protocol is modeled and analyzed in Section 5. Section 6 concludes the paper.

2   Coloured Petri Nets

2.1   Definition of Coloured Petri Nets

The definition follows quite faithfully the definition of Coloured Petri Nets in [12]. CPNs were chosen because they are relatively simple to define while still being high-level Petri nets. Also, the fact that they are well-known contributed to their choice. The results to be presented later can easily be generalized to other high-level Petri net classes.

No concrete syntax and semantics for the net expressions will be given. We will however assume that it exists so that the following concepts are well defined.

- $Type(e)$ - The type of the expression e
- $Var(e)$ - The set of variables in an expression e.
- A *binding* $b(v)$ associates with each variable v a value of the type of the variable.
- $e\langle b\rangle$ - The value obtained by evaluating the expression e with the binding b.

Definition 1. *A tuple* $\Sigma = \langle \Pi, P, T, A, N, C, E, G, M_0\rangle$ *is a **Coloured Petri Net** (CPN) [12] where,*

i.) Π *is a finite set of non-empty types or **colour sets**.*
ii.) P *is a finite set of **places**.*
iii.) T *is a finite set of **transitions**, such that* $P \cap T = \emptyset$.

iv.) A *is a finite set of* **arcs**, *such that* $P \cap A = T \cap A = \emptyset$.

v.) $N : A \to (P \times T) \cup (T \times P)$ *is a* **node function**.

vi.) $C : P \to \Pi$ *is a* **colour function**.

vii.) E *is an* **arc expression function**, *defined from arcs to expressions, such that* $\forall a \in A : E(a)\langle b \rangle$ *is a multi-set over the colour of the place component of the arc, for all legal bindings* b.

viii.) G *is a* **guard function**, *defined from transitions to expressions, such that* $G(t)\langle b \rangle \in \{true, false\}$ *for any legal binding* b *and* $t \in T$.

ix.) M_0 *is an* **initial marking**.

We define the following notations:

- $A(x) = \{a \in A \mid \exists x' \in P \cup T : [N(a) = (x, x') \vee N(a) = (x', x)]\}$
- $A(x_1, x_2) = \{a \in A \mid N(a) = (x_1, x_2)\}$
- $\forall t \in T : Var(t) = \{v \mid v \in Var(G(t)) \vee \exists a \in A(t) : v \in Var(E(a))\}$.
- $\forall (x_1, x_2) \in (P \times T \cup T \times P) : E(x_1, x_2) = \sum_{a \in A(x_1, x_2)} E(a)$.

$A(x)$ returns the set of surrounding arcs, i.e. the arcs that have x as a source or a destination, for a given node x. $A(x_1, x_2)$ return the arcs which are between the nodes x_1 and x_2. $Var(t)$ is the set of variables of t, while $E(x_1, x_2)$ is the expression of (x_1, x_2) and returns the multi-set sum of all expressions connected to the arcs which have x_1 and x_2 as nodes.

Definition 2. *A* **token element** *is a pair* $(p, c) \in P \times C(p)$. *The set of all token elements is denoted by* $\mathcal{TE}$. *A* **marking** *is a multi-set over* $\mathcal{TE}$.

Because each marking defines a unique function $M(p)$, which maps each place to a multi-set over the colour set of the place, a marking is usually presented as a function on P.

Definition 3. *A* **binding** *of a transition* $t \in T$ *is a binding function on* $Var(t)$ *such that* $\forall v \in Var(t) : b(v) \in Type(v)$ *and* $G(t)\langle b \rangle = true$. *We denote the binding* $t\langle b \rangle$ *and call* $t\langle b \rangle$ *an* **instance** *of* t. *A transition instance* $t\langle b \rangle$ *is* **enabled** *in a marking* M *iff* $\forall p \in P : E(p, t)\langle b \rangle \leq M(p)$.

The function $en(M)$ returns the transition instances which are enabled in the marking M. If a transition instance $t\langle b \rangle \in en(M)$ it can **occur** changing M into another marking M' which is given by

$$\forall p \in P : M'(p) = M(p) - E(p, t)\langle b \rangle + E(t, p)\langle b \rangle$$

Hence M' is reachable from M, which we denote by $M \overset{t\langle b \rangle}{\to} M'$.

The behavior of the net is given by the Kripke structure of the net.

Definition 4. *The* **Kripke structure** *of a CPN* Σ *is a triple* $K = \langle S, \rho, s_0 \rangle$, *where* S *is the set of markings,* ρ *is the transition relation, and* s_0 *is the initial marking.* S *and* ρ *are defined inductively as follows.*

1. $s_0 = M_0 \in S$
2. If $M \in S$ and $M \to M'$ then $M' \in S$ and $(M, M') \in \rho$. If $M \in S$ and $en(M) = \emptyset$ then $(M, M) \in \rho$.
3. S and ρ have no other elements.

The executions of the net are infinite sequences of markings $\xi = M_0 M_1 M_2 \ldots$, for which $(M_i, M_{i+1}) \in \rho$ for all $i \geq 0$.

2.2   Petri Nets and Fairness

Fairness can be a useful abstraction, when one does not want to model the details of the scheduling required in the implementation of the system. Using *fairness assumptions* for the transitions is perhaps the most convenient way one can restrict the set of legal executions to the desired ones when verifying liveness properties. Fairness does not affect safety properties. The most common fairness assumptions are known as *weak fairness* and *strong fairness*. In [7] they are defined using the familiar concepts of *enabledness* and *occurrence* of the relevant events. An event is *weakly fair* when continuous enabledness implies that the event occurs infinitely often. For some situations, however, weak fairness is not enough. In this situation *strong fairness* might be appropriate. Strong fairness assumes that if an event is infinitely often enabled then it will occur infinitely often.

There is some related work which has combined model checking and fairness constraints. Latvala and Heljanko presented how P/T nets can be extended with fairness constraints and model checked. In [6] Emerson and Lei presented how to cope with strong fairness constraints when model checking CTL properties of a Kripke structure. A similar method was used to design a BDD based algorithm when the property was given as an automaton in [10] while in [14] a procedure for model checking LTL properties using BDDs with both weak and strong fairness constraints was presented.

The traditional way of incorporating fairness when model checking Petri nets exploits the fact that fairness is expressible by LTL. First, one usually has to add places and transitions to the model so that the occurrence of transitions is explicitly visible in the reachable markings. These modifications in a sense model a scheduler and have to be made because the state based version of LTL can only express properties of markings and thus cannot express properties of the transitions unless they are explicitly visible in the Kripke structure. The model is then verified by checking the formula "*fairness* $\Rightarrow$ *property*". This approach has several drawbacks. The two most obvious ones are that adding places and transitions might increase the size of the state space, and that the size of the Büchi automaton representing the property can grow exponentially in the number of fairness constraints (see e.g. [8]). A more subtle drawback is that adding the scheduler often reduces the concurrency in the model, which may affect the performance of some partial order methods (see e.g. [36]).

Let $\xi = M_0 \overset{t_0\langle b_0 \rangle}{\rightarrow} M_1 \overset{t_1\langle b_1 \rangle}{\rightarrow} \cdots$, with $M_i \overset{t_i\langle b_i \rangle}{\rightarrow} M_{i+1}$, be an execution of a CPN. A fairness function F is a function from transitions to boolean valued expressions. We define $EN_{F,i}(\xi) = true$ if $\exists t\langle b \rangle \in en(M_i) : F(t)\langle b \rangle = true$; otherwise $EN_{F,i}(\xi) = false$. Also let $OC_{F,i}(\xi) = true$ if $F(t_i)\langle b_i \rangle = true$; otherwise $OC_{F,i}(\xi) = false$. These two functions return true if a transition instance is enabled ($EN_{F,i}(\xi)$) or has occured ($OC_{F,i}(\xi)$) w.r.t. a fairness function. Denote the quantifier "there exist infinitely many" by $\exists^\omega$ and then let $InfEN_F(\xi)$ and $InfOC_F(\xi)$ be defined in the following way:

$$InfEN_F(\xi) = \begin{cases} true & \text{if } \exists^\omega i : EN_{F,i}(\xi) = true \\ false & \text{otherwise.} \end{cases}$$

$$InfOC_F(\xi) = \begin{cases} true & \text{if } \exists^\omega i : OC_{F,i}(\xi) = true \\ false & \text{otherwise.} \end{cases}$$

Now strong and weak fairness can be defined w.r.t. a fairness function. An execution is strongly fair if the set of transitions instances, defined by a fairness function F, are infinitely often enabled implies that they occur infinitely often [12].

$$InfEN_F(\xi) \Rightarrow InfOC_F(\xi).$$

The definition for weak fairness is that persistent enabling implies an occurrence [12].

$$\forall i \in \mathbb{N} : EN_{F,i}(\xi) \Rightarrow \exists k \geq i : [\neg EN_{F,k}(\xi) \lor OC_{F,k}(\xi)]$$

The semantics of our fairness constraints are equivalent to those presented by Jensen in [12], but the notation is a little different.

We now extend CPNs with fairness constraints on the transitions.

Definition 5. *A **fair CPN** (FCPN) is a triple $\Sigma_F = \langle \Sigma, WF, SF \rangle$ where Σ is a CPN and $WF = \{wf_1, \ldots wf_k\}$ a set of weak fairness functions, where wf_i is a function from the set of transitions to expressions such that $Var(wf_i(t)) \subseteq Var(t)$ for all $t \in T$ and $wf_i(t)\langle b \rangle \in \{true, false\}$ for any legal binding b of the expression. $SF = \{sf_1, \ldots, sf_m\}$ is the corresponding set of strong fairness functions with similar restrictions. An execution is an infinite sequence of markings and transition instances $\xi = M_0 \overset{t_0\langle b_0 \rangle}{\rightarrow} M_1 \overset{t_1\langle b_1 \rangle}{\rightarrow} \cdots$, such that for all $i \geq 0$ there exists $t_i\langle b_i \rangle$ for which $M_i \overset{t_i\langle b_i \rangle}{\rightarrow} M_{i+1}$ and ξ obeys the fairness constraints defined previously for all weak and strong fairness functions in WF and SF.*

The expression of a fairness function is true for all instances, which should be treated as equivalent; if one of the instances is fair, the fairness requirement has been satisfied.

In the following discussion we only consider finite state and finitely branching net systems. For analysis purposes these execution based semantics are not very convenient. Something similar to a Kripke structure is needed. However, the behavior of a FCPN cannot be described accurately by a Kripke structure because the fairness constraints are not taken into account in any way. Some mechanism is needed so that unfair executions can be rejected, and only those which conform to the fairness constraints are accepted. One way of doing this is by extending the definition of a Kripke structure to a fair Kripke structure. It could have been possible to extend a labeled transition system (LTS), but as they in [14] use a state based approach and the fastest available Streett emptiness algorithm [24] is state based, extending Kripke structures seemed more appropriate.

Let $\rho = s_0 s_1 s_2 \ldots \in S^\omega$ be an infinite sequence of states. The set of states occuring infinitely often in the sequence is given by:

$$Inf(\rho) = \{s \in S^\omega \mid \exists^\omega : \rho(i) = s\}$$

Definition 6. *A **fair Kripke structure** (**FKS**) [14] is a quintuple $K_F = \langle S, \rho, s_0, \mathcal{W}, \mathcal{S} \rangle$, where S is a set of states, $\rho \subseteq S \times S$ is a transition relation and $s_0 \in S$ is the initial state. The fairness requirements are defined by a set of weak fairness requirements[1] $\mathcal{W} = \{J_1, J_2, \ldots, J_k\}$ where $J_i \subseteq S$, and a set of strong fairness requirements, $\mathcal{S} = \{\langle L_1, U_1 \rangle, \ldots, \langle L_m, U_m \rangle\}$ where $L_i, U_i \subseteq S$. An execution is an infinite sequence of states $\sigma = s_0 s_1 s_2 \ldots \in S^\omega$, where s_0 is the initial state, and for all $i \geq 0$, $(s_i, s_{i+1}) \in \rho$. Computations, i.e. fair executions of the system, are sequences that obey the fairness requirements $\bigwedge_{i=1}^{k} Inf(\sigma) \cap J_i \neq \emptyset$ and $\bigwedge_{i=1}^{m} (Inf(\sigma) \cap L_i = \emptyset \vee Inf(\sigma) \cap U_i \neq \emptyset)$.*

An execution σ is a computation if both the weak and strong fairness requirements are satisfied. From each weak fairness set, at least one state occurs infinitely often in the execution and for each strong fairness if a state from a set L_i occurs infinitely often, also state from U_i must occur infinitely often in the execution.

Using the fairness requirements of the FKS, it is possible to only accept the computations which adhere to the fairness constraints on the transitions for FCPN. However, generating a FKS from a FCPN is not completely straightforward. For the same reason that a normal CPN must sometimes be modified in order for the LTL formulas to be able to express the fairness assumptions, a FKS cannot simply be a normal Kripke structure where we have added some fairness sets. The occurrence of transition instances must be made explicit in the FKS. Here, this is done by adding an intermediate state for each occurrence of a transition instance in the FKS. We make the transition "visible" by adding the intermediate state. For instance, if in the normal Kripke structure the marking M_j is followed by M_{j+1} when taking the transition instance $t\langle b \rangle$, in the FKS this

---

[1] In order to have consistent terminology weak and strong fairness are used instead of justice and compassion as in [14].

sequence will have an intermediate state. If the intermediate state is denoted by $M_{t\langle b\rangle}$ the sequence will be $M_j M_{t\langle b\rangle} M_{j+1}$. With the intermediate states added it is now possible to use the weak and strong fairness sets to ensure that only executions which obey the fairness constraints on the transitions are considered legal.

The states of the FKS are defined as pairs $\langle M, t\langle b\rangle\rangle$ so that the intermediate states can be distinguished from "normal" states. The special symbol $\bot$ replaces the transition instance if the state is not an intermediate state. Hence, to obtain a FKS $K_F = \langle S, \rho, s_0, \mathcal{W}, \mathcal{S}\rangle$ from a FCPN $\Sigma_F = \langle \Sigma, WF, SF\rangle$, we define S and ρ inductively as follows:

1. $s_0 = \langle M_0, \bot\rangle \in S$.
2. If $\langle M, \bot\rangle \in S$ and $M \xrightarrow{t\langle b\rangle} M'$ then, $\langle M', t\langle b\rangle\rangle \in S, \langle M', \bot\rangle \in S$ and $(\langle M, \bot\rangle, \langle M', t\langle b\rangle\rangle) \in \rho, (\langle M', t\langle b\rangle\rangle, \langle M', \bot\rangle) \in \rho$. If $\langle M, \bot\rangle \in S$ and $en(M) = \emptyset$ then $(\langle M, \bot\rangle, \langle M, \bot\rangle) \in \rho$.
3. S and ρ have no other elements.

The weak fairness sets and the strong fairness sets are defined as:

1. For each $wf_i \in WF$ the weak fairness set is
   - $J_i = \{\langle M, \bot\rangle \in S \mid \forall t\langle b\rangle \in en(M) : wf_i(t)\langle b\rangle = false\} \cup \{\langle M', t\langle b\rangle\rangle \in S \mid wf_i(t)\langle b\rangle = true\}$.
2. For each $sf_i \in SF$ the strong fairness sets are
   - $L_i = \{\langle M, \bot\rangle \in S \mid \exists t\langle b\rangle : t\langle b\rangle \in en(M) \wedge sf_i(t)\langle b\rangle = true\}$ and
   - $U_i = \{\langle M', t\langle b\rangle\rangle \in S \mid sf_i(t)\langle b\rangle = true\}$.

We are now ready prove that the construction of the FKS is correct, in the sense that the semantics of the fairness constraints are as we wanted.

Theorem 1. *Let Σ_F be a FCPN. $\xi = M_0 \xrightarrow{t_0\langle b_0\rangle} M_1 \xrightarrow{t_1\langle b_1\rangle} \cdots$ is a fair execution of Σ_F, if and only if the the FKS of the FCPN has a computation ξ'. (Under the assumption that Σ_F is finite state and finitely branching.)*

Proof. See [18].

Example. Consider a simple system consisting of a sender and a receiver. The sender can send n different messages to the receiver and the receiver acknowledges the messages. The communication could be modeled so that a sent message can be lost, while acknowledgements always come through. A Petri Net model checker would declare that if a message is ready to be sent, it would not always be received. A closer study of the counterexamples reveals two reasons. A message may not be sent because only one type of messages are sent even if though all types are ready to be sent, or the channel looses all messages. This could be remedied by including scheduler in the model, however, a simpler solution would be to use some fairness constraints on the transitions. By giving the transition "Send" a weak fairness constraint w.r.t the message type and the transition "Receive" a strong fairness constraint w.r.t the message type, the model checker

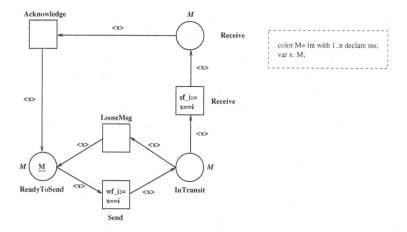

Fig. 1. FCPN model of the sender-receiver system.

would now report that the property holds. The weak fairness constraint ensures progress, so that no message type is left unsent when it is ready. The strong fairness constraint ensures that a message gets through eventually if it is resent. A FCPN model of the fair system can be found in Figure 1.

3   Model Checking LTL

3.1 ω-Automata

The close connection between automata on infinite words and LTL is used by many model checking procedures. Here the necessary automata theory is introduced and the most important terms are defined.

Büchi automata are the basic theoretical construction for every LTL model checker which uses the automata theoretic approach.

Definition 7. *A **labeled generalized Büchi automaton** (LGBA) [3] is a tuple $\mathcal{A} = \langle Q, \Delta, I, \mathcal{F}, \mathcal{D}, \mathcal{P} \rangle$, where Q is a finite set of states, $\Delta \subseteq Q \times Q$ is the transition relation, I is a set of initial states, $\mathcal{F} = \{F_1, F_2, \ldots, F_n\}$ with $F_i \subseteq Q$ is a finite set of acceptance sets, $\mathcal{D}$ some finite domain (in LTL model checking $\mathcal{D} = 2^{AP}$ for some finite set of atomic propositions AP) and $\mathcal{P} : Q \mapsto 2^{\mathcal{D}}$ is a labeling function. A run of $\mathcal{A}$ is an infinite sequence of states $\rho = q_0 q_1 q_2 \ldots$ such that $q_0 \in I$ and for each $i \geq 0$, $(q_i, q_{i+1}) \in \Delta$.*

Let the operator $Inf(\rho)$ be defined similarly for a run as for an execution. A run ρ is accepting if for each acceptance set $F_i \in \mathcal{F}$ there exists at least one state $q \in F_i$ that appears infinitely often in ρ, i.e. $Inf(\rho) \cap F_i \neq \emptyset$ for each $F_i \in \mathcal{F}$. An infinite word $\xi = x_0 x_1 x_2 \ldots \in \mathcal{D}^\omega$ is accepted iff there exists an accepting run $\rho = q_0 q_1 q_2 \ldots$ of $\mathcal{A}$ such that for each $i \geq 0$, $x_i \in \mathcal{P}(q_i)$. If $\mathcal{F} = \{F_1\}$ the LGBA corresponds to an ordinary Büchi automaton.

With Streett automata it is possible to extend the LTL model checking procedure to also cope with strong fairness in an efficient manner.

Definition 8. *A **Streett automaton** (see [35] for an arc labeled version) is a tuple $\mathcal{A} = \langle Q, \Delta, I, \Omega, \mathcal{D}, \mathcal{P} \rangle$, where Q, Δ, I, $\mathcal{D}$ and $\mathcal{P}$ have the same meanings as above. $\Omega = \{(L_1, U_1), \ldots, (L_k, U_k)\}$ with $L_i, U_i \subseteq Q$ is a set of pairs of acceptance sets. A run of a Streett automaton is defined in the same way as for an LGBA. The Streett automaton accepts a run $\rho = q_0 q_1 q_2 \ldots$ if $\bigwedge_{i=1}^{k}(Inf(\rho) \cap L_i = \emptyset \vee Inf(\rho) \cap U_i \neq \emptyset)$.*

We can read the acceptance condition as that the automaton accepts when "for each i, if some state in L_i is visited infinitely often, then some state in U_i is visited infinitely often". We define the set of infinite words accepted by $\mathcal{A}$ analogously to the LGBA case, using the new acceptance condition Ω.

Streett automata and generalized Büchi automata both accept the class of ω-regular languages, however, there is no polynomial translation from a Streett automaton to a Büchi automaton (see e.g. [28]). The converse can easily be done by letting $L_i = Q$ and $U_i = F_i$.

The set of ω-words the automaton $\mathcal{A}$ accepts is denoted by $\mathcal{L}(\mathcal{A})$, and it is called the language of $\mathcal{A}$. $\mathcal{L}(\mathcal{A}) = \emptyset$ denotes that the language accepted by $\mathcal{A}$ is empty. Determining whether $\mathcal{L}(\mathcal{A}) = \emptyset$ is referred to as performing an emptiness check.

3.2   LTL Definitions

Linear temporal logic (LTL) [22] is commonly used for specifying properties of reactive systems. LTL is interpreted over infinite executions which makes it appropriate to specifying properties of the executions of a Kripke structure.

Given a finite non-empty set of atomic propositions AP, LTL formulas are defined inductively as follows:

1. Every member $p \in AP$ is a LTL formula.
2. If φ and ψ are LTL formulas then so are $\neg\varphi, \varphi \vee \psi, X\,\varphi$ and $\varphi\,U\,\psi$.
3. There are no other LTL formulas.

An interpretation for a LTL formula is an infinite word $\xi = x_0 x_1 x_2 \ldots$ over the alphabet 2^{AP}, i.e. a mapping $\xi : \mathbb{N} \to 2^{AP}$. The mapping is interpreted to give the propositions which are true; elements not in the set are interpreted as being false. With ξ_i we mean the suffix starting at index i, namely $x_i x_{i+1} x_{i+2} \ldots$. The semantics of LTL are given by the following:

- $\xi \models p$ if $p \in x_0$, the first index of ξ, for $p \in AP$.
- $\xi \models \neg\varphi$ if $\xi \not\models \varphi$.
- $\xi \models \varphi \vee \psi$ if $\xi \models \varphi$ or $\xi \models \psi$.
- $\xi \models X\,\varphi$ if $\xi_1 \models \varphi$.
- $\xi \models \varphi\,U\,\psi$ if there exists an $i \geq 0$ such that $\xi_i \models \psi$ and $\xi_j \models \varphi$ for all $0 \leq j < i$.

The constants $\mathbf{T} = p \vee \neg p$, for an arbitrary $p \in AP$, and $\mathbf{F} = \neg\mathbf{T}$ denote atomic propositions which are always true and respectively false. Commonly used abbreviations are $\Diamond\varphi = \mathbf{T}\ U\ \varphi$, $\Box\varphi = \neg\Diamond\neg\varphi$ and the usual boolean abbreviations $\wedge$, $\Rightarrow$ and $\Leftrightarrow$.

By mapping the execution of a CPN or a FCPN to a sequence ξ, it is then possible to use LTL to specify properties of the executions.

Definition 9. *The corresponding sequence ξ_K of an execution $\zeta_K = M_0 M_1 \ldots$ of the Kripke structure of a CPN is the sequence generated by the function $seq(\zeta_K)$ which maps each marking in the execution to the corresponding subset of atomic propositions by evaluating the atomic propositions in the marking. The corresponding sequence ξ_F of a computation $\zeta_F = s_0 s_1 s_2 \ldots$ of the FKS of a FCPN is the sequence generated by the function $seq_F(\zeta_F)$ which maps every second state in the computation to the corresponding subset of atomic propositions by evaluating the atomic proposition in the marking component of every second state in the sequence, starting from s_0.*

Thus it is now possible to define two problems related to model checking of LTL, which are especially interesting for verification of systems modeled with Petri nets.

Model Checking Problem: Given a CPN Σ, and an LTL formula φ, does $seq(\zeta_K) \models \varphi$ hold for every execution ζ_K of Σ.

Fair Model Checking Problem: Given a FCPN Σ_F and an LTL formula ψ, does $seq_F(\zeta_F) \models \psi$ hold for every fair execution ζ_F of Σ_F.

3.3   Model Checking

The automata theoretic approach to model checking utilizes the close relationship between LTL and automata on infinite words. Several procedures [8,4,5,32] have been suggested which construct a LGBA that recognizes all the models of a given LTL formula. Most model checking procedures are designed for ordinary Büchi automata but in our case this is not a problem as they are a special case of the LGBA.

Given a LTL property φ and a corresponding Büchi automaton, model checking a system is now possible by interpreting the Kripke structure as a Büchi automaton. This Büchi automaton represents all the possible executions of the system. If this system automaton is intersected with the property automaton, the result is an automaton which accepts all executions which are common to the two automata. Intersecting the system automaton with an automaton corresponding to the negation of the property yields an automaton which has no accepting executions if and only if the system is a model of the LTL property.

Hence, the steps performed to verify that a system has a property given by a LTL formula φ and solve the model checking problem are the following [3,17]:

1. Construct a generalized Büchi automaton $\mathcal{A}_{\neg\varphi}$ corresponding to the negation of the property φ.

2. Generate the Kripke structure of the system and interpret it as a LGBA $\mathcal{K}$, with $\mathcal{F} = \emptyset$.
3. Form the product automaton $\mathcal{B} = \mathcal{A}_{\neg\varphi} \times \mathcal{K}$.
4. Check if $\mathcal{L}(\mathcal{B}) = \emptyset$.

If $\mathcal{L}(\mathcal{B}) = \emptyset$ the model of the system has the desired property. Combining several of these steps into a single algorithm and performing them in an interleaving manner is referred to as "on-the-fly" model checking [3,17]. Naturally the procedure can also be done with a simple Büchi automaton, if the property LGBA is further expanded to a simple Büchi automaton.

```
proc Check (formula φ, System K) ≡
    LGBA-Automaton A := to-automaton (¬φ);              Step 1.
    LGBA-Automaton B := product (A, K);                 Step 2.
    Streett-Automaton S;
    Component mscc;
    forall mscc ∈ MSCC(B) do                            Step 3.
        if (¬modelcheck (mscc)) then ;                  Step 4.
            continue;
        fi
        if (hasWF(mscc) AND
                ¬wf-modelcheck (mscc)) then             Step 5.
            continue;
        fi
        S = ToStreett (mscc);
        if (hasSF(mscc) AND
                ¬sf-modelcheck (S)) then                Step 6.
            continue;
        fi
        counterexample (S);                             Step 7.
        return true;
    od
    return false;
```

Fig. 2. The fair model checking procedure

The afore mentioned procedure is not appropriate for model checking a FKS, from an efficiency point of view, and thus solving the fair model checking problem. What is needed is a procedure which can handle both generalized Büchi acceptance sets and Streett acceptance sets. Of course, the procedure should also avoid using the more time consuming (see e.g. [24]) Streett emptiness checking procedure if possible.

To solve the fair model checking problem the new procedure for high-level Petri Nets, shown in Figure 2, does the following.

1. Constructs a generalized Büchi automaton $\mathcal{A}_{\neg\varphi}$.
2. The *Kripke* structure of the FCPN model is constructed, interpreted as a LGBA with $\mathcal{F} = \emptyset$, and simultaneously the product with $\mathcal{A}_{\neg\varphi}$ is computed.
3. Tarjan's algorithm is used to compute a *maximal strongly connected component* (MSCC) of the product. In graph theoretic terms a MSCC is a maximal subset of vertices C of a directed graph, such that for all $v_1, v_2 \in C$, the vertex v_1 is reachable from v_2 and vice versa. The set is maximal in the sense that if any state is added to this set, it ceases to be a SCC.
4. When a MSCC of the product automaton has been calculated, we check for generalized Büchi acceptance, i.e. whether there is *any* execution which violates the given property. There cannot exist a fair counterexample if there is no failing execution. Hence, if the component does not contain a state from each Büchi acceptance set (LGBA acceptance condition), we return to step 3.
5. If a component is accepted, the component is checked if it is weakly fair. This can be done without generating any intermediate states, which is of course desirable. The memberships of the fairness sets are assigned in the following manner. Let the MSCC be denoted by C. For a state $s = \langle M, P \rangle$, where M is the corresponding marking in the Kripke structure and P the corresponding state in the formula automaton. Then, for all $s \in C$, s is member of F_i if:

    – $\forall t\langle b \rangle \in en(M) : wf_i(t)\langle b \rangle = false$, or

    – $\exists t\langle b \rangle \in en(M), s' \in S : wf_i(t)\langle b \rangle = true$ and $(s, s') \in \Delta, M \overset{t\langle b \rangle}{\to} M', s' = \langle M', P' \rangle$, such that $s' \in C$.

    See Theorem 2 why this works. If the component is accepted, i.e. it contains all weak fairness sets, and has no strong fairness constraints, the intermediate states are added according to the definition of a FKS. However, the generalized Büchi sets are interpreted as Streett acceptance sets U_i with each L_i set initialized to the universal set. The intermediate states are needed for the correctness of the counterexample. The counterexample algorithm must be able to identify fair transitions instances which can occur infinitely often. Now we can generate a counterexample at step 7.
6. We now know that the MSCC contains a weakly fair counterexample. To ensure that there is also a counterexample which is both strongly and weakly fair, we will use a Streett emptiness checking algorithm on this MSCC. (Using the Streett emptiness checking to handle strong fairness constraints goes back to at least [6,20].) However, we cannot yet ignore the property sets and the weak fairness sets. Therefore the weak fairness sets are computed according to the definition of the FKS and both the property sets and the weak fairness sets are again simulated with Streett acceptance sets, using the technique given in step 5. Before the component is given to the Streett emptiness algorithm, also the strong fairness sets L_i and U_i must be computed. These are also computed according to the definition of a FKS. The correctness of this step is proven in Theorem 3. We simulate the FKS with a Streett automaton and if no weakly and strongly fair counterexample is found, we continue from step 3 with the next MSCC of the product automaton.

7. A counterexample is generated using the subset of vertices of the MSCC (the Streett emptiness algorithm possibly deletes some states and edges), which the emptiness checking algorithm gives to the counterexample algorithm.

Theorem 2. *Let C be a MSCC of the product automaton. The component contains a weakly fair counterexample if and only if the component interpreted as an automaton A, using the set assignments done in step 5, is non-empty.*

Proof. If C contains a counterexample which is weakly fair, then by step 3 of the procedure, the sets representing the property must be present in the automata. For all weak fairness functions there are transition instances in such a way that the implicit set generated by the fairness function is infinitely often disabled or can occur infinitely often. A state in A belongs to a weak fairness set, i.e. an acceptance set of the automaton, if no transition of the set is enabled in the state or a transition of the set occurs and the resulting state is also in the component. The first condition handles cases where the weakly fair transition instances never are enabled while the second condition handles the cases where the transition instances occur infinitely often. Thus all weak fairness sets are present if there is a weakly fair counterexample, and thus A is non-empty.

If A is non-empty, we know from step 3 of the procedure that the component contains a counterexample. Any execution respecting the acceptance sets of the property is a counterexample. As all sets are present in the component, and all states are reachable from each other, there must exist an execution which goes through both the property sets and weak fairness sets (a trivial example is an execution which visits all states of the component infinitely often). This execution is a weakly fair counterexample.

Theorem 3. *Let C be a MSCC of the product automaton. The component contains a strongly fair counterexample if and only if the Streett automaton A resulting from transforming C according to the definition of a FKS and simulating the property and weak fairness set with the Streett sets (consider the CPN marking of each product state only, ignoring the product automaton state) is non-empty.*

Proof. If C contains a counterexample which is weakly and strongly fair, then by step 3 of the procedure the property sets must be present in A, as they are simulated by some Streett sets. From Theorem 1 we know that the acceptance sets simulating the weak fairness sets will also be present as the counterexample is weakly fair. As the counterexample is also strongly fair for each strong fairness function there are transition instances, such that if the instances are infinitely often enabled, some instances occur infinitely often. In A each state which has a transition instance enabled that makes a strong fairness function true, belongs to the corresponding L set, thus marking all possible states where some strongly fair transition instances defined by a fairness function are enabled. An intermediate state is generated which will belong to the corresponding U set, which will be belong to the MSCC if the occurrence of the transition instance results in a state which is in the component. Thus the U set marks all the states making it possible for transition instances defined by a fairness function to occur. Remembering that the Streett acceptance condition is similar to the strong fairness constraint, clearly the Streett acceptance will be satisfied if a strongly fair execution is present in the component. Thus the component is non-empty.

Let A be non-empty. Any execution respecting the property sets will be a counterexample. We know from Theorem 1 that any execution respecting the FKS fairness sets must be both strongly and weakly fair. The simulation of the generalized Büchi sets (the property and the weak fairness sets) is done by setting $L_i = S$ and $U_i = F_i$. As each L_i is guaranteed to be present, each U_i must be satisfied which corresponds to the LGBA acceptance condition that each F_i must be satisfied. Since the component respects the fairness sets it is possible to construct an execution which respects the fairness sets (otherwise the component would be empty). Hence the component contains a weakly and strongly fair counterexample.

Corollary 1. *The procedure solves the fair model checking problem.*

The procedure tries to avoid the cost of the more expensive Streett emptiness check, whenever possible, by always first testing for weak fairness and only invoking the Streett check if there are strong fairness constraints enabled. This might result in faster running times compared to always performing the check. Also by performing the verification in an on-the-fly manner, checking one MSCC at a time, the cost of computing all MSCCs of the product automaton might be avoided.

There are several other algorithms for automata theoretic model checking LTL which have been presented in the literature. The nested-depth-first-search algorithm of [3] was designed for (non-generalized) Büchi automata. The algorithm of [14] is similar in the sense it uses both Büchi and Streett acceptance conditions, however their emptiness checking procedure is BDD based, making it unsuitable for the explicit state tool this procedure was designed for. The transitive closure and fixpoint computations the algorithm requires makes it an unwise choice in an explicit state setting. The same argument applies to the algorithm of [10]. An algorithm tailored to handle only generalized Büchi acceptance sets was presented in [4], but due to some optimizations it makes it could not be used here. It is, however, somewhat similar to the procedure presented in this work, as it also is a Tarjan based on-the-fly algorithm.

4   Implementation

The model checking procedure described in this work has been implemented in the MARIA analyzer [21]. The MARIA analyzer is a reachability analyzer for algebraic system nets [15,16,25] and it has been developed at the Laboratory for Theoretical Computer Science at Helsinki University of Technology.

4.1 The MARIA Analyzer

The MARIA analyzer, is a reachability analyzer for Algebraic System Nets. The intention is to develop an analyzer with model checking capabilities for a formalism which is powerful enough to model in a straightforward manner high-level programming languages. By using language specific front ends, the idea is that MARIA can function as the analysis tool for several formalisms. Currently a front end for SDL [1] is under development.

The net class of MARIA is based on an algebra with powerful built-in data types and expressions. MARIA supports leaf types (bool, char, enum, int, unsigned) familiar from high-level programming languages and also complex structured types such as structs, arrays and bounded queues. There are built-in operations for multi-set operations, multi-set sums, etc.

4.2 Implementation

The implementation was programmed in C++, like the rest of the analyzer. The emptiness checking algorithm for Streett automata and counterexample generation follows the algorithms described in [19]. Some optimizations were however performed. Management of the arcs of the product automaton was carefully designed, so that only during the Streett emptiness checking phase were the arcs kept in main memory. The Streett emptiness check was also modified so that not all intermediate states were added to the FKS. Only transition instances related to a strong fairness constraint and some instances related to a weak fairness constraint caused an intermediate state to be added. Specifically the transition instances which belonged to a weak fairness set and could occur so that the resulting state was still in the current MSCC, caused an intermediate state to be generated. An external implementation [26] of the algorithm presented in [8] was used as the translator from an LTL formula to a LGBA.

5 Verifying a Sliding Window Protocol

The sliding window protocol provides reliable transmission over an unreliable communication medium. It is used in several data link control (DLC) protocols and it is serves as a basis for reliable transport in many protocols.

Several papers have been written on the verification of the protocol. Recent works which focus on automatic verification of the protocol include [13,9,31]. In [13] Kaivola combines a compositional approach with property preserving preorders and equivalences to verify both safety and liveness properties for arbitrary channel lengths. Results up to a window size of seven is presented with bounded buffer sizes. Using Queue BDDs Godefroid and Long [9] verify a full duplex version of the protocol with queue sizes up to eleven. Smith and Klarlund [31] use a theorem proving tool to verify safety properties of the protocol for unbounded buffer sizes, window sizes and channel capacities. They also verify safety properties with a fixed window size as large as 256.

5.1   Protocol Description

The specific version of the protocol on which we focus here is a unidirectional version due to [33]. The sliding window protocol consists of four main components. These are the sender, the receiver, the transmission channel and the acknowledgment channel. The structure of the protocol can be seen in Figure 3. Additionally we assume there is a data source which generates messages to be sent and a target which receives them. The protocol has the following important parameters:

- tw The size of the transmission window. This specifies the maximum amount of messages which the sender can send without receiving an acknowledgment.
- rw The size of the receive window. This specifies the maximum amount of messages which the receiver can receive without forwarding them to the target.
- w The maximum value of the sequence numbers.

Only when $tw + rw \leq w + 1$ holds will the protocol function properly [33].

The abstraction made in the following description follow Kaivola [13]. The sender side of the protocol functions in the following way. It receives data to be sent from the data source. It can then choose to send the message containing the data and a sequence number or receive more messages to be sent. The sender can send tw messages without receiving an acknowledgment. If no acknowledgment is received the sender can timeout and resend the messages. Following the timeout model of Kaivola, a timeout can occur only if no further messages can be sent to the transmission channel. When a valid acknowledgment is received the transmission buffer is emptied up till the last acknowledged message and the sender can receive new messages from the data source.

The receiver side of the protocol functions in the following way. When the receiver receives a message it is stored and marked as unacknowledged, if the sequence number of the message is in the reception window. Then the protocol proceeds to forward the messages to the data target and empties the receive

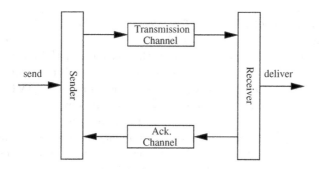

Fig. 3. The main components of the protocol.

buffer. The sequence number of the last message sent to the target is sent as an acknowledgment to the sender.

All channels in the protocol are modeled as queues with a fifo discipline, which can loose any message.

5.2   Properties

The aim of this small case study was to test the performance of the MARIA model checker. Hence, liveness properties are of special interest as they in many cases require fairness assumptions.

The liveness property required is that as many items should be delivered to the data target as are read from the data source. If the system is connected to a data source which generates the ω-regular language (see e.g. [34]) $0^\omega \cup 0^* \cdot 1$, using *data-independence* [38,27] and properties of ω-regular languages Kaivola argues that it is then enough to verify the property:

LIVE Either infinitely many 0:s are delivered to the target or 1 is delivered.

The property can be expressed in LTL in the following way:

$$\Box\Diamond(\mathit{receive}(0)) \vee \Diamond(\mathit{receive}(1)),$$

where $\mathit{receive}(i)$ denotes an atomic proposition indicating that the receiver has received a message with the data i. However, the property does not hold unconditionally. A fairness constraint is needed for the channels. The following fairness constraint is used:

FAIR For all sequence numbers $i \in \{0, \dots, w\}$, if a message with the sequence number i is sent to the receiver infinitely often, it receives a message with the sequence number i infinitely often. The same constraint applies for the acknowledgment channel.

5.3   The MARIA Model

Modeling the protocol in the MARIA Petri Net class is quite straightforward. Using the queue and array types of MARIA, it is easy to model the channels as queues and the internal data buffers of the sender and the receiver as arrays. The complete model consists of 12 places and 9 high-level transitions. Two places are used to make it possible to express the liveness property **LIVE** in LTL. The fairness constraints on the channels are easy to express by putting strong fairness constraints, tied to the sequence number of the message, on the channels which handle the receiving of data and acknowledgments respectively. Hence there is a fairness set for each channel for each sequence number. A weak fairness constraint is needed on the receiver side to guarantee progress in the sequential parts. To obtain the complete model see http://www.tcs.hut.fi/~timo/pn2001.

w	$tw = rw$	states	arcs	|product|	time (s)
1	1	118	250	297	0.1
3	2	2048	6104	5337	2.3
5	3	14952	50982	39659	36.5
7	4	79720	296088	212997	508.2
9	5	356570	1404910	955745	5978.4
11	6	1427184	5871588	3830669	59748.2

Fig. 4. Statistics for the model checking.

5.4 Results

The new procedure was tested by model checking the liveness property **LIVE** for different window sizes with the fairness constraint **FAIR**. The parameters were chosen such that $rw = tw$ and the capacity of the queues modeling the channels were w. To yet decrease the possible values for the parameters the value of w was fixed such that $w = tw + rw - 1$.

All tests were performed on a PC with a 1 GHz AMD Athlon processor and 512 MB of RAM. The system was running the Debian GNU/Linux distribution, version 2.2. Times were measured using the UNIX command "time" and the user plus system time was recorded. In the table of Figure 4 the results can be seen. The "states" and the "arcs" column indicate the number of nodes and the number of arcs respectively in the reachability graph. The |product| column gives the number of states in the product. The same protocol was also modeled using the Spin [11] model checker. However, because the automata translator of Spin could not translate the fairness constraints to an automaton for $w > 1$, due to memory exhaustion, a comparison was not deemed fair and hence omitted. Actually, no algorithm available to the author could translate the fairness constraints, for the window sizes presented here, on the hardware used. As can be seen from the results the MARIA model checker can due to the efficient handling of the fairness constraints cope with quite large window sizes. It should be noted that no reduction methods were used on the state space. Using reduction methods could probably enable model checking of larger window sizes.

6 Conclusions

In this work LTL model checking for high-level Petri nets has been extended with nets, which have fairness constraints on the transitions. For P/T Nets a similar construction was presented, but not implemented, in [19]. The semantics for the fairness constraints is given by an execution based semantics and a fair Kripke structure. Using Streett automata the model checking procedure is extended to handle the fairness constraints in an efficient manner.

The procedure has been implemented in the MARIA analyzer. Using a model of a unidirectional sliding window protocol, the implementation was tested. Results indicate that the procedure seems to scale well, even when there are many strong fairness constraints present. The protocol could not have been verified for so large window sizes, had the fairness constraints been part of the property to be verified, instead of being part of the model.

There are still some open questions concerning the new LTL model checking procedure. It is clear that not all intermediate states have to be added when model checking. It should be possible to formulate a better sufficient condition, which could be statically checked, for transitions in the model which need the intermediate states to be generated. This could reduce the number of intermediate states needed in the procedure. It also could be interesting to generalize this method to encompass the full branching time logic CTL*. As CTL* model checking can be reduced to several calls to a LTL model checker [6] this should be possible. Another interesting question is what kind of effect would the procedure have on partial order methods, such as the stubborn set method [36].

Acknowledgments. Thanks to Keijo Heljanko many fruitful discussions and comments, to Prof. Ilkka Niemelä for suggesting the sliding window protocol example and to the anonymous referee whose comments helped me find a bug in the original Maria model of the protocol.

References

1. CCITT. Specification and description language (SDL). Technical Report Z.100, ITU-T, 1996.
2. E.M. Clarke and E.A. Emerson. Design and synthesis of syncronization of skeletons using branching time temporal logic. In *Proceedings of the IBM Workshop on Logics of Programs*, pages 52–71. Springer-Verlag, 1981. LNCS 131.
3. C. Courcoubetis, M.Y. Vardi, P. Wolper, and M. Yannakakis. Memory efficient algorithms for the verification of temporal properties. *Formal Methods in System Design*, 1:275–288, 1992.
4. J-M. Couvreur. On-the-fly verification of linear temporal logic. In *Proceeding of the World Congress on Formal Methods in the Development of Computing Systems (FM'99)*, volume 1, pages 253–271, Berlin, 1999. Springer-Verlag. LNCS 1708.
5. M. Daniele, F. Giunchiglia, and M.Y Vardi. Improved automata generation for linear temporal logic. In *Proceedings of the International Conference on Computer Aided Verification (CAV'99)*, pages 249–260, Berlin, 1999. Springer-Verlag. LNCS 1633.
6. E.A. Emerson and C-L. Lei. Modalities for model checking: Branching time logic strikes back. *Science of Computer Programming*, 8(3):275–306, 1987.
7. N. Francez. *Fairness*. Springer-Verlag, New York, 1986.
8. R. Gerth, D. Peled, M.Y. Vardi, and P. Wolper. Simple on-the-fly automatic verification of linear temporal logic. In *Proceedings of the 15th Workshop Protocol Specification, Testing, and Verification*, Warsaw, June 1995. North-Holland.
9. P. Godefroid and D.E Long. Symbolic protocol verification with Queue BDDs. *Formal Methods in System Design*, 14(13):257–271, May 1999.

10. R. Hojati, V. Singhal, and R.K. Brayton. Edge-Streett / edge-Rabin automata environment for formal verification using language containment. Memorandum UCB/ERL M94/12, Electronics Research Laboratory, University of California, Cory Hall, Berkley, 1994.

11. G.J. Holzmann. The model checker Spin. *IEEE Transactions on Software Engineering*, 23(5):279–295, May 1997.

12. K. Jensen. *Coloured Petri Nets*, volume 1. Springer-Verlag, Berlin, 1997.

13. R. Kaivola. Using compositional preorders in the verification of sliding window protocol. In *Proceedings of the 9th International Conference on Computer Aided Verification (CAV'97)*, pages 48–59. Springer-Verlag, 1997. LNCS 1254.

14. Y. Kesten, A. Pnueli, and L. Raviv. Algorithmic verification of linear temporal properties. In *Proceedings of the 25th International Colloquium on Automata, Languages and Programming (ICALP 1998)*, pages 1–16. Springer-Verlag, 1998. LNCS 1443.

15. E. Kindler and W. Reisig. Algebraic system nets for modeling distributed algorithms. *Petri Net Newsletter*, 51:16–31, 1996.

16. E. Kindler and H. Völzer. Flexibility in algebraic nets. In *Proceedings of the International Coneference on Application and Theory of Petri Nets 1998 (ICAPTN'98)*, pages 345–364. Springer-Verlag, 1998. LNCS 1420.

17. R.P. Kurshan. *Computer-Aided Verfication of Coordinating Processes: The Automata-Theoretic Approach*. Princeon University Press, Princeton, New Jersey, 1994.

18. T. Latvala. Model checking linear temporal logic properties of Petri nets with fairness constraints. Research Report A67, Helsinki University of Technology, Jan 2001.

19. T. Latvala and K. Heljanko. Coping with strong fairness. *Fundamenta Informaticae*, 43(1–4):175–193, 2000.

20. O. Lichtenstein and A. Pnueli. Checking that finite state programs satisfy their linear specification. In *Proceedings of the 12th ACM Symposium on Principles of Programming Languages*, pages 97–107, 1985.

21. M. Mäkelä. Maria: Modular reachability analyzer for algebraic system nets. On-line documentation, 1999. http://www.tcs.hut.fi/maria.

22. A. Pnueli. The temporal logic of programs. In *Proceedings of 18th IEEE Symposium on Foundation of Computer Science*, pages 46–57, 1977.

23. J.P. Quielle and J. Sifakis. Specification and verification of concurrent systems in CESAR. In *Proceedings of the 5th International Symposium on Programming*, pages 337–350, 1981.

24. M. Rauch Henzinger and J.A. Telle. Faster algorithms for the nonemptiness of Streett automata and for communication protocol pruning. In *Proceedings of the 5th Scandinavian Workshop on Algorithm Theory(SWAT'96)*, 1997.

25. W. Reisig. Petri nets and algebraic specifications. *Theoretical Computer Science*, 80:1–34, March 1991.

26. M. Rönkkö. A distributed object oriented implementation of an algorithm converting a LTL formula to a generalised Büchi automaton. On-line documentation, 1998. http://www.abo.fi/%7Emronkko/PROJECT/LTL2BUCHI/abstract.html.

27. K. Sabnani. An algorithmic technique for protocol verification. *IEEE Transactions on Communications*, 36(8):235–248, 1988.

28. S. Safra. On the complexity of omega-automata. In *29th Annual Symposium on Foundations of Computer Science*, pages 319–327. IEEE, 1988.

29. K. Schmidt. LoLA: A low level analyser. In *Proceedings of the 21st International Conference on Application and Theory of Petri Nets (ICATPN 2000)*, pages 465–474. Springer-Verlag, 2000. LNCS 1825.

30. A.P. Sistla and E.M. Clarke. The complexity of propositional linear temporal logic. *Journal of the Association for Computing Machinery*, 32(3):733–749, July 1985.

31. M.A. Smith and N. Klarlund. Verification of a sliding window protocol using IOA and MONA. In *Proceeding of the 20th Joint International Conference on Formal Methods for Distributed Systems and Communication Protocols (FORTE/PSTV 2000)*, pages 19–34. Kluwer Academic Publishers, 2000.

32. F. Somenzio and R. Bloem. Efficient büchi automata from LTL formulae. In *Proceedings of the International Conference on Computer Aided Verification (CAV2000)*, pages 248–263. Springer-Verlag, 2000. LNCS 1855.

33. N.V Stenning. A data transfer protocol. In *Computer Networks*, volume 11, pages 99–110, 1976.

34. W. Thomas. Automata on infinite objects. In *Handbook of Theoretical Computer Science*, volume B, pages 133–191. Elsevier, 1990.

35. W. Thomas. Languages, automata and logic. In G Rozenberg and A Salomaa, editors, *Handbook of Formal Languages*, volume 3, pages 385–455. Springer-Verlag, New York, 1997.

36. A. Valmari. The state explosion problem. In *Lectures on Petri Nets I: Basic Models*, pages 429–528. Springer-Verlag, 1998. LNCS 1491.

37. K. Varpaaniemi, J. Halme, K. Hiekkanen, and T. Pyssysalo. PROD reference manual. Technical Report 13, Helsinki University of Technology, 1995.

38. P. Wolper. Expressing interesting properties of programs in propositional temporal logic. In *Proceedings of the 13th ACM Symposium on Principles of Programming Languages*, pages 184–194, 1986.

Incremental State Space Construction for Coloured Petri Nets

Glenn Lewis[1] and Charles Lakos[2]

[1] Department of Computing, University of Tasmania
Hobart, Tas, 7001, Australia
Glenn.Lewis@utas.edu.au
[2] Computer Science Department, University of Adelaide
Adelaide, SA, 5005, Australia
Charles.Lakos@adelaide.edu.au

Abstract. State space analysis is a popular formal reasoning technique. However, it is subject to the crippling problem of state space explosion, where its application to real world models leads to unmanageably large state spaces. In this paper we present algorithms which attempt to alleviate the state space explosion problem by taking advantage of the common practice of incremental development, i.e. where the designer starts with an abstract model of the system and progressively refines it. The performance of the incremental algorithm is compared to that of the standard algorithm for some case studies, and situations under which the performance improvement can be expected are identified.

1 Introduction

A major advantage of formal methods is that they allow for formal reasoning. State-based formal reasoning techniques commonly involve examining every possible state of a system. Such techniques are automatic, can be applied by less trained personnel, and can be used for analysis and error detection as well as verification. For these reasons they are seen as one of the most promising formal reasoning techniques [10].

Unfortunately the number of states of a system increases exponentially as the complexity of the system increases. This means that the total number of states is often far too large with respect to time and/or space resources to be fully generated. This growth of the state space is referred to as *state space explosion*, and is the primary obstacle to the practical application of state-based formal reasoning techniques.

Most state space methods investigate the reachable states by constructing a *Reachability Graph* (also known as an *Occurrence Graph* [4]). In its most basic form a reachability graph is a directed graph consisting of all the states the system can reach from its given initial state. Each vertex of the graph represents a state of the system being analysed, and each directed edge is labelled with the action that leads to the next state (that is, the next vertex of the graph). A basic reachability graph represents an *interleaving semantics* of a system. That is, it does not model the possibility of two or more actions occurring simultaneously.

J.-M. Colom and M. Koutny (Eds.): ICATPN 2001, LNCS 2075, pp. 263–282, 2001.
© Springer-Verlag Berlin Heidelberg 2001

State space explosion does not preclude the use of state space analysis in practice, since the great advantages of these methods have motivated many researchers to try to find ways of alleviating the problem [10]. Two popular approaches are *Symmetric Occurrence Graphs* [4], and *Stubborn Sets* [10]. Symmetric occurrence graphs identify sets of symmetric states and store only one representative from each set. Thus, large sections of the state space are omitted, at the cost of additional computation to determine which states are symmetric. Another benefit is that the full state space can be recovered. Stubborn sets, on the other hand, reduce the size of the state space by eliminating a number of interleavings of independent processes. Here, the full state space cannot be recovered, but it is guaranteed that the desirable properties are not affected by the reduction in the state space.

A recent technique by Christensen and Petrucci [1] exploits the modular structure present in many formal specifications to minimise the representation of the interleaving of independent actions and therefore to help alleviate the state space explosion. The current paper has certain points of contact with the approach of Christensen and Petrucci, in attempting to alleviate state space explosion by utilising the incremental structure found in many Coloured Petri Net (CPN) models. This structure arises because designers commonly develop their models incrementally — they start with an abstract model of the system (and possibly verify certain properties of the abstraction) and then progressively refine that model till sufficient detail is included.

This paper is organised as follows: through the use of a simple example Section 2 informally introduces three forms of incremental change that have previously been identified as suitable for use in practice; Section 3 presents algorithms that take advantage of each of the forms of incremental change; Section 4 reports on the implementation of the incremental algorithm; Section 5 examines the performance of the incremental algorithm for some case studies and identifies the situations under which performance improvements are maximised; and finally conclusions and areas for further work are given in Section 6. We assume the reader has a working knowledge of CPNs, as in [3].

2   Incremental Development of Coloured Petri Nets

Incremental change is fundamental to the way people solve complex problems. They tend to first develop a solution to a simpler problem, and then incrementally add detail to change this solution to address the problem at hand.

In the context of Coloured Petri Nets [3], three forms of incremental change or refinement have been identified as being commonly applicable in practice [5, 7]. They have been termed *type refinement*, *subnet refinement*, and *node refinement*. These are all special cases of so-called *system morphisms*, i.e. morphisms (or mappings) $\phi : N \rightarrow N'$, from a refined net, N, to an abstract net, N', which maintain behavioural compatibility. This means that every (complete) action sequence of the refined system corresponds to an action sequence of the abstract system, and every reachable state of the refined system (following a complete

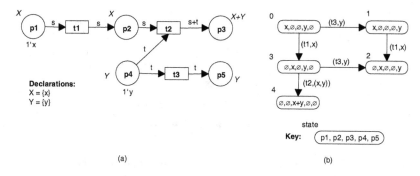

Fig. 1. A simple net (a) and its reachability graph (b)

action sequence) corresponds to a state of the abstract system. This correspondence is achieved by either ignoring the refined action or state components, or projecting them onto abstract components [5].

Using the net of Figure 1 (a) as an example, we now informally present type, subnet and node refinement. (A formal presentation can be found in [5].) In this net, place p_1 initially holds a token x, while place p_4 initially holds a token y. If transition t_1 fires with mode x, it results in transferring a token x from place p_1 to p_2. The subsequent firing of transition t_2 with mode (x, y) would result in token x being consumed from place p_2 and token y from place p_4, with tokens x and y being deposited in place p_3.

The first and simplest form of refinement, *type refinement*, involves incorporating additional information in the tokens and firing modes, while keeping the net structure unchanged. Each value of the refined type can be projected onto a value of the abstract type. For example it may be desirable to introduce further information into the type X of Figure 1 (a). This will simply involve extending the type X, to say $X = \{(x, 1)\}$, extending the corresponding transition firing modes, and changing the initial marking so that the place p_1 contains the token $(x, 1)$. In this refined version of the system, it is certainly the case that if there is a behaviour of the refined system, then there is a corresponding behaviour of the abstract system.

The second form of refinement, *subnet refinement*, involves augmenting a subnet with additional places, transitions, and arcs. We can use subnet refinement on the net of Figure 1 (a) to give the net of Figure 2, where places p_6 and p_7 and transition t_4 have been added. As with type refinement, constraints on subnet refinement ensure that for each behaviour of the refined system there is a corresponding behaviour of the abstract system (but not necessarily vice versa).

The third form of refinement, *node refinement*, is to replace a place (transition) by a place (transition) bordered subnet. Canonical forms of such refinements are as mandated in [5]. The place p_2 and transition t_3 of the net of Figure 1 (a) might be refined as shown in Figure 3. The refined place has one input border place called p_2-*inp1* and one output border place called p_2-*out1*. The p_2-*accept* and p_2-*offer* transitions, together with the internal place p_2-*buf* constitute the

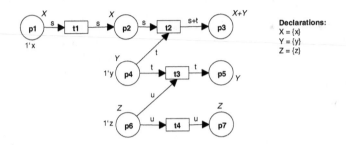

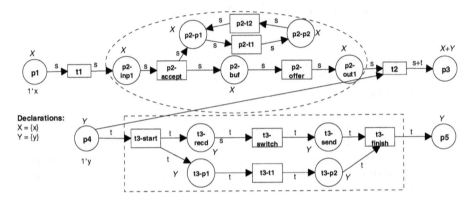

Fig. 2. The net of Figure 1 (a) refined using subnet refinement

Fig. 3. The net of Figure 1 (a) refined using node refinement

basis of the canonical place refinement. It guarantees that tokens are preserved. Further activity is achieved by the subnet refinement which extends transition p_2-*accept*. The transitions t_3-*start*, t_3-*finish*, t_3-*switch*, and the places t_3-*recd*, t_3-*send*, constitute the basis of the canonical transition refinement. This guarantees that the border transitions (t_3-*start* and t_3-*finish*) will fire with matching modes. Again, a behaviour of the refined system will have a corresponding abstract behaviour, though the reverse will not necessarily be the case.

Even though the above three forms of refinement can be identified and analysed in isolation, they will commonly be used in combination in practical applications.

3   Incremental State Space Algorithms

In this section we present reachability graph algorithms that take advantage of type, subnet and node refinement. These can be combined into a single algorithm that takes advantage of a mixture of type, subnet, and node refinement, but we do not do so due to space constraints.

Coloured Petri Nets are defined in the context of a *universe of non-empty colour sets* Σ, the functions over Σ given by $\Phi\Sigma = \{X \to Y \mid X, Y \in \Sigma\}$, the multisets over a colour set X given by $\mu X = \{X \to \mathbb{N}\}$, and the sequences over a colour set X given by $\sigma X = \{x_1 x_2 \ldots x_n \mid x_i \in X\}$.

Definition 1. A *Coloured Petri Net* N is a tuple
$N = (P, T, A, C, E, \mathbb{M}, \mathbb{Y}, M_0)$ where:

   a. P is a set of places
   b. T is a set of transitions, s.t. $P \cap T = \emptyset$
   c. A is a set of arcs, s.t. $A \subseteq (P \times T) \cup (T \times P)$
   d. $C : P \cup T \to \Sigma$ determines the colours of places and (modes) of transitions
   e. $E : A \to \Phi\Sigma$ gives the arc inscriptions, s.t. $E(p,t), E(t,p) : C(t) \to \mu C(p)$
   f. $\mathbb{M} = \mu\{(p,c) \mid p \in P, c \in C(p)\}$ is the set of markings
   g. $\mathbb{Y} = \mu\{(t,c) \mid t \in T, c \in C(t)\}$ is the set of steps
   h. M_0 is the initial marking, $M_0 \in \mathbb{M}$

Note that there is at most one arc in each direction for any (place,transition) pair and that the effect of an arc is given by the arc inscription in conjunction with a particular transition firing mode. We refer to a (place,colour) pair as a *token element*, and a (transition,colour) pair as a *firing element*. We denote the set of all firing elements by FE, i.e. $FE = \{(t,c) \mid t \in T, c \in C(t)\}$. The markings of N are multisets of token elements, and the steps are multisets of firing elements. While markings and steps are derivative quantities, they are included in the definition so that it is clear that system morphisms $\phi : N \to N'$ map markings and steps to markings and steps respectively.

The above definition is, to all intents and purposes, equivalent to the common definition [3]. It does not include a guard function defined on transitions, but the same effect is achieved by limiting the colour set associated with the transition.

Having defined the structure of CPNs, we are now ready to consider their behaviour.

Definition 2. The *incremental effects* $E^+, E^- : \mathbb{Y} \to \mathbb{M}$ of the occurrence of a step Y are given by:

   a. $E^-(Y) = \displaystyle\sum_{(t,m) \in Y} \sum_{(p,t) \in A} \{p\} \times E(p,t)(m)$
   b. $E^+(Y) = \displaystyle\sum_{(t,m) \in Y} \sum_{(t,p) \in A} \{p\} \times E(t,p)(m)$

The enabling and firing of steps and sequences is defined in the usual manner:

Definition 3. For a net CPN, N, a step $Y \in \mathbb{Y}$ is *enabled* in marking $M \in \mathbb{M}$, written $M[Y\rangle$, if $M \geq E^-(Y)$. If a step $Y \in \mathbb{Y}$ of CPN N is enabled in marking $M_1 \in \mathbb{M}$, it may *fire* leading to marking $M_2 \in \mathbb{M}$, written $M_1[Y\rangle M_2$ with $M_2 = M_1 - E^-(Y) + E^+(Y)$. The set of *reachable markings* $\mathbb{M}_R \subseteq \mathbb{M}$ is given by $\mathbb{M}_R = \{M \in \mathbb{M} \mid \exists \, Y^* \in \sigma\mathbb{Y} : M_0[Y^*\rangle M\}$.

The standard way to represent the state space of a system is using a directed graph called a reachability graph (as introduced in Section 1). A reachability graph has a vertex (node) for every reachable state or marking of the system, and a directed edge (arc) for every possible step that occurs (here with a single firing element). As in [4], we allow multiple edges between pairs of vertices.

Definition 4. The *reachability graph* of a net $N = (P, T, A, C, E, \mathbb{M}, \mathbb{Y}, M_0)$ is the directed graph $G = (\mathcal{V}, \mathcal{E}, f)$ where:

$\mathcal{V} = \mathbb{M}_R$, the set of vertices, each of which is a reachable marking

$\mathcal{E} = \left\{ (M_1, (t, c), M_2) \in \mathcal{V} \times \mathbb{Y} \times \mathcal{V} \mid M_1[(t, c)\rangle M_2 \right\}$, the set of edges, each of which identifies an enabled firing element

$\forall\, e = (M_1, (t, c), M_2) \in \mathcal{E} : f(e) = (M_1, M_2)$, the end points of each edge

The reachability graph of the net of Figure 1 (a) is shown in Figure 1 (b) (where the key indicates how the marking of each place contributes to a marking of the net). This reachability graph is built using the standard reachability graph algorithm for CPNs given in Algorithm 1. This is based on that of Jensen [4, p. 5], and differs from it only to simplify the subsequent development of incremental versions. The algorithm determines the reachability graph G for the net N starting from the initial marking M_0. *Waiting* is the set of reachable markings whose successors have not yet been examined. It is therefore initialised to $\{M_0\}$. The algorithm repeatedly examines a marking in *Waiting*, adds edges and associated vertices to the graph for all the immediate successor states. This process continues till all reachable states have been examined.

Functions are defined for building the graph — ADDVERTEX(G, M) adds a vertex representing the marking M to the graph G, ADDEDGE$(G, (M, (t, c), M_1))$ adds an edge from M to M_1 labelled by (t, c) to G. The function SELECT(*Waiting*) returns a marking from the *Waiting* set. We do not indicate how the selection of this state is performed. The nature of this selection will determine whether the graph is constructed in a breadth-first manner, depth-first manner, or some other order.

The variable, *possible*, is a set of candidate transitions to be examined. We do not indicate how this set is calculated. In the worst case, it would be the set of all transitions. In the best case, it would be the set of transitions enabled at M. Unfortunately, there is no known heuristic to efficiently determine exactly those transitions which are enabled. Therefore *possible* will include all enabled transitions plus possibly some which are not enabled at M. The fewer disabled transitions included in the set, the better the performance of the algorithm. The function EDGESFROM$(N, M, possible)$ returns the set of edges that result from the occurrence of a transition in the set *possible* at marking M, namely:

$\left\{ (M_1, (t, c), M_2) \} \mid (t \in possible) \;\wedge\; M_1[(t, c)\rangle M_2 \right\}.$

The EDGESFROM function therefore must determine which firing elements are enabled. This can be a bottleneck in the performance of the reachability graph algorithm. The function MATCH(G, M) returns true if and only if M matches any vertex in G. We do not indicate how matching of markings is performed. It could be something trivial like an equality test, or possibly something more subtle such as allowing for symmetry [4].

Pseudo code for the EDGESFROM function is also presented. The function ENABLEDFIRINGELEMENTS(N, M, t) returns the set of firing elements involving t enabled at marking M. ENABLEDFIRINGELEMENTS could be implemented using the *transition instance analysis* algorithm of Maria [8,9]. The algorithm is rather

---

Algorithm 1 Standard Reachability Graph Algorithm

REACHABILITYGRAPH(G, N, M_0)
begin
  ADDVERTEX(G, M_0)
  $Waiting := \{M_0\}$
  while $Waiting \neq \emptyset$ **do**
    $M_1 :=$ SELECT$(Waiting)$
    for all $(M_1, (t, c), M_2) \in$ EDGESFROM$(N, M_1, possible)$ **do**
      if not MATCH(G, M_2) **then**
        ADDVERTEX(G, M_2)
        $Waiting := Waiting + \{M_2\}$
      end if
      ADDEDGE$(G, (M_1, (t, c), M_2))$
    end for
    $Waiting := Waiting - \{M_1\}$
  end while
end

EDGESFROM$(N, M, possible)$
begin
  $Result := \emptyset$
  for all $t \in possible$ **do**
    for all $(t, c) \in$ ENABLEDFIRINGELEMENTS(N, M, t) **do**
      $M_1 := M - E^-((t, c)) + E^+((t, c))$
      $Result := Result + \{(M, (t, c), M_1)\}$
    end for
  end for
  return $Result$
end

---

complicated, but the basic idea is to bind tokens in the input places, one at a time, to the variables on the input arcs of the transition being analysed.

3.1 Catering for Type Refinement

Given a net that has been derived from an abstract net by the type refinement, we can use the reachability graph of the abstract net to help produce the reachability graph of the refined net. We refer to *reachability graphs, markings* and *firing elements* as either *abstract* or *refined*.

In the previous section, we noted that a time consuming task in reachability graph generation is determining which firing elements are enabled for a given marking. Recall that a system morphism, $\phi : N \to N'$, maps refined net components to abstract net components. For type refinement, the refined firing elements enabled at M project onto abstract firing elements enabled at $\phi(M)$ [5]. Therefore the determination of refined firing elements which are enabled at marking M can be constrained by the knowledge of the abstract firing elements enabled at marking $\phi(M)$, which is already known from the abstract reachability graph. Further, if neither the transition t nor its neighbouring places have been modified by type refinement, then the enabled abstract and refined firing elements coincide, and the follower marking M_1 can simply be determined by applying the changes to $\phi(M)$ to M.

An algorithm which takes advantage of both type refinement and subnet refinement is presented in the following section.

We illustrate the above approach using the net of Figure 1 (a). Suppose that the type $X = \{x\}$ in this net is refined to $X = \{(x, 1)\}$ and that the initial marking is changed so that the place p_1 contains the token $(x, 1)$.

Given the marking $M_0 = (p_1, (x, 1)) + (p_4, y)$ of the type refined net, then the corresponding marking in the abstract net is $\phi(M_0) = (p_1, x) + (p_4, y)$. From the reachability graph of the abstract net, we know that the firing elements (t_1, x), (t_3, y) are enabled at $\phi(M_0)$. Since the neighbouring places to t_1 have been type refined, we must check if firing elements involving t_1 are still enabled in the refined net. On the other hand the neighbouring places to t_3 have not changed, and so the enabled firing elements for t_3 in the refined net are exactly those which were enabled in the abstract net (i.e. (t_3, y)). The successor of firing (t_3, y) from $\phi(M_0)$ in the abstract net is $M_1' = (p_1, x) + (p_5, y)$. We can efficiently find the successor of M_0 with firing element (t_3, y) by applying the changes to M_0' to M_0. Hence the successor of M_0 by firing (t_3, y) is $M_1 = (p_1, (x, 1)) + (p5, y)$.

3.2   Catering for Subnet Refinement

As was the case with type refinement, if a net is refined using subnet refinement, we can use the reachability graph of the abstract net to help determine the firing elements that are enabled at a given refined marking and therefore reduce the time required to construct the reachability graph of the refined net.

In the case of subnet refinement, the system morphism, $\phi : N \to N'$, is a restriction of the net N. In other words, the components of N are either retained or ignored in N'. Thus, if a refined firing element of N is retained in N', then it can only be enabled if the corresponding abstract firing element is enabled at the corresponding marking, a fact which can be determined from the abstract reachability graph. If a refined firing element of N is ignored in N', the abstract reachability graph does not help us and we must determine its enabling in the usual way. Thus, we again have that if an abstract transition t is unchanged in the refined net, then its enabled refined firing elements at marking M are exactly the enabled abstract firing elements at marking $\phi(M)$.

We illustrate the principle using the net of Figure 1 (a), modified by subnet refinement, as in Figure 2. Given the marking $M_0 = (p_1, x) + (p_4, y) + (p_6, z)$, the corresponding abstract marking is $\phi(M_0) = (p_1, x) + (p_4, y)$, where the marking of newly added places is ignored. The abstract firing elements enabled at $\phi(M_0)$ are (t_1, x), and (t_3, y). The transition t_1 has not been changed by the refinement, nor have its neighbouring places, and hence the refined firing element (t_1, x) is enabled in the refined net. Further, the abstract successor can be used to efficiently determine the refined successor. In this case, the successor of the initial

---

Algorithm 2 EDGESFROM modified to cater for type and subnet refinement

---

EDGESFROM-TYPESUBNET($N, N', M, possible$)
begin
  $Result := \emptyset$
  for all $(\phi(M), (t, c'), M_1') \in$ ABSTRACTEDGESFROM($N', \phi(M), \phi(possible)$) **do**
    if not CHANGED(N, N', t) **then**
      $M_1 :=$ UPDATE(N, N', M, M_1')
      $Result := Result + \{(M, (t, c'), M_1)\}$
    else
      for all $(t, c) \in FE \mid \phi((t, c)) = (t, c')$ **do**
        if $M \geq E^-((t, c))$ **then**
          $M_1 := M - E^-((t, c)) + E^+((t, c))$
          $Result := Result + \{(M, (t, c), M_1)\}$
        end if
      end for
    end if
  end for
  for all $(t, c) \in FE \mid (t \in possible) \land$ **not** MAPPED(t, ϕ) **do**
    if $M \geq E^-((t, c))$ **then**
      $M_1 := M - E^-((t, c)) + E^+((t, c))$
      $Result := Result + \{(M, (t, c), M_1)\}$
    end if
  end for
end

---

marking in the abstract net is $M_1' = (p_2, x) + (p_4, y)$. The changes to places p_2 and p_3 must be applied to the initial marking of the refined net, namely M_0, giving $M_1 = (p_2, x) + (p_4, y) + (p_6, z)$.

On the other hand, the transition t_3 has been modified (it has an extra input arc), so firing elements involving t_3 must be examined to determine whether they are enabled in the refined net. It turns out that the firing element $(t_3, (y, z))$ is enabled. Finally, there is a newly added transition, t_4, which must be examined to find out if it has any enabled firing elements.

Algorithm 2 takes advantage of type and subnet refinement to improve the performance of the reachability graph construction. The ABSTRACTEDGESFROM function returns all the edges from the abstract marking $\phi(M)$. That is, it returns all the enabled abstract firing elements at $\phi(M)$ and the corresponding successor markings. We would usually expect the edges simply to be looked up in the reachability graph of the abstract net, but they could be calculated as required. The function CHANGED(N, N', t) determines if the transition t or its neighbouring places have been modified by type or subnet refinement. The function UPDATE(N, N', M, M_1') determines the refined successor of M given the abstract successor, M_1', of $\phi(M)$. The function MAPPED(t, ϕ) returns *true* if the transition t is mapped to a transition in N' by the morphism ϕ (rather than being ignored).

3.3   Catering for Node Refinement

For a net with refined nodes, the state space is developed using a variant of modular analysis as proposed by Christensen and Petrucci [1]. Thus, the state space for each refined node is developed separately, since it is an independent subsystem apart from those points where it interacts with its environment. This leads to several reachability graphs which combine to represent the complete state space of the refined net [7]. We refer to the collection of graphs as the *Refined-Node State Space* (RNSS). Due to space constraints, we only present an informal explanation of the RNSS here. Formal definitions and the derivation of RNSS properties can be found in [7].

The RNSS is composed of a *local graph* for each refined node, and a *global graph*. The local graph of a refined node only contains local information, namely the reachable markings of the (subnet of the) refined node and the associated enabled firing elements[1].

The global graph is similar to the *synchronisation graph* of modular analysis [1], in that each vertex of the global graph refers to strongly connected components (SCCs)[2] of the local graphs, rather than the individual markings. We call such vertices *global vertices*. As with modular analysis, this approach avoids much of the interleaving that would normally be present in the full reachability graph. The full reachability graph can be recovered from the RNSS. However, the various dynamic properties (reachability, dead markings, liveness, home properties, etc.) can be determined directly from the RNSS [7]. This is particularly important since the recovery of the full reachability graph may be computationally expensive.

We explain the construction of the RNSS by considering the node refinement of Figure 1 (a) as in Figure 3. In the following, for an abstract node x'' which is refined, we denote the local graph of x'' by $G_{x''}$, and the subnet of N by $N_{x''}$. We will also use the notation $M|_p$ to refer to the marking M restricted to the place p, and $M|_{x''}$ to refer to the marking M restricted to (the places occurring in) the subnet of the node x'' plus the environment places (if x'' is a transition). This notation is generalised to M_X, where X is a set of nodes (refined or otherwise).

To construct the RNSS, we start with the global graph and add a vertex representing the initial marking, as shown in Figure 4 (a). It is worth noting that:

- The global graph has the rather unusual property that it stores markings for refined transitions. Unlike simple transitions, refined transitions may retain some state information between firings.
- To avoid cluttering the graphs, we have not included the labels on edges of local graphs.

---

[1] For technical reasons, the marking of the neighbouring places of a border transition of the refined transition are also included in its local reachability graph.

[2] Informally, a strongly connected component is a subset, S, of the nodes of a directed graph such that any node in S is reachable from any other node in S and S is not a subset of any larger such set.

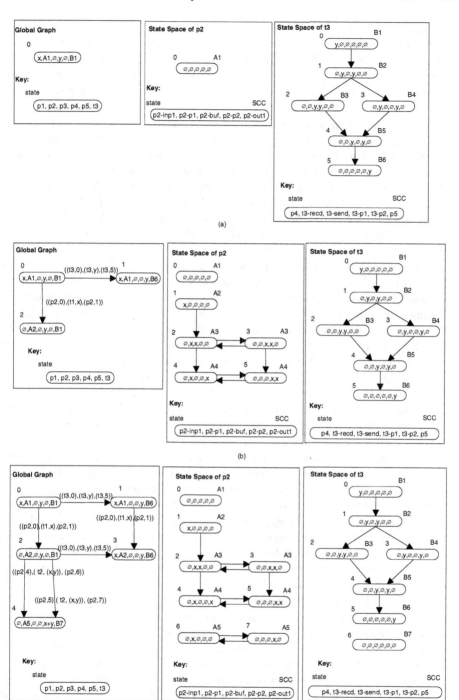

Fig. 4. RNSS generation for the net of Figure 3

The next step in the RNSS generation involves adding successors from M_0 to the global graph. If there is an unrefined transition which only takes input from unrefined places, then its enabling can be determined directly (or from the abstract reachability graph). This is the case with transition t_1. Since it has output to a refined place (here p_2), we need to determine the relevant strongly connected component for the local graph.

If there is a refined transition, then we consider its border terminal step sequences. Such a sequence involves only internal transitions of the refined transition, and ends with the occurrence of an output border transition. This would be the case for an internal sequence of transitions of the refined transition t_3 that ends with the occurrence of $t3$-*finished*. The edge in the global reachability graph simply indicates the sum of abstract firing modes for the refined transition (i.e. the sum of modes for the *switch* transition in the border terminal sequence), together with the local markings before and after the border terminal step sequence.

Finally, if the transition under consideration in the global graph has input from refined places, then we consider successors which are possible following a sequence of internal transitions of those refined places. Naturally, the occurrence of those internal transitions is only recorded in the relevant local graph, but the edge in the global graph is labelled with the actual source and successor markings (and not just the relevant SCCs). In our example, this would be the case for the firing of transition t_2, which is not enabled in M_0.

In our example the result of adding immediate successors from M_0 is shown in Figure 4 (b). At this point, the transition t_2 is enabled. The above process of adding successors is repeated for each vertex of the global graph for which immediate successors have not been examined. The complete RNSS is shown in Figure 4 (c).

If we compare the ordinary state space of the whole system, with the RNSS, we observe that even for this trivial example, the RNSS is smaller than the ordinary state space. The RNSS contains a total of 20 nodes and 21 edges, while the ordinary state space contains 38 nodes and 88 edges.

The structure of the algorithm to develop the RNSS is the same as that of the basic reachability algorithm (Algorithm 1), modified with the changes given in Algorithm 3. Since the global graph stores global vertices then the *Waiting* set of Algorithm 1 will now store global vertices, and the function ADDVERTEX of Algorithm 1 will now add global vertices.

As in Christensen and Petrucci [1], if $v \in \mathcal{V}$, then we use v^c to denote the strongly connected component to which v belongs, and M^ϕ to denote the global vertex corresponding to M. The function GLOBALVERTEX(M) of Algorithm 3 calculates the global vertex corresponding to M (i.e. it calculates M^ϕ). Firstly, this is initialised to $M|_{EP}$, where EP is the set of *external places*, that is, places of the refined net that are not part of a refined node. The state space of each refined node is then developed from the current marking. REFINEDNODES(N') returns the set of places and transitions of the abstract net N' that are refined by node refinement; REACHABILITYGRAPH†(G, N, M) calculates the reachability graph from M, where M may already appear in G; and COMPUTESCCS($G_{x''}, M|_{x''}$)

Algorithm 3 GLOBALVERTEX, and modified EDGESFROM, for node refinement

---

GLOBALVERTEX(M)
begin
   $M^{\notin} := M|_{EP}$
   for all $x'' \in$ REFINEDNODES(N') **do**
      REACHABILITYGRAPH$^\dagger(G_{x''}, N_{x''}, M|_{x''})$
      $scc :=$ COMPUTESCCS($G_{x''}, M|_{x''}$)
      $M^{\notin} := M^{\notin} + (x'', scc)$
   end for
   return $M^{\notin}$
end

EDGESFROM-NODE($N, N', M^{\notin}, possible$)
begin
   $Result := \emptyset$
   for all $t \in ET \cap possible$ **do**
      for all $M_1 \in$ INTERNALLYREACHABLE($N, M^{\notin}, {}^{\circ}t$) **do**
         if $M_1 \geq E^-((t,c))$ **then**
            $M_2 := M_1 - E^-((t,c)) + E^+(t,c)$
            $Result := Result + \{(M^{\notin}, (M_1|_{({}^{\circ}t \cup t^{\circ})}, (t,c), M_2|_{({}^{\circ}t \cup t^{\circ})}), M_2^{\notin})\}$
         end if
      end for
   end for
   for all $t'' \in T''$ **do**
      for all $M_1 \in$ INTERNALLYREACHABLE($N, M^{\notin}, {}^{\circ}t''$) **do**
      REACHABILITYGRAPH$^\dagger(G_{t''}, N_{t''}, M_1|_{t''})$
      for all $(m,c) \in$ BORDERTERMINAL($G_{t''}, M_1|_{t''}$) **do**
         $M_2 := M_1 - M_1|_{t''} + m$
         $Result := Result + \{(M^{\notin}, (M_1|_{({}^{\circ}t'' \cup t''^{\circ} \cup t'')}, (t,c), M_2|_{({}^{\circ}t'' \cup t''^{\circ} \cup t'')}), M_2^{\notin})\}$
      end for
      end for
   end for
   return $Result$
end

---

computes the SCCs of the vertices reachable from $M|_{x''}$ in the graph $G_{x''}$, and returns the SCC index of the marking $M|_{x''}$. Thus the marking $M|_{x''}$ is replaced by its SCC index to form $M^{\notin}$.

Since each refined node has an associated local graph, and since the local state space is developed from different starting points, then it follows that the local graph is not necessarily connected. Also, it may be the case that part or all of the local state space of a given refined node has previously been developed. This is clearly an advantage, saving time and space as the state space does not need to be developed again (as would be the case in the standard algorithm). We also note that the local graphs are independent, and the implementation could therefore develop them in parallel.

The EDGESFROM function has also been changed as in Algorithm 3 to produce the RNSS. It first considers those edges due to *external transitions*. (These are the transitions of the refined net that are not part of a refined node, and are denoted *ET*.) It then finds those edges due to refined transitions. We use $°t$ to denote the set of abstract places that are refined by node refinement and are inputs of the transition t. Similarly $t°$ denotes the set of abstract places that are refined by node refinement and are outputs of the transition t. The function INTERNALLYREACHABLE$(N, M^{\not\epsilon}, °t)$ returns the set of markings reachable from M, after internal activity of refined places in $°t$. This set includes the marking M. The source and successor markings stored with the edge of the global graph are the source and successor markings restricted to the refined places which are adjacent to the transition t. We note that the labelling of edges extends that of Algorithm 1 by incorporating the relevant local markings.

Having considered the external transitions, the function EDGESFROM considers the refined transitions. (We denote by T'' the set of transitions of the abstract net which are refined by node refinement.) For each refined transition, we generate its local reachability graph, $G_{t''}$ starting at marking $M|_{t''}$. We then consider the border terminal firing sequences in these local graphs. (The function BORDERTERMINAL$(G_{t''}, M|_{t''})$ identifies these firing sequences and returns a set of tuples (m, c) where m is a marking reachable by a border terminal sequence, and c is the sum of abstract firing modes that are fired to reach m.) Recall that a *border terminal* firing sequence of a refined transition consists of only internal transitions of the refined transition, and ends with the occurrence of an output border transition.

4   Implementing the Incremental Algorithm

The Maria reachability analyser [8] is a relatively new tool building on the earlier work with PROD [11]. Maria has a modular design, so that different algorithms, front-ends, and state storage mechanisms can easily be incorporated. This modularity, together with its simple text-based input language made it an attractive choice for implementing the above incremental reachability algorithms.

Maria has been modified so that the following methodology can be adopted for analysing incrementally developed models: the abstract and refined nets are parsed and the refinements are detected; the reachability graph of the abstract net is then developed (if it does not already exist) and the incremental algorithm is used to develop the state space of the refined net. Thus, the syntax of Maria was extended to support the specification of the various refinements; and the analyser of Maria was modified to support the incremental algorithms. Full details of these changes can be found in [7].

Currently, the data structures of Maria do not provide optimum support for incremental analysis. For example, the function EDGESFROM-TYPESUBNET from Algorithm 2, requires us to determine those refined firing elements which map to enabled abstract firing elements. In order to do this, we need to be able (within the context of markings) to map from refined token elements to abstract token elements and back again. While the former direction is easily supported, the latter is not. Computing this information on the fly requires the same amount of effort as the transition instance analysis algorithm of Maria. Performance improvements can therefore be expected with special support for these mappings. This is a matter for further research.

However, we are still able to gain some advantage in the implementation from the fact that the net has been refined by type and/or subnet refinement. First and foremost, we do not have to use the transition instance analysis algorithm on those transitions for which there is no corresponding enabled abstract transition. (This does not apply to transitions and/or firing modes which are introduced by subnet refinement.) Second, if the transition has not been changed then we do not have to check if it is enabled in the refined net and can obtain the refined successor marking by updating the abstract successor marking.

5   Performance of the Incremental Algorithm

In this section, we characterise the situations where the incremental algorithm can be expected to give a performance improvement over the standard algorithm, and also apply the algorithm to some case studies. The results quoted have been obtained using a 500MHz Intel 686 machine running Linux (kernel 2.2.15). The machine has 256MB random access memory (RAM), and 2GB of virtual memory (which is the limit for this kernel). The code was compiled using the optimising option of the GNU Compiler Collection (version egcs-2.91.60).

These tests use the reachability graph of the abstract net to determine the enabled abstract firing modes. We indicate the *total time* for the incremental algorithm, which consists of the time required to construct the abstract graph together with the time required to construct the refined graph using the incremental algorithm. In practice, however, we would normally expect the reachability graph for the abstract net to be constructed and analysed before the refined net is constructed, thus making the incremental algorithms even more attractive. As it is, even the total time for the incremental algorithm is, in some situations, less than the time required to construct the full reachability graph using the standard algorithm.

One disadvantage of using the abstract graph to determine the enabled refined firing modes, is that both the abstract and refined graphs must be represented in system memory[3]. This means that if the abstract graph is large then the performance of the incremental algorithm may not be as good. On the other hand, if the refined graph is much larger than the abstract graph, then the amount of extra memory used to store the abstract graph becomes insignificant.

We note that the longer the standard algorithm takes to construct the refined graph relative to the abstract graph, then the greater the likelihood that the incremental algorithm will demonstrate improvements.

One significant advantage gained by the incremental algorithm is that refined firing elements do not have to be considered if the corresponding abstract firing elements are disabled. Therefore we can expect good performance improvement

---

[3] Maria uses a hash table to represent the graph in system memory [9].

for the incremental algorithm if there is a large number of refined firing elements disabled for this reason. Such an example is shown in Figure 5. This net has an initialisation section which generates a number of tokens, and a processing section which consumes them. The initialisation has a lot of disabled firing modes, which do not need to be examined in the refinement. The graph plots the time taken for the refined graph using the standard algorithm, the time for the abstract graph (using the standard algorithm), and the total time for the incremental algorithm as the the value of n is increased in the abstract and refined net of Figure 5 (a) and (b) respectively (where only the changed part of the refined net has been shown).

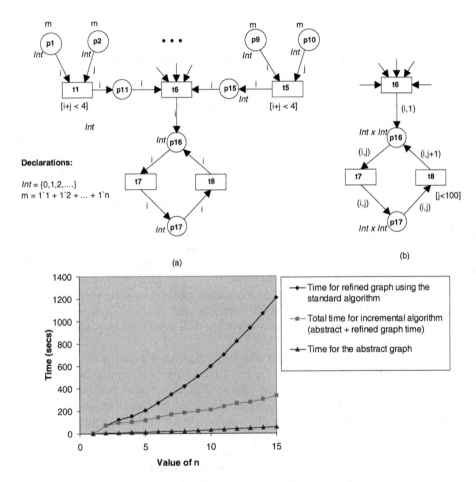

Fig. 5. The effect of increasing the number disabled firing elements

Another significant advantage of the incremental algorithm is that it can save both time and space for a net with refined nodes, because it will not consider all the possible interleavings of the internal activity of the refined nodes. As the extent of interleaving between internal and external transitions of the net increases, so too does the performance improvement of the incremental algorithm compared to that of the standard algorithm. We can demonstrate this using the net of Figure 6. The amount of internal activity is determined by the number of places in the sequence p_1-p_1 to p_1-p_n. The graph shows the performance of the incremental algorithm compared to that of the standard algorithm as the level of internal activity is increased (i.e. as the number of places in this sequence is increased). When the number of transitions in the sequence was greater than 12 the standard algorithm could not complete due to insufficient virtual memory, whereas the incremental algorithm produced the complete RNSS in a little over a minute.

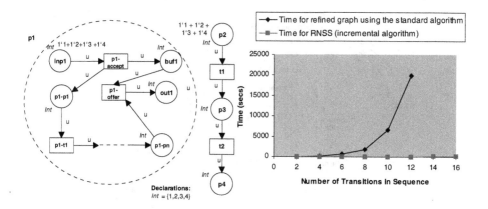

Fig. 6. The effect of increasing the number disabled firing elements

Thus, it is not difficult to produce examples where the incremental algorithm can perform significantly better than the standard algorithm. We have also implemented two separate case studies to assess the performance of the incremental algorithm in practice: *the Z39.50 Protocol for Information Interchange* [6] and *the Distributed Missile Simulator Model* [2]. Both of these case studies have been developed incrementally. Our implementation of these studies together with a more detailed examination of the results obtained is given in [7].

The Z39.50 Protocol model uses subnet refinement to introduce segmentation of responses — a capability added in the 1995 version of the protocol. With an initial marking typical of what we expect to be analysed, the standard algorithm takes 1 273 seconds whilst the time for the incremental algorithm is 676 seconds (plus 159 seconds to construct the abstract graph). We have observed similar results for other refinements of the Z39.50 protocol [7].

The abstract model of the Distributed Missile Simulator is given in Figure 7. The *Outputs* place and the *Simulate* transition of this abstract model are refined to capture more of the detail of the simulation algorithm. The refined *Simulate* transition includes calculations for the target, infrared, radar, and missile control. Type refinement is also used to introduce values for the coordinate data, such as the position of the missile and target. We have implemented a modified version of the refined model in Maria (since Maria only supports integer arithmetic).

The distance to the target can be varied to achieve different numbers of iterations of the basic model. The graph of Figure 7 shows the performance of the incremental algorithm compared to that of the standard algorithm for the refined model as the x-coordinate of distance is increased (while the y and z coordinates remain unchanged). When the x-coordinate of distance between the missile and target was set to 600m, the incremental algorithm constructed a RNSS in a few minutes whereas the standard algorithm exhausted virtual memory and could not complete.

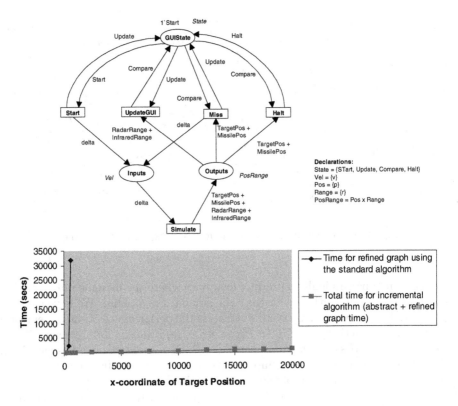

Fig. 7. The effect of increasing distance from the missile to the target

6   Conclusions and Further Work

In this paper we have presented algorithms for generating state spaces of systems exploiting their incremental structure. The incremental algorithms have been shown to give significant improvements in situations where there are a large number of refined firing elements that map to disabled abstract elements, and where there is significant interleaving between the local transitions of refined nodes. These significant benefits have been observed for real world studies — in one case it was the difference between producing the state space in less than an hour, as opposed to the standard algorithm which could not complete.

As we have noted, this approach has points of contact with the modular analysis of Christensen and Petrucci [1]. Their modules are quite general, provided they are built with transition fusion. Place fusion is problematic since finite local reachability graphs can still result in infinite global reachability graphs, as tokens are passed backwards and forwards between the modules. Our modules arise from node refinement, which imposes certain behavioural constraints. Thus, if a superplace is considered to be a module, joined to its environment by place fusion, then the environment cannot extract any tokens which it has not deposited or which were not present initially. Thus, the combination of such modules by place fusion ceases to be a problem.

We have considered transition refinements with multiple input and output border transitions (the distributed input and output of Vogler [12]). We have shown that with distributed input, we can generate the full reachability graph by unfolding the RNSS. With distributed output, unfolding is possible if there is a guarantee that once the first output border transition has fired, then the step sequence can be guaranteed to complete without firing any further input border transitions. In this case, the unfolding results in a stubborn set reduction of the full reachability graph [7]. This optimisation was applied to the Missile Simulator example quoted in Section 5.

We have also shown that it is possible to determine properties of the net — reachability, dead markings, liveness, home properties, and boundedness — directly from the RNSS, without needing to unfold it [7]. If one is prepared to sacrifice the capability of recovering the full reachability graph, then it is possible to optimise the incremental algorithm even further. Thus, for example, there is no point trying to extract tokens from a refined place if the corresponding abstract tokens are not even available in the abstract marking.

There are a number of interesting avenues for future work. An important issue would be the development of a state storage mechanism for Maria which would be tailored to the special needs of incremental analysis, including direct access to place markings. It may even be possible to design a data structure which would support sharing of state information between abstract and refined graphs.

It would also be interesting to investigate the extent to which the incremental algorithms could be combined with other state space reduction techniques. The incremental approach seems to be orthogonal to the Symmetric Occurrence Graphs. Also, while it does eliminate interleaving between the activity of different refined nodes, it may still be improved by combination with the Stubborn Set approach.

Acknowledgements. This work was partially funded under ARC Large Grant A49800926. We thank Nisse Husberg and the other members of the Theoretical Computer Science Laboratory, University of Helsinki for their assistance with the implementation work with Maria. We acknowledge the helpful discussions held with Robert Esser and Vishv Malhotra.

References

1. S. Christensen and L. Petrucci. Modular Analysis of Petri Nets. *The Computer Journal*, 43(3):224–242, 2000.
2. S. Gordon and J. Billington. Analysing a Missile Simulator using Coloured Petri Nets. *International Journal on Software Tools for Technology Transfer*, 2(2):144–159, December 1998.
3. K. Jensen. *Coloured Petri Nets: Basic Concepts, Analysis Methods and Practical Use: Volume 1, Analysis Methods.* Monographs in Theoretical Computer Science. Springer-Verlag, 1992.
4. K. Jensen. *Coloured Petri Nets: Basic Concepts, Analysis Methods and Practical Use: Volume 2, Analysis Methods.* Monographs in Theoretical Computer Science. Springer-Verlag, 1995.
5. C.A. Lakos. Composing Abstractions of Coloured Petri Nets. In *21st International Conference on Application and Theory of Petri Nets (ICATPN'2000) Aarhus, Denmark, June 25-29, 2000*, volume 1825 of *Lecture Notes in Computer Science*, pages 323–342, June 2000.
6. C.A. Lakos and J. Lamp. The Incremental Modelling of the Z39.50 Protocol with Object Petri Nets. In J. Billington, M. Diaz, and G. Rozenberg, editors, *Application of Petri nets to Communication Networks : Advances in Petri Nets*, volume 1605 of *Lecture Notes in Computer Science*, pages 37–68. Springer-Verlag, 1999.
7. G.A. Lewis. *An Incremental Perspective — Incremental Development and Incremental Analysis in the Context of Coloured Petri Nets.* PhD thesis, Department of Computing, University of Tasmania, Hobart, Tasmania, Australia. *(to be submitted)*.
8. M. Mäkelä. *Maria — Modular Reachability Analyzer for Many-Sorted Petri Nets.* Helsinki University of Technology, Espoo, Finland, 1999. version 0.1.
9. M. Mäkelä. A Reachability Analyser for Algebraic System Nets. Licentiate thesis, Helsinki University of Technology, Theoretical Computer Science Laboratory, Espoo, Finland, March 2000.
10. A. Valmari. The State Explosion Problem. In W. Reisig and G. Rozenberg, editors, *Lectures on Petri Nets I: Advances in Petri Nets*, volume 1491 of *Lecture Notes in Computer Science*, pages 429–528. Springer-Verlag, 1998.
11. K. Varpaaniemi, J. Halme, K. Hiekkanen, and T. Pyssysalo. PROD reference manual. Technical Report B13, Helsinki University of Technology, Department of Computer Science and Engineering, Digital Systems Laboratory, Espoo, Finland, August 1995.
12. W. Vogler, W. Brauer, and R. Gold. A Survey of Behaviour and Equivalence Preserving Refinements of Petri Nets. In *Advances in Petri Nets 1990*, volume 483 of *Lecture Notes in Computer Science*, pages 1–46. Springer-Verlag, 1991.

Optimising Enabling Tests and Unfoldings of Algebraic System Nets

Marko Mäkelä[*]

Helsinki University of Technology
Laboratory for Theoretical Computer Science
P.O.Box 9700, 02015 HUT, Finland
msmakela@tcs.hut.fi
http://www.tcs.hut.fi/Personnel/marko.html

Abstract. Reachability analysis and simulation tools for high-level nets spend a significant amount of the computing time in performing enabling tests, determining the assignments under which transitions are enabled. Unlike the majority of earlier work on computing enabled transition bindings, the techniques presented in this paper are highly independent of the algebraic operations supported by the high-level net formalism.

Performing enabling tests is viewed as a unification problem. A unification algorithm is presented and modifications to it are suggested. One variant of the algorithm constructs finite unfoldings for nets with unbounded domains. Some heuristics for optimising the enabling tests are discussed and their usefulness is evaluated based on experiments. The algorithms have been implemented in the reachability analyser MARIA.

Keywords: high-level Petri nets, reachability analysis, unification, unfolding

1 Introduction

Constructing computer-readable models for systems resembles programming in many aspects. High-level languages make it easier to create models or programs, but analysing or executing them involves an overhead, since the operations in the high-level specification have to be transformed to simpler operations that the underlying computing machinery is able to perform. This can be done either in one big preprocessing step that translates the whole input to a simpler language, or in smaller steps that interpret the high-level operations one at a time, or by performing a mixture of preprocessing and interpreting.

There are several approaches to analysing high-level Petri nets. *Structural techniques*, such as determining invariants of a high-level net [10] and proving some properties based on them, are typically applied by humans and therefore

---

[*] This research was financed by the Helsinki Graduate School on Computer Science and Engineering, the National Technology Agency of Finland (TEKES), the Nokia Corporation, Elisa Communications and the Finnish Rail Administration.

J.-M. Colom and M. Koutny (Eds.): ICATPN 2001, LNCS 2075, pp. 283–302, 2001.

only work on relatively small, highly abstracted models. Many computer-aided techniques are based on *exhaustive state space exploration* or *reachability analysis*, generating all states reachable from the initial state of the model.

Some reachability analyser tools work on low-level nets [15]. Such tools can analyse high-level nets if these are *unfolded*, translated to low-level nets in a preprocessing step. A straightforward unfolding, as the one defined for Algebraic system nets in [10, Section 5.1], may yield places not connected to any transition, or transitions whose input places will never become marked.

The unfolded net of a high-level net can be reduced by analysing the high-level net and overestimating the set of reachable markings. The unfolded net needs to contain only such places that can ever become marked according to the estimate. Similarly, only those transitions that are connected to these places need to be included in the unfolded net. Even when such reductions are applied, a high-level model whose variables have large domains may yield an unmanageably large unfolding, even if the full state space of the model is moderate.

Reachability analysis can also be performed on the high level. This is computationally more complex than analysing low-level nets, since the transitions may fire in different *modes*, depending on the values assigned to their variables. Compared to unfolding, analysing models on the high level usually trades execution time for memory space. When a net is unfolded, its high-level transitions are processed only once. When it is analysed on the high level, the transitions must be "unfolded", or interpreted in each state that is explored.

Our approach to the reachability analysis of high-level nets is a mixture of preprocessing and interpreting. We perform a series of translations on the model and set up auxiliary data structures that make it possible to use a simpler and more efficient algorithm for performing enabling tests. The idea is to find efficient static schedulings for input arc inscriptions and to apply computationally cheap heuristics for pruning transitions that are disabled in a marking.

The notations in this paper is based on Algebraic system nets, defined by Kindler and Völzer in [10]. They can be considered as a slightly more formal version of coloured Petri nets [9]. The class of nets we consider is more generic than the well-formed nets used by Chiola et al. [2], Gaeta [4] and Ilié et al. [7,8] and others at least in the following aspects:

- data types are not limited to enumerations and tuples
- algebraic operations may be irreversible
- arcs may have variable-dependent weights or be multiset-valued

The coloured nets used by Sanders [16] are more generic than well-formed nets but less generic than Algebraic system nets. His approach represents input arc expressions as variables with constant multiplicity. Since Sanders performs enabling tests by solving constraint satisfaction problems, it is nontrivial to allow the arcs in his formalism to have variable-dependent weights.

The formalism supported by our tool MARIA is Algebraic system nets with some extensions and limitations. The main limitations are that variables on input arcs may not be multiset-valued, and all data types must have finite domains. These limitations ensure that every model in our formalism can be unfolded to a finite low-level net.

1.1 Example: Changing Money

Figure 1 illustrates a situation that could happen near a coin-operated machine. A customer comes to a cashier with a bank bill in his hand, asking "Could you break this for me?" The cashier then changes the money to an equivalent amount of money in smaller coins. In our algorithm, he always returns one type of coin, e.g. ten ① coins for 10 units of money, and not e.g. one ⑤ coin and five ① coins.

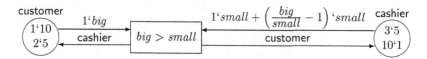

Fig. 1. An algorithm for breaking money.

The model contains two places, customer and cashier, which represent the money held by the two parties. The only transition of the model has two input variables, *big* and *small*, the monetary values. A transition guard specifies that the monetary value of *big* must be greater than that of the change coins *small*.

When the cashier receives a piece of money from the customer, he first chooses one of his coins and then picks enough of them so that the monetary values match. The output arcs of the transitions make use of special multiset-valued variables, which are short-hand notation of our tool for more complex arc inscriptions. These variables refer to the multisets that the input arcs connected to the corresponding places evaluate to. Thus, the cashier receives the coins the customer took from his purse and vice versa.

As we shall see later, the definition of Algebraic system nets allows arbitrary multiset-valued arc inscriptions. We could replace the complex inscription of the arc running from the place cashier to the transition in Figure 1 with a reference to a multiset-valued variable *change*, and replace the transition guard with $big = \sum_m change(m)m$, requiring that no money is made or lost. Alas, this kind of a definition could introduce a combinatorial explosion in the analysis. On input arcs, our approach does not allow multiset-valued *variables*, but it does support more complex multiset-valued terms, provided that they can be evaluated based on variable bindings obtained from other arcs.

2 Basic Concepts

Before presenting our algorithm, we must define some basic concepts. We refer the reader to [10, Section 3.1–3.2] for a more detailed introduction to Algebraic system nets and the underlying mathematical concepts. Our definition of algebras has some extensions to the original. *Evaluation errors* are helpful in tracking modelling errors. Our tool does not silently ignore transitions that cannot be fired due to errors such as arithmetic overflow. Models with variable-weight arcs may benefit from *undefined variables*. If a variable occurs only on arcs whose multiplicity evaluates to zero, it does not need to be defined in order for the transition to be fired. Space limitations prohibit us from formally defining another extension, *short-circuit evaluation* of if-then-else expressions.

Algebras and signatures. A signature $SIG = \langle S, OP \rangle$ consists of a finite set S of *sort symbols* and a pairwise disjoint family $OP = (OP_a)_{a \in S^+}$ of *operation symbols*. A SIG-*algebra* $\mathcal{A} = \langle A, f \rangle$ consists of a family $A = (A_s)_{s \in S}$ of sets and a family $f = (f_{op})_{op \in OP}$ of total functions. Let $\epsilon \notin A$ be an *error symbol* and $A'_s = A_s \cup \{\epsilon\}$. For $op \in OP_{s_1 \ldots s_n s_{n+1}}$, let $f_{op} : A'_{s_1} \times \cdots \times A'_{s_n} \to A'_{s_{n+1}}$ such that the image of the subset $(A'_{s_1} \times \cdots \times A'_{s_n}) \setminus (A_{s_1} \times \cdots \times A_{s_n})$ equals ϵ; that is, whenever an argument equals ϵ, so does the result. A set A_s of an algebra is called a *domain* and a function f_{op} is called an *operation* of the algebra.

In the following we assume that a signature SIG has the sort symbols $bool, nat \in S$ and in each SIG-algebra the corresponding domains are $A_{bool} = \mathbb{B} = \{\text{true}, \text{false}\}$ and $A_{nat} = \mathbb{N} = \{0, 1, \ldots\}$.

Variables and terms. For a signature $SIG = \langle S, OP \rangle$ we call a pairwise disjoint family $X = (X_s)_{s \in S}$ with $X \cap OP = \emptyset$ a *sorted SIG-variable set*. A *term*, associated with a particular sort, is built up from variables and operation symbols. The *set of SIG-terms over X of sort s* is denoted by $\mathbf{T}_s^{SIG}(X)$ and inductively defined by:

1. If $x \in X_s$, then $x \in \mathbf{T}_s^{SIG}(X)$.
2. If $u_k \in \mathbf{T}_{s_k}^{SIG}(X)$ for some $k \in \{1, \ldots, n\}$ and $op \in OP_{s_1 \ldots s_n s_{n+1}}$, then $op(u_1, \ldots, u_n) \in \mathbf{T}_{s_{n+1}}^{SIG}(X)$.

The set of all terms (of any sort) is denoted by $\mathbf{T}^{SIG}(X)$. A term without variables, a *ground term*, of sort s belongs to the set $\mathbf{T}_s^{SIG} = \mathbf{T}_s^{SIG}(\emptyset)$.

Evaluation of terms. For a signature $SIG = \langle S, OP \rangle$, a sorted SIG-variable set $X = (X_s)_{s \in S}$, and a SIG-algebra $\mathcal{A} = \langle (A_s)_{s \in S}, (f_{op})_{op \in OP} \rangle$, a mapping $\beta : X \to A \cup \{\epsilon\}$ is an *assignment* for X iff for each $s \in S$ and $x \in X_s$ holds $\beta(x) \in A_s \cup \{\epsilon\}$ where $\epsilon \notin A$ denotes an *undefined variable*. We canonically extend β to a mapping $\bar{\beta} : \mathbf{T}^{SIG}(X) \to A \cup \{\epsilon\}$ by:

1. $\bar{\beta}(x) = \beta(x)$ for $x \in X$.
2. $\bar{\beta}(op(u_1, \ldots, u_n)) = f_{op}(\bar{\beta}(u_1), \ldots, \bar{\beta}(u_n))$ for $op(u_1, \ldots, u_n) \in \mathbf{T}^{SIG}(X)$.

Let $\beta_\emptyset : \emptyset \to A \cup \{\epsilon\}$ be the unique assignment for the empty variable set.

2.1 Algebraic System Nets

Algebraic system nets are based on a special case of the algebras defined above. We distinguish some *ground-sorts* and assign a *bag-sort* (a finite nonnegative multiset sort) to each ground-sort. The domain associated with a bag-sort must be a multiset over the domain of the corresponding ground-sort.

Definition 1 (Bag-signature, *BSIG*-algebra). *Let $SIG = \langle S, OP \rangle$ be a signature and $BS, GS \subseteq S$. $BSIG = \langle S, OP, bs \rangle$ is a bag-signature iff $bs : GS \to BS$ is a bijective mapping. An element of GS is called a* ground-sort, *an element of BS is called a* bag-sort *of $BSIG$. A SIG-algebra $\mathcal{A} = \langle A, f \rangle$ is a $BSIG$-algebra iff for each $s \in GS$ holds $A_{bs(s)} = \mathrm{BAG}(A_s) = (A_s \to \mathbb{N})$.*

Definition 2 (Algebraic system net). *Let $BSIG = \langle S, OP, bs \rangle$ be a bag-signature with bag-sorts BS. An* algebraic system net *$\Sigma = \langle N, \mathcal{A}, X, i \rangle$ over $BSIG$ consists of*

1. *a finite net $N = \langle P, T, F \rangle$ where $P \cap T = \emptyset$, $F \subseteq (P \times T) \cup (T \times P)$ and P is sorted over BS, i.e., $P = (P_s)_{s \in BS}$ is a bag-valued $BSIG$-variable set,*
2. *a $BSIG$-Algebra $\mathcal{A}$,*
3. *a sorted $BSIG$-variable set X disjoint from P,*
4. *a net inscription $i : P \cup T \cup F \to \mathbf{T}^{BSIG}(X)$ such that*
   a) *for each $p \in P_s : i(p) \in \mathbf{T}_s^{BSIG}$,*
   b) *for each $t \in T : i(t) \in \mathbf{T}_{bool}^{BSIG}(X)$, and*
   c) *for each $t \in T$ and $p \in P_s$ and $f \in F$ with $f = \langle p, t \rangle$ (input arc) or $f = \langle t, p \rangle$ (output arc) holds $i(f) \in \mathbf{T}_s^{BSIG}(X)$.*

For a place $p \in P$, the inscription $i(p)$ is called the symbolic initial marking *of p; for a transition $t \in T$, the term $i(t)$ is called the* guard *of t.*

It is worth noting that Definition 2, replicated from [10, Definition 3] allows multiset-valued operations and variables in arc inscriptions (annotations). The MARIA tool [11,12] makes use of both.[1] Arcs with variable weights are useful when modelling certain types of resource management.

The basic semantics of Algebraic system nets, including the firing rule, have been defined by Kindler and Völzer in [10, Definitions 4–6].

2.2   Unification Concepts

There are at least two ways to construct the set of assignments under which a transition is enabled. One way is to construct all possible assignments for the variables that occur in the arc inscriptions and in the guard of the transition, and to prune those assignments under which the arc inscriptions and the guard fail to fulfil the firing rule. This is the usual way when a net is unfolded; see e.g. [10, Definition 13]. This approach does not work very well if the transitions have a large (or infinite) number of possible assignments (firing modes), and the transitions are enabled in only a few firing modes in the reachable states.

Fortunately, there is a more efficient approach for the case when the input places of a transition are marked sparsely. The process of finding assignments or substitutions under which two algebraic terms are equivalent is often referred to as *unification*, e.g. [1, pp. 74–76]. In algebraic system nets, we can unify input arc

---

[1] MARIA allows multiset-valued variables on output arcs, where they refer to the multisets removed from the input places; see Figure 1.

inscriptions with a marking of the net. In this case, a unifier is an assignment for the transition variables under which the evaluations of the input arc inscriptions are contained in the corresponding input place markings.

If the algebraic operations are not restricted, there might be prohibitively many unifiers. For instance, consider the constant $2 \in \mathbb{N}$ and the expression $x + y$. If the variables x and y are known to be sorted over nat, then three assignments are possible unifiers: $\{\langle x, 0\rangle, \langle y, 2\rangle\}$, $\{\langle x, 1\rangle, \langle y, 1\rangle\}$, and $\{\langle x, 2\rangle, \langle y, 0\rangle\}$. If the constant was n, there would be $n + 1$ different unifiers. If either variable was allowed to be negative, there would be infinitely many unifiers.

In order to avoid a combinatorial explosion, we have to restrict the set of algebraic terms that the unification algorithm examines to find values for variables. A natural way of making this restriction is to limit the set of operations the unification algorithm recognises in such a way that the choice of unifiers is always unique. This rules out the operation $+$ in our previous example.

We distinguish two classes of operations that are recognised by our algorithm. Reversible unary operations, such as taking the successor of an element in a sequence, can be "neutralised" by applying a reverse operation, such as the predecessor operator. Other operations that the algorithm must know are constructors that tie terms together. For instance, we want to be able to unify the variables in the term $\langle x, y \rangle$ with the constants in the ground term $\langle 1, 2 \rangle$.

Definition 3 (Unifier candidate, assignment compatibility). *Let $SIG = \langle S, OP \rangle$ be a signature with the variable set X, and let $\mathcal{A} = \langle A, f \rangle$ be a SIG-algebra with the error symbol $\epsilon \notin A$. Let $OP_c \subseteq OP$ be the set of constructor operations, and let $rop \subseteq (OP \to OP)$ be the map of reversible unary operations such that*

$$\forall op \in \textbf{dom}\, rop : \exists s, s' \in S : op \in OP_{s\,s'} : \forall a \in A_s : f_{rop(op)}(f_{op}(a)) = a.$$

Furthermore, let $s, s' \in S$, $x \in X_s$ and $T \in \mathbf{T}_{s'}^{SIG}(X)$. The variable x is said to be unifiable from T, denoted $x \triangleleft T$, if

1. *$T = x$, or*
2. *for some $op \in OP_c$ and $k \in \{1, \ldots, n\}$, $T = op(T_1, \ldots, T_n)$ and $x \triangleleft T_k$, or*
3. *for some $op \in \textbf{dom}\, rop$, $T = op(T')$ and $x \triangleleft T'$.*

Let $T_\emptyset \in \mathbf{T}_{s'}^{SIG}$ and $x \triangleleft T$. A unifier candidate $x \triangleleft_{T_\emptyset} T$ is inductively defined as follows:

1. *$T_\emptyset$, if $T = x$*
2. *$x \triangleleft_{T_{k\emptyset}} T_k$, if for some $op \in OP_c$, $T = op(T_1, \ldots, T_n)$, $T_\emptyset = op(T_{1\emptyset}, \ldots, T_{n\emptyset})$, $k \in \{1, \ldots, n\}$ and $x \triangleleft T_k$, and there is no $1 \leq j < k$ such that $x \triangleleft T_j$,[2] or*
3. *$x \triangleleft_{T'_\emptyset} T'$, if for some $op \in \textbf{dom}\, rop$, $T = op(T')$ and $x \triangleleft T'$ for $T'_\emptyset = rop(op)(T_\emptyset)$.*

Let $\beta : X \to A$. The terms T and $T_\emptyset$ are compatible under β, denoted $T \sim_\beta T_\emptyset$, if either

---

[2] Requiring the smallest k to be chosen ensures that unifier candidates are unique.

1. $\bar{\beta}(T) = \bar{\beta}_{\emptyset}(T_{\emptyset})$, or
2. for some $op \in OP_c$, $T = op(T_1, \ldots, T_n)$ and $T_{\emptyset} = op(T_{1\emptyset}, \ldots, T_{n\emptyset})$, and $T_k \sim_{\beta} T_{k\emptyset}$ for $k \in \{1, \ldots, n\}$, or
3. for some variable x in T, $\beta(x) = \epsilon$.

Continuing our example, and assuming that the operation $+$ is a constructor, $+ \in OP_c$, the definition yields no unifier candidates for x and y, if T is $x + y$ and $T_{\emptyset}$ is 2. If $T_{\emptyset}$ was $1 + 2$, then we would have $x \triangleleft_{T_{\emptyset}} T$ equal to 1 and $y \triangleleft_{T_{\emptyset}} T$ equal to 2. The terms are compatible under the assignment $\beta = \{\langle x, 1 \rangle, \langle y, 2 \rangle\}$ constructed from these candidates, since $x + y \sim_{\beta} 1 + 2$.

Restrictions of Unification. An analyser implementation can considerably restrict the set of operations supported by unification and the set of reversible operations. MARIA only looks for variables inside so-called constructor terms which construct values of structured data types out of components. From the constructor term $\langle x, \langle y + 1, z \rangle \rangle$, it could find unifier candidates for x and z, but not for y, since $y \ntriangleleft y + 1$. It also performs *constant folding* by replacing ground terms with equivalent nullary operators. It would transform the $T_{\emptyset} = 1 + 2$ in our above example to $\bar{\beta}_{\emptyset}(T_{\emptyset}) = 3$.

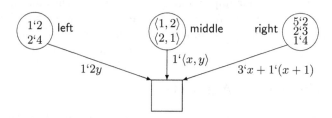

Fig. 2. A simple model for illustrating unification.

The operation collections OP_c and **dom** rop in Definition 3 strongly affect the set of unifiable variables. The more operations are contained in these sets, the more unifier candidates are possible. Consider the model shown in Figure 2. If multiplication by a constant belongs to the set of reversible operations, it is possible to unify y from the term $2y$ by unifying y with the given ground term (2 or 4) divided by 2.

We have not presented any algorithms yet, but we are about to face a somewhat philosophical question. Should a unification algorithm be able to find all possible assignments that enable the transition, or does it suffice for the algorithm to deal with real models, and report errors for cases it cannot handle? An implementation that restricts the sets of supported operations is likely to be more efficient and less prone to errors than one that tries to handle everything. For instance, when making basic arithmetic operations reversible, one must take care of arithmetic precision and exceptional situations.

Variables that are not unifiable by Definition 3 could be handled by nondeterministically picking values for them from their domains and by checking the terms for compatibility, but doing so is computationally expensive if the domains are large or there are many such variables. It is easier to report "variables cannot be unified" even though it might be possible to unify them. According to our experience with practical models, this works pretty well. One can always gain expressive power by replacing problematic terms with new variables and guards.

Splitting the Arcs. In typical models, arc expressions consist of elementary multisets (single-item multiset constructor terms) combined with multiset summation. In order to improve the granularity of our algorithm, we write each input arc inscription as a such combination of terms. Sanders refers to this as *arc unfolding* [16, Section 3]. For instance, the rightmost arc of the model illustrated in Figure 2 is split into two arcs, with the inscriptions $3`x$ and $1`(x+1)$.

The claim "any arc with a non-elementary multi-set may be 'unfolded' into multiple arcs" by Sanders [16, Section 3] is difficult to fulfil if the arcs contain multiset-valued variables or other multiset operations than the two we defined above. Our approach does not restrict the set of multiset operations.

We distinguish three kinds of split arc inscriptions: ones that contain unifiable variables, ones that can be evaluated under a partial assignment incrementally constructed by our algorithm, and others. In Figure 2, the arcs $1`\langle x, y\rangle$ and $3`x$ contain variables that our implementation can unify. Other inscriptions can be arbitrary multiset-valued terms. What matters is that whenever the unification algorithm finds a complete assignment, all arc inscriptions are compatible under it with the ground terms corresponding to the given marking of the model.

3   The Unification Algorithm

Our unification algorithm performs a depth-first search on the input arc inscriptions of the transition, split as described earlier. The algorithm is remarkably simple, since it processes the arcs in a fixed order produced in static analysis. Static analysis also determines which variables will be unified from which arc inscriptions, and verifies that all variables can be unified.[3]

The input arc inscriptions are split into items $S_k = \langle T_k, X_k, p_k, m_k\rangle \in \mathbf{T}^{BSIG}(X) \times \mathcal{P}(X) \times P \times \mathrm{BAG}(A_s)$, $k \in \{1, \ldots, n\}$ for some n, such that

- the variable sets X_k are pairwise disjoint: $X_j \cap X_k = \emptyset$ if $j \neq k$
- no T_k refers to a variable outside $\bigcup_{j=1}^{n} X_j$: $T_k \in \mathbf{T}^{BSIG}(\bigcup_{j=1}^{n} X_j)$
- if $X_k = \emptyset$: $T_k \in \mathbf{T}^{BSIG}(\bigcup_{j=1}^{k-1} X_j)$
- if $X_k \neq \emptyset$: $T_k = T_k' `T_k''$ and $T_k' \in \mathbf{T}^{BSIG}(\bigcup_{j=1}^{k-1} X_j)$ and $\forall x \in X_k : x \triangleleft T_k''$
- the input arcs of the transition and their inscriptions can be constructed from all places p_k and split inscriptions T_k via multiset summation

The last component, m_k, is a place-holder for the multiset the term is supposed to evaluate to. Our algorithm does not refer to it before initialising it; for convenience, we can assign it to the empty multiset here.

---

[3] A variable unified from a variable-multiplicity arc may remain undefined if the multiplicity of the arc evaluates to zero.

The input arcs of the only transition in Figure 2 can be split e.g. so that

$$S_1 = \langle T_1, X_1, p_1, m_1 \rangle = \langle 3`x, \{x\}, \text{right}, \emptyset \rangle$$
$$S_2 = \langle T_2, X_2, p_2, m_2 \rangle = \langle 1`\langle x, y \rangle, \{y\}, \text{middle}, \emptyset \rangle$$
$$S_3 = \langle T_3, X_3, p_3, m_3 \rangle = \langle 1`(x + 1), \emptyset, \text{right}, \emptyset \rangle$$
$$S_4 = \langle T_4, X_4, p_4, m_4 \rangle = \langle 1`2y, \emptyset, \text{left}, \emptyset \rangle.$$

The first components of the tuples are the split arc inscriptions. The second components are the "new" unifiable variables. Let us observe S_2 a bit more closely. We have $X_2 = \{y\}$, although also x could be unified: $x \triangleleft \langle x, y \rangle$. Including x in X_2 would violate the disjointness property, since $x \in X_1$. In a sense, the variable x will "already" be unified from S_1. Also, the terms T_3 and T_4 are "constant" since their variables can be unified from the earlier arcs S_1 and S_2.

3.1 The Basic Algorithm

Analyse arcs $S_1..S_n$ w.r.t. marking M
ANALYSE(S, n, M):
$\beta \leftarrow (\bigcup_{k=1}^{n} X_k) \times \{\epsilon\}$
ANALYSE-ARCS($S, 1, n, M, \beta$)

Analyse arcs $S_k..S_n$ w.r.t. M and β
ANALYSE-ARCS(S, k, n, M, β):
if $k = n$ **then print** β
else ▷ $S_k = \langle T_k, X_k, p_k, m_k \rangle$
  if $X_k = \emptyset$ **then**
    ANALYSE-CONSTANT(S, k, n, M, β)
  else
    ANALYSE-VARIABLE(S, k, n, M, β)

Evaluate arc S_k
ANALYSE-CONSTANT(S, k, n, M, β):
▷ $S_k = \langle T_k, X_k, p_k, m_k \rangle$
$m_k \leftarrow \bar{\beta}(T_k)$
if $m_k = \epsilon$ **then**
  print "undefined arc", β, T_k
else
  if $M(p_k) \geq m_k$ **then**
    $M' \leftarrow M$
    $M'(p_k) \leftarrow M(p_k) - m_k$
    ANALYSE-ARCS($S, k + 1, n, M, \beta$)

Analyse arc S_k, augment β
ANALYSE-VARIABLE(S, k, n, M, β):
▷ $S_k = \langle T_k, X_k, p_k, m_k \rangle$
▷ $T_k = T_k'`T_k''$
$c \leftarrow \bar{\beta}(T_k')$
if $c = \epsilon$ **then**
  print "undefined multiplicity", β, T_k
  return
if $c = 0$ **then**
  $m_k \leftarrow \emptyset$
  ANALYSE-ARCS($S, k + 1, n, M, \beta$)
else
  for each $m : M(p_k) \geq c`m$ **do**
    $m_k \leftarrow c`m$
    $\beta' \leftarrow \beta$
    for each $x \in X_k$ **do**
      $\beta'(x) \leftarrow (x \triangleleft_m T_k'')$
    if $\bigwedge_{j=1}^{k} T_j \sim_{\beta'} m_j$ **then**
      $M' \leftarrow M$
      $M'(p_k) \leftarrow M(p_k) - m_k$
      ANALYSE-ARCS($S, k + 1, n, M', \beta'$)

Fig. 3. The unification algorithm.

Our unification algorithm is presented in Figure 3. The computation is initiated by invoking ANALYSE with the split input arc inscriptions S and their amount n and a marking $M : P \to \mathrm{BAG}(A)$ of the net. The computation step of the depth-first search is divided into two alternatives: processing a "constant" arc (arc with no new bindable variables), and obtaining new variable bindings from an arc.

An Example. Continuing our running example from Figure 2, the call to ANAL-YSE on the initial marking of the model proceeds as follows. The assignment is initialised to $\beta = \{\langle x, \epsilon \rangle, \langle y, \epsilon \rangle\}$, and control is passed to ANALYSE-ARCS and further to ANALYSE-VARIABLE. The multiplicity of $T_1 = 3`x$ evaluates to $c = 3$. Now ANALYSE-VARIABLE loops over all items in the marking of right whose multiplicity is at least 3. It turns out that $m = 2$ is the only choice.

A new assignment with $\beta'(x) = 2$ is computed. Since all terms unified so far are compatible under this assignment, the multiset m_k is reserved from the marking and the control is transferred to ANALYSE-ARCS, which passes it again to ANALYSE-VARIABLE to handle the next term, $1`\langle x, y \rangle$. Both tokens in the place middle are tried, but only $\langle 2, 1 \rangle$ passes the compatibility check with x. Therefore, the assignment is transformed to $\beta' = \{\langle x, 2 \rangle, \langle y, 1 \rangle\}$.

The remaining two arcs are handled by ANALYSE-CONSTANT, which ensures that there are enough tokens for them. Finally, ANALYSE-ARCS prints out the assignment. At this point, the marking passed to it equals the original marking minus the evaluations of the input arcs under the assignment. The algorithm starts to backtrack. Since there were no other feasible choices in either active instance of ANALYSE-VARIABLE, the algorithm terminates.

Some Remarks. For the sake of simplicity, the illustrated procedures do not cover guards. In our implementation, guards are split to terms combined via logical conjunction. Whenever all the variables of a guard term become defined (due to assignments to $\beta'(x)$ in ANALYSE-VARIABLE), the term is evaluated. If the guard evaluates to false, the algorithm backtracks, just like it does in case a term T_k becomes incompatible. If an evaluation error occurs, the algorithm displays the assignment for diagnostics and backtracks.

Procedure ANALYSE-VARIABLE evaluates the multiplicity of a term T_k. When the multiplicity evaluates to zero, the variables in X_k remain undefined. As a result of this, the completed valuations displayed by ANALYSE-ARCS may contain undefined variables. This is not a problem if these variables are never evaluated due to short-circuit evaluation. Otherwise an error may occur when the transition is fired and its output arcs are evaluated. Also, before firing a transition, our implementation ensures that the guard evaluates to true.[4]

Correctness. The calling hierarchy of the algorithm is straightforward. The main procedure ANALYSE invokes ANALYSE-ARCS, which passes control to ei-

---

[4] Traditionally, "don't care" variables are assigned a nondeterministic choice of values from their domains. This generates unnecessary transition instances with identical behaviour. To avoid this, we assign these variables the special value ϵ.

ther ANALYSE-CONSTANT or ANALYSE-VARIABLE, which in turn call ANALYSE-ARCS. Each recursive call to ANALYSE-ARCS increments k, and the recursion terminates at $k = n$.

The assignment passed to ANALYSE-ARCS initially maps each variable to the undefined value. The only place where the assignment is modified is in ANALYSE-VARIABLE, where only previously undefined variables can be assigned.[5]

When ANALYSE-ARCS is invoked with $k = n$, the evaluation of the arc inscriptions under the gathered assignment is a subset of the marking passed to ANALYSE. This follows from two facts. Firstly, whenever the algorithm unifies an inscription and a multiset, it ensures that the marking contains the multiset and removes the multiset from the marking used for unifying further inscriptions.

Secondly, all split arc inscriptions are evaluated and ensured to match the multiset assigned to them. The procedure ANALYSE-CONSTANT evaluates the split arc inscription under the assignment gathered so far. In the procedure ANALYSE-VARIABLE, the relationship between arc inscriptions and markings is restricted by the compatibility check $\bigwedge_{j=1}^{k} T_j \sim_{\beta'} m_j$. At the deepest call to ANALYSE-VARIABLE, all variables have been assigned, and the compatibility check is equivalent to $\bigwedge_{j=1}^{k} \bar{\beta}'(T_j) = \bar{\beta}'(m_j)$.

To be sure that the algorithm finds all assignments or reports errors, we must investigate the conditions under which it backtracks without reporting anything. The procedures ANALYSE and ANALYSE-ARCS do not backtrack. ANALYSE-CONSTANT does backtrack when an input place would have an insufficient marking, when the test $M(p_k) \geq m_k$ fails. ANALYSE-VARIABLE silently backtracks when the arc inscriptions unified so far would be incompatible under the assignment, causing the test $\bigwedge_{j=1}^{k} T_j \sim_{\beta'} m_j$ to fail. Clearly, all assignments under which transitions are enabled must pass these tests. Therefore, the algorithm finds all relevant assignments.

3.2 Firing Transitions

At the moment when ANALYSE-ARCS displays a completed valuation β, the marking M passed to it is exactly the original marking passed to ANALYSE, minus the evaluations of the input arc inscriptions under β. Transition firing can be integrated to our enabling test algorithm by just replacing the **print** statement with something that binds the rest of the variables[6] and adds the evaluations of the output arcs to M.

Our current implementation of the algorithm combines enabling tests with firing. This is useful when all immediate successors of a state are to be generated, since there is no need to explicitly store the assignments. Also, if input and output arcs have similar inscriptions, some output arc inscriptions may have

---

[5] The variables are previously undefined, since the variable sets X_k are required to be pairwise disjoint.

[6] In our implementation, output arc inscriptions may make use of nondeterministically bound variables and multiset-valued variables that represent the tokens removed from the input places.

already been evaluated on the input side, and applying an optimisation technique called common subexpression elimination can save computations.

3.3   Unfolding

With slight modifications, the enabling test algorithm can also be used for unfolding Algebraic system nets to compact Place/Transition nets. Doing so has at least the following advantages:

- simple modifications: easy implementation, small chance of errors
- smaller unfolded net:
  - no unconnected places
  - sometimes finite unfoldings for nets with infinite domains

There are two unfolding options in MARIA: reduced and traditional. The latter option essentially implements the traditional definition of unfolding, e.g. [10, Definition 13], generating all possible assignments for all transitions. It doesn't generate all low-level places, though; only places that are connected to a low-level transition or are initially marked are generated.

The reduced unfolding option works by maintaining a set of low-level places that can ever be marked. This set is represented as a marking M of the high-level net. For all places $p \in P$ that contain tokens m in the marking, $M(p)(m) \geq 0$, there exists a low-level place $\langle p, m \rangle$. This marking is constructed incrementally, starting from the initial marking of the net.

The multiset containment comparisons $M(p_k) \geq m_k$ in ANALYSE-CONSTANT and ANALYSE-VARIABLE are modified so that they ignore the exact multiplicities: $M(p_k) \succeq m_k$ if and only if for each d such that $m_k(d) > 0$, it holds that $M(p_k)(d) > 0$.

Once the modified algorithm completes an assignment β of a transition t, it must unfold the input and output arcs of the transition. Our implementation accomplishes this by constructing two collections of multisets for the high-level input and output arcs: $M_-(p) := \bar{\beta}(i(\langle p, t \rangle))$ and $M_+(p) := \bar{\beta}(i(\langle t, p \rangle))$. For each value d such that $M_-(p)(d) > 0$, it constructs a low-level input arc of weight $M_-(p)(d)$ from the low-level place $\langle p, d \rangle$ to the low-level transition $\langle t, \beta \rangle$. The output arcs are constructed in similar way. The marking M is augmented with M_+. The algorithm keeps unfolding the high-level transitions in different modes until no new items are introduced in M.

When applied to the net illustrated in Figure 1, this algorithm yields the place/transition system illustrated in Figure 4, no matter how big domains the high-level places have. It can be easily seen that if the initial marking of this net contains n different tokens, the unfolded net can have at most $2n$ places and $\frac{1}{2}(n^2 + n)$ transitions.

It should be noted that also the reduced unfolding may be unmanageably large even if the high-level system has a small state space. A minimal unfolding (with no dead places or transitions) could be extracted from the full reachability graph of the high-level system by constructing only those low-level places that ever become marked and those transitions that ever fire. In the case of

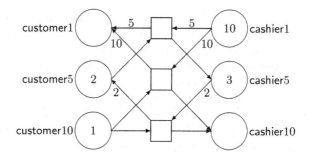

Fig. 4. A reduced unfolding of the net presented in Figure 1.

Figure 4, the reduced unfolding is also a minimal unfolding, since all transitions are enabled in the initial state and there cannot be dead places or transitions.

4   Optimisation Techniques

The first unification algorithm implemented in MARIA was very dynamic. It even rewrote input arc expressions on the fly, expanding multiset summations whose limits may depend on other variables. The first optimisation step was to expand quantifications in the parsing stage, transforming dynamic limits to variable-dependent multiplicities, so that the arc expressions remained static during the analysis. This improved the overall performance of the tool by 15–20 percent. At the same time, we started to make experiments with executable code generation [14]. Using the C program code generated by our current implementation usually shortens the analysis times to less than a third of the times consumed by the interpreter written in C++.

Previously presented algorithms, such as [7,8,16], dynamically schedule the input arcs. When we replaced dynamic scheduling with static scheduling in our implementation, we noticed a significant performance boost, 20–30 percent. This is mainly due to less bookkeeping, as the variables are always unified in the same order.[7] Dynamic scheduling may provide shortcuts for simulators that randomly pick one assignment for firing a transition without generating all assignments.

4.1   Representing Multisets

It is important to choose the right data structures for representing multisets in the enabling test algorithm. It appears that we are not the first ones to come up with binary search trees. Haagh and Hansen [6, Chapter 3] suggest using a form of balanced binary trees.

Our implementation uses two representations for markings: encoded (used for storing states) and expanded (for computations). The expanded representation is

---

[7] Variables that are unifiable from several variable-multiplicity arcs and no constant-multiplicity arcs form an exception. Our implementation attempts to unify them from each arc having nonzero multiplicity.

a binary search tree whose keys are multiset items and values are multiplicities. When a multiset is decoded from the state storage, only items with nonzero multiplicity are added. When items are removed from the tree, their multiplicities are set to zero. Since single items are never removed from the multiset, there is no need for costly balancing operations.

According to our experiments, using unbalanced trees is faster than using red-black trees if the places contain a small number of distinct tokens in the reachable markings. Even though unbalanced trees easily degenerate to linked lists, and searches may have to process n nodes instead of $\lceil \log_2 n \rceil$, the savings in insertions dominate for small values of n. Our executable code generator contains an option for enabling or disabling red-black trees.

4.2   Static Heuristics: Sorting the Arcs

We use multisets in two remarkably different ways. ANALYSE-CONSTANT (Figure 3) performs one containment comparison on a multiset and calls ANALYSE-ARCS zero or one times. ANALYSE-VARIABLE may iterate through all items in a multiset and invoke ANALYSE-ARCS for any number of them.

Let us assume that our enabling test algorithm is invoked on a sequence of n arcs, k of which are constant. Furthermore, let us assume that all multisets contain m distinct items. If the constant arcs are processed first, the search tree will consist of a linear sequence of k calls to ANALYSE-CONSTANT followed by a tree of ANALYSE-VARIABLE invocations. There will be at most $k + m^{n-k}$ recursive calls to ANALYSE-ARCS. The other extreme, analysing constant arcs as late as possible, yields at most $(k + 1)m^{n-k}$ recursive calls.

The proportion of these numbers of iteration steps is

$$\frac{k + m^{n-k}}{(k+1)m^{n-k}} = \frac{k}{k+1}\frac{1}{m^{n-k}} + \frac{1}{k+1} \approx \frac{1}{k+1},$$

and the approximation is pretty good already for $m = 2$. Thus, if there are k constant arcs, it is about k times slower to analyse them at the leaves of the search tree than at the root. The difference becomes even more significant if the transition only has a few enabled instances in each marking and most instances of ANALYSE-CONSTANT backtrack. The earlier this can happen, the better.

The problem, finding an optimal static scheduling that minimises the number of ANALYSE-ARCS calls, becomes more complicated when we consider the fact that ANALYSE-CONSTANT can handle non-ground terms that can be evaluated under the assignment generated so far. Intuitively, an optimal scheduling should

- minimise the number of arcs from which variables are unified, and
- minimise the worst-case number of multiset iterations, and
- schedule the remaining arcs as early as possible in such an order that the algorithm is most likely to backtrack early.

Especially the last requirement is difficult to fulfil in static analysis. We apply Gaeta's "Less Different Tokens First" policy [4, Section 5.1] and compute the maximum numbers of distinct items in the input places. A multiset associated with a place whose domain is $BAG(A_s)$ can have at most $|A_s|$ distinct items.[8]

Let us shortly return to our example from Figure 2. If we assume that the domain sizes of the places left, middle and right are d, d^2 and d for some $d > 1$, then the scheduling we presented in the beginning of Section 3 is not very optimal. In the worst case, it iterates through d items in right and d^2 items in middle, at most d of which can pass the compatibility requirements. ANALYSE-ARCS can be invoked $1 + d(1 + d(1 + 2)) = 3d^2 + d + 1$ times, and ANALYSE-VARIABLE may scan up to $d + d^3$ multiset items. Scheduling the term $1'(x + 1)$ before $1'\langle x, y \rangle$ would reduce the maximum number of invocations to $1 + d(1 + 1 + d(1 + 1)) = 2d^2 + 2d + 1$. The same number of multiset items need to be scanned in the worst case, but if analysing the term $1'(x + 1)$ fails every time, the d^3 scans for $1'\langle x, y \rangle$ can be avoided.

Gaeta divides input arc expressions to three categories: simple, complex and guarded. We use four categories: closed arcs (arcs that may only depend on already unified variables), constant-multiplicity arcs with unifiable variables, variable-multiplicity arcs with unifiable variables, and other arcs.

We have implemented a depth-first search algorithm for splitting the input arc inscriptions as described in the beginning of Section 3. Since the algorithm has exponential complexity with regard to the number of split arcs containing unifiable variables, we programmed a special condition that terminates the search when a solution is found with more than five arcs containing unifiable variables.

Our algorithm uses three cost functions. The primary cost function is the number of variables that will be unified from variable-multiplicity arcs. The secondary cost function is a sum of costs $c_2(S_k)$ for each arc $S_k = \langle T_k, X_k, p_k, m_k \rangle$ defined as

$$c_2(S_k) = \begin{cases} 0 & \text{if } X_k \neq \emptyset \\ \sum_{j=1}^{k-1}[X_k \neq \emptyset] & \text{if } X_k = \emptyset \end{cases}$$

where the square brackets map truth values to 0 and 1. Thus, for closed arcs, the secondary cost is the number of preceding non-closed arcs. Minimising this cost ensures that all closed arcs will be scheduled as early as possible.

As a shortcut, our algorithm prioritises closed arcs over arcs with unifiable variables. Every time the algorithm picks an arc with unifiable variables, some of the remaining arcs may become closed. Only after the closed arcs run out, the algorithm picks the next arc with unifiable variables. If the only arcs left are in the "other" category, the search backtracks. If no complete schedulings are found, a unification error will be reported.

The third and last cost function is the maximum number of iterations possible with the scheduling. For each split arc $S_k = \langle T_k, X_k, p_k, m_k \rangle$, we define

$$c_3(S_k) = \begin{cases} m(p_k) & \text{if } X_k \neq \emptyset \\ 1 & \text{if } X_k = \emptyset \end{cases}$$

---

[8] If the multiset is associated with a maximum cardinality, then it is another limit.

where $m(p_k)$ denotes the maximum possible number of distinct tokens in the place p_k. The total cost is defined as

$$c_3(S_1) \cdot (1 + c_3(S_2) \cdot (1 + c_3(S_3) \cdot (1 + \cdots)))$$

where the term $(1 + \cdots)$ after the last cost $c_3(S_n)$ is replaced with 1.

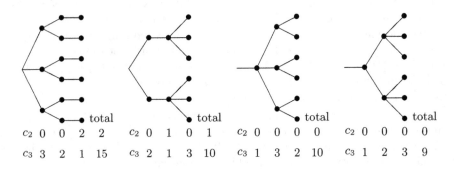

| c_2 | 0 | 0 | 2 | 2 |
| c_3 | 3 | 2 | 1 | 15 |

| c_2 | 0 | 1 | 0 | 1 |
| c_3 | 2 | 1 | 3 | 10 |

| c_2 | 0 | 0 | 0 | 0 |
| c_3 | 1 | 3 | 2 | 10 |

| c_2 | 0 | 0 | 0 | 0 |
| c_3 | 1 | 2 | 3 | 9 |

Fig. 5. Minimising Search Trees with Cost Functions

Figure 5 demonstrates the relation between the secondary and ternary cost functions. It illustrates the maximal search trees imposed by four different orderings for three split arcs, one of which is constant. The constant arc is the only contributor to the secondary cost, and scheduling it first minimises the secondary cost. The total ternary cost is the maximum number of recursive invocations to ANALYSE-ARCS, denoted in the figure with opaque circles. The scheduling presented on the right implies the smallest tree, 9 nodes.

The algorithm selects a scheduling that has the minimum primary cost, number of variables unified from variable-multiplicity arcs. If there are several such schedulings, then the one with the least secondary cost is selected. If that selection is not unique, then one with the smallest ternary cost is chosen.

As a finishing touch for the found static scheduling, our algorithm sorts contiguous sequences of closed terms in such a way that arcs whose places have the smallest number of distinct tokens are scheduled first. This enforces Gaeta's "Less Different Tokens First" policy.

4.3   Dynamic Heuristics

Caching. Ilié and Rojas suggest in [8, Section 3.4] that to speed up a simulator, one could build up a cache that maps input place markings of transitions to sets of enabling assignments. In our experiments, it turned out that in exhaustive reachability analysis, this kind of cache is only useful when there are a large number of states in which the input places of a transition are marked in exactly the same way.

In simulations of timed nets, where exactly the same states are visited over and over again, using such caches may pay off if the bookkeeping overhead (comparing and duplicating input place markings and copying associated sets of enabling assignments) is smaller than the cost of computing the enabling assignments from the scratch. This depends on the implementation, on the size and the scheduling policy of the cache and on the model.

We experimented with an artificial model, one of whose transitions had n input arcs from a place holding n tokens, with a total of $n!$ enabled assignments. For $n = 7$, generating the 5040 assignments took several thousands of times longer than a cache look-up. In more realistic models, we witnessed differences of at most a few percent. In many models, such as the dining philosophers [3] or the distributed data base management system [5], all cache look-ups failed. Due to this experience, we decided to eliminate the cache altogether and to integrate transition firings with enabling tests. This improved the execution times by about ten percent.

Cardinality Tests. Gaeta [4, Section 4.3] has implemented heuristics for detecting when a transition is disabled. He keeps track of the number of tokens in each input place. If a place contains less tokens than a transition would consume, the transition cannot be enabled and the search for enabling assignments can be avoided.

Our implementation of the cardinality test needs to consider arcs with variable multiplicity. Their multiplicities are assumed to be zero. Since the heuristics is implemented in generated code, the comparisons can be omitted if they are known to hold. In MARIA models, it is possible to speed up analysis by specifying conditions on the amounts of tokens places may hold in reachable markings.

4.4   Some Experimental Results

MARIA uses an explicit technique for maintaining the set of reachable states and the transition instances leading from one state to another. Everything related to the reachability graph is kept in disk files [13]. Because of this, the analyser spends most of its execution time checking whether an encoded state exists in the reachability graph. In our tests, MARIA has generated full state spaces of converted PROD [18] models in 0.5 to 1 times the PROD speed. One explanation for the slowness is that MARIA detects evaluation errors and supports much more powerful algebraic operations than PROD, which makes optimisations in the C code generation difficult. Also, it is possible to use probabilistic verification with MARIA. When no arcs are stored and a reachability set is maintained in memory, the tool performs an order of magnitude faster.

Analysing the biggest state space so far with MARIA, a translation of a radio link control protocol specified in SDL consisting of 15,866,988 states and 61,156,129 events, took 5 megabytes of memory and 1.55 gigabytes of disk space, most of which was consumed by the arc inscriptions and double links stored with the reachability graph. The analysis was completed in less than nine hours on a 266 MHz Pentium II system.

Table 1 lists some models we have analysed and unfolded in MARIA. All models except "rlc" are distributed with the tool. There are three figures for

Table 1. Unfoldings and State Spaces of Selected Models

Model	Folded $\|P\|$	$\|T\|$	Unfolded $\|P\|$	$\|T\|$	Reduced $\|P\|$	$\|T\|$	Minimal $\|P\|$	$\|T\|$	State Space states	arcs
dining(10)	2	3	40	30	40	30	40	30	6,726	43,480
dbm(5)	8	4	111	60	96	50	96	50	406	1,090
dbm(10)	8	4	421	220	391	200	391	200	196,831	1,181,000
sw(1,1)	12	9	41	422	35	288	35	54	164	352
sw(2,2)	12	9	2,048	1,087,382	729	239,478	129	688	2,640	7,716
sw(6,6)	12	9	–	–	–	–	9,805	145,464	1,774,716	7,127,688
resource	4	3	336	8,158	193	4,610	111	45	538,318	4,136,459
rlc	18	104	708	14,736	114	1,429	–	–	15,866,988	61,156,129

unfolded net sizes. The first column is for "traditional" unfoldings, excluding unconnected places; the second is for unfoldings reduced with our method, and the third is for minimal unfoldings obtained from the reachability graph.

The model named "resource" solves a resource allocation problem. On our system, MARIA generates its full reachability graph in 26 minutes, using 3 megabytes of memory and 85.7 megabytes of disk space. With the default capacity limit, LOLA consumes 4 minutes less time but about 530 megabytes more memory on the reduced unfolding of this model. We tried to tighten the capacity limit to save memory but failed, because the limit is global for all places.

Reduced unfoldings work best for models with a sparse state space, i.e. only a fraction of the possible states are actually reachable. For some theoretically pleasing symmetric models, our reduced unfolding does not gain much.

We have the feeling that models of communication protocols, especially those translated from a high-level programming language, have sparse state spaces. The sliding window protocol model we experimented with ("sw" in Table 1) is a good example of this. Already with very small window sizes its traditional unfoldings become unmanageably large. Even the reduced and minimal unfoldings are not very helpful for larger window sizes. This is because all buffer reads and writes in the model are atomic, which reduces the reachability graph but makes the unfolding explode.

5   Conclusion and Future Work

Earlier work on performing enabling tests for high-level nets appears to be limited to nets whose arc inscriptions have constant weights. According to Kindler and Völzer [10], it is difficult to model distributed network algorithms under such limitations. They present Algebraic system nets as a solution, but do not define any algorithms for analysing these nets on the high level.

We viewed enabling tests for high-level nets—constructing the set of assignments under which a transition is enabled in a given marking—as a unification problem, matching multiset-valued terms and subsets of constant multisets. Our approach avoids the combinatorial explosion inherent in this problem by disallowing multiset-valued variables on input arc inscriptions, by restricting the set of operations recognised by the unification algorithm, and by requiring that variable-dependent weights and arbitrary multiset-valued terms can be evaluated based on variable bindings gathered from other terms.

Our reachability analyser MARIA supports queues and stacks on the data type level. Powerful operations, such as removing multiple items from the middle of a queue, make it easy to construct compact high-level models. Experiments show that it is often infeasible to unfold such models in the traditional way. Special constructs for translating large blocks of atomic operations in high-level nets into behaviour-equivalent compact low-level nets are subject to further research.

The presented unification algorithm processes terms in a fixed order. A method for statically ordering the terms in a close to optimal way was presented, and some optimisations to the algorithm were discussed. Some of the presented techniques may be best suited for exhaustive analysis tools; their applicability in simulators was not tested.

A new method for unfolding high-level nets based on a kind of "coverable marking" was presented. The method often produces considerably smaller unfoldings than the common approach of iterating over all domains.

Acknowledgements. The author would like to thank Dr. Kimmo Varpaaniemi, who implemented the skeleton of the first unification algorithm in MARIA, for fruitful discussions and ideas for improving the algorithm, and the referees for their constructive comments.

References

1. Chin-Liang Chang and Richard Char-Tung Lee. *Symbolic Logic and Mechanical Theorem Proving.* Academic Press, New York, NY, USA, 1973.
2. Gianfranco Chiola, Giuliana Franceschinis and Rossano Gaeta. A symbolic simulation mechanism for well-formed coloured Petri nets. In Philip Wilsey, editor, *Proceedings, 25th Annual Simulation Symposium,* pages 192–201, Orlando, FL, USA, April 1992. IEEE Computer Society Press, Los Alamitos, CA, USA.
3. Edsger Wybe Dijkstra. Hierarchical ordering of sequential processes. *Acta Informatica,* 1:115–138, 1971.
4. Rossano Gaeta. Efficient discrete-event simulation of colored Petri nets. *IEEE Transactions on Software Engineering,* 22(9):629–639, September 1996.
5. Hartmann J. Genrich and Kurt Lautenbach. The analysis of distributed systems by means of Predicate/Transition-Nets. In Gilles Kahn, editor, *Semantics of Concurrent Computation,* volume 70 of *Lecture Notes in Computer Science,* pages 123–146, Evian, France, July 1979. Springer-Verlag, Berlin, Germany, 1979.
6. Torben Bisgaard Haagh and Tommy Rudmose Hansen. Optimising a Coloured Petri Net Simulator. Master's thesis, University of Århus, Denmark, December 1994. http://www.daimi.au.dk/CPnets/publ/thesis/HanHaa1994.pdf.
7. Jean-Michel Ilié, Yasmina Maïzi and Denis Poitrenaud. Towards an efficient simulation based on well-formed Petri nets, extended with test and inhibitor arcs. In Tuncer I. Oren and Louis G. Birta, editors, *1995 Summer Computer Simulation Conference,* pages 70–75, Ottawa, Canada, July 1995. Society for Computer Simulation International.

8. Jean-Michel Ilié and Omar Rojas. On well-formed nets and optimizations in enabling tests. In Marco Ajmone Marsan, editor, *Application and Theory of Petri Nets 1993*, volume 691 of *Lecture Notes in Computer Science*, pages 300–318, Chicago, IL, USA, June 1993. Springer-Verlag, Berlin, Germany.

9. Kurt Jensen. *Coloured Petri Nets: Basic Concepts, Analysis Methods and Practical Use: Volume 1, Basic Concepts*. Monographs in Theoretical Computer Science. Springer-Verlag, Berlin, Germany, 1992.

10. Ekkart Kindler and Hagen Völzer. Flexibility in algebraic nets. In Jörg Desel and Manuel Silva, editors, *Application and Theory of Petri Nets 1998: 19th International Conference, ICATPN'98*, volume 1420 of *Lecture Notes in Computer Science*, pages 345–364, Lisbon, Portugal, June 1998. Springer-Verlag, Berlin, Germany.

11. Marko Mäkelä. *Maria—Modular Reachability Analyser for Algebraic System Nets*. On-line documentation, http://www.tcs.hut.fi/maria/.

12. Marko Mäkelä. *A Reachability Analyser for Algebraic System Nets*. Licentiate's thesis, Helsinki University of Technology, Department of Computer Science and Engineering, Espoo, Finland, March 2000.

13. Marko Mäkelä. Condensed storage of multi-set sequences. In Kurt Jensen, editor, *Practical Use of High-Level Petri Nets*, DAIMI report PB-547, pages 111–125. University of Århus, Denmark, June 2000.

14. Marko Mäkelä. Applying compiler techniques to reachability analysis of high-level models. In Hans-Dieter Burkhard, Ludwik Czaja, Andrzej Skowron and Peter Starke, editors, *Workshop on Concurrency, Specification & Programming 2000*, Informatik-Bericht 140, pages 129–142. Humboldt-Universität zu Berlin, Germany, October 2000.

15. Wolfgang Reisig. *Petri Nets: An Introduction*. Springer-Verlag, Berlin, Germany, 1985.

16. Michael J. Sanders. Efficient computation of enabled transition bindings in high-level Petri nets. In *2000 IEEE International Conference on Systems, Man and Cybernetics*, pages 3153–3158, Nashville, TN, USA, October 2000.

17. Karsten Schmidt. LoLA: a low level analyser. In Mogens Nielsen and Dan Simpson, editors, *Application and Theory of Petri Nets 2000: 21st International Conference, ICATPN 2000*, volume 1825 of *Lecture Notes in Computer Science*, pages 465–474, Århus, Denmark, June 2000. Springer-Verlag, Berlin, Germany.

18. Kimmo Varpaaniemi, Jaakko Halme, Kari Hiekkanen and Tino Pyssysalo. PROD reference manual. Technical Report B13, Helsinki University of Technology, Department of Computer Science and Engineering, Digital Systems Laboratory, Espoo, Finland, August 1995.

Extending the Petri Box Calculus with Time

Olga Marroquín Alonso and David de Frutos Escrig

Departamento de Sistemas Informáticos y Programación
Universidad Complutense de Madrid, E-28040 Madrid, Spain
{alonso,defrutos}@sip.ucm.es

Abstract. PBC (*Petri Box Calculus*) is a process algebra where real parallelism of concurrent systems can be naturally expressed. One of its main features is the definition of a denotational semantics based on Petri nets, which emphasizes the structural aspects of the modelled systems. However, this formal model does not include temporal aspects of processes, which are necessary when considering real-time systems. The aim of this paper is to extend the existing calculus with those temporal aspects. We consider that actions are not instantaneous, that is, their execution takes time. We present an operational semantics and a denotational semantics based on timed Petri nets. Finally, we discuss the introduction of other new features such as time-outs and delays. Throughout the paper we assume that the reader is familiar with both Petri nets and PBC.

1 Introduction

Formal models of concurrency are widely used to specify concurrent and distributed systems. In this research field, process algebras and Petri nets are well-known. Each of them has its own advantages and drawbacks: In Petri nets and their extensions one can make assertions about events, even if causality relations are not given explicitly. Emphasis is put on the partial order of events and on structural aspects of the modelled systems. However, their algebraic basis is poor, thus modelling and verification of systems are affected. On the other hand, the main feature of process algebras is the simple algebraic characterization of the behaviour of each of the syntactic operators, although it is true that in most of the cases, we obtain it by losing important information concerning event causality.

Recently, a new process algebra, PBC (*Petri Box Calculus*) [1,2,3], has arisen from the attempts to combine the advantages of both Petri nets and process algebras. When defining PBC the starting point were Petri nets and not any well known process algebra, so their authors looked for a suitable one whose operators could be easily defined on Petri nets. As a consequence, they obtained a Petri net's algebra which can be seen as the denotational semantics of PBC [1, 2,3] and since Petri nets are endowed with a natural operational semantics, we also also derive an operational semantics of the given process algebra which can be also directly defined by means of Plotkin-like syntax-guided rules.

J.-M. Colom and M. Koutny (Eds.): ICATPN 2001, LNCS 2075, pp. 303–322, 2001.

Nevertheless, these models do not include any temporal information, and it is obvious that some kind of quantitative time representation is needed for the description of real-time systems. Many timed process algebras have been proposed. Among them we will mention: timed CCS [11], temporal CCS [7], timed CSP [10] and timed Observations [8]. Besides, since long time ago we have several timed extensions of Petri nets like Petri nets with time [6], and timed Petri nets [9].

The aim of this paper is to extend PBC with time maintaining its main properties and the basic concepts in which it is based. In this way, we propose TPBC (*Timed Petri Box Calculus*). Our model differs from the previous approach by M. Koutny [4] in several relevant aspects, among which we mention the duration of actions and the nonexistence of illegal action occurrences. Both models correspond to different ways of capturing time information, and as a consequence they are not in competition but are complementary.

Here we consider a discrete time domain. Most of the systems which we are interested in can be modelled under this hypothesis, and some of the definitions and results in the paper can be presented in a more simple way. Our results cannot be generalized to the case of continuous time, because of the fact that discrete time cannot be considered just as a simplification of continuous time.

In order to improve the readability of the paper, we will layer the presentation: first we present a simple extension in which actions have a minimal duration, that is, we allow that their executions will take more time than expected. We can find an intuitive justification of this fact considering the behaviour in practice of real-time systems. One usually knows which is the minimum time needed to execute a task, but depending on how it interacts with the environment, it could take more time than expected. Besides, even if we consider that the duration of each action is fixed, when we execute an action α whose minimal duration is d but the real duration is $d' > d$, we could consider that we are just representing a delayed execution of the action, which would start after $d' - d$ time units, in such a way that d would still be the effective duration of the action.

In this simplified timed extension there is no limit to the time a process can be idle without changing its state. So, we cannot model in it any kind of urgency or time-out. In order to do it, we present in Section 6 of the paper a more elaborate extension where these characteristics are introduced.

As we said before, we introduce time information by means of duration of actions instead of combining delays and instantaneous actions. There are several reasons that justify our choice. First, one can find in the literature both kind of timed models; since Koutny has already investigated the other case, by studying here the case of actions with duration we are somehow completing the picture. It is true that similar results to these in the paper could be obtained for the case of delays and instantaneous actions, although the corresponding translation could be non-immediate in all the cases. But it is not the consideration of actions with duration what makes our models rather more complicated than Koutny's one: The introduced complications are necessary in order to avoid *illegal action occurrences*. We will compare more in depth the two timed models in the last sections of the paper.

2   The TPBC Language

Although probably it would be more than desirable, by lack of space we cannot give here a fast introduction to PBC. Unfortunately, this is a technically involved model whose presentation require several pages [1,2,3]. But you can also infer from our paper the main characteristics of PBC, just by abstracting all the references to time in it.

Throughout the paper we use standard mathematical notation. In particular, the set of finite multisets over a set $\mathcal{S}$ is denoted by $\mathcal{M}(\mathcal{S})$, and defined as the set of functions $\alpha : \mathcal{S} \longrightarrow I\!N$ such that $\{ s \in \mathcal{S} \mid \alpha(s) \neq 0 \}$ is finite. $\varnothing$ denotes the empty multiset, and $\{a\}$ is a unitary multiset containing a.

To define the syntax of TPBC, we consider a countable alphabet of *labels* $\mathcal{A}$, which will denote atomic actions. In order to support synchronization we assume the existence of a bijection $\hat{\ } : \mathcal{A} \longrightarrow \mathcal{A}$, called *conjugation*, by means of which we associate to each label $a \in \mathcal{A}$ a corresponding one $\hat{a} \in \mathcal{A}$. This function must satisfy the following property: $\forall a \in \mathcal{A} \quad \hat{a} \neq a \wedge \hat{\hat{a}} = a$.

As in plain PBC, the basic actions of TPBC are finite multisets of labels, called *bags*. In the prefix operator of the language each bag $\alpha \in \mathcal{M}(\mathcal{A})$ carries a duration $d \in I\!N^+$, thus we obtain the *basic action* $\alpha : d$. In the following, we will denote by $\mathcal{BA} = \mathcal{M}(\mathcal{A}) \times I\!N^+$ the set of those basic actions.

The rest of the syntactic operators in TPBC are those in PBC. Although recursive processes have not been included, the calculus is not finite: We have infinite behaviours due to the presence of the iteration operator.

Definition 1 (Static expressions). *A **static expression** of TPBC is any expression generated by the following BNF grammar:*

$$E ::= \alpha : d \mid E; E \mid E \square E \mid E \| E \mid [E * E * E] \mid E[f] \mid E \operatorname{sy} a \mid E \operatorname{rs} a \mid [a : E]$$

where $d \in I\!N^+$ and $f : \mathcal{A} \longrightarrow \mathcal{A}$ is a conjugate-preserving function. The set of static expressions of TPBC is denoted by $Expr^s$, and we use letters E and F to denote its elements.

3   Operational Semantics

The operational semantics of TPBC is defined by means of a labelled transition system including two types of transitions: instantaneous transitions, which relate equivalent processes, and non-instantaneous transitions, which express how processes evolve due to the execution of actions and the progress of time. The set of states of this transition system corresponds to a new class of expressions, the so called *dynamic expressions*.

Definition 2 (Dynamic expressions). *A **dynamic expression** of TPBC is any expression generated by the following BNF grammar:*

$$G ::= \overline{E} \mid \underline{E} \mid \widetilde{\alpha : d}, d' \mid G; E \mid E; G \mid G \square E \mid E \square G \mid G \| G \mid$$
$$[G * E * E] \mid [E * G * E] \mid [E * E * G] \mid G[f] \mid G \operatorname{sy} a \mid G \operatorname{rs} a \mid [a : G]$$

where $d \in \mathbb{N}^+$, $d' \in \mathbb{N}$ and f is a conjugate-preserving function from $\mathcal{A}$ to $\mathcal{A}$. The set of dynamic expressions of TPBC will be denoted by $Expr^d$, and we use letters G and H to represent its elements.

In the definition above the static expressions of the calculus are marked with three types of barring: *overlining*, *underlining* and *executing barring*. The first two have the same meaning as in PBC: $\overline{E}$ denotes that E has been activated and it offers all the behaviours E represents, whereas $\underline{E}$ denotes that the process E has reached its final state and the only move it can perform is letting time pass. Finally, the dynamic expression $\widetilde{\alpha : d}, d'$ represents that the action $\alpha \in \mathcal{M}(\mathcal{A})$ has begun its execution some time ago with a minimum duration of d time units, and from now on its execution will take d' time units until it terminates.

Since we are interested in expressing locally the passage of time, and it is at the level of basic actions where this can be done in a proper way, the overline operator will be distributed over the current expression until a basic action is reached. Then the *executing bar* $\frown$ will only be applied to this kind of actions.

As we already mentioned, non-instantaneous transitions represent the execution of actions and their labels indicate which bags have just begun to execute together with their durations. This information is expressed by means of *timed bags*, $\alpha_{d'}$, which consist of a bag α and a temporal annotation $d' \in \mathbb{N}^+$. The set of timed bags will be denoted by $\mathcal{TB}$.

Timed bags express the first level of concurrency we can distinguish in a concurrent system. They represent the simultaneous execution of atomic actions in the same component of the system. To cover also a second level, which represents the concurrent evolution of the different components of the system, we introduce *timed multibags*, which are finite multisets of timed bags.

Thus non-instantaneous transitions have the form $G \stackrel{\Gamma}{\longrightarrow} G'$, where $\Gamma \in \mathcal{M}(\mathcal{TB})$. This can be interpreted as follows: process G has changed its state to G' by starting to execute the multiset of timed bags Γ during one unit of time. More in general, we assume that each labelled transition represents the passing of one time unit. As a consequence, the execution of a multibag Γ is only observable (at the level of labels of the transition system) at the first instant of it. Afterwards, we will let time to progress until the execution of Γ terminates, although in between the execution of some other actions whose performance do not need the termination of Γ could be initiated.

The passage of one time unit without starting the execution of any new bag is represented by transitions labelled by $\varnothing$. Since for any overlined or underlined process we let time pass without any change in the state, we have the rules:

$$\overline{\overline{E} \stackrel{\varnothing}{\longrightarrow} \overline{E}} \; \textbf{(V1)} \qquad \underline{E} \stackrel{\varnothing}{\longrightarrow} \underline{E} \; \textbf{(V2)}$$

Instantaneous transitions take the form $G \longleftrightarrow G'$, and their intended meaning is that the involved process has these two different syntactic representations: G and G'. Therefore this kind of transitions relates expressions with the same operational behaviour.

3.1 Transition Rules

In this section we first present those transition rules that represent the timed aspects of our model. They are basic actions rules and synchronization rules. We also provide the operational semantics of iteration, since it is not common in most of process algebras. The rest of the operators behave as in the untimed model [1,2,3]. A complete set of rules will appear in the PhD thesis of the first author (see [5] for a partial preliminary version).

$$\overline{\alpha : d} \xrightarrow{\{\alpha_{d'}\}} \widetilde{\alpha : d, d'} - 1 \text{ if } d' \geq d \textbf{ (B1)} \quad \widetilde{\alpha : d, d'} \xrightarrow{\varnothing} \widetilde{\alpha : d, d'} - 1 \text{ if } d' > 0 \textbf{ (B2)}$$

$$\widetilde{\alpha : d, 0} \longleftrightarrow \underline{\alpha : d} \quad \textbf{(B3)}$$

Operational semantics of *basic actions*

$$\overline{E} \text{ sy } a \longleftrightarrow \overline{E \text{ sy } a} \quad \textbf{(S1)}$$

$$\frac{G \longleftrightarrow G'}{G \text{ sy } a \longleftrightarrow G' \text{ sy } a} \quad \textbf{(S2a)} \qquad \frac{G \xrightarrow{\Gamma} G'}{G \text{ sy } a \xrightarrow{\Gamma} G' \text{ sy } a} \quad \textbf{(S2b)}$$

$$\frac{G \text{ sy } a \xrightarrow{\{\{\alpha+\{a\}\}_{d'}\}+\{\{\beta+\{\hat{a}\}\}_{d'}\}+\Gamma} G' \text{ sy } a}{G \text{ sy } a \xrightarrow{\{\{\alpha+\beta\}_{d'}\}+\Gamma} G' \text{ sy } a} \quad \textbf{(S2c)}$$

$$\underline{E} \text{ sy } a \longleftrightarrow \underline{E \text{ sy } a} \quad \textbf{(S3)}$$

Operational semantics of *synchronization*

$$\overline{[E * F * E']} \longleftrightarrow [\overline{E} * F * E'] \quad \textbf{(It1)}$$

$$\frac{G \longleftrightarrow G'}{[G * F * E] \longleftrightarrow [G' * F * E]} \quad \textbf{(It2a)} \qquad \frac{G \xrightarrow{\Gamma} G'}{[G * F * E] \xrightarrow{\Gamma} [G' * F * E]} \quad \textbf{(It2b)}$$

$$[\underline{E} * F * E'] \longleftrightarrow [E * \overline{F} * E'] \quad \textbf{(It2c)} \qquad [\underline{E} * F * E'] \longleftrightarrow [E * F * \overline{E'}] \quad \textbf{(It2d)}$$

$$\frac{G \longleftrightarrow G'}{[E * G * E'] \longleftrightarrow [E * G' * E']} \quad \textbf{(It3a)} \qquad \frac{G \xrightarrow{\Gamma} G'}{[E * G * E'] \xrightarrow{\Gamma} [E * G' * E']} \quad \textbf{(It3b)}$$

$$[E * \underline{F} * E'] \longleftrightarrow [E * \overline{F} * E'] \quad \textbf{(It3c)} \qquad [E * \underline{F} * E'] \longleftrightarrow [E * F * \overline{E'}] \quad \textbf{(It3d)}$$

$$\frac{G \longleftrightarrow G'}{[E * F * G] \longleftrightarrow [E * F * G']} \quad \textbf{(It4a)} \qquad \frac{G \xrightarrow{\Gamma} G'}{[E * F * G] \xrightarrow{\Gamma} [E * F * G']} \quad \textbf{(It4b)}$$

$$[E * F * \underline{E'}] \longleftrightarrow \underline{[E * F * E']} \quad \textbf{(It5)}$$

Operational semantics of *iteration*

Rule **(B1)** states that basic processes can only leave its initial state by starting the execution of the corresponding bag. The duration of this execution will be greater or equal than the annotated minimum duration. Once this execution starts, the process will let time pass until its termination (rule **(B2)**). Then the basic action has finished its execution, what is represented by rule **(B3)**.

The synchronization is activated whenever its first argument becomes active (rule **(S1)**), but synchronization is not forced, so that G sy a can mimic all the behaviours of G (rule **(S2b)**). Rule **(S2c)** shows what happens when the process synchronizes with itself: The timed bags involved in the operation must have the same real duration, and they join together in a new timed bag in which we have removed a pair of labels $(a, \hat{a})$ such that each component is in a different multiset. Finally, the process finishes when its argument terminates, as indicated by rule (**S3**). All the rules can be applied on any equivalent state of G (rule (**S2a**)).

Iteration rules define the behaviour of this syntactic operator. Control is transmitted from each argument to some other (either the next, or the same in the case of the second argument), until the last one terminates. More in detail, rules (**It2c**) and (**It2d**) state that once the entry condition has finished (first argument) we can choose between executing the loop body (second argument) or the exit condition (third argument). Rules (**It3c**) and (**It3d**) state that each time the second argument terminates, we can choose between executing it again or we advance to execute the last argument.

4 Denotational Semantics

The denotational semantics of any language is defined by means of a function which maps the set of its expressions into the adequate semantic domain. In this way we associate to each expression of the language an object which reflects its structure and behaviour. The denotational semantics of TPBC is based on timed Petri nets, whose transitions have a duration. A *labelled timed Petri net*, denoted by TPN, is a tuple $(P, T, F, W, \delta, \lambda)$ such that P and T are disjoint sets of *places* and *transitions*; $F \subseteq (P \times T) \cup (T \times P)$ is the set of *arcs* of the net; W is a *weight function* from F to the set of positive natural numbers $I\!N^+$; δ is a function from the transition set T to $I\!N^+$ that defines the duration of each transition; and λ is a function from $P \cup T$ into a set of labels $\mathcal{L}$. In our case, λ maps elements in P into the set $\{\mathsf{e}, \mathsf{i}, \mathsf{x}\}$ and transitions in T into the set $\mathcal{C} = \mathcal{P}(\mathcal{M}(\mathcal{BA}) \backslash \{\varnothing\} \times \mathcal{BA})$. Its intended meaning is the same as in PBC, that is, we consider that a net has three types of places: *entry places* (those with $\lambda(p) = \mathsf{e}$), *internal places* (those with $\lambda(p) = \mathsf{i}$), and *exit places* (those with $\lambda(p) = \mathsf{x}$). Moreover, each transition v is labelled with a binary relation $\lambda(v) \subseteq \mathcal{M}(\mathcal{BA}) \backslash \{\varnothing\} \times \mathcal{BA}$, whose elements are pairs of the form $(\{\alpha_1 : d_1, \alpha_2 : d_2, \ldots, \alpha_n : d_n\}, \alpha : d)$. The informal meaning of such a pair is that the behaviour represented by the multiset $\{\alpha_1 : d_1, \alpha_2 : d_2, \ldots, \alpha_n : d_n\}$ will be substituted by the execution of the bag α with a minimum duration of d time units. The most usual binary relations are the following:

1. Constant relation: $\rho_{\alpha:d} = \{ (\{\beta : d\}, \alpha : d) \}$.
2. Identity: $\rho_{id} = \{ (\{\alpha : d\}, \alpha : d) \}$.
3. Synchronization: $\rho_{\mathsf{sy}\, a}$ is defined as the smallest relation satisfying:
   - $\rho_{id} \subseteq \rho_{\mathsf{sy}\, a}$,
   - $(\Gamma, \alpha + \{a\} : d_1), (\Delta, \beta + \{\hat{a}\} : d_2) \in \rho_{\mathsf{sy}\, a}$
     $\implies (\Gamma + \Delta, \alpha + \beta : \max\{d_1, d_2\}) \in \rho_{\mathsf{sy}\, a}$.
4. Basic relabelling: $\rho_{[f]} = \{ (\{\alpha : d\}, f(\alpha) : d) \mid f(\hat{a}) = \widehat{f(a)}\, \forall a \in \mathcal{A} \}$.
5. Restriction: $\rho_{\mathsf{rs}\, a} = \{ (\{\alpha : d\}, \alpha : d) \mid a, \hat{a} \notin \alpha \}$.
6. Hiding: $\rho_{[a:\_]} = \{ (\Gamma, \alpha : d) \in \rho_{\mathsf{sy}\, a} \mid a, \hat{a} \notin \alpha \}$.

As it is done in PBC, we distinguish two kinds of nets: *plain nets*, whose transitions are labelled with a constant relation; and *operator nets*, in which transitions are labelled with non-constant relations.

Only plain nets will be marked, and therefore they are the only ones able to fire transitions. Due to this fact, it is not necessary to include time information in operator nets. The formal definitions are the following:

Definition 3 (Timed plain net). *A **timed plain net** $N = (P, T, F, W, \delta, \lambda)$ is a timed Petri net such that* $\lambda : P \cup T \longrightarrow \{e, i, x\} \cup \{\rho_{\alpha:d} | \alpha \in \mathcal{M}(\mathcal{A}), d \in I\!N^+\}$ *where* $\forall p \in P \; \lambda(p) \in \{e, i, x\}$ *and* $\forall v \in T \; \lambda(v) = \rho_{\alpha:d}$ *with* $\delta(v) = d$.

Definition 4 (Operator net). *An **operator net** N is a labelled Petri net (P, T, F, W, λ) such that* $\lambda : P \cup T \longrightarrow \{e, i, x\} \cup \mathcal{C}$ *where* $\forall p \in P \; \lambda(p) \in \{e, i, x\}$ *and* $\forall v \in T \; \lambda(v) \in \mathcal{C} \setminus \{ \rho_{\alpha:d} \mid \alpha \in \mathcal{M}(\mathcal{A}), d \in I\!N^+ \}$.

To support the duration of transitions, markings of timed plain nets will be constituted by two components, M^1 and M^2. M^1 represents the available marking of the net, that is, where the tokens are and how many of them are. M^2 is the multiset of transitions currently in execution, each one carrying the time units its execution will still take from now on. Formally speaking, if $N = (P, T, F, W, \delta, \lambda)$ is a timed plain net, a *marking* M of N is a pair (M^1, M^2) where $M^1 \in \mathcal{M}(P)$ and M^2 is a finite multiset of tuples in $T \times I\!N^+$. We say that a transition v is in M^2, $v \in M^2$, iff there is some $d' \in I\!N^+$ such that $(v, d') \in M^2$. In order to visualize the set of transitions in execution, and also simplify some results, it is useful to introduce the derived concept of *frozen token*. A place $p \in P$ is occupied by a *frozen token* iff there is a transition $v \in T$ such that $(p, v) \in F$ and $v \in M^2$. In this case, if there exists only one value d' for v, we denote it by $rem(v)$. Then we define the *token-marking* $T(M)$ associated to $M = (M^1, M^2)$ by $T(M) = (M^1, \overline{M}^2)$ where $\overline{M}^2$ is the multiset of places defined by $\overline{M}^2(p) = \sum\limits_{(p,v)\in F, \, d\in I\!N^+} M^2(v, d)$. In the following we will call *available tokens* to the ordinary tokens in a marking in order to avoid confusion with frozen tokens.

A marking $M = (M^1, M^2)$ is *safe* if each place is occupied, at most, by one token, either ordinary or frozen. That is, $\forall p \in P$

$$M^1(p) + \overline{M}^2(p) \leq 1$$

A safe marking $M = (M^1, M^2)$ is *clean* if the following conditions hold:

- $(\forall p \in {}^\bullet N \quad M^1(p) + \overline{M}^2(p) \neq 0 \implies (\forall p \in P \quad \lambda(p) \neq e \Rightarrow M^1(p) + \overline{M}^2(p) = 0)$
- $(\forall p \in N^\bullet \quad M^1(p) \neq 0 \,) \implies N^\bullet = M^1$

In order to define the firing of transitions we need to know their real durations. So, we consider *timed transitions*, which are pairs of the form (v, d') where v is a transition and $d' \in I\!N^+$. Then we say that a multiset of timed transitions RT is *enabled* at a marking M if the following conditions are satisfied:

- $\forall (v, d') \in RT \quad d' \geq \delta(v)$
- $\forall p \in P \quad M^1(p) \geq \sum\limits_{v \in T} RT(v) \cdot W(p, v)$ where $RT(v) = \sum\limits_{d' \in I\!N^+} RT(v, d')$.

Once we know when a multiset of timed transitions RT is enabled at a marking, the following firing rule defines the effect of its firing:

Definition 5 (Firing rule). *Let $N = (P, T, F, W, \delta, \lambda)$ be a TPN, and $M = (M^1, M^2)$ be a marking of N at some instant $\beta \in \mathbb{N}$. If a multiset of timed transitions RT is enabled at M and its transitions are fired, then the marking $M' = (M'^1, M'^2)$ reached at the instant $\beta + 1$ is defined as follows:*

- $M'^1 = M^1 - \sum\limits_{v \in C_0} RT(v)W(-, v) + \sum\limits_{v \in C_1} RT(v)W(v, -) + \sum\limits_{(v,1) \in C_2} M^2(v, 1)W(v, -)$

  where
  $C_0 = \{\, v \in T \mid \exists d' \in \mathbb{N}^+, RT(v, d') > 0 \,\}$, $C_1 = \{\, v \in T \mid (v, 1) \in RT \,\}$, *and*
  $C_2 = \{\, (v, 1) \in T \times \mathbb{N}^+ \mid M^2(t, 1) > 0 \,\}$.
- $M'^2 : T \times \mathbb{N}^+ \longrightarrow \mathbb{N}$

  with
  $$M'^2(v, \beta') = \begin{cases} RT(v, d') & \text{if } (v, d') \in RT \wedge \beta' = d' - 1 \\ M^2(v, \beta' + 1) & \text{otherwise} \end{cases}$$

The step generated by the firing of a set of transitions RT is denoted by $M[RT\rangle M'$. Step sequences are defined as usual, and the set of reachable markings in N from M is denoted by $Reach(N, M)$.

So, frozen tokens are those consumed by a transition in execution. Whenever that execution finishes they become available tokens in the postconditions of the fired transitions.

4.1   A Domain of Timed Boxes

Timed Petri boxes are equivalence classes of labelled timed Petri nets. A suitable equivalence relation should allow, at least, the derivation of identities such as those induced by associativity and commutativity of several operators, such as parallel composition or choice. Besides, this relation must allow us to abstract away the names of places and transitions; also it must provide a mechanism to identify duplicate elements in the nets. The relation that we propose is a natural extension of the one used in plain PBC, by adequately considering the temporal aspects of the nets, which means to preserve the duration of related transitions.

Definition 6 (Structural equivalence). *Being $N_1 = (P_1, T_1, F_1, W_1, \lambda_1)$ and $N_2 = (P_2, T_2, F_2, W_2, \lambda_2)$ two operator nets, they are said to be **structurally equivalent** (or just equivalent) iff there is a relation $\varphi \subseteq (P_1 \cup T_1) \times (P_2 \cup T_2)$ such that:*

1. *$\varphi(P_1) = P_2$ and $\varphi^{-1}(P_2) = P_1$,*
2. *$\varphi(T_1) = T_2$ and $\varphi^{-1}(T_2) = T_1$,*
3. *$\forall (p_1, p_2), (v_1, v_2) \in \varphi \quad W_1(p_1, v_1) = W_2(p_2, v_2), W_1(v_1, p_1) = W_2(v_2, p_2)$,*
4. *If $(x_1, x_2) \in \varphi$ then $\lambda_1(x_1) = \lambda_2(x_2)$,*
5. *$\forall v_1 \in T_1, v_2 \in T_2 \quad |\varphi(v_1)| = 1$ and $|\varphi^{-1}(v_2)| = 1$.*

*Being $N_1 = (P_1, T_1, F_1, W_1, \delta_1, \lambda_1, M_1)$ and $N_2 = (P_2, T_2, F_2, W_2, \delta_2, \lambda_2, M_2)$ two marked timed plain nets, they are said to be **structurally equivalent** (or just equivalent) iff there is a binary relation $\varphi \subseteq (P_1 \cup T_1) \times (P_2 \cup T_2)$ such that:*

1-4. As before,

5. *If $(v_1, v_2) \in \varphi$ then $\delta_1(v_1) = \delta_2(v_2)$,*
6. *If $(p_1, p_2) \in \varphi$ then $M_1^1(p_1) = M_2^1(p_2)$,*
7. *If $(v_1, v_2) \in \varphi$ then $\forall d \in \mathbb{N}^+ \quad M_1^2(v_1, d) = M_2^2(v_2, d)$.*

There are several conditions that a net N has to satisfy in order to generate a timed box. First, we impose *T-restrictedness* which means that the pre-set and post-set of each transition are non-empty sets. Besides, we impose *ex-restrictedness* (there is at least one entry place and one exit place) and *ex-directedness* (pre-sets of entry places and post-sets of exit places are empty). All these assumptions are inherited from the untimed version of the calculus.

Remark 7. Next we will only consider nets N satisfying the following conditions:

- N is T-restricted: $\forall v \in T \quad {}^\bullet v \neq \varnothing \neq v^\bullet$,
- There are no side conditions : $\forall v \in T \quad {}^\bullet v \cap v^\bullet = \varnothing$,
- N has at least one entry place: ${}^\bullet N \neq \varnothing$,
- N has at least one exit place: $N^\bullet \neq \varnothing$,
- There are no incoming arcs to entry places and no outgoing arcs from exit places: ${}^\bullet({}^\bullet N) = \varnothing \wedge (N^\bullet)^\bullet = \varnothing$,
- N is simple: $\forall p \in P \; \forall v \in T \quad W(p, v), W(v, p) \in \{0, 1\}$.

Definition 8 (Plain and operator timed boxes).

- *A marked **timed plain box** B is an equivalence class $B = [N]$ induced by the structural equivalence over labelled nets, where N is a marked plain net.*
- *An **operator box** Ω is an equivalence class $\Omega = [N]$ induced by the structural equivalence over labelled nets, where N is an operator net.*

Plain timed boxes will be the semantic objects to be associated with syntactic expressions, that is, the denotational semantics of an expression will be always a plain box. Its structural construction relies on operator boxes. Each semantic operator has a certain number of arguments (the same as the corresponding syntactic operator has). By applying them to a tuple of arguments we can obtain a new plain box, using the refinement procedure which we will explain later.

Next we define *static* and *dynamic* boxes. A plain box B is *static* if the marking of its canonical representative is empty ($M^1 = \varnothing \wedge M^2 = \varnothing$), and the reachable markings from the initial one (${}^\bullet B, \varnothing$) are safe and clean. A plain box $B = [(P, T, F, W, \lambda)]$ is *dynamic* if the following conditions are satisfied:

- The marking of its canonical representative is non-empty,
- The plain box $[(P, T, F, W, \delta, \lambda, (\varnothing, \varnothing))]$ is static,
- The reachable markings from M are safe and clean.

The set of static boxes is Box^s, and the set of dynamic boxes is Box^d.

Plain boxes are classified in several classes depending on the type of tokens that they contain, and the labels of the places they are in.

Definition 9 (Classes of plain boxes). *Let* $B = [(P, T, F, W, \delta, \lambda, (M^1, M^2))]$ *be a plain box. We say that B is a **stable box** if $M^2 = \varnothing$; otherwise we say that B is **unstable**. If B is a stable box, then we say that it is an **entry-box** if $M^1 = {}^\bullet B$; an **exit-box** if $M^1 = B^\bullet$; and an **intermediate-box**, otherwise. All these classes are denoted by Box^{st}, Box^{ust}, Box^e, Box^x, Box^i, respectively.*

For operator boxes we need the additional property of being *factorisable*. In order to extend this notion to the timed case we first present the concept of (reachable) marking of an operator box. In this case it is enough to know how available and frozen tokens are distributed. By means of them we define which arguments of the operator are stable and which ones are in execution.

Definition 10 (Markings of operator boxes). *Being $\Omega = [(P, T, F, W, \lambda)]$ an operator box, a **marking** M of Ω is a pair $(M^1, M^2) \in \mathcal{M}(P) \times \mathcal{M}(P)$.*

In the definition above M^1 represents the multiset of available tokens, while M^2 defines the set of places where we have frozen tokens.

Definition 11 (Reachable markings of operator boxes). *Let $\Omega = [(P, T, F, W, \lambda)]$ be an operator box. We say that a multiset of transition $RT \in \mathcal{M}(T)$ is **enabled** at a marking M if the following condition is satisfied:*
$$\forall p \in P \quad M^1(p) \geq \sum_{v \in T} RT(v) \cdot W(p, v)$$
*The set of **reachable markings** of Ω after the firing of RT is defined as the set of markings (M'^1, M'^2) such that:*

- $M'^1 = M^1 - \sum_{v \in T} RT(v) \cdot W(-, v) + \sum_{v \in C} W(v, -)$
- $M'^2 = M^2 + RT - C$ *where* $C \subseteq M^2 \cup RT$.

Available tokens indicate the arguments of the connective which are in stable form (that is, they have no executing transition), while frozen tokens say us the ones that are currently in execution. The condition of factorisability tries to capture these distributions of tokens. Basically, it means that when a postcondition (precondition) of a transition is marked, all its postconditions (preconditions) must be also marked with tokens of the same type. To define the condition three sets of transitions are considered, one for frozen tokens and two for available tokens. This distinction is necessary because frozen tokens are always placed in the preconditions of the transitions in execution, while available tokens can either be consumed by the firing of a transition or obtained as a consequence of the execution of another transition.

Definition 12 (Factorisability). *Let $\Omega = [(P, T, F, W, \lambda)]$ be an operator box and $M = (M^1, M^2)$ be a marking of Ω. A **factorisation** of M is a triple of sets of transitions $\Phi = (\Phi_1, \Phi_2, \Phi_3)$ such that the following conditions hold:*

$$\bullet M^1 = \left(\biguplus_{v \in \Phi_1} {}^\bullet v \right) \uplus \left(\biguplus_{v \in \Phi_3} v^\bullet \right) \quad \bullet M^2 = \left(\biguplus_{v \in \Phi_2} {}^\bullet v \right)$$

*We say that Ω is **factorisable** iff every safe marking $M \in Reach(\Omega, ({}^\bullet\Omega, \varnothing))$ satisfies that for every set U of transitions enabled at M, there is a factorisation*

$\Phi = (\Phi_1, \Phi_2, \Phi_3)$ of M such that $U \subseteq \Phi_1$. In the following, the set of factorisations of Ω is denoted by $fact_\Omega$, and we will use $\tilde{\Phi}$ to denote $\Phi_1 \cup \Phi_2 \cup \Phi_3$.

If factorisability were violated, the marking of an operator box could not be distributed over its arguments. Indeed, when factorisability is violated there must be a token (available or frozen) which neither can have been produced by the firing of any transition, nor enables by itself any marking; otherwise all the preconditions or postconditions of the involved transition would be marked.

Now we can finally define operator boxes:

Definition 13 (Acceptable operator boxes). *Let (P, T, F, W, λ) be an operator net satisfying the requirements in Remark 7. The equivalence class $\Omega = [(P, T, F, W, \lambda)]$ is an **acceptable operator box** if it is factorisable, and all the markings reachable from $(\bullet\Omega, \varnothing)$ are safe and clean.*

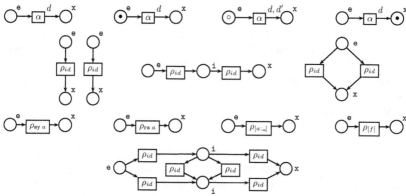

Fig. 1. Denotational semantics of TPBC

The denotational semantics of the algebra is defined in Figure 1. From left to right and top to bottom, we show the semantics of: basic actions ($\alpha : d$, $\widetilde{\alpha : d}$, $\alpha : d, d'$ and $\underline{\alpha : d}$); disjoint parallelism ($\_\|\_$), sequential composition ($\_;\_$), and choice ($\_\Box\_$); synchronization ($\_\mathbf{sy}\,a$), restriction ($\_\mathbf{rs}\,a$), hiding ($[a : \_]$), basic relabelling ($\_[f]$) and iteration ($[\_ * \_ * \_]$). In this figure, frozen tokens only appear in the semantics of $\alpha : d, d'$, where the corresponding marking is given by $M^1 = \varnothing$ and $M^2 = \{(\alpha, d')\}$.

4.2 Refinement

Refinement is the mechanism to obtain a plain box $op_\Omega(B_1, B_2, \ldots, B_n)$ from a given operator box Ω and a tuple of plain boxes $(B_1, B_2, \ldots, B_n)$. The basic idea is that every transition of the operator box is replaced by one element of the tuple $(B_1, B_2, \ldots, B_n)$, where we have previously done the changes indicated by the label of the transition.

In the static case, when the involved boxes are unmarked, the mechanism followed is the same as in plain PBC. The same happens when frozen tokens are not involved. We will just comment how it works in the remaining case.

When frozen tokens appear the refinement basically involves the same structural changes as those in plain PBC [1,2,3]. However this procedure has been slightly modified with two purposes. First, we avoid the existence of arcs with a weight greater than one, which in fact were useless, since they only increase the size of the obtained box, without allowing new firings. The second and more important aim is the inclusion of time. The underlying intuition is that any transition in execution in a box remains the same when this box is "mixed" with some others, unless it proceeds from a synchronization, in which case the synchronized transitions can be considered in execution.

The following definition of the refinement domain reflects the fact that each operator box is a partial operation from plain boxes to plain boxes.

Definition 14 (Refinement domain). *Let $\Omega = (P, T, F, W, \lambda)$ be an operator box with n arguments. The **domain** of application of Ω, denoted by dom_Ω, is defined by the following conditions:*

1. *It comprises all the tuples $(B_{v_1}, B_{v_2}, \ldots, B_{v_n})$ of static plain boxes,*
2. *For every factorisation of Ω, $\Phi = (\Phi_1, \Phi_2, \Phi_3) \in fact_\Omega$, it comprises all tuples of boxes $\mathbb{B} = (B_{v_1}, B_{v_2}, \ldots, B_{v_n})$ such that:*
   - $\forall v \in \Phi_1 \quad B_v \in Box^e \cup Box^i,$
   - $\forall v \in \Phi_2 \quad B_v \in Box^{ust},$
   - $\forall v \in \Phi_3 \quad B_v \in Box^x,$
   - $\forall v \in T\backslash\tilde{\Phi} \quad B_v \in Box^s.$

Φ_1 indicates which transitions will be replaced by boxes that contribute with stable non-final markings; Φ_2 represents the transitions corresponding to unstable boxes; Φ_3 denotes the transitions instantiated by boxes with terminal markings, and $T\backslash\tilde{\Phi}$ are the remaining transitions, that is, those which contribute with empty markings to the generated plain box.

In order to simplify the formal definition of refinement, we assume that in operator boxes all place and transition names of their canonical representatives are primitive names from P_{op} and T_{op}, respectively. We shall further assume that in basic plain boxes, that is, in the denotational semantics of basic actions, all places and transitions have primitive names from, respectively, P_{box} and T_{box}. As we will see, when we apply this mechanism, we have as names of the newly constructed places and transitions labelled trees in two sets which are denoted by P_{tree} and T_{tree}, respectively. These trees are defined in a recurrent way and then each subtree of such a tree is also in the corresponding set.

Next we define a collection of auxiliary concepts which will be later used in order to adequately define the frozen tokens of any net obtained by refinement.

Definition 15. *If $q = l(q_1, \ldots, q_n) \in T_{tree}$, where l is the label of its root and $q_1, \ldots, q_n$ are the root's sons, we say that each q_i is a **component** of q. In the following, we will denote the set of components of q, $\{q_1, \ldots, q_n\}$, by $comp(q)$.*

Definition 16. *Given $T_1, T_2 \subseteq T_{tree}$, we say that $dec : T_1 \longrightarrow T_2 \times \mathbb{N}$ is a **decomposition** of T_1 iff the following condition holds:*

$$\forall q \in T_1 \quad dec(q) = \langle l(q_1, \ldots, q_n), k \rangle \text{ with } q_k = q$$

*dec is a **consistent decomposition** iff the following condition is satisfied:*

$$\forall q' = l(q_1, \ldots, q_n) \in T_2 \quad (\exists k \in I\!N \; \exists q \in T_1, dec(q) = \langle q', k \rangle) \implies$$
$$\implies (\forall k \in 1..n \; \exists q \in T_1, dec(q) = \langle q', k \rangle)$$

*Finally, if $B = [(P, T, F, W, \delta, \lambda, (M^1, M^2)]$ is a timed plain box, $T_1 \subseteq T$, and dec $: T_1 \longrightarrow T_2 \times I\!N$ is a decomposition of T_1, we say that dec is a **timed consistent decomposition** iff the following condition holds:*

$$\forall q' = l(q_1, \ldots, q_n) \in T_2 \quad (\exists k \in I\!N \; \exists q \in T_1, dec(q) = \langle q', k \rangle) \implies$$
$$\implies (\exists d \in I\!N^+ \; \forall k \in 1..n \; \exists q \in T_1, (q, d) \in M^2 \wedge dec(q) = \langle q', k \rangle)$$

The formal definition of refinement is the following:

Definition 17 (Refinement). *Let $\Omega = [(P, T, F, W, \lambda)]$ be an operator box, and for each $v \in T$ $B_v = [(P_v, T_v, F_v, W_v, \delta_v, \lambda_v, (M_v^1, M_v^2))]$ be a timed plain box. Under the assumptions above on Ω and $\mathbb{B} = (B_{v_1}, B_{v_2}, \ldots, B_{v_n})$, the result of the simultaneous substitution of the nets B_{v_i} for the transitions in Ω is any plain timed box whose canonical representative is a timed plain net*

$$op_\Omega(\mathbb{B}) = (P_0, T_0, F_0, W_0, \delta_0, \lambda_0, (M_0^1, M_0^2))$$

defined as follows:

1. ***Places, their labels and markings:*** *The set of places, P_0, is given by*

$$P_0 = \left(\bigcup_{v \in T} IP_{new}^v \right) \cup \left(\bigcup_{p \in P} OP_{new}^p \right)$$

where the sets IP_{new}^v and OP_{new}^p are defined as follows:
   - *For each $v \in T$, IP_{new}^v is the set of places $\{v.p_v | p_v$ internal place in $P_v\}$. The label of all these places is $\mathtt{i}$ and their marking is given by $M_v^1(p_v)$.*
   - *Let $p \in P$ with $^\bullet p = \{v_1, v_2, \ldots, v_k\}$ and $p^\bullet = \{v_{k+1}, v_{k+2}, \ldots, v_{k+m}\}$. Then, OP_{new}^p is the set of places $p(v_1 \lhd p_1, \ldots, v_{k+m} \lhd p_{k+m})$ where*

$$\begin{cases} p_i \in (P_{v_i})^\bullet \; \forall i \in \{1, \ldots, k\} \\ p_i \in \,^\bullet(P_{v_i}) \; \forall i \in \{k+1, \ldots, k+m\}. \end{cases}$$

   Each one of these places will be labelled by $\lambda(p)$ and its marking is

$$M_{v_1}^1(p_1) + M_{v_2}^1(p_2) + \ldots + M_{v_{k+m}}^1(p_{k+m}).$$

2. ***Transitions, their labels and durations:*** *The set of transitions T_0 is defined by*

$$T_0 = \bigcup_{v \in T} T_{new}^v$$

where for each $v \in T$ the set T_{new}^v is obtained as follows: Whenever we have a pair of the form $(\{\lambda(q_1) : d_1, \lambda(q_2) : d_2, \ldots, \lambda(q_n) : d_n\}, \alpha : d) \in \lambda(v)$ for $q_i \in T_v$, and the following condition holds:

$$\forall i, j \in \{1, \ldots, n\} \; i \neq j \implies (^\bullet q_i \cap \,^\bullet q_j = \varnothing) \wedge (q_i^\bullet \cap q_j^\bullet = \varnothing),$$

a new transition $v.\alpha(q_1, \ldots, q_n)$ is generated. Its duration is d and its label is α. In the following, we will denote the transition v by $root(v_0)$.

3. **Set of arcs:** For each transition v_0 in T_0, the set of arcs leaving or reaching v_0 in F_0 are those obtained as follows:

- $p_0 \in IP_{new}^v$
$$\begin{cases} \text{if } \exists q_i \in comp(v_0), (p_v, q_i) \in F_v \text{ then } (p_0, v_0) \in F_0 \\ \text{if } \exists q_i \in comp(v_0), W_v(q_i, p_v) = 1 \text{ then } (v_0, p_0) \in F_0 \end{cases}$$

- $p_0 \in OP_{new}^p$
$$\begin{cases} \text{if } \exists i \in \{1, \dots, k+m\} \wedge \exists q_j \in comp(v_0), (p_i, q_j) \in F_v \text{ then } (p_0, v_0) \in F_0 \\ \text{if } \exists i \in \{1, \dots, k+m\} \wedge \exists q_j \in comp(v_0), (q_j, p_i) \in F_v \text{ then } (v_0, p_0) \in F_0 \end{cases}$$

The weight function returns 1 for all the arcs in F_0.

4. **Transitions in execution (frozen tokens):** The frozen marking M_0^2 will be defined as the union of a collection of submarkings $M_{0,v}^2$ with $v \in T$. As a matter of fact, each M_v^2 will be not defined in a unique way and therefore we will have several possible frozen markings M_0^2. This is justified for technical reasons, since once we prove the equivalence between the operational and the denotational semantics, we obtain as a consequence that all the different boxes defining the denotational semantics of an expression are in fact equivalent. Thus, we could also select any of the possibilities to obtain a function instead of a relation, but if we would do it in this way the definition would be much less readable.

Then, in order to get a value for each $M_{0,v}^2$ we consider any timed consistent decomposition $dec : M_v^2 \longrightarrow T_0 \times \mathbb{N}$, and we take as $M_{0,v}^2$

$$M_{0,v}^2 = \{ q' \in T_0 \mid \exists q \in M_v^2, dec(q) = q' \}$$

where for each $q' \in M_{0,v}^2$ we take $rem(q') = rem(q)$.

Theorem 18.

1. Every step sequence of the operational semantics of G is also a step sequence of any of the timed plain boxes corresponding to G.
2. Every step sequence of any of the timed plain boxes corresponding to G is also a step sequence of the operational semantics of G.

Proof. It follows the lines of that for plain PBC [1,2,3] with the changes needed to cope with time information (see [5]). Unfortunately, even the original proof is so long and involved that even it is not possible to sketch it here. It would be nice to comment at least the necessary changes to extend the proof, but this is neither possible since they are both local and distributed all along the proof. Therefore, we will only justify here why we need nondeterminism in the definition of the denotational semantics. This is needed in order to preserve a one to one relation between the evolution of the expressions and that of their denotational semantics, since there are some expressions which evolve to the same one after the execution of different multibags, but the same is not true for the evolution of a single box.

5   Example: A Producer/Consumer System

In this section we present an example that models the producer/consumer system with a buffer of capacity 1. The system describes the behaviour of a simple production line which involves two workers. A conveyor belt is between them, so it can hold any item that has been produced by the first worker and has to be consumed by the second.

The given specification consists on three components that are combined to obtain the TPBC expression of the system: the producer system, the consumer system and the conveyorBelt system. The timing assigned to them reflects a set of hypotheses about the production and consumption speed and the characteristics of the conveyor belt.

- producer $= [(\{a, b\} : 1) * ((p : 2); (p_b : 1)) * (\{\neg a, \neg b\} : 1)]$
- consumer $= [(\{\hat{a}, b\} : 1) * ((g_b : 1); (c : 1)) * (\{\neg \hat{a}, \neg b\} : 1)]$
- conveyorBelt $= [G_1 * [(\hat{p_b} : 1) * ((\hat{g_b} : 1; \hat{g_b} : 1)\Box(\hat{g_b} : 1; \hat{p_b} : 1)) * (\hat{g_b} : 1)] * G_2]$
  where $G_1 = (\{b, b\} : 1)$ and $G_2 = (\{\neg b, \neg b\} : 1)$
- system $= (((\text{producer}\|\text{consumer})\text{sy } A\|\text{conveyorBelt})\text{sy } B)\text{rs}(A \cup B)$
  where $A = \{a, \neg a\}$ $B = \{b, \neg b, p_b, g_b\}$

The meaning of the actions is the following:

p: Produce an item.
c: Consume an item.
p_b: Put an item into the conveyor belt.
g_b: Get an item from the conveyor belt.
b, ¬b: Activate and deactivate the running of the conveyor belt.
a, ¬a: Activate and deactivate the production/consumption of items.

Figure 2 shows the denotational semantics of producer and consumer expressions. The construction of the boxes corresponding to conveyorBelt and the full system expressions can be completed in the same way, but due to lack of space we can not show the obtained nets. You can check by hand the equivalence between the operational semantics of the given expressions and those of the obtained nets.

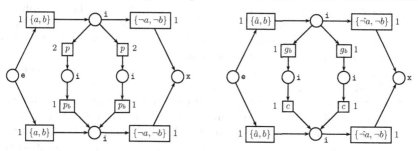

Fig. 2. The producer/consumer problem

6   The Time-Constraining Operators

The previous sections describe a simple timed extension of the basic model where we need not to introduce any new operator. Note that due to rules **(V1)** and

(V2), processes can let time progress indefinitely. More precisely, for any over-lined or underlined process time passes without causing any change of the state, out of the fact that processes become stable after finishing their pending actions.

This property, called *unlimited waiting*, is not too useful from a practical point of view. We need to introduce some kind of urgency in the formalism; otherwise, the expressive power of TPBC would be limited to specify very simple systems. Therefore, in this section we will add to our model some new features that force systems to evolve, namely *timed choice* and *time-out with exception handler*.

6.1   Timed Choice

Timed choice constraints an action to occur at some instant of a given interval $[t_0, t_1]$. That is, a so restricted action cannot begin its execution either before t_0 or after t_1. The static expression used to implement this behaviour is $\alpha^{[t_0, t_1]} : d$, where $\alpha \in \mathcal{M}(\mathcal{A})$, $t_0 \in I\!N$, $d \in I\!N^+$, and $t_1 \in I\!N^+ \cup \{\infty\}$.

Depending on the values of t_0 and t_1 there are three particular cases of timed choice. A rough description of each one is as follows:

1. **Finite delay:** It corresponds to expressions of the form $\alpha^{[t_0, \infty]} : d$, where $t_0 \neq 0$ (note that processes of the form $\alpha^{[0, \infty]} : d$ will be just equivalent to $\alpha : d$.) When such an expression gets the control of the system, it delays the execution of the action $\alpha : d$ for at least t_0 time units.
2. **Time-out:** It is obtained when $t_0 = 0$ and $t_1 \neq \infty$. If a process of the form $\alpha^{[0, t_1]} : d$ is activated, the beginning of the execution of $\alpha : d$ cannot be postponed more than t_1 units of time.
3. **Time-stamped actions:** We say that an action is *time-stamped* when its execution is enforced to start at a given instant. The corresponding static expressions are $\alpha^{[t_0, t_0]} : d$ with $t_0 \neq \infty$, which we will usually syntactic sugar by $\alpha^{t_0} : d$.

Now, activated timed choices need to have information about the passage of time over the system components. More exactly, it is necessary to know for how long a timed choice has been activated, in order to restrict the beginning of the execution of the corresponding action. Until now, dynamic expressions only give us information about the global state of processes, in such a way that an external observer is able to determine whether a process has pending actions or not, but there is no way to obtain quantitative information about time aspects, unless the observer knows the full history of the process.

With the purpose of avoiding this limitation, we introduce a temporal an-notation over both overlined and underlined expressions. In the first case, the new clock tells us for how long the process has been activated, while in the sec-ond case it tells us how long ago the process finished. This annotation will be included as an index close to the corresponding bars.

Due to these modifications, the operational semantics previously defined must be changed. In particular, any clock of the system needs to be updated when time progresses. This is done by means of new rules **(V1)** and **(V2)**:

$$\overline{E}^k \xrightarrow{\varnothing} \overline{E}^{k+1} \text{ (V1)} \qquad \underline{E}_k \xrightarrow{\varnothing} \underline{E}_{k+1} \text{ (V2)}$$

Table 1. New transition rules with time information

$$\overline{\alpha : d}^{k} \xrightarrow{\{\alpha_{d'}\}} \alpha : \widetilde{d, d'} - 1 \;\; d' \geq d \;\; \textbf{(B1)} \qquad \widetilde{\alpha : d, 0} \longleftrightarrow \underline{\alpha : d}_{0} \qquad \textbf{(B3)}$$

$$\overline{E \| F}^{k} \longleftrightarrow \overline{E}^{k} \| \overline{F}^{k} \qquad \textbf{(Pc1)} \qquad \overline{E}_{k} \| \overline{E}_{k'} \longleftrightarrow \underline{E \| F}_{min\{k,k'\}} \qquad \textbf{(Pc3)}$$

$$\overline{E; F}^{k} \longleftrightarrow \overline{E}^{k}; F \qquad \textbf{(Sc1)} \qquad \underline{E}_{k}; F \longleftrightarrow E; \overline{F}^{k} \qquad \textbf{(Sc3)}$$

$$E; \underline{F}_{k} \longleftrightarrow \underline{E; F}_{k} \qquad \textbf{(Sc5)}$$

$$\overline{E \Box F}^{k} \longleftrightarrow \overline{E}^{k} \Box F \qquad \textbf{(Ch1a)} \qquad \overline{E \Box F}^{k} \longleftrightarrow E \Box \overline{F}^{k} \qquad \textbf{(Ch1b)}$$

$$\underline{E}_{k} \Box F \longleftrightarrow \underline{E \Box F}_{k} \qquad \textbf{(Ch2a)} \qquad E \Box \underline{F}_{k} \longleftrightarrow \underline{E \Box F}_{k} \qquad \textbf{(Ch2b)}$$

$$\overline{E \operatorname{sy} a}^{k} \longleftrightarrow \overline{E}^{k} \operatorname{sy} a \qquad \textbf{(S1)} \qquad \underline{E}_{k} \operatorname{sy} a \longleftrightarrow \underline{E \operatorname{sy} a}_{k} \qquad \textbf{(S3)}$$

$$\overline{E \operatorname{rs} a}^{k} \longleftrightarrow \overline{E}^{k} \operatorname{rs} a \qquad \textbf{(Rs1)} \qquad \underline{E}_{k} \operatorname{rs} a \longleftrightarrow \underline{E \operatorname{rs} a}_{k} \qquad \textbf{(Rs3)}$$

$$\overline{[a : E]}^{k} \longleftrightarrow [a : \overline{E}^{k}] \qquad \textbf{(Sp1)} \qquad [a : \underline{E}_{k}] \longleftrightarrow \underline{[a : E]}_{k} \qquad \textbf{(Sp3)}$$

$$\overline{E[f]}^{k} \longleftrightarrow \overline{E}^{k}[f] \qquad \textbf{(Rl1)} \qquad \underline{E}_{k}[f] \longleftrightarrow \underline{E[f]}_{k} \qquad \textbf{(Rl3)}$$

$$\overline{[E * F * E']}^{k} \longleftrightarrow [\overline{E}^{k} * F * E'] \qquad \textbf{(It1)}$$

$$[\underline{E}_{k} * F * E'] \longleftrightarrow [E * \overline{F}^{k} * E'] \quad \textbf{(It2c)} \qquad [\underline{E}_{k} * F * E'] \longleftrightarrow [E * F * \overline{E'}^{k}] \quad \textbf{(It2d)}$$

$$[E * \underline{F}_{k} * E'] \longleftrightarrow [E * \overline{F}^{k} * E'] \quad \textbf{(It3c)} \qquad [E * \underline{F}_{k} * E'] \longleftrightarrow [E * F * \overline{E'}^{k}] \quad \textbf{(It3d)}$$

$$[E * F * \underline{E'}_{k}] \longleftrightarrow \underline{[E * F * E']}_{k} \qquad \textbf{(It5)}$$

In addition to this, we have to modify most of the rules concerned with control transmission. The resulting rules are those shown in Table 1, and the reasoning supporting them is straightforward. Next the definition of the operational semantics of timed choice, which is shown in Table 2. Rule **(TCh1)** states that such a process can perform the empty timed bag at any time. In order to restrict the execution of the action $\alpha : d$, the value k of the system clock is compared with the limits of the interval $[t_0, t_1]$. If $t_0 \leq k \leq t_1$ we can apply rule **(TCh2)**, which establishes that the corresponding (activated) timed choice, $\overline{\alpha^{[t_0, t_1]} : d}^{k}$, can perform the timed bag $\{\alpha_{d'}\}$. Afterwards, rules **(TCh3)** and **(TCh4)** define the expected behaviour: The remaining execution of $\{\alpha_{d'}\}$ is hidden to the observer, who only sees the passage of time (rule **(TCh3)**). When the action terminates, the process is underlined and the value of its clock is reset to zero (rule **(TCh4)**).

6.2   Time-Out with Exception Handler

By means of the timed choice operator just introduced, we are able to limit the time at which a process will be able to start its execution. If this time is exceeded, the process dies, and there is no way to express that some alternative continuation will be activated. In order to get this, we introduce a new operator called *time-out with exception handler*. This operator has two arguments, $E, F \in Expr^s$, and a parameter $t_0 \in \mathbb{N}^+$. E is called the *body* and F the *exception handler* of the time-out.

Table 2. Operational semantics of *timed choice*

$$\overline{\alpha^{[t_0,t_1]}:d}^k \xrightarrow{\ \varnothing\ } \overline{\alpha^{[t_0,t_1]}:d}^{k+1} \tag{TCh1}$$

$$\overline{\alpha^{[t_0,t_1]}:d}^k \xrightarrow{\ \{\alpha_{d'}\}\ } \widetilde{\alpha^{[t_0,t_1]}:d}, d'-1 \ \text{if}\ k \geq t_0,\ k \leq t_1 \ \text{and}\ d' \geq d \tag{TCh2}$$

$$\widetilde{\alpha^{[t_0,t_1]}:d}, d' \xrightarrow{\ \varnothing\ } \widetilde{\alpha^{[t_0,t_1]}:d}, d'-1 \ \text{if}\ d' > 0 \tag{TCh3}$$

$$\widetilde{\alpha^{[t_0,t_1]}:d}, 0 \longleftrightarrow \widetilde{\alpha^{[t_0,t_1]}:d_0} \tag{TCh4}$$

When a time-out with exception handler gets the control, it behaves as its body whenever it begins to perform actions before the instant t_0; otherwise, after time t_0, the process behaves as defined by its exception handler.

The static form of a time-out with exception handler is $\lfloor E \rfloor^{t_0} F$. The possible dynamic forms are equivalent to either $\lfloor G \rfloor^{t_0;t} F$, with $t \geq -1$, or $\lfloor E \rfloor^{t_0;-1} G$. In both cases, the expressions include two temporal annotations. The former is the parameter of the time-out and it is immutable, that is, it does not change during the execution of the process. If t is the value of the latter, this means that the exception handler F will be activated within t time units, unless the body of the time-out begins its execution before. t is set to -1 when either the body of the time-out has begun to execute its first action before t_0 or that limit has been exceeded without performing any non-empty transition of E. The operational semantics reflecting these ideas is shown in Table 3.

Rules **(Te1a)** and **(Te1b)** state that we have two cases depending on the elapsed time from the activation of the process: If $k \leq t_0$ then the body gets the control of the system, and it has $t_0 - k$ time units to execute its first action (rule **(Te1a)**). Otherwise, the exception handler of the time-out is activated, just as rule **(Te1b)** states. Rule **(Te2a)**[1] shows how the passage of time affects this class of processes. Notice carefully that control is transmitted according to the second temporal annotation, so the first argument remains active as long as $t \geq 0$. Otherwise, the body has totally consumed its disposal time and the exception handler is activated. The performance of actions is formalized by means of rules **(Te2b)**, **(Te3b)**, and **(Te4b)**, whose explanation is straightforward. They can be applied on any equivalent form of the arguments (rules **(Te3a)** and **(Te4a)**). Finally, a time-out with exception handler finishes its execution when either its body or its exception handler terminates (rules **(Te5a)** and **(Te5b)**).

6.3   Remarks on Denotational Semantics

Denotational semantics of the time-constraining operators can be found in [5]. Plain nets have been modified by adding temporal restrictions $[t_0, t_1]$ to transitions, and labelling the available tokens in the net with their age. Then, a modified firing rule uses this information, in such a way that a transition v restricted by $[t_0, t_1]$ can only consume tokens younger than t_1.

---

[1] Predicate Init(G) (see [5]) returns **true** if and only if G is equivalent to some overlined expression H.

Table 3. Operational semantics of the *time-out with exception handler*

$$\lfloor \overline{E} \rfloor^{t_0} F^k \longleftrightarrow \lfloor \overline{E}^k \rfloor^{t_0; t_0 - k} F \text{ if } k \le t_0 \qquad \textbf{(Te1a)}$$

$$\lfloor \overline{E} \rfloor^{t_0} F^k \longleftrightarrow \lfloor E \rfloor^{t_0; -1} \overline{F}^{(k-t_0-1)} \text{ if } k > t_0 \quad \textbf{(Te1b)}$$

$$\frac{G \xrightarrow{\varnothing} G' \quad \text{Init}(G) \wedge t \ge 1}{\lfloor G \rfloor^{t_0; t} F \xrightarrow{\Gamma} \lfloor G' \rfloor^{t_0; t-1} F} \qquad \textbf{(Te2a)}$$

$$\frac{G \xrightarrow{\Gamma} G' \quad \Gamma \ne \varnothing \quad \text{Init}(G) \wedge t \ge 0}{\lfloor G \rfloor^{t_0; t} F \xrightarrow{\Gamma} \lfloor G' \rfloor^{t_0; -1} F} \qquad \textbf{(Te2b)}$$

$$\frac{G \longleftrightarrow G'}{\lfloor G \rfloor^{t_0; t} F \longleftrightarrow \lfloor G' \rfloor^{t_0; t} F} \qquad \textbf{(Te3a)}$$

$$\frac{G \xrightarrow{\Gamma} G' \quad \neg\text{Init}(G)}{\lfloor G \rfloor^{t_0; -1} F \xrightarrow{\Gamma} \lfloor G' \rfloor^{t_0; -1} F} \qquad \textbf{(Te3b)}$$

$$\frac{G \longleftrightarrow G'}{\lfloor E \rfloor^{t_0; -1} G \longleftrightarrow \lfloor E \rfloor^{t_0; -1} G'} \qquad \textbf{(Te4a)}$$

$$\frac{G \xrightarrow{\Gamma} G'}{\lfloor E \rfloor^{t_0; -1} G \xrightarrow{\Gamma} \lfloor E \rfloor^{t_0; -1} G'} \qquad \textbf{(Te4b)}$$

$$\lfloor E_k \rfloor^{t_0; -1} F \longleftrightarrow \lfloor E \rfloor^{t_0} F_k \qquad \textbf{(Te5a)}$$

$$\lfloor E \rfloor^{t_0; -1} F_k \longleftrightarrow \lfloor E \rfloor^{t_0} F_k \qquad \textbf{(Te5b)}$$

As a consequence, we obtain that a transition generated by synchronization is enabled at a marking M if and only if all its component transitions would be enabled. This is also true even if any of these components have been removed by application of a restriction operator. This is probably the most important difference between our model and that in [4]. There the author had to introduce *illegal action occurrences* in order to reflect some (intuitively undesirable) synchronizations whenever they become executable in the box defining the denotational semantics of an expression. This happens when the application of a restriction operator removes some timing restrictions which made that the synchronization was non-executable before the application of the restriction operator.

Although in that paper the author does not explain why he decided to get the equivalence between both semantics by allowing the execution of *impossible* synchronizations, we assume that he wants to maintain that equivalence result without changing too much the ideas of untimed PBC. As a consequence, it is not possible to avoid the fireability of those *impossible* synchronizations. To do it, it is mandatory the introduction of age information in the tokens, as we have done in our model. So, we have to pay the price of a more complicated model, but as a reward we avoid those undesirable transitions.

7 Conclusions and Work in Progress

In this paper we have presented a new proposal to introduce time in PBC, preserving the main ideas of this model as far as possible. In the discussed model, called TPBC (*Timed PBC*), processes execute actions which have a duration.

The work by Koutny [4] has been somehow completed by considering an alternative model. More importantly, we have defined a more involved model in order to avoid the generation of undesired transitions. This cannot be done if we just work with plain time Petri nets since these are not prepared to support the correct definition of some time operators such as urgency.

As work in progress, currently we are introducing in the calculus new features enhancing its time characteristics, mainly *maximal parallelism* and *urgency*. Another line of research concerns not only TPBC but also the original PBC. In order to exploit these Petri box calculi, we are working on the axiomatization of the semantics, as it is done in the framework of process algebras. Once we will get the equations in this axiomatization, we will try to interpret them at the level of Petri nets, obtaining a collection of basic correct transformations. We are sure that this will be a fine way to exploit the facilities of both PBC and TPBC.

References

1. E. Best, R. Devillers and J. Hall *The Petri Box Calculus: A New Causal Algebra with Multi-label Communication.* Advances in Petri Nets 1992, LNCS vol.609, pp.21-69. Springer-Verlag, 1992.
2. E. Best and M. Koutny *A Refined View of the Box Algebra.* Petri Net Conference'95, LNCS vol.935, pp.1-20. Springer-Verlag, 1995.
3. M. Koutny and E. Best *Operational and Denotational Semantics for the Box Algebra.* Theoretical Computer Science 211, pp.1-83, 1999.
4. M. Koutny *A Compositional Model of Time Petri Nets.* Application and Theory of Petri Nets 2000, LNCS vol.1825, pp.303-322. Springer-Verlag, 2000.
5. O. Marroquín Alonso and D. Frutos Escrig. *TPBC: Timed Petri Box Calculus.* Technical Report, Dept. Sistemas Informáticos y Programación. UCM, 2000. In Spanish.
6. P. Merlin *A Study of the Recoverability of Communication Protocols.* PhD. Thesis, University of California, 1974.
7. F. Moller and C. Tofts *A Temporal Calculus of Communicating Systems.* CONCUR'90: Theories of Concurrency: Unification and Extension, LNCS vol.458, pp.401-415. Springer-Verlag, 1990.
8. Y. Ortega Mallén *En Busca del Tiempo Perdido.* PhD. Thesis, Universidad Complutense de Madrid. 1990.
9. C. Ramchandani *Analysis of Asynchronous Concurrent Systems by Timed Petri Nets.* Technical Report 120. Project MAC. 1974.
10. G.M. Reed and A.W.Roscoe *Metric Spaces as Models for Real-time Concurrency.* Mathematical Foundations of Programming, LNCS vol.298, pp.331-343. Springer-Verlag, 1987.
11. Wang Yi *A Calculus of Real Time Systems.* PhD. Thesis, Chalmers University of Technology, 1991.

Abstractions and Partial Order Reductions for Checking Branching Properties of Time Petri Nets[*]

Wojciech Penczek[1] and Agata Półrola[2]

[1] ICS PAS, Ordona 21, 01-237 Warsaw, Poland
penczek@ipipan.waw.pl
[2] Faculty of Mathematics, University of Lodz, Banacha 22, 90-238 Lodz, Poland
polrola@math.uni.lodz.pl

Abstract. The paper deals with verification of untimed branching time properties of Time Petri Nets. The atomic variant of the geometric region method for preserving properties of CTL* and ACTL* is improved. Then, it is shown, for the first time, how to apply the partial order reduction method to deal with next-time free branching properties of Time Petri Nets. The above two results are combined offering an efficient method for model checking of $ACTL^*_{-X}$ and CTL^*_{-X} properties of Time Petri Nets.

1 Introduction

Model checking is one of the most popular methods of automated verification of concurrent systems, e.g., hardware circuits, communication protocols, and distributed programs. However, the practical applicability of this method is strongly restricted by the state explosion problem, which is mainly caused by representing concurrency of operations by their interleaving. Therefore, many different reduction techniques have been introduced in order to alleviate the state explosion. The major methods include application of partial order reductions [Pel96,Val89, WG93], symmetry reductions [ES96], abstraction techniques [DGG94], BDD-based symbolic storage methods [Bry86], and SAT-related algorithms [BCCZ99].

Recently, the interest in automated verification is moving towards concurrent real-time systems. Two main models for representing such systems are usually exploited: timed automata [AD94], and Time Petri Nets [Sta90]. The properties to be verified are expressed in either a standard temporal logic like LTL and CTL*, or in its timed version like MITL [AFH96], and TCTL [ACD90].

Most of the efficient reduction techniques exist for linear time formalisms. For verification of concurrent systems this is maybe sufficient, but evidently is not for verification of timed and multi-agent systems. If one reviews the existing verification methods for multi-agent systems [Rou01], then it is easy to notice that most of them rely on translating the formalisms to branching time temporal

[*] Partly supported by the State Committee for Scientific Research under the grant No. 8T11C 01419

J.-M. Colom and M. Koutny (Eds.): ICATPN 2001, LNCS 2075, pp. 323–342, 2001.
© Springer-Verlag Berlin Heidelberg 2001

logics (see description of Nepi, KARO, HSTS in [Rou01]). This is one of new important motivations for considering reduction methods for branching time properties.

The present paper deals with verification of untimed branching time temporal properties of Time Petri Nets. Since Time Petri Nets have usually infinite state spaces, abstraction techniques [DGG94] are used to represent these by finite ones. To reduce the sizes of abstract state spaces partial order reductions are used. The main contribution of the paper relies on:

- improving the variant of the geometric region method [YR98] for defining abstract state spaces preserving properties of CTL* and ACTL* such that the structure of a verified formula is exploited,
- showing, **for the first time**, how to extend the po-reduction methods of [DGKK98,GKPP99,PSGK00] to deal with next-time free branching properties of Time Petri Nets,
- combining the above two results offering an efficient method for model checking of ACTL$^*_{-X}$ and CTL$^*_{-X}$ properties of Time Petri Nets.

The rest of the paper is organized as follows. Section 2 reviews the existing results. In Section 3 Time Petri Nets and concrete state spaces are introduced. Temporal logics of branching time are defined in Section 4. Atomic and pseudo-atomic state spaces are described in section 5. A partial order reduction method is presented in Section 6. The next two sections contain experimental results and conclusions.

2   Related Work

Our approach to abstract state spaces of Time Petri Nets improves the one of [YR98], which is a refinement of the method of [BD91,Dil89]. As far as we know this approach gives much smaller state spaces than the region graph method of [ACD90]. So far partial order reductions have been defined only for linear time properties either for Time Petri Nets [YS97,Lil99] or for Timed Automata of standard [Pag96,DGKK98] or local semantics [BJLW98,Min99]. Our approach is closely related to [DGKK98], from which we draw the idea of the covering relation, and to [GKPP99,PSGK00], from which we take the general method of partial order reductions preserving branching time properties.

The partial order approach presented in [VP99] works for Time Petri Nets and TCTL, but due to a very restrictive notion of visibility (all the transitions easily get visible) is of a rather restricted practical value.

3   Time Petri Nets

Let Q^+ denote the set of non-negative rational numbers.

Definition 1. A Time Petri Net *(TPN, for short) is a six-element tuple* $N = (P, T, F, Eft, Lft, m_0)$, *where*

- $P = \{p_1, p_2, \ldots, p_m\}$ *is a finite set of* places,
- $T = \{t_1, t_2, \ldots, t_n\}$ *is a finite set of* transitions,
- $F \subseteq (P \times T) \cup (T \times P)$ *is the* flow relation,
- $Eft, Lft : T \longrightarrow Q^+$ *are functions describing the* earliest *and the* latest firing times *of the transitions such that* $Eft(t) \le Lft(t)$ *for each* $t \in T$, *and*
- $m_0 \subseteq P$ *is the* initial marking *of* N.

We need also the following notations and definitions:

- $pre(t) = \{p \in P \mid (p, t) \in F\}$ *is the* preset *of* $t \in T$,
- $post(t) = \{p \in P \mid (t, p) \in F\}$ *is the* postset *of* $t \in T$,
- A *marking* of N is any subset $m \subseteq P$,
- A transition t is *enabled* at m ($m[t >$, for short) if $pre(t) \subseteq m$ and $post(t) \cap (m \setminus pre(t)) = \emptyset$; and leads from marking m to marking m' ($m[t > m'$), where $m' = (m \setminus pre(t)) \cup post(t)$.
- Let $en(m) = \{t \in T \mid m[t >\}$.

For the ease of the presentation we consider TPNs without self-loops, i.e., we require that $pre(t) \cap post(t) = \emptyset$ for all $t \in T$.

3.1 Concrete State Spaces of TPNs

A concrete state σ of N is an ordered pair $(m, clock)$, where m is a marking and $clock$ is a function $T \longrightarrow Q^+$, which for each transition t enabled at m gives the time elapsed since t became enabled most recently. Therefore, the initial state of N is $\sigma_0 = (m_0, clock_0)$, where m_0 is the initial marking and $clock_0(t) = 0$ for each $t \in T$. The states of N change when time passes or a transition fires. In state $\sigma = (m, clock)$, time $r \in Q^+$ can pass leading to new state $\sigma' = (m, clock')$ provided $clock(t) + r \le Lft(t)$ for all $t \in en(m)$. Then, $clock'(t) = clock(t) + r$ for all $t \in T$ (denoted $clock' = clock + r$).

As well, in state $\sigma = (m, clock)$ transition $t \in T$ can fire leading to new state $\sigma' = (m', clock')$ (denoted $t(\sigma)$) if $t \in en(m)$ and $Eft(t) \le clock(t) \le Lft(t)$. Then, $m' = m[t >$ and for all $u \in T$:

- $clock'(u) = 0$, for $u \in en(m') \setminus en(m \setminus pre(t))$,
- $clock'(u) = clock(u)$, otherwise.

Notice that firing of transitions takes no time. Let $\sigma \overset{(r,t)}{\rightarrow} \sigma'$ denote that σ' is obtained from σ by passing time r and firing transition t. We write $\sigma \overset{t}{\rightarrow} \sigma'$ or $\sigma \rightarrow \sigma'$ if there is r and t such that $\sigma \overset{(r,t)}{\rightarrow} \sigma'$. A *run* of N is a maximal sequence of states and transitions $\rho = \sigma_0 \overset{(r_0,t_0)}{\rightarrow} \sigma_1 \overset{(r_1,t_1)}{\rightarrow} \sigma_2 \ldots$. A state σ' is *reachable* from σ_0 if there is a run ρ and $i \in N$ s.t. $\sigma' = \sigma_i$.

Definition 2. *Let C be the set of all the states reachable from σ_0. The* concrete state space *of N is the structure $C_N = (C, \rightarrow, \sigma_0)$.*

Notice that the structure C_N can be infinite, but all its runs are discrete, i.e., each state contains at most one successor within a run. This property allows to use these structures as frames for our branching time temporal logics (subsets of CTL*).

Separating passing time and firing a transition leads to a different notion of (dense) concrete state spaces, which could not be directly taken as frames for CTL*. It would be, however, possible to reinterpret CTL* over these frames, but this problem goes beyond the scope of the paper.

4   Branching Time Logics: CTL* and ACTL*

Syntax of CTL*

Let PV be a finite set of propositions. First, we give a syntax of CTL* and then restrict it to standard sublanguages. The set of state formulas and the set of path formulas of CTL* are defined inductively:

S1. every member of PV is a state formula,
S2. if φ and ψ are state formulas, then so are $\neg\varphi$, $\varphi \vee \psi$ and $\varphi \wedge \psi$,
S3. if φ is a path formula, then $A\varphi$ is a state formula,
P1. any state formula φ is also a path formula,
P2. if φ, ψ are path formulas, then so are $\varphi \wedge \psi$, $\varphi \vee \psi$, and $\neg\varphi$,
P3. if φ, ψ are path formulas, then so are $X\varphi$, $G\varphi$, $\text{Until}(\varphi,\psi)$, and $\overline{\text{Until}}(\varphi,\psi)$.

The modal operator A has the intuitive meaning "for all paths". Until denotes the standard Until and $\overline{\text{Until}}$ is the operator dual to Until. CTL* consists of the set of all state formulae. The following abbreviations will be used: $E\varphi \overset{def}{=} \neg A\neg\varphi$, $F\varphi \overset{def}{=} \text{Until}(true, \varphi)$.

Sublogics of CTL*.

ACTL*. Negation can be applied only to subformulas that do not contain modalities.
ACTL. The sublogic of ACTL* in which the state modality A and the path modalities X, Until, $\overline{\text{Until}}$, and G may only appear paired in the combinations AX, $A\text{Until}$, $A\overline{\text{Until}}$, and AG.
CTL. The sublogic of CTL* in which the state modality A and the path modalities X, Until, $\overline{\text{Until}}$, and G may only appear paired in the combinations AX, $A\text{Until}$, $A\overline{\text{Until}}$, and AG.
LTL. Restriction to formulas of the form $A\varphi$, where φ does not contain A. We usually write φ instead of $A\varphi$ if confusion is unlikely.
L_{-X}. The sublogic of $L \in \{\text{CTL*, ACTL*, ACTL, CTL, LTL}\}$ without the operator X.

Semantics of CTL*

Let Σ be a finite set of labels and PV be a set of propositions. A *model* for CTL* is a pair (F, V), where $F = (S, R, \iota^0)$ is a directed, rooted, edge-labeled graph with node set S, and initial node $\iota^0 \in S$, $R \subseteq S \times \Sigma \times S$, while V is a valuation function $V : S \longrightarrow 2^{PV}$. The edge relation R is assumed to be total; i.e. $\forall u \, \exists v, a \ u \xrightarrow{a} v$. The labels on the edges in the definition of the graph are only used in the sequel for the benefit of the description of the suggested algorithm, but are ignored by the interpretation of the temporal logics. We assume that F is deterministic, i.e., $\forall u, v, v', a$ if $u \xrightarrow{a} v$ and $u \xrightarrow{a} v'$, then $v = v'$. For finite graphs the determinism is imposed by renaming all the copies of each non-deterministic transition. If the original set of labels was Σ, the new set is called Σ_r. Totality is ensured by adding cycles at the end nodes.

Let $M = (F, V)$ be a model and let $\pi = s_0 a_0 s_1 \cdots$ be an infinite path of F. Let π_i denote the suffix $s_i a_i s_{i+1} \cdots$ of π and $\pi(i)$ denote the state s_i. Satisfaction of a formula φ in a state s of M, written $M, s \models \varphi$ or just $s \models \varphi$, is defined inductively as follows:

S1. $s \models q$ iff $q \in V(s)$, for $q \in PV$,
S2. $s \models \neg\varphi$ iff not $s \models \varphi$, $\quad s \models \varphi \wedge \psi$ iff $s \models \varphi$ and $s \models \psi$,
$\quad\;\; s \models \varphi \vee \psi$ iff $s \models \varphi$ or $s \models \psi$,
S3. $s \models A\varphi$ iff $\pi \models \varphi$ for every path π starting at s,
P1. $\pi \models \varphi$ iff $s_0 \models \varphi$ for any state formula φ,
P2. $\pi \models \neg\varphi$ iff not $\pi \models \varphi$, $\quad \pi \models \varphi \wedge \psi$ iff $\pi \models \varphi$ and $\pi \models \psi$,
$\quad\;\; \pi \models \varphi \vee \psi$ iff $\pi \models \varphi$ or $\pi \models \psi$,
P3. $\pi \models X\varphi$ iff $\pi_1 \models \varphi$
$\quad\;\; \pi \models G\varphi$ iff $\pi_j \models \varphi$ for all $j \geq 0$.
$\quad\;\; \pi \models \text{Until}(\varphi, \psi)$ iff there is an $i \geq 0$ such that $\pi_i \models \psi$ and $\pi_j \models \varphi$ for all $0 \leq j < i$.
$\quad\;\; \pi \models \overline{\text{Until}}(\varphi, \psi)$ iff for all $i \geq 0$ ($\pi_i \models \psi$ or there is $j < i$ such that $\pi_j \models \varphi$).

4.1 Distinguishing Power of ACTL* and CTL*

Let $M = ((S, R, \iota), V)$ and $M' = ((S', R', \iota'), V')$ be two models.

Definition 3 ((Simulation [GL91])). *A relation $\leadsto_s \subseteq S' \times S$ is a simulation from M' to M if the following conditions hold:*

1. $\iota' \leadsto_s \iota$,
2. *if $s' \leadsto_s s$, then $V'(s') = V(s)$ and for every s_1 such that $s \xrightarrow{a} s_1$, there is s'_1 such that $s' \xrightarrow{a} s'_1$ and $s'_1 \leadsto_s s_1$.*

Model M' simulates model M ($M' \leadsto_s M$) if there is a simulation from M' to M. Two models M and M' are called simulation equivalent if $M \leadsto_s^1 M'$ and $M' \leadsto_s^2 M$. Two models M and M' are called bisimulation equivalent if $M \leadsto_s M'$ and $M' \leadsto_s' M$, where $\leadsto_s'$ is the inverse of $\leadsto_s$.

Theorem 1 (([GL91],[BCG88])). *Let M and M′ be two (bi-)simulation equivalent models[1], where the range of the labeling function V and V′ is 2^{PV}. Then, $M, \iota \models \varphi$ iff $M', \iota' \models \varphi$, for any ACTL* (CTL*, resp.) formula φ over PV.*

The reverse of this theorem, essentially stating that no coarser (bi-)simulation preserves the truth value of ACTL* (CTL*, resp.) formulas, also holds and can be proved by an easy induction, when M and M′ are finitely branching [GKP92].

5 Abstract State Spaces of TPNs

Since the concrete state space C_N of a TPN can be infinite, it cannot be directly used for model checking. We need to define its finite abstraction, i.e., another structure, which is finite and preserves all the formulas of our logic. This task is performed in the following way. First, we give a general definition of abstract state spaces. Next, we recall the notion of geometric state regions [BD91], which are finite abstractions preserving only linear time properties. Then, we define atomic state classes [YR98], which are obtained by splitting some of the geometric state regions in order to preserve all the branching time properties. Consequently, we improve on the algorithm generating atomic state classes by restricting the logic to ACTL* and the splitting to the regions, for which this it is really necessary. Then, we go for further reductions of the abstract state spaces by applying partial order reductions.

Let $\equiv \subseteq C \times C$ be an equivalence relation on the set of concrete states C. By *abstract states* we mean equivalence classes of $\equiv$, i.e., $A = \{[\sigma]_\equiv \mid \sigma \in C\}$. We assume that $\equiv$ satisfies at least the condition: if $(m, clock) \equiv (m', clock')$, then $m = m'$. The other conditions depend on the properties to be preserved.

Definition 4. *An* abstract state space *of N is a structure $A_N = (A, \rightarrow_A, \alpha_0)$, where $\alpha_0 = [\sigma_0]_\equiv$ and $\rightarrow_A \subseteq A \times T \times A$ satisfies the condition:*

EE) $\alpha \rightarrow_A \alpha'$ iff $(\exists \sigma \in \alpha) (\exists \sigma' \in \alpha')$ s.t. $\sigma \rightarrow \sigma'$.

5.1 Geometric State Regions

Geometric state regions are defined by sets of inequalities representing different clock functions. For an inequality $\theta ::= x - y \sim c$, where x, y are variables, c is a rational constant, and $\sim \in \{<, \leq, >, \geq\}$, let $\overline{\theta}$ denote the inequality $x - y \sim' c$, where $\sim'$ is the complement of the relation $\sim$.

Let I be a set of inequalities over $\{x_1, \ldots, x_n\}$. A *solution set $Sol(I)$ of* I is the set of all the vectors $(c_1, \ldots, c_n)$ of rational constants, which make every inequality of I valid in the theory of rational numbers. I is *consistent* if $Sol(I) \neq \emptyset$. Now, following [YR98] we introduce the notion of geometric state classes.

---

[1] Formulated for finite models, but easily extends to infinite ones.

A *(geometric) state class* is a tuple $\alpha = (m, I, w)$, where m is a marking, I is a set of inequalities, and $w \in T^*$, such that m is obtained by firing from m_0 the sequence w of transitions satisfying the timing constraints represented by I.

The initial state class is $\alpha_0 = (m_0, \emptyset, \epsilon)$. Since I needs to be defined over the variables representing the firing times of transitions, we distinguish between the different firings of the same transition. A transition t is *i-times fired* in $\alpha = (m, I, w)$ if t occurs i-times in w. A transition t is *enabled* in α (denoted $en(\alpha)$), if $t \in en(m)$.

In the following discussion we assume some fixed transition sequence w. A variable t^i represents the time of the i-th firing of $i - 1$-times fired transition t. Let $\alpha = (m, I, w)$ and $t^i \in w$ denote that t occurs i-times in w.

If transition t which is $i - 1$-times fired in α became enabled most recently by the jth firing of a transition u such that $u^j \in w$, then we say that u^j is the parent of t^i in α, denoted $u^j = parent(t^i, \alpha)$. For v being a special variable representing the time when the net started we assume that $v \in w$, for each sequence of transitions w. The following notions are defined:

- $Parents(\alpha) = \bigcup_{t \in en(\alpha)} parent(t^i, \alpha)$.
  For example $parent(t^1, \alpha_0) = v$ for $t \in en(\alpha_0)$.
- $ParTrans(\alpha) = \{(u, t) \mid u^j = parent(t^i, \alpha) \in Parents(\alpha)\}$,
- An $i-1$-times fired transition t is called *firable* in α if $t \in en(m)$ and for any $j-1$-times fired transition $u \in en(m)$:
  $I \cup \{"parent(t^i, \alpha) + Eft(t) \leq parent(u^j, \alpha) + Lft(u)"\}$ is consistent,
- A successor α' is obtained from α by firing a firable transition t (denoted $\alpha' = t(\alpha)$), as follows: if t is $i-1$-times fired in α, then $m' = m[t>$, $w' = wt$ and $I' = I \cup J$, where $J = J_1(\alpha, t^i) \cup J_2(\alpha, t^i)$, with
  - $J_1(\alpha, t^i) = \{"Eft(t) \leq t^i - parent(t^i, \alpha) \leq Lft(t)"\}$,
  - $J_2(\alpha, t^i) = \{"t^i \leq parent(u^j, \alpha) + Lft(u)" \mid u$ is enabled and $j-1$-times fired in $\alpha\}$.

Using the above successor relation, we can define the state class graph, where the nodes are state classes and the edges are represented by this relation. The future behaviour of N from α depends only on marking m and the firing times of the enabled transitions [BD91]. If the firing times of the parents of the corresponding enabled transitions (i.e., these which differ by upper indices only) are the same in α and α', then the firing times of these transitions are also the same. Therefore, two state classes are equivalent (denoted $\alpha \equiv \alpha'$) iff $m = m'$, $ParTrans(\alpha) = ParTrans(\alpha')$, and the projections of the solutions $Sol(I)|_{Parents(\alpha)} = Sol(I')|_{Parents(\alpha')}$. The number of equivalence classes of state classes is finite [BD91] and therefore these equivalence classes constitute the carrier of the abstract state spaces, $\mathcal{A} = (A, \to_A, \alpha_0)$ with $\to_A \subseteq A \times T \times A$, of geometric state regions. When we need to refer to the original state class graph, we call it $Unfold(\mathcal{A})$.

Note that upper indices of variables can be unbounded if there exists a cycle in behaviour of N. From $ParTrans(\alpha) = ParTrans(\alpha')$, we can consider a fixed ordering of the variables used in $ParTrans(\alpha)$ and $ParTrans(\alpha')$. Assuming that all the elements in the tuple $Sol(I)|_{Parents(\alpha)}$ and $Sol(I')|_{Parents(\alpha')}$ are arranged in this way, we can correctly find the equivalence of these classes.

Connection between concrete and abstract states. For any run $\rho = \sigma_0 \overset{(r_0,t_0)}{\to} \sigma_1 \overset{(r_1,t_1)}{\to} \sigma_2 \ldots$ of N define $\rho^i = \sigma_i$ and $time(\rho, i) = \Sigma_{k=0}^{i} r_k$ (represents the time t_i fires in ρ). For each state σ and a state class $\alpha = (m, I, w)$ we define $\sigma \in \alpha$ iff there exists a run $\rho = \sigma_0 \overset{(r_0,t_0)}{\to} \sigma_1 \overset{(r_1,t_1)}{\to} \sigma_2 \ldots$ of N and $n \in \mathcal{N}$:

- $\sigma = \rho^n$,
- $w = t_0 t_1 \cdots t_{n-1}$,
- $(time(\rho, 0), time(\rho, 1), \ldots, time(\rho, n - 1))$ and $v = 0$ satisfy I, i.e., the set of inequalities obtained from I by replacing v by 0 and the variables corresponding to t_i (i.e., t_i with upper indices) by $time(\rho, i)$, for $i \leq n - 1$, holds.

In what follows we consider only these TPNs, which satisfy the *progressiveness* condition, i.e., $\Sigma_{k=0}^{\infty} r_k \longrightarrow \infty$ for each infinite run ρ of N. This means that TPNs cannot have cycles of transitions for which the earliest firing times are equal to 0.

5.2   Atomic State Classes

The problem with geometric state graphs A_N, reported in [YR98], is that they do not satisfy condition AE) (see definition) and therefore do not preserve CTL formulas.

Definition 5. *A state class α satisfies condition* **AE)** *if: $\alpha \to_A \alpha'$ iff $(\forall \sigma \in \alpha) (\exists \sigma' \in \alpha')$ s.t. $\sigma \to \sigma'$, for each successor α' of α in A.*

This means that there is a formula, which holds at σ_0, but does not hold at α_0. Therefore, the geometric region approach cannot be used for CTL model checking of TPN. One of the ways out is to split the state classes into *atomic state classes* to make the condition AE) hold.

Definition 6. *A state class $\alpha = (m, I, w)$ is* atomic *if it satisfies condition AE). A region graph is* atomic *if all its state classes are atomic.*

Splitting of the state classes. Atomic classes are obtained by splitting the original classes. When the region $\alpha = (m, I, w)$ is split, then two new regions $(m, I \cup \xi, w)$ and $(m, I \cup \overline{\xi}, w)$ are created, where $\xi = IMPCON(\delta)$ is a proper constraint defined below.

Let $\alpha' = (m', I', w')$ be a successor of $\alpha = (m, I, w)$ obtained by the i-th firing of t. An *edge constraint* between α and α' is an inequality $\delta \in I'$ such that $variables(\{\delta\}) = \{t^i, u^j\}$ and $u^j \in Parents(\alpha)$. For an edge constraint $\delta = t^i + c \sim u^j$ and $x = parent(t^i, \alpha)$, define

- $IMPCON(\delta) = x + (Eft(t) + c) \sim u^j$ if $\sim \in \{<, \leq\}$.

- $IMPCON(\delta) = x + (Lft(t) + c) \sim u^j$, otherwise.
- $IMPCON(\delta)$ is *a proper constraint* if $I \cup \overline{IMPCON(\delta)}$ is consistent.

Intuitively, the existence of a proper constraint means that t is not firable from all $\sigma \in \alpha$.

Theorem 2 ([YR98]). *A state class α is not atomic if there exists a proper constraint $IMPCON(\delta)$ for some edge constraint δ between α and α'.*

The algorithm building atomic state classes [YR98] combines the depth-first search together with the partitioning. It starts with generating the initial geometric region and then, recursively, computes the successor classes of each class $\alpha = (m, I, w)$ obtained by firing transitions firable at α. Before firing a transition t, α is split if it is not atomic. In this case, for the class "$\alpha + \overline{IMPCON}$", i.e., α modified by adding $\overline{IMPCON}$ to I, all its successors are generated (instead of copying the subgraph). For α and its successors the recursive splitting procedure is called, which modifies the sets of inequalities of that classes by adding $IMPCON$s. Splitting of a class can also make it necessary, to split some of its predecessors, as every inequality we add in this process is a new edge constraint between a class and its parent. Then, the new class $t(\alpha)$ obtained by firing t at α is computed.

In [YR98] it is proved that the above algorithm always terminates and produces a finite atomic state class graph $\mathcal{A}$, such that $Unfold(\mathcal{A})$ is also atomic. In the following example we display a TPN with its geometric state graph and unfolded atomic state graph.

Example 1. The TPN in Figure 1 is taken from [YR98]. The sets of inequalities of state classes are defined as follows:

$$I_{\alpha_0} = \emptyset$$

$$I_{\alpha_1} = \left\{ \begin{array}{l} 0 \le t_1 - v \le 1 \\ t_1 \le v + 6 \end{array} \right\}$$

$$I_{\alpha_2} = \left\{ \begin{array}{l} 0 \le t_1 - v \le 1 \\ 0 \le t_2 - t_1 \le 1 \\ t_1 \le v + 6 \\ t_2 \le v + 6 \end{array} \right\}$$

$$I_{\alpha_3} = \left\{ \begin{array}{l} 0 \le t_1 - v \le 1 \\ 0 \le t_2 - t_1 \le 1 \\ 0 \le t_3 - t_2 \le 1 \\ t_1 \le v + 6 \\ t_2 \le v + 6 \\ t_3 \le v + 6 \end{array} \right\}$$

$$I_{\alpha_4} = \left\{ \begin{array}{l} 0 \le t_1 - v \le 1 \\ 0 \le t_2 - t_1 \le 1 \\ 0 \le t_3 - t_2 \le 1 \\ 1 \le t_4 - t_3 \le 3 \\ t_1 \le v + 6 \\ t_2 \le v + 6 \\ t_3 \le v + 6 \\ t_4 \le v + 6 \end{array} \right\}$$

$$I_{\alpha_5} = \left\{ \begin{array}{l} 0 \le t_1 - v \le 1 \\ 0 \le t_2 - t_1 \le 1 \\ 0 \le t_3 - t_2 \le 1 \\ 1 \le t_4 - t_3 \le 3 \\ 5 \le t_a - v \le 6 \\ t_1 \le v + 6 \\ t_2 \le v + 6 \\ t_3 \le v + 6 \\ t_4 \le v + 6 \end{array} \right\}$$

$$I_{\alpha_6} = \left\{ \begin{array}{l} 0 \le t_1 - v \le 1 \\ 0 \le t_2 - t_1 \le 1 \\ 0 \le t_3 - t_2 \le 1 \\ 5 \le t_a - v \le 6 \\ t_1 \le v + 6 \\ t_2 \le v + 6 \\ t_3 \le v + 6 \\ t_a \le t_3 + 3 \end{array} \right\}$$

$$I_{\alpha_7} = \left\{ \begin{array}{l} 0 \le t_1 - v \le 1 \\ 0 \le t_2 - t_1 \le 1 \\ 0 \le t_3 - t_2 \le 1 \\ 5 \le t_a - v \le 6 \\ 1 \le t_4 - t_3 \le 3 \\ t_1 \le v + 6 \\ t_2 \le v + 6 \\ t_3 \le v + 6 \\ t_a \le t_3 + 3 \end{array} \right\}$$

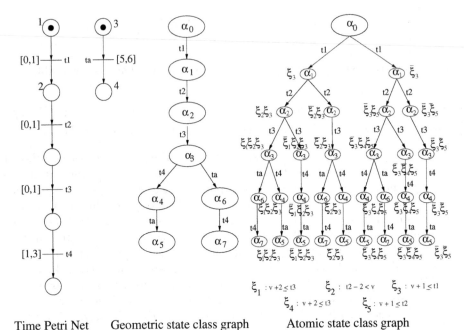

| Time Petri Net | Geometric state class graph | Atomic state class graph |

Fig. 1. TPN and its state graphs with IMPCON's and $\overline{IMPCON}$'s added

5.3   The State Spaces of TPN as Models for CTL*

Let each proposition $q_i \in PV$ correspond to exactly one place $p_i \in P$. For simplicity, we assume the same names for the propositions and places. Notice that the concrete state space $C_N = (C, \rightarrow_C, \sigma_0)$ of a TPN N as well as the abstract ones $A_N = (A, \rightarrow_A, \alpha_0)$ extended by valuation functions:

- $V_C : C \longrightarrow 2^P$, where $p \in V_C((m, clock))$ iff $p \in m$, for each $(m, clock) \in C$,
- $V_A : A \longrightarrow 2^P$, where $p \in V_A((m, I, w))$ iff $p \in m$, for each $(m, I, w) \in A$.

define models for the formulas of CTL*.

   Abstract model $M_{A_N} = (A_N, V_A)$ *preserves a formula* φ if for each $\alpha \in A_N$: $M_{A_N}, \alpha \models \varphi$ iff $(\forall \sigma \in \alpha)\ M_{C_N}, \sigma \models \varphi$. This implies that $M_{A_N}, \alpha_0 \models \varphi$ iff $M_{C_N}, \sigma_0 \models \varphi$.

Theorem 3 (([BD91,YR98])). *Let* $A_N = (A, \rightarrow_A, \alpha_0)$ *be an abstract state space. Then, the following conditions hold:*

- *if* A_N *is a geometric class graph, then* M_{A_N} *preserves LTL formulas,*
- *if* A_N *is a atomic class graph, then* M_{A_N} *preserves CTL formulas.*

Notice that if A_N is an atomic class graph, then it satisfies condition AE) and therefore M_{C_N} and M_{A_N} are bisimulation equivalent. Then, M_{A_N} preserves also CTL* formulas (see Theorem 1).

5.4   Pseudo-Atomic State Classes

Now, our first contribution starts. The aim is to relax the conditions on the atomic class graphs such that only ACTL* formulas remain preserved. We start with weakening the condition AE) to condition U), defined below.

U) For each $\alpha \in A$ there is $\emptyset \neq cor_\alpha \subseteq \alpha$ such that $\sigma_0 \in cor_{\alpha_0}$, and $\to_A$ satisfies:
$\alpha \to_A \alpha'$ iff $(\forall \sigma \in cor_\alpha)(\exists \sigma' \in cor_{\alpha'})$ s.t. $\sigma \to \sigma'$.

If an abstract state space A_N satisfies U, we say that A_N is *pseudo-atomic*. Notice that condition U) is equivalent to AE) when there is $cor_\alpha = \alpha$ for each $\alpha \in A$.

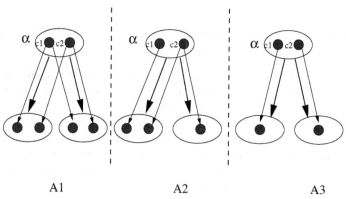

A1   A2   A3

Fig. 2. Fragments of abstract structures formed from sets of concrete states

Example 2. Figure 2 shows three fragments of abstract state spaces, where

A1: AE) is satisfied and $cor_\alpha = \{c_1, c_2\}$,
A2: U) is satisfied and $cor_\alpha = \{c_2\}$,
A3: Neither AE) or U) is satisfied; $cor_\alpha = \emptyset$.

Next, we formulate the theorem, which connects condition U) with ACTL*.

Theorem 4. *If A_N is a pseudo-atomic class graph, then M_{A_N} preserves the formulas of ACTL*.*

Proof. (Sketch) First notice that the following conditions hold:

1. $(\forall \sigma, \sigma' \in cor_\alpha)(\forall \varphi \in ACTL^*)$ $\sigma \models \varphi$ iff $\sigma' \models \varphi$,
2. $\alpha \models \varphi$ iff $(\exists \sigma \in cor_\alpha)$ $\sigma \models \varphi$.
3. $(\forall \sigma \in cor_\alpha)$ $(\forall \sigma' \in \alpha \setminus cor_\alpha)$ $(\forall \varphi \in ACTL^*)$ $\sigma \models \varphi \Rightarrow \sigma' \models \varphi$.

1) and 2) follow directly from the proof that condition AE) preserves CTL* (remember that ACTL* is a sublanguage of CTL*). 3) follows from the fact that by condition U) each $\sigma \in cor_\alpha$ simulates each other $\sigma' \in \alpha$.
   Then, we show that $\alpha \models \varphi$ iff $(\forall \sigma \in \alpha)$ $\sigma \models \varphi$.

($\Rightarrow$) It follows from 1) and 2) that if $\alpha \models \varphi$, then $(\forall \sigma \in cor_\alpha) \ \sigma \models \varphi$. Since (by condition 3) $(\forall \sigma \in cor_\alpha)(\forall \sigma' \in \alpha \setminus cor_\alpha) \ \sigma \models \varphi$ implies $\sigma' \models \varphi$, we get that $(\forall \sigma \in cor_\alpha) \ \sigma \models \varphi$ implies $(\forall \sigma \in \alpha) \ \sigma \models \varphi$. So, we are done.

($\Leftarrow$) If $(\forall \sigma \in \alpha) \ \sigma \models \varphi$, then $(\forall \sigma \in cor_\alpha) \ \sigma \models \varphi$. Thus, by condition 1) and 2) $\alpha \models \varphi$.

Thus, the lemma follows from the fact that $\sigma_0 \in cor_{\alpha_0}$.

The condition U) will be used for improving the size of atomic state spaces preserving ACTL* formulas.

Example 3. Notice that in Fig. 2 A_1 preserves CTL*, A_2 does not preserve CTL*, but preserves ACTL*, whereas A_3 does not preserve ACTL, but preserves LTL.

Example 4. Notice that the geometric state class graph of Fig. 1 satisfies the condition U) and therefore preserves ACTL*.

Now, we discuss how to modify the algorithm generating atomic state spaces in order to get pseudo-atomic state spaces. The classes are now represented by $\alpha = (m, I, I^{cor}, w)$, where $I^{cor} \subseteq I$ represents the *cor* of α. Firing a transition t at α gives the new class $\alpha' = (m[t >, I \cup J, I^{cor} \cup J, wt)$, where $(m[t >, I \cup J, wt)$ is the t-successor of α, described in section 5.1. Notice that splitting is not necessary when I^{cor}_α is consistent. Thus, the algorithm is modified such that rather than splitting α, it computes and memorizes I^{cor}_α. This is performed in the following way. The algorithm starts with assigning $I^{cor}_{\alpha_0} = I_{\alpha_0}$ for the initial state class. Then, assume for simplicity, that $IMPCON(\delta)$ is the only proper constraint between α and its t-successor α'. Then,

i) $I^{cor}_\alpha := I^{cor}_\alpha \cup IMPCON(\delta)$ and ii) $I^{cor}_{\alpha'} := I^{cor}_{\alpha'} \cup IMPCON(\delta)$.

i) follows from the fact that transition t is only firable at α with $I_\alpha \cup IMPCON(\delta)$ to lead to α'. ii) propagates $IMPCON(\delta)$ to α'.

As long as I^{cor}_α is consistent the algorithm does not need to split. Otherwise, the algorithm starts splitting in a similar way as described in [YR98] handling the splitting of I^{cor}'s as well. In this case, class α and its successors are recursively modified by adding $\overline{IMPCON}$ to their sets of inequalities I and I^{cor}, and the new class $\alpha_1 = \alpha + IMPCON$ is created. $I^{cor}_{\alpha_1}$ is a union of I_{α_1} and the set of inequalities obtained by firing a transition t, which was fired to generate the class α, at I^{cor} of the parent class of α. The successors of α_1 are computed, instead of copying the subgraph. A detailed description of this algorithm can be found in [Pół01].

6   Partial Order Reductions for TPN

In this section we show that the covering relation of [DGKK98] can be exploited for defining partial order reductions of abstract state spaces (pseudo-atomic and atomic) of Time Petri Nets preserving branching time properties.

Let $A_N = (A, \rightarrow_A, \alpha_0)$ be a region graph, $\alpha \in A$, and $t(\alpha)$ denotes the class α', obtained by firing t at α. Recall that A_N is deterministic, so the transitions are labelled by T. $\alpha^+ = \{\sigma' \in C \mid \sigma' = \sigma + r,$ for some $\sigma \in \alpha$ and some time $r \in Q^+$, which can pass at $\sigma\}$.

Definition 7. *Let $t, t' \in T$. Transition t covers transition t' (written, $t \leadsto_c t'$) if for all classes $\alpha \in A$ it holds $t(t'(\alpha)) \subseteq (t'(t(\alpha)))^+$.*

Intuitively, any class that can be reached from α by firing t' and then t can also be reached from α by first firing t then t' and then possibly passing some time.

Example 5. In Figure 3 transition t covers transition t'. The components I_i of the abstract states α_i are as follow:

- $I_1 = \{"t - v \leq 2"\}$,
- $I_2 = \{"1 \leq t' - v \leq 3", "t' - v \leq 2"\}$,
- $I_3 = \{"t - v \leq 2", "1 \leq t' - v \leq 3"\}$,
- $I_4 = \{"t - v \leq 2", "1 \leq t' - v \leq 2"\}$.

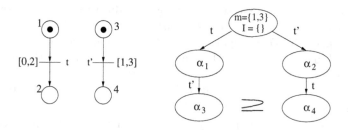

Fig. 3. Transition t covers transition t' as $\alpha_4 \subseteq \alpha_3$

We now give a simple criterium which implies that t covers t'.

Theorem 5. *If transitions $t, t' \in T$ satisfy the following conditions:*

- $(pre(t) \cup post(t)) \cap (pre(t') \cup post(t')) = \emptyset$,
- $Eft(t) = 0$.

Then, $t \leadsto_c t'$.

Proof. (Sketch) We have to show that $t(t'(\alpha)) \subseteq t'(t(\alpha))^+$. Let $\sigma = (m, clock) \in \alpha$ and $\sigma'_2 = (m[t't >, clock + r_{t'} + r_t) \in t(t'(\sigma))$. Since t' can fire at $(m, clock + r_{t'})$, t can fire as well, due to $Eft(t) = 0$. Let $\sigma_2 = (m[t't >, clock + r_{t'})$. From $m[t't >= m[tt' >$ it follows that $\sigma_2 \in t'(t(\sigma))$. Thus, $\sigma'_2 \in (t'(t(\sigma)))^+$.

Below, we show that using the covering relation, we can extend the partial order reduction method of branching time properties from untimed systems [GKPP99] to TPNs. There are several problems with proving the correctness of this approach, which follow from the fact that $t(t'(\alpha)) \neq t'(t(\alpha))$ and the fact that the pseudo-atomic state spaces are generated by the algorithm, which is a combination of DFS and splitting. The following discussion explains how we solve these problems.

First, observe that this is not equality of $t(t'(\alpha))$ and $t'(t(\alpha))$, which is necessary in the proof of correctness, but the fact that $t(t'(\alpha))$ simulates (or bisimulates) $t'(t(\alpha))$. This is provided by the following lemma.

Lemma 1. *If* $t \leadsto_c t'$ *and* t, t' *are firable at* α, *then* $\alpha_1 = t'(t(\alpha))$ *simulates* $\alpha_2 = t(t'(\alpha))$ *(denoted* $\alpha_1 \leadsto_s \alpha_2$), *for each* $\alpha \in A$.

Unfortunately, the proof of Theorem 6 using the above lemma becomes quite involved.

The second problem is connected with the way the po-reduced (pseudo-)atomic state spaces are generated. First of all, DFS together with the partitioning is used and moreover splitting of classes can introduce non-determinism and the need for renaming the copies of transitions of T to T_r.

Now, we prove that for a given (pseudo-)atomic state space, our method of po-reductions gives a reduced state space, which preserves the formulas of (ACTL$^*_{-X}$) CTL$^*_{-X}$. This result allows us to reduce the abstract state spaces, but it does not say how to perform partial order reductions while generating the state spaces by DFS + partitioning. Such a combination is possible and it is discussed in the last section. For the detailed description of the algorithm the reader is referred to [Pół01].

Next, we discuss the above algorithms. Let $A_N = (A, \rightarrow_A, \alpha_0)$ be a pseudo-atomic state space, and $en(\alpha) = \{t \mid \exists \alpha' \in A : \alpha \xrightarrow{t}_A \alpha'\}$

The standard po-reduction algorithm is based upon a straightforward depth-first-search (DFS) algorithm (see [GKPP99,Pel96]) to generate a reduced state graph.

DFS is a recursive algorithm, which starts at the initial pseudo-atomic region of the Time Petri Net N. In current state α, DFS selects an unexpanded (i.e., not yet used for the generation of a successor) enabled transition, say $t \in en(\alpha)$, and generates the successor state α' such that $\alpha \xrightarrow{t}_A \alpha'$. Then, DFS continues from successor state till no more enabled transition is present; then backtracks to the nearest predecessor, which contains unexpanded enabled transition.

PO-reduction algorithm differs from DFS such that whenever the po-reduction algorithm visits a new state α' and the set of enabled transitions $en(\alpha)$ is examined, only a *subset (Ample set)* of it, denoted $E(\alpha')$, is used to generate the successors. The choice of Ample sets is constrained by the following conditions, introduced and precisely discussed in [Pel96,Val89,Val96,WG93,GKPP99, PSGK00]. Let **Vis** denote the set of renamed transitions, which change valuations of the propositions used in φ on model M_{A_N}, and **Invis** $= T_r \setminus Vis$. Recall that for models corresponding to TPN's the places play role of the propositions.

C1 No transition $t \in T_r \setminus E(\alpha)$ that is not covered by a transition in $E(\alpha)$ can be executed in N before a transition of $E(\alpha)$ is executed.
C2 On every cycle in the constructed state graph, there is a node α with $E(\alpha) = en(\alpha)$,

C3 $E(\alpha) \cap \mathbf{Vis} = \emptyset$ or $E(\alpha) = en(\alpha)$,

C4 there is an unrenamed $t \in T$ s.t. $E(\alpha) = \{t\}$ or $E(\alpha) = en(\alpha)$,

CD $(\mathbf{Vis} \times \mathbf{Vis}) \cap \leadsto_c = \emptyset$.

An algorithm for computing sets of transitions satisfying **C1** w.r.t. independency can be found in [HP94]. Its adaptation w.r.t. covering is straightforward (see [DGKK98]).

Remark: for reductions preserving only $ACTL^*_{-X}$ there is an alternative way of defining the Ample sets such that the conditions **C3,C4** are relaxed [PSGK00].

Below, we show that the reduced pseudo-atomic model preserves $ACTL^*_{-X}$, i.e., that the pseudo-atomic and the reduced pseudo-atomic model are simulation equivalent. A similar proof can be obtained for preservation of CTL^*_{-X}, but then we have to reduce atomic models using the definition of Ample sets based on the symmetric relation $\leadsto_c$. To this aim we use the idea of the proof in [GKPP99], but modify the definition of simulation. We start with defining a visible simulation, which has been shown to be stronger than simulation [PSGK00].

Definition 8 ((visible simulation) [Pel96,GKPP99]). *A relation* $\leadsto_{vs} \subseteq S' \times S$ *is a* visible simulation *from* $M' = ((S', R', \iota'), V')$ *to* $M = ((S, R, \iota), V)$ *if (i)* $\iota' \leadsto_{vs} \iota$, *and (ii) whenever* $s' \leadsto_{vs} s$, *the following conditions hold:*

1. $V'(s') = V(s)$.

2. *If* $s \xrightarrow{b} t \in R$, *then either* b *is invisible and* $s' \leadsto_{vs} t$, *or there exists a path* $s' = s_0 \xrightarrow{a_0} s_1 \xrightarrow{a_1} \cdots \xrightarrow{a_{n-1}} s_n \xrightarrow{b} t'$ *in* M' *such that* $s_i \leadsto_{vs} s$ *for* $i \leq n$, a_i *is invisible for* $i < n$ *and* $t' \leadsto_{vs} t$.

3. *If there is an infinite path* $s = t_0 \xrightarrow{b_0} t_1 \xrightarrow{b_1} \cdots$, *where* b_i *is invisible and* $s' \leadsto_{vs} t_i$ *for* $i \geq 0$, *then there exists an edge* $s' \xrightarrow{c} s''$ *such that* c *is invisible and* $s'' \leadsto_{vs} t_j$ *for some* $j > 0$.

Model M' v-simulates *model* M $(M' \leadsto_{vs} M)$ *if there is a visible simulation from* M' *to* M. *Two models* M *and* M' *are called* v-simulation equivalent *if* $M \leadsto^1_{vs} M'$ *and* $M' \leadsto^2_{vs} M$. *Two models* M *and* M' *are called* v-bisimulation equivalent *if* $M \leadsto_{vs} M'$ *and* $M' \leadsto'_{vs} M$, *where* $\leadsto'_{vs}$ *is the inverse of* $\leadsto_{vs}$.

Theorem 6. *Let* M *be a pseudo-atomic model and* M' *be its po-reduction generated by our algorithm. Then,* M *and* M' *are v-simulation equivalent.*

Proof. (Sketch) The proof is a generalization of the corresponding proof of [GKPP99]. Let $M = ((S, R, \iota), V)$ be the pseudo-atomic model of N and $M' = ((S', R', \iota'), V')$ be the reduced one generated for M by our partial order reduction algorithm.

Remember that $\iota' = \iota$ and $V' = V|S'$. Since M' is a sub-model of M, it is obvious that M v-simulates M'. In order to show the opposite, we define a new visible simulation relation:

Let $s \leadsto_s s'$ denote that s simulates s' in the sense of Lemma 1. We use also the notation: $s'(\leadsto_s)^{-1}s$ for $s \leadsto_s s'$. In order to obtain a v-simulation between the reduced model M' and the model M, define the following relation:

Definition 9. *Let* $\sim \subseteq S \times S$ *be such that* $s \sim s'$ *iff there exists a sequence of states* $s = s_0 (\leadsto_s)^{-1}w_0 \xrightarrow{a_0} s_1(\leadsto_s)^{-1}w_1 \xrightarrow{a_1} \cdots \xrightarrow{a_{n-1}} s_n(\leadsto_s)^{-1}w_n = s'$ *such that* a_i *is invisible and* $\{a_i\}$ *satisfies condition* **C1** *from state* w_i *for* $0 \le i < n$. *Such a sequence will be called a* generalized forming sequence *(gf-sequence, for short). A gf-sequence of the form* $s = s_0 \xrightarrow{a_0} s_1 \xrightarrow{a_1} \cdots \xrightarrow{a_{n-1}} s_n = s'$ *is called a* standard forming path *(sf-path).*

The number of $\longrightarrow$ in a gf-sequence is called its *length*. Let $\approx = \sim \cap (S \times S')$. Now, our goal is to show that $\approx$ is a v-simulation. We use a number of simple lemmas (2 - 5) to prove the main theorem 7:

Lemma 2. *Let* $s \xrightarrow{a} r$ *be an edge of* M *such that* $\{a\}$ *satisfies Condition* **C1** *from the state* s. *Let* $s \xrightarrow{b} s'$ *be another edge of* M, *with* $a \ne b$. *Then* $\{a\}$ *satisfies Condition* **C1** *from* s'.

Proof. See [GKPP99].

Lemma 3. *Let* $s (\leadsto_s)^{-1}r$ *and* $s \xrightarrow{b} s'$ *be an edge of* M. *Then, there is* $r' \in S$ *and an edge* $r \xrightarrow{b} r'$ *such that* $s' (\leadsto_s)^{-1}r'$.

Proof. Follows directly from Lemma 1.

Lemma 4. *Let* $s = s_0(\leadsto_s)^{-1}w_0 \xrightarrow{a_0} s_1(\leadsto_s)^{-1}w_1 \xrightarrow{a_1} \cdots \xrightarrow{a_{n-1}} s_n(\leadsto_s)^{-1}w_n = r$ *be a gf-sequence and* $s \xrightarrow{b} s'$. *Then there are exactly two possibilities:*

1. *For all* $0 \le i < n$ *transition* a_i *covers* b. *Then, there exists a gf-sequence:*
   $s' = s_0' (\leadsto_s)^{-1} w_0' \xrightarrow{a_0} x_1 (\leadsto_s)^{-1} s_1'(\leadsto_s)^{-1}w_1' \xrightarrow{a_1} \cdots \xrightarrow{a_{n-1}} x_n(\leadsto_s)^{-1}s_n'$
   $(\leadsto_s)^{-1}w_n' = r'$, *with* $s_j \xrightarrow{b} s_j'$ *and* $w_j \xrightarrow{b} w_j'$ *for* $0 \le j \le n$.
2. *There exists* $j < n$ *such that* b *is covered by* a_i *for* $0 \le i < j$, *and* $b = a_j$. *There exists a gf-sequence* $s' = s_0'(\leadsto_s)^{-1}w_0' \xrightarrow{a_0} x_1(\leadsto_s)^{-1}s_1'(\leadsto_s)^{-1}w_1' \xrightarrow{a_1}$
   $\cdots \xrightarrow{a_{j-1}} x_j(\leadsto_s)^{-1}s_j'(\leadsto_s)^{-1}w_j' = r'$, *where* $w_j' = s_{j+1}$, $s_i \xrightarrow{b} s_i'$, *and* $w_i \xrightarrow{b}$
   w_i' *for* $0 \le i \le j$. *Therefore, there is a gf-sequence of length* $n - 1$ *from* s' *to* r.

Proof. Notice that b is covered by a_i for $0 \le i < n$ in Item *1*, since $\{a_i\}$ satisfies **C1** in w_i. The same holds for $0 \le i < j$ in Item *2*. We can now apply a simple induction using Lemma 2 and Lemma 3.

Corollary 1. *Let* $s \sim r$ *and* $s \xrightarrow{b} s'$. *Then there exists an edge* $r \xrightarrow{b} r'$ *such that* $s' \sim r'$ *in each of the following two cases:*

1. b does not appear on some gf-sequence from s to r (in particular this must be the case when b is visible), or
2. $s' \nsim r$.

It is easy to see that the reduction algorithm guarantees the following:

Lemma 5. *Let s be a state in the reduced model M'. Then, there is a sf-path in M' from s to a fully expanded (i.e., with all the successors generated) state s'.*

Theorem 7. *The relation $\approx$ is a v-simulation.*

Proof. First, observe that $\iota = \iota'$ and $\iota \in S'$. Hence $\iota \approx \iota'$. Next, let $s \approx r$. Thus, $s \sim r$. Item *1* of Definition 8 is satisfied since the invisible operations and $\leadsto_s$ preserve valuation. Hence, $V(s) = V(r)$. Thus, also $V(s) = V'(r)$.

We show that Item *2* of Definition 8 holds. Let $s \xrightarrow{b} s' \in M$. We argue by cases:

Case 1. $s' \sim r$ and b is invisible. Then Item 2 follows immediately.

Case 2. $s' \nsim r$ or b is visible. According to Corollary 1, in both cases there is an edge $r \xrightarrow{b} r'$ in M such that $s' \sim r'$. Notice that by the definition of $\approx$, $r \in S'$, but it is not necessarily the case that $r' \in S'$. By Lemma 5, there is an sf-path in M' from r to some fully expanded state t. Hence, $s \sim r \sim t$, which implies by transitivity of $\sim$ that $s \sim t$. Since $t \in S'$, also $s \approx t$. Again there are two cases:

Case 2.1. $r' \sim t$ and b is invisible. Then, $s' \sim r' \sim t$, hence $s' \sim t$ and also $s' \approx t$. Thus, the path required by Item 2 consists of the sf-path from r to t.

Case 2.2. $r' \nsim t$ or b is visible. Then, according to Corollary 1, there is an edge $t \xrightarrow{b} t'$, with $r' \sim t'$. Thus, $s' \sim r' \sim t'$, hence $s' \sim t'$. Since t is fully expanded, $t' \in S'$, thus $s' \approx t'$. Thus, the path required in Item 2 consists of the sf-path from r to t, followed by the edge $t \xrightarrow{b} t'$.

For proving item *3* of Definition 8, let $s_0 \xrightarrow{a_0} s_1 \xrightarrow{a_1} \ldots$ be an infinite path, where $s_0 = s$ and with a_i invisible and $s_i \approx r$ for $i \geq 0$. Consider now two cases. In the first case, there is a single edge $r \xrightarrow{c} r'$, with c invisible, in M', with $\{c\}$ satisfying Condition **C1** from r. In this case, $r \sim r'$, and since $s_1 \sim r$, we have $s_1 \sim r'$. Since r' is in S' then $s_1 \approx r'$.

In the second case, r is fully expanded. We will show that there exists a gf-sequence from some s_j to r, where $j \geq 0$, such that a_j does not occur on it. To show that, we will construct a sequence of gf-sequences l_i from s_i to r, for $0 \leq i \leq j$, with l_0 a path from $s = s_0$ to r. Observe that by Lemma 4, if a_i appears on l_i, then we can construct a path l_{i+1} from s_{j+1} that is shorter than l_i. Since there are infinitely many states s_i, and l_0 has a finite length, this construction must terminate with some j as above. Now, according to Corollary 1, there is an edge $r \xrightarrow{a_j} r' \in M$ such that $s_{j+1} \sim r'$. Since r is fully expanded, also $s_{j+1} \approx r'$. End of proof.

It is important to notice that if $\leadsto_c$ is symmetric, then it follows easily from Lemma 3 that $\approx$ is a v-bisimulation. In this case to preserve CTL^*_{-X} the PO-reduction algorithm should be applied to atomic-models. An easy criterium for $\leadsto_c$ to be symmetric is to require that $Eft(t) = Eft(t') = 0$.

6.1   Combining DFS, Partitioning, and Partial Order Reductions

Now, we describe the (non-sequential) combination of the DFS+partitioning algorithm generating the (pseudo)-atomic state classes and partial order reductions.

The following changes to the algorithm generating the (pseudo-)atomic state classes are made. The set of transitions to be fired at α (an Ample set) is computed w.r.t a *formula* to be checked. Since splitting of a class can imply splitting of some of its predecessors, the original choice of a single transition by **C4** to be fired from the class can be invalidated (the transition gets copies, which have to be renamed). This requires recomputing the ample set and rebuilding the subgraph. A similar problem has to be handled if a class we get, when splitting propagates, is equivalent to some of its predecessors, which can invalidate the condition **C2**. This again requires rebuilding the subgraph. Moreover, all the edge constraints, which could potentially cause splitting of class α even if the po-reduction has been applied to transitions enabled at α, are computed in order to build the graph being a substructure of the unreduced one. A detailed description of the algorithm can be found in [Pół01].

Our proof of correctness shows that the reduced state space generated by the above algorithm is exactly the same as it was generated by the standard PO-reduction algorithm, applied to the already generated (pseudo-)atomic state graph. So, correctness follows again from Theorem 6.

7   Experimental Results

Our algorithm has been implemented on the base of the program described in [YR98] we have received by courtesy of Tomohiro Yoneda. The sets of inequalities describing state classes are represented by DBM matrices [Dil89]. The present implementation does not contain any compressing procedure and therefore should be optimized for practical purposes.

A small net is presented in Figure 4. We display the sizes of its (detailed) region graph of [ACD90] (taken from [YR98]), geometric region, atomic and pseudo-atomic graphs and the graphs obtained using our PO-reduction algorithm (for **Vis**=$\{t_a\}$).

It is obvious that running our implementation on bigger examples (with more concurrent processes) would give much more substantial po-reductions as these depend on the number of independent operations.

	States	Edges	Total	CPU
region graph	15011	25206	40217	
geometric region	16	25	41	0.02
AE) satisfied	53	95	148	0.05
AE)+po(ACTL*_X)	53	76	129	0.04
U) satisfied	31	50	81	0.03
U)+po(ACTL*_X)	31	44	75	0.04

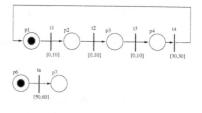

Fig. 4. A comparison between the sizes of graphs for a simple TPN

8   Conclusions

We have improved the variant of the geometric region method [YR98] for defining abstract state spaces preserving properties of ACTL*. Moreover, we have shown how to extend the po-reduction methods to deal with next-time free branching properties of time Petri Nets. Finally, the above two results have been combined offering an efficient method for model checking of $ACTL^*_{-X}$ and CTL^*_{-X} properties of time Petri Nets. So far our method covers only untimed properties of TPN. However, we believe that a similar approach can be used for efficient verification of TCTL properties using the region graph approach of [ACD90].

References

[ACD90]   R. Alur, C. Courcoubetis, and D. Dill, *Model Checking for Real-Time Systems*, Proc. of LICS'90, IEEE, 1990, pp. 414 – 425.

[AD94]   R. Alur and D. Dill, *Automata for Modelling Real-Time Systems*, Theoretical Computer Science **126(2)** (1994), 183–236.

[AFH96]   R. Alur, T. Feder, and T. Henzinger, *The Benefits of Relaxing Punctuality*, Journal of ACM **43(1)** (1996), 116–146.

[BCCZ99] A. Biere, A. Cimatti, E. Clarke, and Y. Zhu, *Symbolic Model Checking without BDDs*, Proc. of DAT'99, 1999.

[BCG88]   M.C. Browne, E.M. Clarke, and O. Grumberg, *Characterizing Finite Kripke Structures in Propositional Temporal Logic*, Theoretical Computer Science **59** (1988), 115–131.

[BD91]   B. Berthomieu and M. Diaz, *Modelling and Verification of Time Dependent Systems Using Time Petri Nets*, IEEE Trans. on Software Eng. **17(3)** (1991), 259–273.

[BJLW98] J. Bengtsson, B. Jonsson, J. Lilius, and Y. Wang, *Partial Order Reductions for Timed Systems*, LNCS **1466** (1998), 485–500.

[Bry86]   R. Bryant, *Graph-based Algorithms for Boolean Function Manipulation*, IEEE Transaction on Computers **35(8)** (1986), 677–691.

[DGG94]   D. Dams, O. Grumberg, and R. Gerth, *Abstract Interpretation of Reactive Systems: Abstractions Preserving ACTL*, ECTL* and CTL**, Proceedings of the IFIP Working Conference on Programming Concepts, Methods and Calculi (PROCOMET), North-Holland, 1994.

[DGKK98]   D. Dams, R. Gerth, B. Knaack, and R. Kuiper, *Partial-Order Reduction Techniques for Real-Time Model Checking*, Proc. of 3rd Int. Workshop on Formal Methods for Industrial Critical Systems, 1998, pp. 157 – 169.

[Dil89]   D. Dill, *Timing Assumptions and Verification of Finite State Concurrent Systems*, LNCS **407** (1989), 197 – 212.

[ES96]   A. Emerson and A.P. Sistla, *Symmetry and Model Checking*, Formal Methods in System Design **7** (1996), 105–131.

[GKP92]   U. Goltz, R. Kuiper, and W. Penczek, *Propositional Temporal Logics and Equivalences*, LNCS **630** (1992), 222–236.

[GKPP99]   R. Gerth, R. Kuiper, D. Peled, and W. Penczek, *A Partial Order Reductions to Branching Time Logic Model Checking*, Information and Computation **150** (1999), 132–152.

[GL91]   O. Grumberg and D.E. Long, *Model Checking and Modular Verification*, LNCS **527** (1991), 250–265.

[HP94]   G.J. Holzmann and D. Peled, *An Improvement in Formal Verification*, Proc. of FORTE'94, Formal Description Techniques, Chapman and Hall, 1994, pp. 197–211.

[Lil99]   J. Lilius, *Efficient State Space Search for Time Petri Nets*, Proc. MFCS'98 Workshop on Concurrency, ENTCS, vol. 18, Springer-Verlag, 1999, p. 21.

[Min99]   M. Minea, *Partial Order Reductions for Model Checking of Timed Automata*, LNCS **1664** (1999), 431–446.

[Pag96]   F. Pagani, *Partial Orders and Verification of Real Time Systems*, LNCS **1135** (1996), 327–346.

[Pel96]   D. Peled, *Partial Order Reduction: Linear and Branching Temporal Logics and Process Algebras*, POMIV'96, Partial Order Methods in Verification, American Mathematical Society, DIMACS, 1996, pp. 79–88.

[Pół01]   A. Półrola, *Generation of Reduced Abstract State Spaces for Time Petri Nets*, Report ICS PAS (2001), to appear.

[PSGK00]   W. Penczek, M. Szreter, R. Gerth, and R. Kuiper, *Improving Partial Order Reductions for Branching Time Properties*, Fundamenta Informaticae **43** (2000), 245–267.

[Rou01]   Ch. Rouff, *editor, Proc. of Formal Approaches to Agent-Based Systems*, Springer-Verlag, 2001, to appear.

[Sta90]   P. Starke, *Analyse von Petri-Netz-Modellen*, Teubner Verlag, 1990.

[Val89]   A. Valmari, *Stubborn Sets for Reduced State Space Generation*, Proc. of the 10th International Conference on Application and Theory of Petri Nets, LNCS, vol. 483, Springer-Verlag, 1989, pp. 491–515.

[Val96]   A. Valmari, *Stubborn Set Methods for Process Algebras*, Proc. POMIV'96, Partial Order Methods in Verification, Americal Mathematical Society, 1996, pp. 213–222.

[VP99]   I.B. Virbitskaite and E.A. Pokozy, *A Partial Order Method for the Verification of Time Petri Nets*, LNCS **1684** (1999), 547–558.

[WG93]   P. Wolper and P. Godefroid, *Partial-Order Methods for Temporal Verification*, LNCS **715** (1993), 233–246.

[YR98]   T. Yoneda and H. Ryuba, *CTL Model Checking of Time Petri Nets Using Geometric Regions*, IEICE Trans. Inf. and Syst. **3** (1998), 1–10.

[YS97]   T. Yoneda and B.H. Schlingloff, *Efficient Verification of Parallel Real-Time Systems*, Formal Methods in System Design **11(2)** (1997), 197–215.

Pr/T–Net Based Seamless Design of Embedded Real-Time Systems

Carsten Rust[1], Jürgen Tacken[1], and Carsten Böke[2]

[1] C-LAB, Fürstenallee 11, D-33094 Paderborn, Germany
{theo, car}@c-lab.de
http://www.c-lab.de/
[2] Paderborn University, Fürstenallee 11, D-33094 Paderborn, Germany
boeke@upb.de
http://www.upb.de/

Abstract. During the last years we have been working towards a complete design method for distributed embedded real-time systems. The main characteristic of the methodology is that within the critical phases of analysis and synthesis the system under development is available in one unique model, that of extended Pr/T–Nets. Among several other reasons we have chosen a high-level Petri Net model in order to benefit from the multitude of analysis and synthesis methods for Petri Nets. Even though the methodology is based upon one common model, it nevertheless supports the modeling of heterogeneous systems using different specification languages. The methodology was introduced and described in several former publications. In this paper we therefore only give a brief overview and afterwards go into details of our recent work, namely the transformation of proper Pr/T–Net-models into synchronous languages, the partitioning of Pr/T–Nets and an OS-integrated execution engine for Pr/T–Nets.

1 Introduction

In recent years the design of embedded systems has gained increasing importance. This development is on the one hand rooted in the growing number of embedded systems. On the other hand, due to the complexity of embedded systems, handmade design of these systems without support of well-elaborated methodologies is not practicable. The reasons for complexity are manifold: First of all many embedded systems are safety-critical. Thus many requirements, logical as well as temporal, have to be assured during design. However, the effort - in terms of hardware resources - for implementing the system should be minimized, since normally embedded systems are series products. Complexity is often raised through the fact that complex embedded systems are distributed and contain concurrent behavior. Finally, the target architectures for their realization are typically heterogeneous. Summarized, sophisticated methodologies and tools for the seamless design of embedded systems are needed. Within our group we have been working towards a complete design methodology for embedded systems

J.-M. Colom and M. Koutny (Eds.): ICATPN 2001, LNCS 2075, pp. 343–362, 2001.
© Springer-Verlag Berlin Heidelberg 2001

for several years now. It covers the whole design flow, reaching from *modeling* of embedded systems via *analysis and partitioning* down to *synthesis*. Analysis methods are used in order to guarantee compliance of the implementation with specified requirements as well as to provide information for an efficient realization of the system. One main problem when analyzing the system is that the modeling of a complex embedded real-time system typically results in a heterogeneous specification. The resulting model consists of several components specified using application-specific languages from the respective domains. With regard to reliability it is not sufficient to analyze each single component on its own. Many functional errors only expose themselves when all individual components work together in the whole context, observed over time. For this reason, one main characteristic of our methodology is that within the critical phases of analysis and synthesis, the system under development is available in one unique formal model. Thereby, our basic model is that of extended Pr/T–Nets.

We have chosen a Petri Net model, since for Petri Net models a lot of analysis methods and tools are already available. Moreover, Petri Nets are well-suited for modeling reactive, local, and concurrent behavior, which is typical for embedded real-time systems. In this context, local behavior means that the global behavior of a model is the sum of its local behaviors. Despite their advantages, a methodology based solely on Petri Nets would gain only little acceptance, since most engineers have their favorite specification language they are used to work with. But as we have already shown in [1], our underlying model of extended Pr/T–Nets is able to integrate other specification languages like diagrams for differential equations or token-based dataflow graphs.

Several tools supporting particular parts of embedded system design are already available within the commercial area. For the purpose of modeling, tools based on StateCharts are very popular. Examples are STATEMATE [2] and the StateChart extension StateFlow of Matlab SIMULINK [3]. Likewise widespreadly used is SDL, for instance within the tool SDT [4]. With regard to verification, an environment based on Statecharts is described in [5]. Within the environment, timing diagrams [6] are used to formulate properties of the specified Statecharts. These can be checked with the help of a symbolic model checker [7]. However, existing commercial tools do not provide a means for a complete seamless design, including for instance a modeling environment as well as support for analysis and verification. Moreover, for commercial tools, attention is rather put to pragmatic aspects, like comfortable generation of code and simple operability, whereas conceptual bases are not well-elaborated.

Some of the first non-commercial approaches towards a complete environment for the design of embedded systems were the projects AutoFocus [8], PTOLEMY [9], POLIS [10], and CodeSign [11]. Just as our approach CodeSign is based on a high-level Petri Net model. In contrast to our work the scope of the project was focused on modeling and analysis. It did not address the generation of code for distributed target architectures as our methodology does. The other approaches mentioned are based on StateCharts and Finite State Machines respectively. They have in common that the underlying formal model is basically synchronous.

However, one main idea behind the POLIS system is that the synchronous subsystems - specified as so called Codesign Finite State Machines - communicate asynchronously. This is similar to our approach. As will be described later on in the paper, we transform suitable parts of a basically asynchronous Petri Net model into synchronous specifications.

The rest of the paper is organized as follows. First, we give an overview of our methodology in Section 2. In Section 3 we describe an approach for integrating synchronous languages into our methodology. Section 4 contains the description of a so called prepartitioning method. Within the prepartitioning information about the Petri Net model is exploited in order to configure a graph partitioning tool. The latter is used for dividing the model into parts, that can be realized separately on the distributed target platform. For the realization of a modeled embedded system we developed a concept, that uses standard services of an operating system to execute extended Pr/T–Nets. A description of this approach is given in section 5.

2 Design Methodology

In this section, we first describe our underlying formal model with a small example. In the second subsection, an overview of the design flow proposed by our methodology is given.

2.1 Formal Model

As a common formal model for our design methodology we chose extended Pr/T–Nets. They are based on the basic model introduced by Genrich and Lautenbach. The tokens are tuples of individuals of basic data types like integer, float or string. Accordingly, the edges are annotated with tuples of variables. Transitions carry conditions as well as actions using these variables and defining them respectively. Figure 1 shows a simple example. Some extensions were made to the standard Pr/T–Net model in order to model heterogeneous systems and real-time systems. The extensions include a means for integration of textual languages, hierarchy, and a concept for graphical abstraction. For details of these extensions we refer to [1].

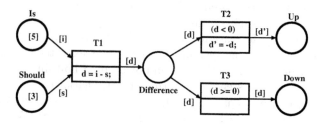

Fig. 1. Pr/T–Net example

A further extension are timing annotations. Extended Pr/T–Nets allow the definition of an *enabling delay* and a *firing delay* for each transition. The enabling delay specifies the time, at which a transition starts firing after having been enabled for some substitution. The time interval is not bound to the substitution, that is the substitution may change during the delay period. The firing delay determines the duration of a transition firing, that is the time between demarking of input places and marking of output places. For both delays an interval (*min, max*) may be specified describing the minimal delay and its maximal value respectively. For the firing delay values with $0 \leq min \leq max < \infty$ are valid, whereas for the enabling delay also $max = \infty$ is allowed.

2.2   Design Flow

Our methodology for the design of parallel embedded real-time systems is based on a design flow first introduced in [12]. The design flow is divided into the three stages *modeling, analysis and partitioning*, and finally *synthesis* (cf. Figure 2). We assume that the process of *modeling* starts with a heterogeneous system consisting of several components with different characteristics. The components may either be modeled using our formalism of extended Pr/T–Nets directly or by using a typical specification language of the respective application domain. Models specified using other languages than Pr/T–Nets are transformed into equivalent extended Pr/T–Nets already within the modeling phase. Thus, at the end of this phase, all components can be coupled together into one hierarchical Pr/T–Net model as shown in [1].

For several specification languages concepts have been developed in order to transform them into a Pr/T–Net model: diagrams for differential equations, token-based dataflow graphs, asynchronous hardware on gate level, Software Circuits, and SDL. The transformation from SDL into Pr/T–Nets was described in [12]. The specification of diagrams for differential equations and token-based dataflow graphs based on Pr/T–Nets was described in [1], the specification of software circuits in [13]. In each case, the specification is based on a Pr/T–Net library for the respective language. Based on the library for diagrams of differential equations, the modeling of hybrid systems is also supported by our methodology [14,15]. As recently shown, even modeling paradigms for autonomous robots may be transformed into extended Pr/T–Nets [16].

After the process of modeling has been finished the design continues with the stage of *analysis and partitioning*. In this phase the extended Pr/T–Net model resulting from the modeling phase (cf. Figure 2: `Hierarchical Pr/T--Net`) is enriched with meta information. The analysis usually starts with simulations in order to check whether the system's behavior is as expected (`Simulation and Animation`). Running simulations is supported by a built-in component of our modeling tool SEA[1] (System Engineering and Animation)

Later on, Petri Net analysis methods are used for verifying functional requirements on the specified system (`Petri Net Analysis`). We therefore try

---

[1] available at: `http://www.c-lab.de/sea/`

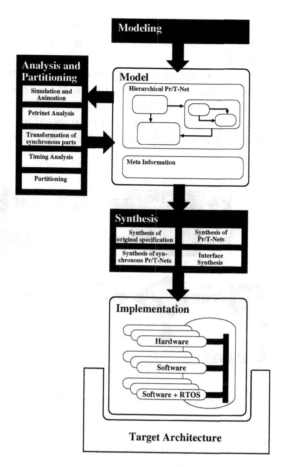

Fig. 2. Design Flow

to determine net properties like reachability of states, deadlock freeness, liveness, and boundness via the skeleton of the extended Pr/T–Net model. From an extended Pr/T–Net we obtain the skeleton by stripping each token of its contents and eliminating the annotations of the transitions and edges. Markings and transition inputs and outputs in the skeleton then involve only numbers of tokens.

The strategy for analyzing Pr/T–Nets based on skeletons is depicted in Figure 3. First, the abstraction of the Pr/T–Net as well as of the investigated property – e.g. the reachability of a certain state – is performed. The application of a Place/Transition Net analysis method yields some information about the respective property. Concerning the reachability analysis, one may get information, whether the abstracted state is reachable in the skeleton and occasionally the firing sequences leading to this state. In some cases, the analysis results are directly valid within the Pr/T–Net. For instance the non-reachability of an abstracted state in a skeleton net implies that the respective state is also not reachable in the original Pr/T–Net. The underlying ideas are similar to those described in [17] and [18].

If the analysis results are not directly applicable, they have to be validated manually. However, the information yielded by the Place/Transition Net analysis may still be useful. Considering the reachability problem it is for instance sufficient to examine only the firing sequences corresponding to those provided by the skeleton analysis. For a detailed description of our analysis approach we refer to [19].

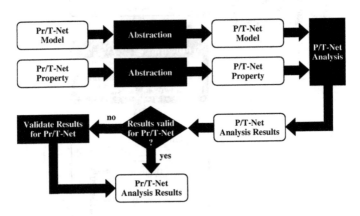

Fig. 3. Skeleton based analysis of Pr/T-Nets

Petri Net analysis methods are not only used for verifying functional requirements. They may also provide information for further steps within our design flow. For instance, the boundness of places is crucial for some synthesis methods as well as for the transformation of synchronous subnets into synchronous languages. The latter becomes obvious in Section 3, in which this step (`Transformation of synchronous parts`) will be further explained. The idea behind the transformation is that appropriate model parts, that is parts with characteristics of synchronous models, are transformed into the world of synchronous languages like ESTEREL, LUSTRE, and SIGNAL [20]. Here, a multiplicity of academical and in particular also commercial design tools exists, which support for instance verification or generation of efficient code for synchronous models.

In order to ensure real-time restrictions we have to check whether distinct components of the system under development react to an input within specified time limits (`Timing Analysis`). For this purpose, a key component within our design methodology is CHaRy [21]. Basically CHaRy is a software synthesis tool for hard real-time systems. It is capable of generating C-code for embedded microcontrollers (target code), whereby a tight estimation of the code's worst case execution time is computed. CHaRy was primarily designed for processing straight-line hard real-time programs, basically a subset of C. For complex parallel systems the timing analysis and synthesis methods provided by CHaRy are combined with Petri Net based methods. The idea for combining them is as follows. In a first step, Petri Net analysis methods yield possible computation paths of a given model. According to these paths, CHaRy afterwards synthesizes target code with a predictable behavior concerning the execution time. For further details we refer to [15]. Timing analysis methods are not only a means for ensuring real-time restrictions, but also for conceiving meta information needed within the partitioning and synthesis phase.

Partitioning is one of the crucial design steps on the way from a high-level Petri Net to an implementation on a distributed target architecture. Before the Petri Net model can be transformed into target code and into dedicated hardware components respectively, it must be divided into corresponding model parts. For partitioning, we use the library PARTY [22] together with the prepartitioning method described in Section 4. PARTY provides min cut methods for computing a balanced partitioning of an undirected weighted graph. This fits well to our problem, if the Petri Net graph partitioning problem is converted into a problem for PARTY as follows. Each unit of the prepartitioned net is mapped to a node. The communication among nodes is specified by arcs. The execution time of the units as well as the communication effort is modeled via weights. In the final partitioning yielded by PARTY, each partition contains at least one unit of the prepartitioned net. Since different partitions may be mapped to different execution units in the final implementation, min cut methods are appropriate in order to keep the communication effort small.

For units of the prepartitioned net, that were merged together during the final partitioning, the respective C-code fragments also have to be merged. Thereby, the units of the final implementation emerge, which may be mapped to a controller or processor of the target hardware. However, this is only one possibility of generating code for one partition of the model, designated as Synthesis of Pr/T-Nets in Figure 2. Whether there are more possibilities, depends on the results of analysis and partitioning. It may for instance be feasible and also reasonable to generate the final implementation of a model part from its original specification in another specification language (Synthesis of original specification). In this case the respective model part was only mapped into a Pr/T–Net in order to enable analysis of the complete model. If synchronous model parts were found during analysis and transformed into synchronous languages as it will be described in Section 3, tools for these languages may be used for the implementation of these parts (Synthesis of synchronous Pr/T-Nets). In any case, for executing the various parts of the implementation, usually a real-time operating system (RTOS) is needed. For executing the units synthesized from Pr/T–Nets, we propose to use a customized RTOS, which is solely dedicated to the execution of Pr/T–Nets (cf. Section 5).

Before finishing the design, it has to be checked whether all real-time constraints will be met by the implementation. If the constraints cannot be ensured, we either repartition the system or replace the software solution for critical parts by special purpose hardware. For the latter synthesizable VHDL–code is generated using the tool PMOSS[2]. The iterative process of repartitioning and/or replacing software by hardware components has to be repeated until an implementation fulfilling all requirements has been found.

3   Mapping Pr/T–Nets to Synchronous Automata

It is generally accepted that Petri Nets are well suited for modeling, analysis, and synthesis of asynchronous systems. Basically they are likewise suitable for synchronous systems. However, during the last decade synchronous languages like ESTEREL, LUSTRE, and SIGNAL [20] have gained great popularity. For these languages a multiplicity of tools supporting the design of embedded systems exist. With respect to the acceptance of our design methodology it therefore appears advisable to transfer suitable model parts into synchronous languages. A linkage to all prevalent synchronous languages is possible via synchronous automata, which were introduced in order to provide a formal model for synchronous languages [23]. Hence, our approach for a linkage to synchronous languages is to determine inherently synchronous parts of a specified Pr/T–Net model and transform them into synchronous automata.

3.1   Synchronous Automata

The operation of synchronous automata is as follows. In each execution cycle an automaton evaluates a set of signals. Depending on these signals as well as on an internal state, output signals are generated. Furthermore the following state is determined. Consequently, the main components of a synchronous automaton are a set of signals, a set of control registers for representing the internal state and a reaction as well as a transition function. The reaction function is used for emitting signals, the transition function for manipulating the control registers.

In Figure 4 a simple example for a synchronous automaton together with a graphical representation is depicted. The element sums up a signal c. In each cycle with c bearing a value, this value is added to the control register h, which is sent to the environment via the signal s. The signal r provides a means for resetting the sum to zero.

The reaction function is specified by statements of the form $s \Leftarrow v$ **if** ϕ, which means that signal s is emitted with value v if the condition ϕ is true. Otherwise the signal is not emitted. The transition function, which is evaluated after the reaction function, is defined by statements like $h \leftarrow v$ **if** ϕ, which means that the value v is assigned to the control register h if the condition ϕ is true. Otherwise the value of h remains unchanged. In order to facilitate initializations a specific signal α is emitted once at the starting point of executing a synchronous automaton. The conditions in the statements are each a logical formula containing con- or disjunctions over signals and control registers. In order to distinguish an instance from an assigned value, we use $V(s)$ for denoting the value of a signal s and $V(h)$ for the value of a control register h.

---

[2] http://www.uni-paderborn.de/cs/ag-hardt/Forschung/Pmoss/pmoss.html

$$S = \{c, r, s\}$$
$$H = \{h\}$$

$s \Leftarrow V(h) + V(c)$ **if** $c \wedge \neg r \wedge \neg \alpha$
$s \Leftarrow 0$ **if** $r \vee \alpha$

$h \leftarrow V(s)$ **if** s

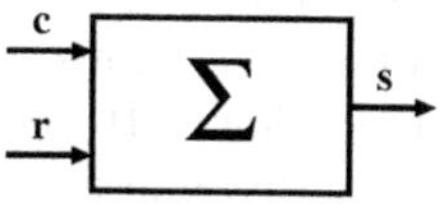

Fig. 4. Example for a synchronous automaton

3.2 Synchronous Pr/T–Nets

We now characterize the class of Pr/T-Nets that are suitable for a transformation into an equivalent synchronous automaton. A Pr/T–Net is called synchronous, if the following conditions are fulfilled:

(1) All transitions of the net have the same timing annotation.
(2) The net is k–safe.
(3) The net is deterministic.

Synchronous languages usually have no explicit notion of physical time. Only simultaneity and precedence between events can be expressed. Accordingly Pr/T–Nets with all transition delays set to zero fulfill one condition to be called synchronous[3]. But the zero delay is no prerequisite. We rather demand in the first condition that all transitions have the same delay. This implies that all enabled transitions fire within the same interval of time which is already a synchronous execution. The second condition guarantees that the state space of the derived synchronous automaton is finite. Finally, deterministic behavior is a basic property of synchronous languages. In the area of Pr/T–Nets deterministic behavior requires that a net contains no input conflicts. It may of course contain structural conflicts, but these have to be solved by the conditions of the involved transitions (cf. Figure 1).

3.3 Transformation

In this subsection we describe the transformation of a synchronous Pr/T–Net into a behavior–equivalent synchronous automaton. Behavior–equivalent means that for each reachable state in the extended Pr/–Net a reachable state in the synchronous automaton exists. One may argue that mapping a finite-state Petri Net to an automaton is a well-known and of course solved problem. However, synchronous automata are not only a particular flavor of finite automata. They have on the contrary several extensions. It seems to be reasonable to exploit these extensions for the realization of appropriate Pr/T–Nets, for instance in order to

---

[3] We thereby suppose a Petri Net execution semantics enforcing each enabled transition to fire immediately.

express parallelism. To our knowledge currently no mapping from a Petri Net-model to synchronous automata exists. We therefore defined the transformation depicted below with the basic ideas described in the following.

The actual marking of a given Pr/T–Net is stored in the actual state of the synchronous automaton, that is in the control registers (cf. (i) in the description below). The initial marking is created by means of statements within the reaction and transition functions, that use the special signal α (ii). Furthermore, the firing of transitions has to be modeled in the synchronous automaton. First of all, the concession to fire is modeled by a signal for each transition. The signal is emitted each time the transition is enabled (iii). In order to model the firing of transitions, additional signals for storing new place markings are needed (i). The new marking of each place after an execution step is stored into these signals by means of the reaction function (iv). At the end of an execution step the place markings are copied from the mentioned signals to control registers by means of the transition function (v). To illustrate the transformation, the Pr/T–Net shown in Figure 1 is used. For a complete description and proof of the transformation see [19].

(i) **Definition of control registers and signals for actual marking**

For each place p with the upper bound k in the extended Pr/T–Net create the control registers $h_{p_1}, \ldots, h_{p_k}$ in the synchronous automaton as well as the signals $s_{p_1}, \ldots, s_{p_k}$.

For the example in Figure 1 the following control registers and signals are created:

$$\{h_{Is_1}, h_{Should_1}, h_{Difference_1}, h_{Up_1}, h_{Down_1}\}$$
$$\{s_{Is_1}, s_{Should_1}, s_{Difference_1}, s_{Up_1}, s_{Down_1}\}$$

(ii) **Initial marking**

For each token m_i within the initial marking $\{m_1, \ldots, m_j\}$ of a place p create the following statement within the reaction function:

$$s_{p_i} \Leftarrow v_{m_i} \text{ if } \alpha$$

Thereby v_{m_i} denotes the value of m_i.

The initial marking in our example is created by:

$$s_{Is_1} \Leftarrow 5 \text{ if } \alpha \qquad and \qquad s_{Should_1} \Leftarrow 3 \text{ if } \alpha$$

(iii) **Signal emission for enabled transitions**

For each transition t create a signal s_t and the following statement within the reaction function:

$$s_t \Leftarrow \lambda \text{ if } c_t$$

Thereby c_t is a logical formula for deciding whether t is enabled.[4]

---

[4] There is no need to assign a value to s_t; hence it is set to λ.

For the example in Figure 1 this step creates three signals with the following statements emitting them:

$$s_{T1} \Leftarrow \lambda \text{ if } h_{Is_1} \wedge h_{Should_1}$$
$$s_{T2} \Leftarrow \lambda \text{ if } h_{Difference_1} \wedge V(h_{Difference_1}) < 0$$
$$s_{T3} \Leftarrow \lambda \text{ if } h_{Difference_1} \wedge V(h_{Difference_1}) \geq 0$$

(iv) **Signal emission for token resulting from transition firing**
For each token m_i in the marking $\{m_1, \ldots, m_k\}$ of a place p with the upper bound k, the preset $\{t_1, \ldots, t_m\}$, and the postset $\{t_1, \ldots, t_n\}$ create the following statements within the reaction function:

$$s_{p_i} \Leftarrow V(h_{p_i}) \text{ if } \bigwedge_{j=1}^{m} \neg s_{t_j} \wedge \bigwedge_{j=1}^{n} \neg s_{t_j}$$

$$\text{for each } j \in \{1, \ldots, m\} : s_{p_i} \Leftarrow v_{t_j} \text{ if } s_{t_j} \wedge \bigwedge_{j=1}^{n} \neg s_{t_j}$$

Due to the first statement, the token m_i remains unchanged, if no transition within the pre- and postset of p fires. The creation of tokens by firing of transition t_j is realized with the other statements.[5]

For the place $Difference$ in the example the following statements are created by this step:

$$s_{Difference_1} \Leftarrow V(h_{Difference_1}) \text{ if } \neg s_{T1} \wedge \neg s_{T2} \wedge \neg s_{T3}$$
$$s_{Difference_1} \Leftarrow V(h_{Is_1}) - V(h_{Should_1}) \text{ if } s_{T1} \wedge \neg s_{T2} \wedge \neg s_{T3}$$

(v) **Storing of new marking**
For each token m_i in the marking $\{m_1, \ldots, m_k\}$ of a place p with the upper bound k create the following statements within the transition function:

$$h_{p_i} \leftarrow V(s_{p_i}) \text{ if } s_{p_i}.$$

The new marking of the place $Difference$ in the example (which may be its old marking) is set using the following statement within the transformation function.

$$h_{Difference_1} \leftarrow V(s_{Difference_1}) \text{ if } s_{Difference_1}.$$

4   Partitioning

In this section we describe an approach for the fine-granular partitioning of Pr/T–Nets. As already described in Section 2, this method is used as a prepartitioning in the overall design flow. The step of clustering the partitions resulting

---

[5] The demarking of places due to firing of transitions is realized implicitly without further effort, since a signal for a token will only be emitted, if the token is created or remains unchanged.

from prepartitioning into the units of the final implementation is left to the standard graph partitioning tool PARTY. By all means the prepartitioning aims at identifying model parts that are well-suited for parallel execution. Our main point of concern thereby is the handling of conflicts. Encapsulating them into partitions avoids that the conflict handling is distributed over several partitions. This could lead to communication overhead (cf. Figure 5b), since different partitions may be mapped to different nodes of the target architecture. Due to the handling of conflicts, our method ensures that instead each communication between partitions can be realized with a simple send and forget mechanism (cf. Figure 5a). As can be seen in the figure, our algorithm cuts Pr/T–Nets at places. This is the natural way for partition Pr/T–Nets, since transitions are their active elements, whereas places are just data buffers. Cutting a place p stands for dividing the net into two partitions, that are disjoint except for p, which is duplicated and part of both partitions.

Besides encapsulating conflicts the prepartitioning also aims at a fine-granular partitioning in order to provide a wide scope for the graph partitioning tool used for clustering the resulting partitions. Unfortunately, a straightforward partitioning, that encapsulates conflicts, usually leads to large partitions. For this reason, the prepartitioning first performs a transformation of the given Petri Net, which does not change the semantics of the net, but facilitates the partitioning.

In order to specify our ideas more precisely we first need some basic definitions.

Let N be an extended Pr/T–Net and t a transition of N.

(1) t is called exclusively triggered iff
   - The pre–set of t contains at most one place.
   - The post–set of each place in the pre–set of t contains one transition (namely t).
(2) t is called exclusively producing iff
   - The post–set of t contains at most one place.

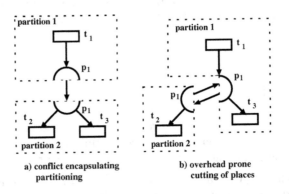

a) conflict encapsulating
partitioning

b) overhead prone
cutting of places

Fig. 5. Partitioning alternatives for an input conflict

– The pre–set of each place in the post–set of t contains one transition (namely t).

Each Pr/T–Net can easily be transformed into an equivalent one containing only transitions that are either exclusively triggered or exclusively producing. The transformation is done by splitting transitions as illustrated in Figure 6. In

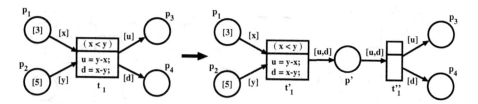

Fig. 6. Transition splitting

this figure the transition t_1 is split into the exclusively producing transition t_1' and the exclusively triggered transition t_1''. As can be seen, the annotations of t_1 remain unchanged at t_1'. The edges between t_1' and t_1'' are annotated with a tuple containing the output variables of t_1.

By applying this transformation to each transition of a Pr/T–Net our prepartitioning algorithm ensures, that no transition is involved in both an input conflict and output conflict. Thereby it becomes easy to completely cover the net with well-formed subnets, which are characterized in the next definition. In the definition $Part(N, T')$ denotes the subnet of a given net N, that contains the transitions T', the places within the pre–set and the post–set of these transitions, and the edges between these sets of nodes. The annotation of all subnet elements is left unchanged.

Let N be an extended Pr/T–Net and T' a subset of its transitions.

(1) The subnet $Part(N, T')$ is called output conflict complete iff
   – Each transition $t \in T'$ is exclusively triggered.
   – For each place p within the post–set of T' the pre–set of p is a subset of T'.
(2) The subnet $Part(N, T')$ is called input conflict complete iff
   – Each transition $t \in T'$ is exclusively producing.
   – For each place p within the pre–set of T' the post–set of p is a subset of T'.

For a Pr/T–Net containing only exclusively triggered and exclusively producing transitions, a partitioning into subnets with the just defined characteristics is computed in a straightforward way. Until all transitions are assigned to a partition, the algorithm fetches one of the not processed transitions and builds a new partition with all (directly or indirectly) conflicting transitions. The resulting well-formed subnets are the units produced by our prepartitioning step. An example for applying our prepartitioning method to a Pr/T–Net is depicted in Figure 7. The transitions t_1 and t_2 of the original net have been split. In the

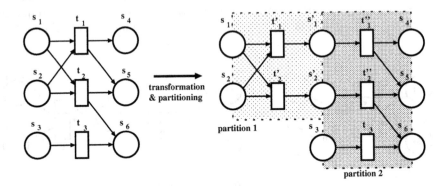

Fig. 7. Partitioning of a Pr/T–Net

resulting net partition 1 is an input conflict complete subnet whereas partition 2 is output conflict complete.

With the example the main advantage of our partitioning technique becomes visible. The interface places belonging to both partitions have the property that all incoming edges originate from the same partition and all outgoing edges lead to the same partition. This is generally true for interface places. If for instance a place p has incoming edges from different transitions these transitions will have an output conflict and are contained in the same partition. If on the other hand p has outgoing edges to different transitions these have an input conflict and again are contained in the same partition. Due to this property of the potential interface places, communication can be realized with a simple send and forget mechanism, as it was promised in the introduction of this section. As long as the partitions computed by our algorithm are considered as minimal ones it is guaranteed that no additional overhead for a complicated protocol is necessary to realize the communication between partitions. Furthermore, the structure of the partitions is quite simple. This is due to the fact that no partition contains both input conflicts and output conflicts and the partitions contain no inner places (there is no sequence of more than one transition).

5   Pr/T–Net Engine

In the following, we describe our approach for the execution of Pr/T–Nets using standard services of an operating system, which is part of the software implementation of extended Pr/T–Nets on a network of interconnected microcontrollers and processors respectively. More precisely, the approach accompanies the step called Synthesis of Pr/T-Nets in Figure 2. Since it is part of the synthesis stage within the overall design flow, the approach described in this section presumes the steps performed in the second stage, in particular a partitioning of the specified model and a mapping of the partitions to nodes of the target architecture.

A canonical strategy for generating a software implementation of a Pr/T–Net is as follows. For each transition, code for its execution is produced. Depending on the semantics of our underlying Petri Net model as well as on the structure of the specific model additional simulation code is produced. For each execution step, the latter code (simulator) selects one of the currently enabled transitions, initiates its execution and determines modifications on the set of enabled transitions afterwards. Both the transition code and the simulation code are usually produced in one of the usual programming languages (e.g. C, C++, Java). From this code a program for the respective execution platform is produced.

The approach just described, which is well-established for the automatic generation of simulation code from Petri Net models, leads to some overhead in the implementation of the system. Some functions performed for the execution of Pr/T–Nets are also inherently present in the operating system. An example is the scheduling of enabled transitions by the simulator and the scheduling of executable threads by the operating system respectively. This existence of two schedulers, which cannot be avoided in the area of workstations, is dispensable when implementing embedded systems. Here, the Pr/T–Net execution is the only task of the operating system on each microcontroller. Hence, the scheduler can be used directly for scheduling transitions, reducing the overhead and thereby the costs of the execution. These considerations lead to the main idea of our software synthesis approach. The execution of Pr/T–Nets is integrated into the operating system, that is we realize a mapping of Pr/T–Net functionality onto OS services as depicted in Figure 8. When realizing this idea, which was already proposed in [24], mainly two questions arise: how may Pr/T–Net functionalities be mapped to OS services, and how have these services to be adapted for Pr/T–Net execution.

The realization of the presented approach is based on our customizable library based real-time operating system kit DREAMS. It seems to be a drawback that we are using a proprietary solution instead of a standard RTOS. But realizing our concepts is not possible based on a fixed RTOS kernel. We rather need a tool allowing for the flexible generation of an RTOS, that is customized to a specific application. However, a trend towards RTOS kits like DREAMS seems to be established in the field of operating systems research, since several approaches emerged during the last years, e.g. [25,26].

DREAMS [27] is a construction set for real-time operating systems. From the DREAMS library an optimized and to the requests of the Pr/T–Net highly adapted RTOS can be derived by customization. After selection, configuration, and compilation of appropriate parts of DREAMS an RTOS is created out of DREAMS's basic components. The thereby defined RTOS may support memory management, device access, multithreading (scheduling), resource allocation, mutual exclusion, synchronization, and communication for embedded real-time

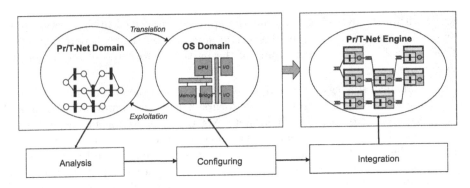

Fig. 8. Integration of Pr/T–Net Domain and OS Domain

applications. DREAMS is an object-oriented system. The basic components are written completely in C++ (except some very few lines in assembler for context switching). The class structure of DREAMS is the basis for its customization. The configurable classes are defined in terms of so called *Skeletons*. For providing configuration facilities C++ was extended by a *Skeleton Customization Language* (SCL). Customization within DREAMS is applied during compilation of a concrete execution platform (kernel-like virtual machine). Therefore SCL is translated into preprocessor commands which are finally handled in the compilation process. By applying customization, several class properties may be configured, for instance the base classes and the components of a class.

In the following, we first deal with the question, how Pr/T–Net functionalities may be mapped to OS services, and afterwards how these services have to be adapted for Pr/T–Net execution.

The active elements of Pr/T–Nets are the transitions, whereas in OS's processes are the active units. Moreover, the lifecycle of a transition with being disabled, enabled, firing, and again enabled or disabled is similar to that of processes, which is depicted in Figure 9. Processes change their states between blocked (waiting for a message), ready (ready for execution because a message was received), and running (the process is executing). Similarly, the token flow in a Pr/T-Net can be compared with message passing in a process system. Thus we identify transitions with processes, tokens with messages and, consequently, places with mailboxes.

The required global instance (simulator) that controls the execution order of transitions can be identified with the process scheduler (with its ready queue) of a multi-tasking operating system. Nevertheless, the operating system requires some special adaptions according to the special semantics of the (transition) process system. These adaptions especially include changes in the scheduler function which selects the next ready process for execution, the scheduling policy itself, and the handling of mailboxes.

The execution cycle of a periodic process consuming and producing messages usually is as follows. A process is blocked when trying to receive a message from

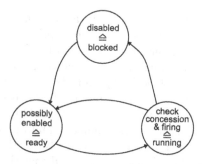

Fig. 9. Assignment of transition and process states

its incoming mailbox. Once a message arrives, the operating system sets the state of the process from *blocked* to *ready*. Therefore, the process is entered into the ready queue, from which the process scheduler fetches the processes to execute. Having been started by the scheduler, the process changes into the *running* state, consumes the available message, and produces output messages. These are sent to the mailboxes of other processes awaiting the data. After doing this the process starts a new cycle. The sending procedure is handled by the operating system, which means the data are inserted into the mailbox and the awaiting process is set to *ready* state.

There are some differences to the execution cycle of transitions in Pr/T–Nets. In contrast to a process which is waiting for the arrival of a single message in its mailbox, a transition can wait for multiple tokens on several input places. Furthermore, a process has exclusive access to its mailbox, whereas token on a place may be consumed by several transitions. Finally, the receiving of a message is already a sufficient condition for a process to become ready. The tokens available for a transition must fulfill further conditions, before the transition can fire.

To embrace these differences the following adaptions have to be implemented. First, mailboxes must be separated from the processes. Therefore, a mailbox ID must be specified in order to receive a message. Second, it must be possible to await for the reception of messages in different mailboxes. The state of the awaiting process may only change from *blocked* to *ready* if a message is available in every mailbox. Furthermore the message's values must fulfill the enabling condition of the respective transition. In order to achieve this it must be possible to read the values contained in messages without consuming them, which is no standard functionality of operating systems. The step of reading and consuming all messages must be atomic because otherwise side effects violating the Petri Net model occur. This can easily and without any overhead be achieved if a *run-to-completion* or *co-operative* policy is applied to the scheduling mechanism instead of a preemptive version.

All required changes have been implemented in an RTOS using DREAMS. An appropriate configuration of the class library was created so that all prerequisites are fulfilled. In order to reveal the benefits of this approach we compared it to the traditional implementation with a simulator process on top of an operating system. The operating system was a preemptive 'normal' configured RTOS which uses semaphores for serialization. As an input for both implementations we used 41 Pr/T–Nets each consisting of one transition. They were connected in a cycle. The execution of both experiments was done on a single PowerPC processor running at 40 MHz. The results of their execution confirmed our expectations. The time used for a transition cycle could be reduced from 526.38 μs to 270.90 μs. The time used for the simulator and operating system to switch from one transition per net (process respectively) to the next one was reduced from 665.60 μs to 215.40 μs. It should be stated that the reduction of execution time highly depends on the net structure. But experiments with other nets have shown that there is an overall benefit of using our approach. In the worst case the reduction only reached 30%. In the above mentioned experiment there is a reduction up to 59% for the execution of a distributed Pr/T–Net using our approach.

6   Conclusion

In this paper we presented our recent work towards a complete methodology for the Pr/T–Net based design of complex embedded real-time systems. We first described an approach for mapping Pr/T–Nets to synchronous languages. It enables us to exploit the numerous design methods and tools available for synchronous languages also for the design based on Pr/T–Nets. Secondly a technique was presented to configure a standard graph partitioning tool in a way that Petri Net graphs are partitioned with respect to their execution semantics. This is valuable for instance in order to reduce the communication overhead in a distributed implementation of the partitioned system. Finally a new approach for a software implementation of high-level Petri Nets was presented. The main idea is to integrate the execution of a Pr/T–Net into an RTOS, which leads to an efficient implementation of a specified model on a target architecture of interconnected microcontrollers. Up to now, we realized the basic implementations for our approaches, integrated them into our methodology, and made several qualitative investigations, whether the developments are valuable for our design methodology. Currently we are performing additional quantitative experiments in order to measure the effect of our techniques on factors like code-size or performance for several different models. In particular we thereby investigate the potential of some improvements we have in mind for the presented approaches. An intended enhancement of the prepartitioning method is for instance to take into consideration not only structural properties of the given Pr/T–Net models, but also information about their dynamic behavior, that can be provided by analysis methods. Similarly, in the future analysis results shall be used to configure the Pr/T–Net engine generated for the execution of a specified model as described in this paper. If for instance due to analysis results a place is known to be safe, a static mailbox could be generated for this place, which is much more efficient than the generic implementation of a mailbox.

In order to evaluate our methodology we have also studied several application examples for the last years. The methodology was for instance used for modeling a series hybrid drive for vehicles [15] as well as an anti blocking system for a car [13]. Our main focus within these projects was on modeling, simulation, and timing analysis. In [12] the modeling and analysis of a decentralized traffic management system using our methodology was presented. The analysis in particular included the verification of essential system properties like deadlock-freeness. More recently, we applied our tools for designing a control for a small mobile robot [16]. This robot, the so called C-LAB Pathfinder[6], is currently our main target platform. It is equipped with several standard microcontrollers and a communication system, that are widespreadly used in the area of embedded systems. This platform enables us to evaluate our tools in the complete design flow reaching from modeling down to the implementation on embedded controllers.

References

1. B. Kleinjohann, E. Kleinjohann, and J. Tacken. The SEA Language for System Engineering and Animation. In *Applications and Theory of Petri Nets*, LNCS 1091, pages 307–326. Springer Verlag, 1996.
2. D. Harel. STATEMATE: A Working Environment for the Development of Complex Reactive Systems. In *IEEE Transactions on Software Engineering*, volume 16(4), pages 403–414, 1990.
3. The MathWorks Inc. *Stateflow User's Guide*, 1999.
4. Telelogic AB. *SDT 3.1 Reference Manual*, 1996.
5. W. Damm, B. Josko, H. Hungar, and A. Pnueli. A compositional real-time semantics of STATEMATE designs. In *Proceedings of the COMPOS'97*, volume 1536 of *LNCS*. Springer-Verlag, 1998.
6. W. Damm, B. Josko, and R. Schloer. Specification and verification of VHDL-based system-level hardware designs. In E. Börger, editor, *Specification and Validation Methods*, pages 331 – 409. Oxford University Press, UK, 1995.
7. T. Filkorn. Applications of Formal Verification in Industrial Automation and Telecommunication. In *Proc. of the Workshop on Formal Design of Safety Critical Embedded Systems*, 1997.
8. F. Huber, B. Schätz, A. Schmidt, and K. Spies. AutoFocus - A Tool for Distributed Systems Specification. In *Proceedings FTRTFT'96 - Formal Techniques in Real-Time and Fault-Tolerant Systems, P. 467-470*. Springer Verlag, LNCS 1135, 1996.
9. J. Davis II et al. Overview of the Ptolemy Project. ERL Technical Report M99/37, Dept. EECS, University of California, Berkeley, CA 94720, July 1999.
10. F. Balarin et al. *POLIS A design environment for control-dominated embedded systems, version 0.4.* University of California, Berkeley, http://www-cad.eecs.berkeley.edu/ polis/, 1999.
11. R. Esser. *CodeSign, Version 1.0, Concepts and Tutorial.* Computer Engineering and Networks Laboratory, Swiss Federal Institute of Technology, Zürich, 1996.
12. B. Kleinjohann, J. Tacken, and C. Tahedl. Towards a Complete Design Method for Embedded Systems Using Predicate/Transition-Nets. In *Proc. of the XIII IFIP WG 10.5 Conference on Computer Hardware Description Languages and Their Applictions (CHDL-97)*, pages 4–23, Toledo, Spain, April 1997. Chapman & Hall.

---

[6] http://www.c-lab.de

13. C. Rust, F. Stappert, P. Altenbernd, and J. Tacken. From High–Level Specifications down to Software Implementations of Parallel Embedded Real–Time Systems. In *Proceedings of DATE 2000*, Paris, France, March 2000.

14. G. Lehrenfeld, R. Naumann, R. Rasche, C. Rust, and J. Tacken. Integrated Design and Simulation of Hybrid Systems. In *International Workshop on Hybrid Systems: Computation and Control*, Berkeley, California, April 1998.

15. C. Rust, J. Stroop, and J. Tacken. The Design of Embedded Real-Time Systems using the SEA Environment. In *Proc. of the 5th Annual Australasian Conference on Parallel And Real-Time Systems*, Adelaide, Australia, September 1998.

16. B. Kleinjohann, C. Rust, and J. Tacken. Entwurf von autonomen Systemen mit High Level Petrinetzen. In *16. Fachgespräch Autonome Mobile Systeme (AMS'2000)*, Karlsruhe, Germany, September 2000.

17. G. Findlow. Obtaining Deadlock-Preserving Skeletons for Coloured Nets. In *International Conference on Applications and Theory of Petri Nets*, volume 616 of *Lecture Notes in Computer Science*, Sheffield, UK, 1992. Springer-Verlag.

18. J. Vautherin. Parallel system specifications with Coloured Petri Nets and algebraic specifications. In G. Rozenberg, editor, *Advances in Petri Nets 1987*, volume 266 of *Lecture Notes in Computer Science*. Springer-Verlag, 1987.

19. J. Tacken. *Eine Pr/T-Netz basierte, durchgängige Entwurfsmethodik für eingebettete Realzeitsysteme*. PhD thesis, Universität Paderborn, FB 17, 2000.

20. N. Halbwachs. *Synchronous programming of reactive systems*. Kluwer Academic Pub., 1993.

21. P. Altenbernd. CHaRy: The C-LAB Hard Real-Time System to Support Mechatronical Design. In *Proc. of the International Conference on Engineering of Computer Based Systems (ECBS-97)*, Monterey, California, March 1997.

22. R. Preis and R. Diekmann. PARTY - A Software Library for Graph Partitioning. In B.H.V. Topping, editor, *Advances in Computational Mechanics with Parallel and Distributed Processing*, pages 63 – 71. Civil-Comp Press, 1997.

23. Olivier Maffeïs and Axel Poigné. Synchronous automata for reactive, real-time and embedded systems. *Arbeitspapiere der GMD, No. 967*, January 1996.

24. C. Böke, M. Hübel, F. J. Rammig, and C. Rust. Zero-Overhead Pr/T-Net Execution. In *Proc. of the SCS International European Simulation Multi-Conference (ESM)*, Gent, Belgium, May 23-26 2000.

25. W. Schröder-Preikschat, U. Spinczyk, F. Schön, and O. Spinczyk. Design rationale of the pure object-oriented embedded operating system. In *Proc. of the Workshop on Distributed and Parallel Embedded Systems*. IFIP, 1999.

26. B. Ford, K. Van Maren, J. Lepreau, S. Clawson, B. Robinson, and J. Turner. The Flux OS Toolkit: Reusable Components for OS Implementation. In *Proc. of the 6th Workshop on Hot Topics in Operating Systems*, May 1997.

27. C. Ditze. *Towards Operating System Synthesis*. Ph.d. thesis, Paderborn University, FB17, Paderborn, Germany, 1999.

Rewriting Logic and Elan: Prototyping Tools for Petri Nets with Time

L. Jason Steggles

Department of Computer Science, University of Newcastle, UK
l.j.steggles@ncl.ac.uk
http://www.cs.ncl.ac.uk/people/l.j.steggles/

Abstract. Rewriting logic (RL) is an extension of standard algebraic specification techniques which uses rewrite rules to model the dynamic behaviour of a system. In this paper we consider using RL and the associated support tool Elan as an environment for rapidly prototyping and analysing Petri nets with time. We link these algebraic tools to the existing Petri net tool PEP which we use to provide a user–friendly front end to our framework. Our flexible approach allows the wide range of possible time extensions presented in the literature to be investigated and thus overcomes one of the major drawbacks of the current hardwired tools. We demonstrate our ideas by considering *time Petri nets* in which transitions are associated with a time interval within which they can fire. The flexibility of our approach is illustrated by modelling a range of semantic alternatives for time Petri nets taken from the literature.

1   Introduction

The theory of Petri nets (see for example [16], [18] and [15]) provides a graphical notation with a formal mathematical semantics for modelling and reasoning about concurrent, distributed systems. One shortcoming of basic Petri nets is that they do not provide any insight into the time behaviour of systems. For real–time systems such as protocols with timeouts such timing information is extremely important (see for example [21]). To address this a variety of Petri net extensions with time have been proposed in the literature (see the surveys [4] and [19]). One problem however is the sheer number of different semantic interpretations that can be made: timing information can be assigned either to places, transitions, arcs or tokens; time durations or intervals can be used; specified time can represent a period of inhibition or a period when an activity can occur. The tools currently available are unable to cope with this wide range of choices and tend to be hardwired to one specific time approach. This makes investigating different time extensions extremely difficult.

In this paper we consider using *rewriting logic* (RL) (see [13]) and the associated support tool *Elan* (see [7]) to rapidly prototype and analyse Petri nets with time. RL is an algebraic formalism that extends the standard algebraic specification techniques by allowing the dynamic behaviour of systems to be modelled using rewrite rules. The idea in RL is to define the static and functional aspects

J.-M. Colom and M. Koutny (Eds.): ICATPN 2001, LNCS 2075, pp. 363–381, 2001.

of a system using a standard algebraic specification and to then view terms over this specification as system states. Rewrite rules are then used to specify the dynamic transitions between these states.

As a case study we consider prototyping and analysing *time Petri nets* in which an interval for firing is associated with each transition (see [12]). We present an RL model for this time extension and consider how the support tool Elan can be used to simulate and analyse the RL model. We consider what it means for our model to be correct and provide a formal argument to show that the model we have given *correctly* simulates time Petri nets. Even for this standard approach to extending Petri nets with time there are a range of possible semantic interpretations that can be considered (see [19]). We illustrate the flexibility of our approach by considering how to adapt our model to represent some of these semantic alternatives.

In order to make our approach practical and take advantage of existing Petri net tool support we have linked our approach with the widely used PEP tool (see [3]). We use PEP as a front end to our framework, using it to create the initial Petri net graphs and to animate the firing sequences that result from our RL simulations using Elan. An overview of how we integrate the two tools Elan and PEP is given in Figure 1. This work illustrates how existing modelling tools can be combined to address new problems.

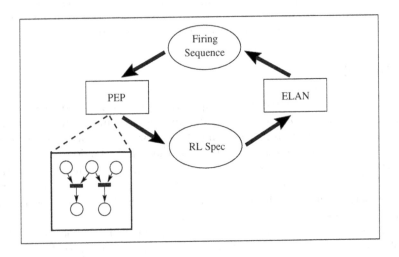

Fig. 1. Integration of Elan and PEP

The paper is organized as follows. In Section 2 we introduce the essential background definitions and results concerning RL and the support tool Elan. In Section 3 we introduce *time Petri nets* and consider how to model and analyse such nets using RL and Elan. This case study demonstrates how different semantic choices can be explored and we present a correctness argument to show that our model correctly simulates a time Petri net. Finally in Section 4 we present some concluding remarks.

We note that we assume the reader is familiar with the basic notation and definitions of Petri nets (see for example [16], [18] and [15]).

2   Background: Rewriting Logic and Elan

In this section we briefly present the background material on rewriting logic (RL) and its associated support tool Elan needed for this paper. We present a small illustrative example of how RL can be used to model simple P/T nets (see [14]).

2.1   Rewriting Logic

Rewriting logic (RL) is an extension of standard algebraic specification techniques which is able to model dynamic system behaviour. In RL the functional and static properties of a system are described by a standard algebraic specification, whereas the dynamic behaviour of the system is modelled using rewrite rules. Terms over a given signature Σ represent the global states of a system and rewrite rules model the dynamic transitions between these states. We now present a brief introduction to RL; for a more detailed introduction to RL see [13].

A standard algebraic specification (Σ, E) is a pair consisting of a signature Σ and a set of equations E over Σ and a set of variables X (see for example [10] and [11]). In RL such a specification is seen as defining the states of a system with each equivalence class (with respect to the equations E) of terms $[t]_E$ being a particular state. We can then define rewrite rules $t \longrightarrow t'$, for terms t, t' over Σ and variables X, which define the dynamic transitions that can occur between states.

Definition 1. *A Rewriting logic specification Spec $= (\Sigma, E, R)$ is a triple consisting of: an algebraic signature Σ which defines a set of sorts S and a set of function symbols Σ; a set of equations E over Σ and a set of variables X; and a set of (labelled) rewrite rules R over Σ and X.* $\square$

As an example of an RL specification let us consider how we might model the simple Petri net depicted in Figure 2 (we follow the approach given in [14] and [13]). The basic idea will be to model a token being present on a place p_i by a constant $p(i)$. A marking can then be modelled as a multi-set of these constants, for example the marking which contains two tokens on place p_1, one token on place p_3 and three tokens on place p_4 could be represented by the term

$$p(1) \otimes p(1) \otimes p(3) \otimes p(4) \otimes p(4) \otimes p(4),$$

where $\otimes$ is the symbol used to denote multi-set union. Note that since places p_2 and p_5 don't contain any tokens they do not appear in the multi-set. Each transition will be represented by a rewrite rule which consumes tokens and produces

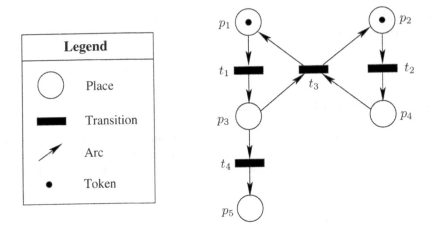

Fig. 2. A simple example of a Petri net

new tokens. For example, transition t_3 would be modelled by the following rule:

$$p(3) \otimes p(4) \longrightarrow p(1) \otimes p(2).$$

The complete RL specification $SpecPN = (\Sigma, E, R)$ for the Petri net depicted in Figure 2 is defined below.

(i) **Signature** Σ: Let $S = \{pnet\}$ be a sort set and let Σ be an S-sorted signature which contains the following function symbols:

$$p(1), p(2), p(3), p(4), p(5) : pnet,$$

$$empty : pnet, \quad \otimes : pnet\ pnet \rightarrow pnet.$$

(ii) **Equations** E: Define the set of equations E to contain the following three equations which axiomatize the properties of $\otimes$:

$$m1 \otimes empty = m1, \quad m1 \otimes m2 = m2 \otimes m1,$$

$$m1 \otimes (m2 \otimes m3) = (m1 \otimes m2) \otimes m3.$$

Note that these equations allow the elements within a multi-set to move around and that the rewrite rules defined below will be applied modulo these equations.

(iii) **Rewrite rules** R: Finally, define the set of rewrite rules R to contain the following rules which axiomatize the transitions in the Petri net:

$$p(1) \longrightarrow p(3), \qquad p(2) \longrightarrow p(4),$$

$$p(3) \otimes p(4) \longrightarrow p(1) \otimes p(2), \qquad p(3) \longrightarrow p(5).$$

Let $p(1) \otimes p(2)$ be the multi-set representing the initial marking in Figure 2. Then the firing sequence $t_1\ t_2\ t_3$ in the Petri net can be simulated by the following sequence of rewrites:

$$p(1) \otimes p(2) \longrightarrow p(3) \otimes p(2) \longrightarrow p(3) \otimes p(4) \longrightarrow p(1) \otimes p(2).$$

2.2   The Support Tool *Elan*

A number of advanced support tools have been developed to allow RL specifications to be simulated and analysed including Maude (see the tutorial available from the Maude website[1]), Elan [7] and CafeObj [9]. These tools provide the means of performing fast AC rewriting, modular structuring mechanisms, and powerful user definable rewrite strategies. We have chosen to use the Elan system here (see [6], [7] and the tools web site[2]). This choice was motivated mainly by the author's experience with the tool and the fact that Elan has a simple built-in strategy language.

As an example of the syntax of Elan we formulate the RL specification presented in the previous section as an Elan module. Note that we use the ASCII symbol # to represent multi-set union $\otimes$, E the empty multi-set and for generality replace the constants $p(1), \ldots, p(5)$ by a function symbol $p : int \rightarrow pnet$.

```
module PNet
   import global int;      end
   sort pnet;              end
   operators global
      E                    : pnet;
      @ # @                : (pnet pnet) pnet    (AC) ;
      p(@)                 : (int) pnet;
   end
   rules for pnet
      pn : pnet;
    global
    []   pn # E => pn                                    end
    []   p(1)  => p(3)                                   end
    []   p(2)  => p(4)                                   end
    []   p(3) # p(4)  => p(1) # p(2)                     end
    []   p(4)  => p(5)                                   end
   end
end
```

The symbol @ is used to denote the position of an argument to a function symbol allowing a mix-fix notation. Each rule can be given an optional label by including text within the square brackets at the start of the rule (all the rules above are unlabelled). The equations of the RL specification have not been explicitly given in the Elan specification. Instead for reasons of efficiency the built-in associativity and commutativity facility of Elan has been used by flagging # as an (AC) operator.

One key feature of Elan is that it provides a built-in *strategy language* for controlling the application of rewrite rules. It allows the user to specify

[1] http://maude.csl.sri.com/tutorial/
[2] http://elan.loria.fr/

a general order in which rewrite rules are to be applied and the possible choices that can be made. The result of applying a strategy to a term is the set of all possible terms that can be produced according to the strategy. A strategy is said to fail if, and only if, it can not be applied (i.e. produces no results). The following is a brief overview of Elan's *elementary strategy language*:

(i) **Basic strategy**: 1   Any label used in a labelled rule [1] t => t' is a strategy. The result of applying a basic strategy 1 is the set of all terms that could result from one application of any rule labelled 1. The strategy is said to fail if, and only if, no rule labelled 1 can be applied.

(ii) **Concatenation strategy**: s1;s2   The concatenation strategy allows two strategies s1 and s2 to be sequentially composed, i.e. s2 is applied to the results from s1. The strategy fails if, and only if, either s1 or s2 fails.

(iii) **Don't know strategy**: dk(s1,...,sn)   The don't know strategy takes a list of strategies s1,...,sn and returns the union of all possible sets of terms that can result from these strategies. This strategy fails if, and only if, all the strategies s1,...,sn fail.

(iv) **Don't care strategy**: dc(s1,...,sn)   The don't care strategy takes a list of strategies s1,...,sn and chooses nondeterministically to apply one of these strategies si which does not fail. Thus the strategy can only fail if all of s1,...,sn fail. The strategy dc one(s1,...,sn) works in a similar way but chooses a *single* result term to return. One final variation is the strategy first(s1,...,sn) which applies the first successful strategy in the sequence s1,...,sn.

(v) **Iterative strategies**: repeat*(s)   The repeat*(s) strategy repeatedly applies s, zero or more times, until the strategy s fails. It returns the last set of results produced before the strategy s failed. The repeat+(s) version works in a similar way but insists that s must be successfully applied at least once.

Elan also provides the so called *defined strategy language* which extends the above elementary language by allowing recursive strategies. For a detailed discussion of Elan's strategy language see [7]. Examples of the application of Elan's strategy languages will be presented in the case study that follows.

3   Modelling Time Petri Nets Using RL

In this section we consider modelling and analysing *time Petri nets*, a time extension in which transitions are associated with firing intervals. We begin by introducing the general ideas and basic definitions of time Petri nets. We then consider how to model time Petri nets using RL and their analysis using the Elan tool. We conclude by presenting a correctness argument for our RL model and by demonstrating the flexibility of our approach by modelling a range of semantic alternatives for time Petri nets.

3.1   Time Petri Nets

Time Petri nets were introduced in [12] and have since become one of the most popular Petri net time extensions (see for example [2] and [1]). Time Petri nets are based on associating a firing interval $[e, l]$ with each transition, where e is referred to as the *earliest firing time* and l is referred to as the *latest firing time*. The idea is that a transition is only allowed to fire if it has been continuously enabled for at least e units of time and is forced to fire once it has been enabled for l units of time (unless a conflicting transition fires first). Firing a transition (i.e. consuming enabling tokens and producing output tokens) is assumed to be instantaneous.

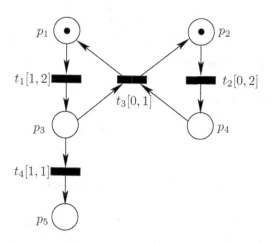

Fig. 3. Example of a time Petri net

As an example consider the time Petri net depicted in Figure 3. In this example both t_1 and t_2 are enabled but only transition t_2 can fire since its earliest firing time is zero. Transition t_1 needs to be enabled for at least 1 clock cycle before it can fire. Both transitions will be forced to fire when they have been enabled for 2 clock cycles.

For simplicity, we assume we are dealing with discrete intervals (see [1]) and let $\mathbf{I} = \{[e, l] \mid e \in \mathbf{N}, l \in \mathbf{N} \cup \{\infty\}, e \leq l\}$. Note that a latest firing time of ∞ indicates that a transition will never be forced to fire. We can formally define a time Petri net as follows.

Definition 2. *A Time Petri Net* $TPN = (P, T, F, m_0, SI)$ *is a 5-tuple where:*
$P = \{p_1, p_2, \ldots, p_n\}$ *is a finite set of places;*
$T = \{t_1, t_2, \ldots, t_k\}$ *is a finite set of transitions, such that* $P \cap T = \emptyset$;
$F \subseteq (P \times T) \cup (T \times P)$ *is a set of arcs (called a flow relation);*
$m_0 : P \rightarrow \mathbf{N}$ *is the initial marking of the Petri net; and*

$SI : T \rightarrow \mathbf{I}$ *is a static interval function that assigns a firing interval to each transition.*   □

Let $TPN = (P, T, F, m_0, SI)$ be a time Petri net. When TPN is clear from the context we let $Eft(t)$ and $Lft(t)$ denote the earliest and latest firing times respectively for any transition $t \in T$. A *state* (m, c) in TPN is a pair consisting of a marking $m : P \rightarrow \mathbf{N}$ and a clock function $c : T \rightarrow \mathbf{N}$ indicating the state of each transition's local clock. For each transition $t \in T$ its local clock records the amount of time $0 \le c(t) \le Lft(t)$ that t has been continuously enabled. We let $States(TPN)$ denote the set of all possible states in TPN. For any transition $t \in T$, we let $\bullet t = \{p \mid (p, t) \in F\}$ denote the set of *input places* to t and $t\bullet = \{p \mid (t, p) \in F\}$ denote the set of *output places* to t. A transition $t \in T$ is said to be *enabled* in a state (m, c) if, and only if, $m(p) > 0$, for each $p \in \bullet t$. We let $Enabled(m)$ denote the set of all enabled transitions in a state (m, c).

Next we define the conditions necessary for a transition to be able to fire.

Definition 3. *A transition $t \in T$ is fireable in state (m, c) after delay d if, and only if,*

(i)   *t is enabled in (m, c);*
(ii) *$Eft(t) \le c(t) + d \le Lft(t)$; and*
(iii) *for all other enabled transitions $t' \in Enabled(m)$ we have $c(t') + d \le Lft(t')$.*
                                                                                        □

We denote by $Fireable((m, c), d)$ the set of all transitions that may be fired in a state (m, c) with delay d. We can define what happens to a state when a transition fires as follows.

Definition 4. *Given a transition $t \in Fireable((m, c), d)$ we can fire t after a delay d to produce a new state (m', c'), denoted $(m, c)[t, d\rangle(m', c')$, which is defined as follows:*

$$m' = m'' \cup t\bullet \text{ and } m'' = m \backslash \bullet t;$$

$$c'(t_i) = \begin{cases} c(t_i) + d, & \text{if } t_i \in Enabled(m'') \text{ and } t_i \neq t; \\ 0, & \text{otherwise;} \end{cases}$$

for all $t_i \in T$.   □

Note that in the above definition the new tokens produced by a transition are discounted when considering whether or not to reset a transition's local clock (this is the reason for defining the intermediate marking m''). In other words, we are able to distinguish newly produced tokens (see [1]). Other semantic approaches exist for resetting local clocks (see [19]) and we will consider modelling some of these in Section 3.4.

A *firing sequence* for a time Petri net is a sequence of firing steps (i.e. pairs of transitions and delay values):

$$\sigma = (t_{f_1}, d_1), (t_{f_2}, d_2), \dots, (t_{f_k}, d_k),$$

where $t_{f_i} \in T$ and d_i is some delay. A firing sequence σ is said to be fireable from a state s_1 if, and only if, there exist states $s_2, \dots, s_{k+1}$ such that $s_i[t_{f_i}, d_i\rangle s_{i+1}$, for $i = 1, \dots, k$.

3.2   Modelling Time Petri Nets Using RL

We now consider how to construct an RL model of a time Petri net that correctly simulates its behaviour. We build on the multi–set approach introduced in Section 2 and introduce new terms $t(i)[n, e, l]$ to represent transitions, where e is the earliest firing time, l the latest firing time and n is the amount of time a transition has been continuously enabled. For example, the initial state of the time Petri net depicted in Figure 3 would be represented by the following RL term:

$$p(1) \otimes p(2) \otimes t(1)[0, 1, 2] \otimes t(2)[0, 0, 2] \otimes t(3)[0, 0, 1] \otimes t(4)[0, 1, 1].$$

In order to allow an in-depth analysis of a time Petri net we enhance this term structure for a state to include information about what action resulted in the state:

$$\langle\ pn\ \rangle[ts],$$

where pn is a multi–set representing the current state (see above example) and ts is a multi–set recording the action that produced the state.

Next we define the rewrite rules that will be used to model the dynamic behaviour of a time Petri net. In our RL model we choose to simulate time progression by single clock ticks. We show in Section 3.3 that this approach is equivalent to allowing arbitrary time progression as defined in Definition 4. Let $n, e, l : nat$ and $pn, ts : pnet$ be variables. For each transition $t_i \in T$ with input places $p_{in_1}, \ldots, p_{in_j}$ and output places $p_{o_1}, \ldots, p_{o_k}$ we have four distinct types of labelled rules:

(1) A *must fire rule* that forces a transition to fire when it has reached its latest firing time:

$$[mr]\ \langle p(in_1) \otimes \cdots \otimes p(in_j) \otimes t(i)[l, e, l] \otimes pn\rangle[ts] \longrightarrow$$
$$\langle t(i)[0, e, l] \otimes N(p(o_1)) \otimes \cdots \otimes N(p(o_k)) \otimes pn\rangle[t(i) \otimes ts]$$

(2) A *firing rule* that allows a transition to choose to fire if it is within its firing interval:

$$[fr]\ \langle p(in_1) \otimes \cdots \otimes p(in_j) \otimes t(i)[n, e, l] \otimes pn\rangle[ts] \longrightarrow$$
$$\langle t(i)[0, e, l] \otimes N(p(o_1)) \otimes \cdots \otimes N(p(o_k)) \otimes pn\rangle[t(i) \otimes ts]\ \ if\ n \geq e \wedge n < l$$

(3) A *time progression rule* to allow an enabled transition to progress in time:

$$[tr]\ \langle p(in_1) \otimes \cdots \otimes p(in_j) \otimes t(i)[n, e, l] \otimes pn\rangle[ts] \longrightarrow$$
$$\langle p(in_1) \otimes \cdots \otimes p(in_j) \otimes D(t(i)[n + 1, e, l]) \otimes pn\rangle[D(t(i)) \otimes ts]$$

(4) Finally we have an *enabling rule* which distinguishes all enabled transitions (ignoring newly produced tokens):

$$[sr]\ \langle p(in_1) \otimes \cdots \otimes p(in_j) \otimes t(i)[n, e, l] \otimes pn\rangle[ts] \longrightarrow$$
$$\langle p(in_1) \otimes \cdots \otimes p(in_j) \otimes D(t(i)[n, e, l]) \otimes pn\rangle[ts]$$

The $[fr]$ and $[mr]$ rules allow a transition to fire if it is within its firing interval. The new tokens produced are distinguished by a marker N which allows them to be temporarily ignored when considering whether or not a transition is enabled (as required by the semantics defined in Definition 4). Note the use of the variable pn in the above rules which is used to represent the remaining part of a Petri net state which we are not currently interested in. The distinction between a can fire $[fr]$ and a must fire $[mr]$ rule is needed to allow an appropriate rewrite strategy to be formulated (see below) to capture the semantics of a time Petri net. As long as no $[mr]$ rule can be applied the $[tr]$ rule can be used to allow time to progress by one unit. These rules use markers D to synchronize time progression and prevent multiple clock ticks. The enabling rules $[sr]$ are used to distinguish those transitions which are still enabled after a transition has fired (ignoring newly produced tokens). All transition terms which are not surrounded by a D marker will have their local clocks reset to zero after a state step by a reset function.

We note that the rewrite theory defined above does not on its own capture the intended semantics of a time Petri net as we have defined it. In order to do this we will combine the above foundation rewrite theory with a rewriting strategy. We have chosen to take this approach (rather than producing a hard-coded rewrite theory) since it allows us to understand and investigate changes to our time Petri net semantics by simply changing our high-level rewrite strategy (see Section 3.4 for an illustration of this point).

As an example of the above rewrite rules, consider the following partial Elan specification for the time Petri net in Figure 3 (for brevity we have only included the rules for transitions $t1$ and $t3$). The specification is built on top of a module **basic** which specifies the basic components needed such as multi–sets and place/transition terms.

```
module timePN
  import global   basic;              end
  rules for state
    pn, pn2, ts    : pnet;
    m, e, l        : int;
  global
  //** Rules for Transition 1  **//
  [mr]  <p(1)#t(1)[1,e,l]#pn>[ts] =>
          <t(1)[0,e,l]#N(p(3))#pn>[t(1)#ts]    end
  [fr]  <p(1)#t(1)[m,e,l]#pn>[ts] =>
          <t(1)[0,e,l]#N(p(3))#pn>[t(1)#ts]  if m>=e and m<l   end
  [tr]  <p(1)#t(1)[m,e,l]#pn>[ts] =>
          <D(t(1)[m+1,e,l])#p(1)#pn>[D(t(1))#ts]   end
  [sr]  <p(1)#t(1)[m,e,l]#pn>[ts] =>
          <D(t(1)[m,e,l])#p(1)#pn>[ts]   end
```

```
//** Rules for Transition 3  **//
[mr]  <p(3)#p(4)#t(3)[1,e,1]#pn>[ts] =>
           <t(3)[0,e,1]#N(p(1))#N(p(2))#pn>[t(3)#ts]    end
[fr]  <p(3)#p(4)#t(3)[m,e,1]#pn>[ts] =>
           <t(3)[0,e,1]#N(p(1))#N(p(2))#pn>[t(3)#ts]
              if m>=e and m<1    end
[tr]  <p(3)#p(4)#t(3)[m,e,1]#pn>[ts] =>
           <D(t(3)[m+1,e,1])#p(3)#p(4)#pn>[D(t(3))#ts]    end
[sr]  <p(3)#p(4)#t(3)[m,e,1]#pn>[ts] =>
           <D(t(3)[m,e,1])#p(3)#p(4)#pn>[ts]    end
    end
end
```

Given a term representing a state in a time Petri net we can either choose to fire a transition or allow time to progress by one unit. However, time progression is not allowed if there exists a transition which has reached its latest firing time; in this case we are forced to fire such a transition. In order to correctly model the semantics outlined above we define a rewrite strategy step which will control the application of the rewrite rules. This strategy will be used to represent a single state step in a time Petri net. We will see in Section 3.4 that alternative time semantics can be easily incorporated into our model by simple changes to the step strategy.

```
strategies for state
  implicit
      []  fire  => fr;repeat*(dc one(sr))         end
      []  must  => mr;repeat*(dc one(sr))         end
      []  time  => repeat+(dc one(tr))            end
      []  step  => dk(fire, first(must, time) )   end
  end
```

The rewrite strategy fire is used to choose non–deterministically a transition to fire which is enabled and within its firing interval. It applies the [fr] rule to produce a set of possible terms and then applies the strategy repeat*(dc one(sr)) to these results to mark all those transitions which are still enabled (the other transitions will have their local clocks reset). The strategy must is similar to fire but chooses transitions to fire which have reached their latest firing time. The time progression strategy time performs one complete time step, incrementing the local clocks of all enabled transitions by one unit. These three strategies are combined into a strategy step which calculates the set of possible next states. It uses the strategy first(must, time) to ensure that time can only progress if no transition is at its latest firing time (i.e. it tries to apply must but if this fails then it applies time).

After applying step we "clean up" the resulting state term by applying a reset function which removes all N and D markers, and resets local clocks to zero:

```
[]   reset(D(t(n)[m,e,l])) => t(n)[m,e,l]            end
[]   reset(t(n)[m,e,l]) => t(n)[0,e,l]               end
[]   reset(N(p(n))) => p(n)                          end
[]   reset(p(n)) => p(n)                             end
[]   reset(E) => E                                   end
[]   reset(pn # pn2) => reset(pn) # reset(pn2)       end
```

where we have the variables pn, pn2 : pnet and m,e,l : int.

Thus a state step in our RL model, denoted $s \longrightarrow s'$ using step, involves applying the strategy step and then the function reset. Given we can now define a state step it is interesting to consider how to explore the resulting state space. We begin by defining the *exit states* we wish to find by specifying exit conditions as rewrite rules, e.g.

```
[exit] <p(4) # pn>[ts] => <p(4) # pn>[ts]
```
Place P4 contains at least one token;

```
[exit] <pn>[t(4) # ts] => <pn>[t(4) # ts]
```
Transition T4 has fired;

```
[exit] <pn>[ts] => <pn>[ts] if length(pn) > 10
```
The size of the state (tokens plus transitions) has exceeded 10.

A state is said to be an exit state if, and only if, one of the exit rules can be successfully applied to it. Using the defined strategy language of Elan (see [8]) we can then define various search strategies that look for exit states. For example, we could define a strategy search which given an initial state performs a depth first search (possibly bounded) until it finds an exit state. This strategy can be generalized to a strategy searchall which finds all exit states. (For a detailed discussion of search strategies see [5] and [20]).

As an example, consider using Elan and the strategy search to find a firing sequence for the time Petri net in Figure 3 which results in a state with a token on place $p5$, i.e. we have the exit rule:

```
[exit] <p(5) # pn>[ts] => <p(5) # pn>[ts]          end
```

The following is an excerpt from the Elan tool:

```
[] start with term :
    [search]<p(1)#p(2)#t(1)[0,1,2]#t(2)[0,0,2]#t(3)[0,0,1]
            #t(4)[0,1,1]>[E]

[] result term:
    (<p(4)#p(5)#t(3)[0,0,1]#t(4)[0,1,1]#t(2)[0,0,2]#t(1)[0,1,2]>
    [t(4)]).([D(t(4))#D(t(3))]).([t(1)]).([D(t(1))]).([t(2)]).
    ([t(3)]).([t(1)]).([D(t(1))]).([t(2)]).([E])

[] end
```

The above result term indicates one possible firing sequence that produces an exit state from the initial marking. It has been displayed using a *display strategy* (see [20]) that outputs only the final state reached, and the steps involved in reaching that state (i.e. [t(i)] indicates transition t_i has fired and [D(t(i)) # D(t(k))] indicates that transitions t_i and t_k have progressed by one unit in time). The above term represents a firing sequence involving nine state steps (the initial [E] represents the initial state). It corresponds to the following time Petri net firing sequence:

$$(t_2, 0), (t_1, 1), (t_3, 0), (t_2, 0), (t_1, 1), (t_4, 1).$$

3.3 Correctness Argument

In this section we consider the correctness of our RL model (rewrite theory plus step strategy) for time Petri nets. We show that our model is both *sound* (each step in our RL model has a corresponding state step in the time Petri net) and *complete* (every state step possible in a time Petri net has a corresponding step in our RL model).

In the sequel let $TPN = (P, T, F, m_0, SI)$ be an arbitrary time Petri net and let $RL(TPN)$ be the corresponding RL model as defined in Section 3.2.

It turns out that not all terms of type *pnet* in $RL(TPN)$ represent valid states in TPN. Thus we define $ValidRL(TPN)$, the set of all state terms s in $RL(TPN)$ such that: (1) if $p(i)$ in s then $p_i \in P$; and (2) if $t(i)[n, e, l]$ in s then $t_i \in T$, $[e, l] = SI(t_i)$, and $0 \le n \le l$. If $s \in ValidRL(TPN)$ then we say s is a *valid state term for TPN*.

Proposition 5. *The rewrite strategy* step *is well–defined with respect to valid state terms, i.e. for any $s \in ValidRL(TPN)$, if $s \longrightarrow s'$ using* step *then $s' \in ValidRL(TPN)$.*

Proof. Suppose $s \in ValidRL(TPN)$ and $s \to s'$ using step. By definition of the strategy step it follows there are three cases to consider:
Case (1): *The strategy* fire *was used (followed by the* reset *function).* The application of a [fr] rule simply consumes some token terms and produces some new token terms which must by definition correspond to places in TPN. The application of the [sr] rules taken together with the reset function will just reset the local clock of some transition terms to zero. Thus if the original state term s was valid then the resulting state term s' after applying fire will also be valid.
Case (2): *The strategy* must *was used (followed by the* reset *function).* Similar argument to above based on the [mr] rules.
Case (3): *The strategy* must *failed and so the strategy* time *was used (followed by the* reset *function).* If the must strategy fails then we know there are no enabled transition terms in s which have reached their latest firing time. This means that performing a time step, that is incrementing by one the local clock of all enabled transitions (which is exactly what the strategy time does), will not result in any local clock going past its transitions latest firing time. Thus the resulting state term s' must be valid. □

Recall that our model is said to be *sound* if, and only if, each step in our RL model has a corresponding state step in the time Petri net; and *complete* if, and only if, every state step possible in a time Petri net has a corresponding (sequence of) step(s) in our RL model. We can now define what we mean by correctness: *a RL model is correct with respect to a time Petri net if, and only if, it is both sound and complete with respect to the time Petri net.*

We show that for any time Petri net TPN, the corresponding RL model $RL(TPN)$ is correct. We begin our correctness proof by defining a mapping between states of a time Petri net and valid state terms in the corresponding RL model as follows.

Definition 6. *The term mapping* $\sigma : States(TPN) \rightarrow ValidRL(TPN)$ *is defined on each state* (m, c) *to return the multi–set term* $\sigma(m, c)$ *which contains only the following:*
(i) for each place $p_i \in P$, *the multi-set term* $\sigma(m, c)$ *will contain* $m(p_i)$ *occurrences of the place term* $p(i)$;
(ii) for each transition $t_i \in T$, *the multi-set term* $\sigma(m, c)$ *will contain the transition term* $t(i)[c(t_i), Eft(t_i), Lft(t_i)]$. $\qquad\qquad\square$

By the definition of $ValidRL(TPN)$ it is straightforward to show that σ is a well–defined, bijective mapping with an inverse $\sigma^{-1} : ValidRL(TPN) \rightarrow States(TPN)$.

Suppose a state step $(m, c)[t, d\rangle(m', c')$ can occur in a time Petri net. Then observe that we can break such a state step down into a series of intermediate steps consisting of a series of time ticks which allow the delay d to pass, followed by the transition t firing. In other words we can represent $(m, c)[t, d\rangle(m', c')$ by the following sequence of events:

$$(m, c)[tick\rangle(m, c_1)[tick\rangle(m, c_2)[tick\rangle \cdots [tick\rangle(m, c_d)[t\rangle(m', c'),$$

where $(m, c_i)[tick\rangle(m, c_{i+1})$ represents a clock tick (i.e. increments the local clocks of all enabled transitions by one unit, resetting all other local clocks to zero) and $(m, c_d)[t\rangle(m', c_d)$ fires transition t and resets to zero the local clocks of all transitions which are not enabled (i.e. corresponds to $(m, c_d)[t, 0\rangle(m', c')$).

Using the above observation we can show that for any time Petri net TPN, the corresponding RL model $RL(TPN)$ is sound and complete.

Theorem 7. *(Soundness) Let* $s \in ValidRL(TPN)$ *be any state term. If* $s \rightarrow s'$ *using the strategy* step *then either* $\sigma^{-1}(s)[tick\rangle\sigma^{-1}(s')$ *or there must exist a transition* $t \in T$ *such that* $\sigma^{-1}(s)[t\rangle\sigma^{-1}(s')$, *i.e. the diagram in Figure 4.(a) commutes.*

Proof. Suppose $s \in ValidRL(TPN)$ and $s \rightarrow s'$ using step. By definition of the strategies step it follows that there are three cases to consider:
Case (1): *The strategy* fire *was used (followed by the* reset *function).* This strategy begins by applying a [fr] rule which corresponds to some transition, say $t \in T$. We will show that the result of this strategy s' corresponds to the

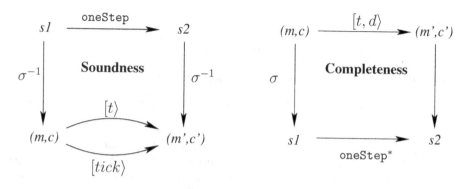

Fig. 4. (a) Soundness; (b) Completeness.

result of $\sigma^{-1}(s)[t\rangle\sigma^{-1}(s')$. Clearly if this rule can be applied then it follows by definition that the local clock for t must be within the $Eft(t)$ and $Lft(t)$ firing times and that for each $p_i \in \bullet t$ there must exist a token term $p(i) \in s$. Thus it follows that the transition t must be fireable in state $\sigma^{-1}(s)$. Applying the [fr] rule will result in the token terms corresponding to the input places $\bullet t$ being removed and token terms corresponding to the output places $t\bullet$ being produced. But this is exactly what will happen to the marking in state $\sigma^{-1}(s)$ when t is fired. Finally, the strategy `fire` will use the [sr] rules and `reset` function to set all unenabled transition clocks (discounting the newly produced token terms) to zero. Again, by definition of firing a transition (see Definition 4) this is what happens when t is fired.

Case (2): *The strategy* `must` *was used (followed by the* `reset` *function)*. Similar argument to above using [mr] rules.

Case (3): *The strategy* `must` *failed and so the strategy* `time` *was used (followed by the* `reset` *function)*. If the `must` strategy fails then we know there are no enabled transition terms in s which have reached their latest firing time. This means that the strategy `time` will be applied and a time step will be performed, i.e. the local clocks of all enabled transitions will be incremented by one and all other local clocks will be reset. Note that allowing time to progress is a valid action since no transition has reached its latest firing time. In this case it will be valid to apply the *tick* action to the state $\sigma^{-1}(s)$ which performs exactly the time update detailed above. Thus we will have $\sigma^{-1}(s)[tick\rangle\sigma^{-1}(s')$ as required.

$\square$

Theorem 8. *(Completeness) Let* $(m, c) \in States(TPN)$ *be any state in TPN. If* $(m, c)[t, d\rangle(m', c')$ *for some transition* $t \in T$ *and some duration* d, *then it follows that* $\sigma(m, c) \to \sigma(m', c')$ *by a series of applications of the strategy* `step`, *i.e. the diagram in Figure 4.(b) commutes.*

Proof. Suppose that $(m, c)[t, d\rangle(m', c')$ for some transition $t \in T$ and some duration d. Then by the observation above we can represent this by the following sequence of events:

$$(m, c)[tick\rangle(m, c_1)[tick\rangle(m, c_2)[tick\rangle \cdots [tick\rangle(m, c_d)[t\rangle(m', c').$$

Thus it suffices to show that the following two facts hold.

Fact 1: If $(m, c)[tick\rangle(m, c')$ then it follows that $\sigma(m, c) \to \sigma(m, c')$ using the strategy step.

Proof. Suppose that $(m, c)[tick\rangle(m, c')$. Then it follows that for all $t_i \in T$ we have $c(t_i) < Lft(t_i)$ (otherwise the *tick* event would not be valid). Therefore we know by definition of σ that no [mr] rule can be applied to the state term $\sigma(m, c)$ and so the strategy must will fail. This means the time strategy can be applied which (along with reset function) will increment by one the local clocks of all enabled transition terms in $\sigma(m, c)$, resetting all unenabled transition clocks to zero. But this is exactly what the *tick* event will do. Thus we have $\sigma(m, c) \to \sigma(m, c')$ as required.

Fact 2: If $(m, c)[t\rangle(m', c')$, for some $t \in T$, then $\sigma(m, c) \to \sigma(m', c')$ using step.

Proof. Suppose $(m, c)[t\rangle(m', c')$. Then we have two cases to consider.
(i) Suppose $c(t) < Lft(t)$. By definition of $RL(TPN)$ there is a [fr] rule for transition t. Given that t is fireable (i.e. enabled in m and $c(t) \geq Eft(t)$) it follows that its corresponding [fr] rule can be applied to the state term $\sigma(m, c)$ as part of the strategy fire and that this will remove the enabling token terms $p(i)$, for each $p_i \in \bullet t$ and add the new token terms $p(i)$ to $\sigma(m, c)$, for each $p_i \in t\bullet$. Applying the [sr] rules and reset function will then reset the local clocks of all unenabled transitions (ignoring the newly produced token terms). But this corresponds to the event of firing t to produce (m', c') and thus we have $\sigma(m, c) \to \sigma(m', c')$.
(ii) Suppose $c(t) = Lft(t)$. The proof follows along similar lines to (i) above but uses the [mr] rules. □

The above two theorems prove that for any time Petri *TPN*, the corresponding RL model $RL(TPN)$ is correct with respect to *TPN*.

Theorem 9. *(Correctness) Given any time Petri net TPN we have that $RL(TPN)$ is correct RL model with respect to TPN.*

Proof. Follows directly from definition of correctness and Theorems 7 and 8. □

3.4   Modelling Alternative Semantic Choices

In the preceding sections we have constructed an RL model for the standard semantic interpretation of time Petri nets (see [2] and [1]). In this section we demonstrate the flexibility of our approach by considering some alternative semantic choices and showing how our RL model can be easily adapted to represent these alternatives. We note that for each of these new semantics a corresponding correctness proof along the lines of that given in Section 3.3

would be needed to ensure the new RL model is correct. For brevity we omit these proofs here and leave them as an instructive exercise for the reader.

(1) *Giving priority to latest firing transitions.*

One possible change to the standard semantics is to give priority to those transitions which have reached their latest firing times. This would mean that a transition which is within its firing interval but not yet at its latest firing time would only be allowed to fire if no transition had reached its latest firing time. Such a change in semantics is easy to represent in our model; we simply change the step strategy to reflect this change in priority as follows:

```
[]  step  => first(must, dk(fire, time) )         end
```

The strategy step now reflects the fact that we start by considering only the must fire transitions. If the strategy must fails then clearly no transition has reached its latest firing time; in such a case we can either choose to fire a transition or perform a time step.

(2) *Resetting local clocks of conflicting transitions.*

In [19] an alternative semantic approach is proposed for resetting the local clocks of conflicting transitions. The idea is that when a transition t fires any transition which shares an input place with t has its local clock reset to zero. For example, in the time Petri net depicted in Figure 3 if transition $t4$ fires then the local clock for transition $t3$ will be reset. This alternative semantics is straightforward to incorporate into our RL model for time Petri nets; we simple change the firing rules ([fr] and [mr]) so that they include all transition terms for conflicting transitions and then reset their clocks when the transition fires. As an illustrative example the following would be the new firing rules for transition $t4$ in the time Petri net depicted in Figure 3:

```
[mr]  <p(3) # t(4)[1,e,1] # t(3)[n,e',1'] # pn>[ts] =>
        <t(4)[0,e,1] # t(3)[0,e',1'] # N(p(5)) # pn>[t(4)#ts]   end
[fr]  <p(3) # t(4)[m,e,1] #  t(3)[n,e',1'] # pn>[ts] =>
        <t(4)[0,e,1] #  t(3)[0,e',1'] # N(p(5)) # pn>[t(4)#ts]
          if m >= e and m < 1      end
```

(3) *Maximal step semantics.*

The final alternative we consider is imposing a *maximal step* semantics on time Petri nets: at each step a maximal set of concurrent fireable transitions are allowed to fire [19]. In practice, the implications of such a semantics would represent a major restriction; transitions would be forced to fire whenever they reach their earliest firing time, which contradicts assigning a firing interval to a transition. However, we consider the maximal step semantics as an illustrative example of the flexibility of our framework. To model the maximal step semantics we need only redefine the fire and step strategies; the time strategy remains the same and of course the must strategy is no longer required.

```
[]  fire  => repeat+(dk(fr,mr));repeat*(dc one(sr))    end
[]  step  => first(fire, time)                          end
```

4   Conclusions

In this paper we have considered using RL and the support tool Elan to model and analyse time Petri nets. We discussed the important issue of correctness and showed that our RL model correctly simulates a time Petri net. We demonstrated the flexibility of our approach by considering several alternative semantics for time Petri nets. We showed that these alternatives could be straightforwardly represented in our model by making small adjustments either to the **step** strategy or to the basic RL rules. This case study illustrates how RL and Elan can be used to prototype and analyse Petri net extensions with time. Furthermore, by coupling our approach with an existing Petri net tool such as PEP we have shown how several different tools can be combined to produce a practical and usable new approach.

We have performed a similar analysis to the one presented here using *timed Petri nets* (see [17] and [22]), in which a duration is associated with transitions. This work is presented in [20].

The aim of this work has been to: (i) provide a flexible formal framework for defining semantic models of Petri net extensions with time which are succinct and easily communicated; (ii) provide tools to allow a range of different Petri net extensions with time to be simulated and practically investigated, thus overcoming the problems associated with the current hardwired tools. We note that we are not proposing that our approach should replace efficient hardwired tools for large scale verification tasks. We see our approach as allowing a tool developer to formally specify the semantics they have chosen and to expediently prototype their ideas before committing themselves to the development of a practical tool. Our framework could also be seen as a design aid, allowing a developer to test out their ideas before committing to a particular Petri net time model.

In future work we intend to investigate extending our approach to prototyping verification algorithms, such as the finite prefix construction. We also intend to perform a variety of verification case studies to illustrate the application of our methods and investigate its limitations.

Acknowledgements. It is a pleasure to thank Maciej Koutny for many valuable discussions during the preparation of this paper. We are very grateful to Hans Fleischhack and Christain Stehno who provided helpful advice and technical assistance. We would also like to thank the anonymous referees whose comments were extremely helpful. Finally, we would like to gratefully acknowledge the financial support provided by the British Council and DAAD during this work.

References

1. T. Aura and J. Lilius. Time Processes for Time Petri Nets. In: *18th Int. Conf. on App. and Theory of Petri Nets*, LNCS 1248, pages 136–155, Springer–Verlag, 1997.
2. B. Berthomieu and M. Diaz. Modeling and Verification of Time Dependent Systems using Time Petri Nets. *IEEE Trans. on Software Engineering*, 17(3):259–273, 1991.

3. E. Best and B. Grahlmann. PEP – more than a Petri Net Tool. In: T. Margaria and B. Steffen (eds), *Proc. of TACAS'96: Tools and Algorithms for the Construction and Analysis of Systems*, LNCS 1055, pp. 397-401, Springer–Verlag, 1996.
4. I. I. Bestuzheva and V. V. Rudnev. Timed Petri Nets: Classification and Comparative Analysis. *Automation and Remote Control*, 15(10):1303–1318, 1990.
5. P. Borovanský. The Control of Rewriting: Study and Implementation of a Strategy Formalism. In: C. Kirchner and H. Kirchner (eds), Proceedings of WRLA '98, *Electronic Notes in Theoretical Computer Science*, Vol. 15, 1998.
6. P. Borovanský, C. Kirchner, H. Kirchner, P.–E. Moreau and M. Vittek. ELAN: A Logical Framework Based on Computational Systems. In: J. Meseguer (ed), Proceedings of WRLA'96, *Electronic Notes in Theoretical Computer Science*, Vol. 4, 1996.
7. P. Borovanský, C. Kirchner, H. Kirchner, P.–E. Moreau and C. Ringeissen. An overview of ELAN In: C. Kirchner and H. Kirchner (eds), Proceedings of WRLA '98, *Electronic Notes in Theoretical Computer Science*, Vol. 15, 1998.
8. P. Borovanský, H. Cirstea, H. Dubois, C. Kirchner, H. Kirchner, P.–E. Moreau, C. Ringeissen and M. Vittek. *ELAN: User Manual*. December, 1998. Available from: http://elan.loria.fr.
9. Diaconescu, R., Futatsugi, K., Ishisone, M., Nakagawa, A. T. and Sawada, T. An Overview of CafeObj. In: C. Kirchner and H. Kirchner (eds), Proceedings of WRLA '98, *Electronic Notes in Theoretical Computer Science*, Vol. 15, 1998.
10. H. Ehrig and B. Mahr. *Fundamentals of Algebraic Specification 1 – Equations and Initial Semantics*. EATCS Monographs on Theoretical Computer Science 6. Springer-Verlag, Berlin, 1985.
11. J. Loeckx, H–D. Ehrich and M. Wolf. *Specification of Abstract Data Types*. Wiley, 1996.
12. P. Merlin and D. J. Faber. Recoverability of communication protocols. *IEEE Transactions on Communications*, COM-24(9):1036–1049, 1976.
13. J. Meseguer. Conditional rewriting logic as a unified model of concurrency. *Theoretical Computer Science*, 96:73–155, 1992.
14. J. Meseguer and U. Montanari. Petri nets are monoids. *Information and Computation*, 88:105-155, 1990.
15. T. Murata. Petri nets: properties, analysis and applications. *Proceedings of the IEEE*, 77(4):541–580, 1989.
16. J. L. Peterson. Petri Nets. *Computing Surveys*, 9(3):223–252, 1977.
17. C. Ramchandani. Analysis of asynchronous concurrent systems by timed Petri nets. Project MAC Technical Report MAC-TR-120, Massachusetts Institute for Technology, Cambridge MA, 1974.
18. W. Reisig. *Petri nets – an introduction*. EATCS Monographs in Theoretical Computer Science, 4, Springer–Verlag, 1985.
19. P. Starke. A Memo on Time Constraints in Petri Nets. Informatik–Bericht Nr. 46, Institut für Informatik, Humboldt Universität, Berlin, 1995.
20. L. J. Steggles. Prototyping and Analysing Petri Nets with time: A Formal Investigation using Rewriting Logic and Elan. Technical Report No. 722, Department of Computing Science, University of Newcastle upon Tyne, 2001.
21. B. Walter. Timed Petri-Nets for Modelling and Analyzing Protocols with Real-Time Characteristics. In: H. Rudin and C. H. West (eds), *Protocol Specification, Testing and Verification III*, North–Holland, 1983.
22. W. M. Zuberek. Timed Petri Nets: Definitions, Properties and Applications. *Microelectronics and Reliability*, 31(4):627–644, 1991.

Partial S-Invariants
for the Verification of Infinite Systems Families

Walter Vogler

Institut für Informatik, Universität Augsburg, Germany
vogler@informatik.uni-augsburg.de

Abstract. We introduce partial S-invariants of Petri nets, which can help to determine invariants and to prove safety if large nets are built from smaller ones using parallel composition with synchronous communication. Partial S-invariants can support compositional reduction and, in particular, the fixed-point approach, used for verifying infinite parameterized families of concurrent systems. With partial S-invariants and the fixed-point approach we prove the correctness of two solutions to the MUTEX-problem based on token rings; for this, we only have to prove liveness of a simplified version due to previous results.

1 Introduction

For the verification of infinite parameterized families of concurrent systems the so-called behavioural fixed-point approach is advocated in [11]. The members of such families are composed of an increasing number of components. If one can show that the composition of, say, two of these components is equivalent to just one, then one can reduce each member of the family to an equivalent small one, and it suffices to prove this small system correct. This approach is a specific case of compositional reduction, for which the equivalence under consideration has to be a congruence for composition – and, of course, it must be strong enough to support verification. We will model systems with Petri nets, and we will use parallel composition with synchronization of common actions, which corresponds to merging transitions; also renaming and hiding of actions are important.

In this paper, we will apply the fixed-point approach to two token-ring based solutions for the problem of mutual exclusion (MUTEX). Such a ring has a component for each user that needs access to the shared resource, and each component has a separate interface that allows the user to *request* access, *enter* its critical section – in which the resource is used –, and *leave* this section again. To verify such a ring, one has to show the safety property that there are never two users in their critical sections at the same time, i.e. that enter- and leave-actions alternate properly, and one has to show the liveness property that to each requesting user access is granted eventually. Modelling token-rings with Petri nets, MUTEX-safety is usually easy to show applying an S-invariant. Hence, we want to apply the fixed-point approach to prove MUTEX-liveness.

An immediate problem is that each ring component has an external interface to its user with actions of its own, and thus two components can hardly be

J.-M. Colom and M. Koutny (Eds.): ICATPN 2001, LNCS 2075, pp. 382–401, 2001.

equivalent to one. In [2], we have shown that under some symmetry assumption it is sufficient to check MUTEX-liveness for one user and hide the other user-actions from the interface. In the modified net, only the actions of one user are visible and two components may be equivalent to one; hence, and this is the first point to be made, our symmetry result opens the door for applying the fixed-point approach.

In fact, one encounters another problem: the composition of two components may not be equivalent to one, because in isolation these nets exhibit behaviour that is not possible in the complete ring. To show the equivalence, one has to restrict their behaviour in a suitable way; this is somewhat similar e.g. to the interface descriptions presented in [5]. The main contribution of this paper is the development of what we call *partial S-invariants* in order to restrict the behaviour suitably. We show how partial S-invariants can support compositional reduction in general, and we will apply them specifically in the fixed-point approach.

Partial S-invariants of components can also be used to obtain S-invariants of composed systems. Another notion of partial S-invariants – for a setting where nets are composed by merging places – has been defined in [8] where it has also been shown how to combine these to obtain S-invariants of composed systems.

The equivalence we use is based on fairness in the sense of the progress assumption, i.e. weak fairness. In [13], one can find such a semantics that is compositional for safe nets. Here, we have to deal with components of a safe net that are not safe themselves, and we show that compositionality for general nets can be achieved very similarly to [13], if one uses a suitable generalization of weak fairness from safe to general S/T-nets. Since we are really interested in safe nets, we are free to choose any generalization that is convenient.

The Petri nets of this paper may have so-called read arcs, which are somewhat similar to loops. If transition t and place s form a loop, then firing t removes a token from s and returns it at the end; hence, this token is not available while t is firing. If t and s are connected by a read arc instead, then t checks for a token on s without actually using it; thus, other transitions might do the same while t is firing. For example, read arcs can model the concurrent reading of data. When we consider firing sequences only, read arcs and loops are interchangeable; when we consider concurrent behaviour or the progress assumption, they make a difference. It is shown in [13] that ordinary nets without read arcs cannot solve the MUTEX-problem. Read arcs have found quite some interest recently, see e.g. [9,6,14,1] and we include them for generality (in particular in the treatment of fairness) and because we need them in our applications.

Section 2 defines Petri nets with read arcs and the operations on nets we will use to construct the MUTEX-solutions. Section 3 gives our definition of (weak) fairness in the sense of progress assumption, refines the resulting semantics to a precongruence for our operations and shows full abstraction. Section 4 introduces partial S-invariants, shows how to combine them to S-invariants and presents the essential result for applying them for compositional reduction and, hence, in the fixed-point approach. Section 5 quotes from [2] the correctness definition for MUTEX-solutions and the symmetry result mentioned above. Section

6 shows how to use partial S-invariants in the fixed-point approach and proves two families of nets correct. For the second family, we use the tool FastAsy that compares the performance of asynchronous systems; the respective performance preorder is closely related to the precongruence we use in the present paper. We close with a discussion of related work in Section 7.

Proofs often had to be omitted in this extended abstract; they will be presented in a forthcoming report.

2   Basic Notions and Operations for Petri Nets with Read Arcs

In this section, we introduce Petri nets with read arcs, as explained in the introduction, and the basic firing rule. Then we define parallel composition, renaming and hiding for such nets and give some laws for these operations. The transitions of our nets are labelled with actions from some infinite alphabet Σ or with the empty word λ; the latter represents internal, unobservable actions.

Thus, a *labelled Petri net with read arcs* $N = (S, T, F, R, l, M_N)$ (or just a *net* for short) consists of finite disjoint sets S of *places* and T of *transitions*, the *flow relation* $F \subseteq S \times T \cup T \times S$ consisting of (ordinary) *arcs*, the set of *read arcs* $R \subseteq S \times T$, the *labelling* $l : T \to \Sigma \cup \{\lambda\}$, and the *initial marking* $M_N : S \to I\!N_0$; we require that $(R \cup R^{-1}) \cap F = \emptyset$. The net is called *ordinary*, if $R = \emptyset$.

We draw transitions as boxes, places as circles, arcs as arrows (as usual), and read arcs as lines (sometimes dashed) without arrow heads. As usual, nets N_1 and N_2 are *isomorphic*, written $N_1 = N_2$, if there is some function that bijectively maps the places (transitions) of N_1 to the places (transitions) of N_2 such that arcs, read arcs, labelling and initial marking are preserved. The *alphabet* $\alpha(N)$ of a net N is the set of all actions from Σ that occur as labels in N.

For each $x \in S \cup T$, the *preset* of x is ${}^\bullet x = \{y \mid (y, x) \in F\}$, the *postset* of x is $x^\bullet = \{y \mid (x, y) \in F\}$, and the *read set* of x is $\hat{x} = \{y \mid (y, x) \in R \cup R^{-1}\}$. If $x \in {}^\bullet y \cap y^\bullet$, then x and y form a *loop*. A *marking* is a function $S \to I\!N_0$. We sometimes regard sets as characteristic functions, which map the elements of the sets to 1 and are 0 everywhere else; hence, we can e.g. add a marking and a postset of a transition or compare them componentwise.

Our basic firing rule extends the firing rule for ordinary nets by regarding read arcs as loops: a transition t is *enabled* under a marking M, denoted by $M[t\rangle$, if ${}^\bullet t \cup \hat{t} \leq M$. If $M[t\rangle$ and $M' = M + t^\bullet - {}^\bullet t$, then we denote this by $M[t\rangle M'$ and say that t can *occur* or *fire* under M yielding the marking M'.

This definition of enabling and occurrence can be extended to finite sequences as usual by repeated application. An infinite sequence w of transitions is *enabled* under a marking M, denoted as above, if all its finite prefixes are enabled under M. We denote the set of finite sequences over a set X by X^*, the set of infinite sequences by X^ω, and their union by X^∞. If $w \in T^\infty$ is enabled under the initial marking, then it is called a *firing sequence*.

We can extend the labelling to sequences of transitions as usual, i.e. homomorphically, which automatically deletes internal actions. With this, we lift the enabledness and firing definitions to actions: a sequence v of actions from Σ is *enabled* under a marking M, denoted by $M[v\rangle\rangle$, if there is some transition sequence w with $M[w\rangle$ and $l(w) = v$; for finite v, $M[v\rangle\rangle M'$ is defined analogously. If $M = M_N$, then v is called a *trace*.

A marking M is called *reachable* if $M_N[w\rangle M$ for some $w \in T^*$. The net is *safe* if $M(s) \leq 1$ for all places s and reachable markings M and if all transitions t satisfy ${}^\bullet t \neq \emptyset$; the latter can be ensured by adding a new marked loop to t.

We are mainly interested in safe nets, but since we will construct safe nets from components that, considered in isolation, violate one or both of the required conditions, we develop our approach for general nets.

Safe nets are without self-concurrency: A transition t is *enabled self-concurrently* under a marking M, if ${}^\bullet t \cup \hat{t} \leq M - {}^\bullet t$, i.e. if there are enough tokens to enable two copies of t at the same time. A net is *without self-concurrency*, if no transition t is enabled self-concurrently under a reachable marking M.

Our parallel composition $\|$, where synchronization is over common actions, is not much different from TCSP-like composition used in [13], but makes notation lighter, and it is also used in [11]. If nets N_1 and N_2 with $A = \alpha(N_1) \cap \alpha(N_2)$ are combined using $\|$, then they run in parallel and have to synchronize on actions from A. To construct the composed net, we have to combine each a-labelled transition t_1 of N_1 with each a-labelled transition t_2 from N_2 if $a \in A$; i.e., we take the disjoint union of N_1 and N_2, and then for each such pair, we introduce a new a-labelled transition (t_1, t_2) that inherits the pre-, post- and read set from both, t_1 and t_2; in the end, we delete all original transitions with label in A.

Other important operators for the modular construction of nets are hiding and renaming, which bind stronger than parallel composition. *Hiding* $A \subseteq \Sigma$ in N means changing all labels $a \in A$ to λ; it results in N/A; we write N/a instead of $N/\{a\}$ for a single $a \in \Sigma$. Similarly, w/A is obtained from a finite or infinite sequence w over Σ by removing all occurrences of actions from A. Clearly, $N/A/B = N/(A \cup B)$; we will freely combine several applications of hiding into one or split one into several.

Just as [11], we use a relabelling that is a bit more general than usual; a *relabelling function* f maps actions from Σ to nonempty subsets of Σ and λ to $\{\lambda\}$. For a relabelling function f, let $dom(f) = \{a \in \Sigma \mid f(a) \neq \{a\}\}$, $cod(f) = \bigcup_{a \in dom(f)} f(a)$ and $\alpha(f) = dom(f) \cup cod(f)$.

The *relabelling* $N[f]$ of N is obtained from N by replacing each transition t with $l(t) = a$ by as many copies as $f(a)$ has elements and labelling each copy by the respective element; the copies are connected to the places just as t.

For $X \subseteq \Sigma$, we set $f(X) = \bigcup_{a \in X} f(a)$, and we set $f^{-1}(X) = \{a \in \Sigma \mid f(a) \cap X \neq \emptyset\}$. We can extend f homomorphically to finite or infinite sequences over Σ such that each sequence is mapped to a set of sequences.

Usually, a relabelling will map almost all $a \in \Sigma$ to $\{a\}$ – we say it is the identity for these a; then, we will only list the exceptions together with their respective images in the form $N[a_1 \to A_1, \ldots, a_n \to A_n]$. Again, we omit the braces of A_i, if it has only one element. Thus, $N[a \to b, b \to \{c, d\}]$ is the net N with each a changed to b and each b-labelled transition duplicated to a c- and a d-

labelled copy; for this relabelling function f, $dom(f) = \{a, b\}$, $cod(f) = \{b, c, d\}$ and $\alpha(f) = \{a, b, c, d\}$; furthermore, $f(aceb) = \{bcec, bced\}$.

We now give some laws for our operations; basically the same were stated e.g. in [11], but for transition systems. These laws are based on isomorphism and should therefore hold whatever more detailed semantics one may choose, and they are true for the fairness based preorder introduced in the next section.

Law 1 $(N_1 \parallel N_2) \parallel N_3 = N_1 \parallel (N_2 \parallel N_3)$

Law 2 $N_1 \parallel N_2 = N_2 \parallel N_1$

These laws will also be used freely without referencing them explicitly.

Law 3 $N[f][g] = N[f \circ g]$ where $(f \circ g)(a) = \bigcup_{b \in f(a)} g(b)$

Law 4 $N/A = N/(A \cup B)$ provided $\alpha(N) \cap B = \emptyset$

Law 5a $N[f]/A = N/A[f]$ provided $A \cap \alpha(f) = \emptyset$

Law 5b $N[a \to B]/A = N/(A \cup \{a\})$ provided $B \subseteq A$

Law 6 $(N_1 \parallel N_2)[f] = N_1 \parallel N_2[f]$ provided $\alpha(N_1) \cap \alpha(f) = \emptyset$

Law 7 $(N_1 \parallel N_2)/A = N_1 \parallel N_2/A$ provided $\alpha(N_1) \cap A = \emptyset$

Law 8 $(N_1 \parallel N_2)[f] = N_1[f] \parallel N_2[f]$ provided f only renames some actions to fresh actions, i.e. $f(a)$ is a singleton with $f(a) \cap (\alpha(N_1) \cup \alpha(N_2)) = \emptyset$ for all $a \in dom(f)$, and for different $a, b \in dom(f)$, $f(a) \neq f(b)$

We can now derive a law for α-conversion, i.e. the renaming of actions that are bound by hiding (apply Law 4 for $B = \{b\}$, Law 5b for $A = \{b\}$ and Law 8):

Law 9 $(N_1 \parallel N_2)/a = (N_1[a \to b] \parallel N_2[a \to b])/b$ provided $b \notin \alpha(N_1) \cup \alpha(N_2)$

3   Fair Semantics

A semantics for specifying and checking liveness properties ('something good eventually happens') usually has to consider some sort of fairness. We will define a semantics that incorporates the progress assumption, also called weak fairness, i.e. the assumption that a continuously enabled activity should eventually occur.

In [13], we defined a fair semantics and determined the coarsest compositional refinement of it for *safe* nets. This result does not directly carry over to general nets, but it can be generalized when we use a slightly peculiar definition of fairness; we will discuss this peculiarity in detail after the definition.

Fig. 1.

But first, we have to discuss the impact of read arcs on the progress assumption. Classically, an infinite firing sequence $M_N[t_0\rangle M_1[t_1\rangle M_2 \ldots$ would be called fair if we have: if some transition t is enabled under all M_i for i greater than some j, then $t = t_i$ for some $i > j$. With this definition, the sequence t^ω of infinitely many t's would not be fair in the net of Figure 1, since t' is enabled under all states reached, but never occurs. But, in fact, t' is not continuously enabled, since every occurrence of t disables it momentarily, compare [10,12]. Thus, t^ω should be fair. On the other hand, if t were on a read arc instead of a loop, t^ω should not be fair: t would only repeatedly check the presence of a resource without actually using it. To model this adequately, we will require in the definition of fairness that a continuously enabled t is enabled also *while* each t_i with $i > j$ is firing, i.e. enabled under $M_i - {}^\bullet t_i$.

Definition 1. For a transition t, a finite firing sequence $M_N[t_0\rangle M_1$ $[t_1\rangle M_2 \ldots M_n$ is called *t-fair*, if M_n does not enable t. An infinite one $M_N[t_0\rangle M_1[t_1\rangle M_2 \ldots$ is called *t-fair*, if we have: For no j, t is enabled under all $M_i - {}^\bullet t_i$ for all $i > j$. If a finite or infinite firing sequence w violates the respective requirement, we say that t is *eventually enabled permanently* in w.

A finite or infinite firing sequence is *fair*, if it is t-fair for all transitions t of N; we denote the set of these sequences by $FairFS(N)$. The *fair language* of N is the set $Fair(N) = \{v \mid v = l(w) \text{ for some } w \in FairFS(N)\}$ of *fair traces*.

What we require in the case of an infinite sequence is stricter than the more usual requirement that, if t is enabled under all $M_i - {}^\bullet t_i$ for i greater than some j, then $t = t_i$ for some $i > j$. For safe nets, these requirements coincide:

Proposition 2. *Let N be a net without self-concurrency (or, in particular a safe net), t a transition and $M_N[t_0\rangle M_1[t_1\rangle M_2 \ldots$ an infinite firing sequence. Assume further that if t is enabled under all $M_i - {}^\bullet t_i$ for i greater than some j, then $t = t_i$ for some $i > j$. Then this sequence is t-fair.*

In our constructions, we have to work with nets that may not be safe; but in the end, we are only interested in safe nets. Thus, it is of no particular importance what a fair firing sequence of an unsafe net is, and we can choose a definition that is technically convenient. This means here that we can obtain a fairness-respecting precongruence easily, i.e. in the same way as for safe nets.

Next, we will determine the coarsest precongruence for parallel composition that respects fair-language-inclusion; this is just the right relation if we want to build systems compositionally and are interested in the fair language. Theorems 4 and 5 generalize respective results of [13] from safe nets to general nets.

Definition 3. A net N_1 is a *fair implementation* of a net N_2, if $\alpha(N_1) = \alpha(N_2)$ and $Fair(N_1\|N) \subseteq Fair(N_2\|N)$ for all nets N.

For a net N, the *fair failure* semantics is the set of the *fair refusal pairs* defined by $\mathcal{FF}(N) = \{(v, X) \mid X \subseteq \Sigma \text{ and } v = l(w) \text{ for some, possibly infinite, firing sequence } w \text{ that is } t\text{-fair for all transitions } t \text{ with } l(t) \in X \cup \{\lambda\}\}$.

We write $N_1 \leq_{\mathcal{FF}} N_2$, if N_1 and N_2 have the same alphabet and $\mathcal{FF}(N_1) \subseteq \mathcal{FF}(N_2)$. If $N_1 \leq_{\mathcal{FF}} N_2$ and $N_2 \leq_{\mathcal{FF}} N_1$, we write $N_1 =_{\mathcal{FF}} N_2$ and call the nets *fair-congruent*.

The motivation for this definition is as follows: assume N_1 is a fair implementation of the specification N_2, N_2 is a component of a parallel system and we replace this component by N_1; then we will get only fair behaviour that is allowed by N_2, i.e. that is possible when N_2 is used. The intuition for $(v, X) \in \mathcal{FF}(N)$ is that all actions in X can be *refused* when v is performed – in the sense that fairness does not force performance of these actions; yet in other words, these actions are treated correctly w.r.t. fairness.

For finite or infinite sequences u and v over Σ and $A \subseteq \Sigma$, $u \parallel_A v$ is the set of all sequences w over Σ such that we can write u, v and w as sequences $u = u_1 u_2 \ldots$, $v = v_1 v_2 \ldots$ and $w = w_1 w_2 \ldots$ of equal finite or infinite length such that for all suitable $i = 1, 2, \ldots$ one of the following cases applies:

- $u_i = v_i = w_i \in A$
- $u_i = w_i \in (\Sigma - A)$ and $v_i = \lambda$
- $v_i = w_i \in (\Sigma - A)$ and $u_i = \lambda$

In this definition, λ's are inserted into the decomposition of u and v to describe the interleaving of actions from $\Sigma - A$, while actions from A are synchronized.

Theorem 4. *For nets N_1 and N_2 with $\alpha(N_1) \cap \alpha(N_2) = A$ we have*

$$\mathcal{FF}(N_1 \parallel N_2) = \{(v, X) \mid \exists (v_i, X_i) \in \mathcal{FF}(N_i), \; i = 1, 2 :$$
$$v \in v_1 \parallel_A v_2 \text{ and } X \subseteq ((X_1 \cup X_2) \cap A) \cup (X_1 \cap X_2)\}$$

The proof of this theorem is similar to proofs that can be found e.g. in the full version of [13], and the proof is simpler than the proof of the corresponding Theorem 5.10 there. The crucial point, where the subtlety of our unusual fairness definition comes into play, concerns inclusion: when we construct (v, X) from (v_1, X_1) and (v_2, X_2) as described on the right hand side of the above equation, we combine firing sequences w_1 of N_1 and w_2 of N_2 to a firing sequence w of $N_1 \parallel N_2$. Now consider a 'combination' transition (t_1, t_2) of $N_1 \parallel N_2$, and assume that t_1 occurs infinitely often in w_1 and t_2 in w_2, i.e. they are both treated fairly according to the more standard definition. This does not ensure that (t_1, t_2) occurs in w at all. But if t_1 is repeatedly disabled in w_1, then (t_1, t_2) is repeatedly disabled in w.

Theorem 5. *i) For nets N_1 and N_2, N_1 is a fair implementation of N_2 if and only if $N_1 \leq_{\mathcal{FF}} N_2$.*

ii) For nets with some fixed alphabet, inclusion of $\mathcal{FF}$-semantics is fully abstract w.r.t. fair-language inclusion and parallel composition of nets, i.e. it is the coarsest precongruence for parallel composition that refines fair-language inclusion.

Theorem 6. *The relation $\leq_{\mathcal{FF}}$ is a precongruence w.r.t. relabelling and hiding.*

In [13], it is shown that for safe nets $\leq_{\mathcal{FF}}$ is decidable. Further, an operation is considered that will be of interest later. We define this operation, state that it preserves safety and quote a result from [13].

Definition 7. An *elongation* of N is obtained by choosing a transition t, adding a new unmarked place s and a new λ-labelled transition t' with ${}^{\bullet}t' = \{s\}$ and $t'{}^{\bullet} = t^{\bullet}$ and, finally, redefining $t^{\bullet}$ by $t^{\bullet} := \{s\}$.

Theorem 8. *If a net N_2 is an elongation of a net N_1, then one of the nets is safe if and only if the other one is; in this case, $N_1 =_{\mathcal{FF}} N_2$.*

We close this section with a notion, also taken from [13] (and similar to one used for ordinary failure semantics), that will make Definition 17 more suggestive. $(v, X) \in \mathcal{FF}(N)$ means that N can perform v in such a way that all internal actions and all actions in X are treated fairly. Hence, $(v, X) \notin \mathcal{FF}(N)$ means that either N cannot perform v in such a way that all internal actions are treated fairly or it can, but whichever way it performs v, it treats some action in X unfairly. The latter means that some $x \in X$ is continuously enabled from some point onward; if N is on its own, it certainly performs such an x – but as a component of a larger system, N simply offers such an x. We therefore define:

Definition 9. If for a net N and some $(v, X) \in \Sigma^{\infty} \times \mathcal{P}(\Sigma)$ we have $(v, X) \notin \mathcal{FF}(N)$, then we say that N **surely offers** (some action of) X **along** v.

If N **surely offers** X **along** v and, in a run of a composed system, N as a component performs v while the environment offers in this run each action in X, then some action in X will be performed in this run.

4   Partial S-Invariants

Corresponding to the interests of this paper, we will only consider a restricted form of S-invariants (and consequently of partial S-invariants) defined as follows.

Definition 10. Let N be a net; a set P of places has *value* n (under a marking M) if the places in P carry together n tokens under the initial marking (under M resp.). An *S-invariant* of N is a set $P \subseteq S$ of value 1 such that for every transition t we have $|P \cap {}^{\bullet}t| = |P \cap t^{\bullet}|$.

A *partial S-invariant* of N with input $I \subseteq \alpha(N)$ and output $O \subseteq \alpha(N)$ is a set P such that for every transition t we have: if the label of t is in I, then $|P \cap t^{\bullet}| - |P \cap {}^{\bullet}t| = 1$; if the label of t is in O, then $|P \cap {}^{\bullet}t| - |P \cap t^{\bullet}| = 1$; if the label of t is neither in I nor in O, then $|P \cap {}^{\bullet}t| = |P \cap t^{\bullet}|$. We call such a P an (I, O, n)-invariant, if additionally P has value n.

N is *covered* by S-invariants if each place is contained in an S-invariant. N is *covered* by partial S-invariants $P_1, P_2, \ldots, P_n$, if $n = 0$ and N is covered by S-invariants or $n > 0$ and each place is contained in an S-invariant or some P_i.

For $I \subseteq \Sigma$, $O \subseteq \Sigma$ and $n \in \mathbb{N}_0$, the (I, O, n)-*component* $C(I, O, n)$ is a net consisting of a place s with n tokens, an i-labelled transition t with $t^{\bullet} = \{s\}$ and ${}^{\bullet}t \cup \hat{t} = \emptyset$ for each $i \in I$ and an o-labelled transition t with ${}^{\bullet}t \cup \hat{t} = \{s\}$ and $t^{\bullet} = \emptyset$ for each $o \in O$.

We only use S-invariants where we just count the places in the pre- and postset of a transition. In general S-invariants, each place has a weight and, instead of counting as in $|P \cap {}^\bullet t|$, one adds up the respective weights; thus, in *our* S-invariants this weight is always one. Furthermore, the value does not have to be 1 in general S-invariants. Similarly, we just count places in the definition of partial S-invariants, and we restrict attention to the case where firing of a transition changes the value by 1, 0 or -1. There is no problem in generalizing the definition to weighted places and general changes, except that one would need arc weights to define the analogue of (I, O, n)-components. The following result is well-known and easy:

Theorem 11. *If P is an S-invariant of a net N, then P has value 1 under all reachable markings. If N is covered by S-invariants, then N is safe.*

Next, we state a number of properties regarding our new notions; their proofs are easy. (In particular, 4 follows from 3.)

Proposition 12.
1. *If P_i is an (I_i, O_i, n_i)-invariant of N_i, $i = 1, 2$, such that $I_1 \cap I_2 = \emptyset = O_1 \cap O_2$, then $P_1 \cup P_2$ is an $((I_1 \cup I_2) - (O_1 \cup O_2), (O_1 \cup O_2) - (I_1 \cup I_2), n_1 + n_2)$-invariant of $N_1 \| N_2$. In particular, P_1 or P_2 can be $\emptyset$, which is an $(\emptyset, \emptyset, 0)$-invariant.*
2. *If N_i is covered by partial S-invariants $P_{i1}, P_{i2}, \ldots, P_{in}$, $i = 1, 2$, then $N_1 \| N_2$ is covered by the partial S-invariants $P_{11} \cup P_{21}, P_{12} \cup P_{22}, \ldots, P_{1n} \cup P_{2n}$.*
3. *An $(\emptyset, \emptyset, 1)$-invariant is an S-invariant.*
4. *If N is covered by partial S-invariants $P_1, P_2, \ldots, P_n$, $n > 0$ and P_1 is an $(\emptyset, \emptyset, 1)$-invariant, then N is covered by partial S-invariants $P_2, \ldots, P_n$.*
5. *$C(I, O, n)$ is covered by an (I, O, n)-invariant.*

Corollary 13. *If N is covered by an (I, O, n)-invariant and $m \in \mathbb{N}_0$ with $m + n = 1$, then $N \| C(O, I, m)$ is safe.*

Proof. By 12.5 and .2, $N \| C(O, I, m)$ is covered by a partial S-invariant that according to Part 1 is an $(\emptyset, \emptyset, 1)$-invariant. By Part 4 $N \| C(O, I, m)$ is covered by S-invariants and, thus, safe by 11.   □

Partial S-invariants and the notion of covering are also consistent with relabelling and hiding in the following sense.

Proposition 14.
1. *Let N be a net, $a \in \Sigma$ and $A \subseteq \Sigma$, such that $\alpha(N) \cap A \subseteq \{a\}$; let the relabelling f map a to A and be the identity otherwise.*
   If P is an (I, O, n)-invariant of N, then P is an $(f(I), f(O), n)$-invariant of $N[f]$. If N is covered by partial S-invariants $P_1, P_2, \ldots, P_n$, then $N[f]$ is covered by the same partial S-invariants.
2. *Let P be an (I, O, n)-invariant of N that does not meet $A \subseteq \Sigma$, i.e. $I \cap A = O \cap A = \emptyset$. Then P is an (I, O, n)-invariant of N/A. If N is covered by partial S-invariants $P_1, P_2, \ldots, P_n$ that do not meet A, then N/A is covered by the same partial S-invariants.*

We close this section with the key result that will allow to insert some sort of interface description into a parallel composition; applications will be shown in Section 6. If N has an (I, O, n)-invariant, then $C(I, O, n)$ is a sort of abstraction of N. Hence, adding it to N as parallel component does not change the behaviour: one can show that the firing sequences stay the same and, in such a sequence, a transition is eventually enabled permanently before if and only if it is after the addition.

Proposition 15. *If N has an (I, O, n)-invariant P, then $N \| C(I, O, n) =_{\mathcal{FF}} N$.*

This proposition can be used to reduce some $N \| N'$ to a smaller fair-congruent net. It might even be that – while N' is quite manageable, in particular not too large – the precise definition of N and its size depend on a parameter, i.e. its size may be arbitrarily large. If we know at least that N has an (I, O, n)-invariant, then $N \| N' =_{\mathcal{FF}} N \| C(I, O, n) \| N'$ by 14 and 5, i.e. we can insert $C(I, O, n)$ into the parallel composition; now we might be able to reduce $C(I, O, n) \| N'$ in a way that would not be possible for N' in isolation; if the component $C(I, O, n)$ perseveres (like a catalyst), we can remove it after the reduction. Corollary 16 describes this compositional reduction, also for the more general case of several partial S-invariants. Observe that 15 is also valid for language-equivalence or bisimilarity in place of fair-congruence; hence, partial S-invariants also support the reduction method if these congruences are used.

In this method, $C(I, O, n)$ is an interface description; it restricts the behaviour of N' to what is relevant in $N \| N'$. Important is that this interface description is verified on N syntactically, while the reduction deals with the behaviour of N' only, and not with that of N. This is in contrast to [5], where some interface description is guessed, used in the reduction of N' to N'' and then verified during the further reduction of $N \| N''$. The latter considers the behaviour of N, which is not possible in the fixed-point approach where N is parametric, i.e. not completely known.

Corollary 16.

1. *Assume that N has an (I, O, n)-invariant and $C(I, O, n) \| N' =_{\mathcal{FF}} C(I, O, n) \| N''$ or $C(I, O, n) \| N' =_{\mathcal{FF}} N''$. Then $N \| N' =_{\mathcal{FF}} N \| N''$.*
2. *Assume N has some partial S-invariants, C is the parallel composition of the respective (I, O, n)-components and C' the parallel composition of just some of these. If $C \| N' =_{\mathcal{FF}} C' \| N''$, then $N \| N' =_{\mathcal{FF}} N \| N''$.*

5 Liveness Properties of MUTEX-Solutions

This section repeats necessary material from [2] regarding the correctness of MUTEX-solutions and introduces our first application example. First, based on the $\mathcal{FF}$-semantics, we will specify correctness with a safety and a liveness requirement. *Safety* requires that no two users are in their critical sections simultaneously; if one user enters, then he must leave before another can enter. Our definition of *liveness* is explained after the definition; we only remark at this point that a MUTEX-solution can only guarantee that each requesting user can enter his critical section, if in turn each user e.g. guarantees to leave after entering.

Definition 17. We call a finite or infinite sequence over $I_n = \{r_i, e_i, l_i \mid i = 1, \ldots, n\}$ *legal* if r_i, e_i and l_i only occur cyclically in this order for each i. An *n-MUTEX net* is a net N with $l(T) \subseteq \{r_i, e_i, l_i \mid i = 1, \ldots, n\} \cup \{\lambda\}$.

Such a net is a *correct n-MUTEX-solution*, if N satisfies *MUTEX-safety*, i.e. e- and l-transitions occur alternatingly in a legal trace, and satisfies *MUTEX-liveness* in the following sense. Let $w \in I_n^* \cup I_n^\omega$ be legal and $1 \leq i \leq n$; then:

1. Each e_i in w is followed by an l_i, or N **surely offers** $\{l_i\}$ **along** w.
2. Assume each e_j is followed by l_j in w. Then either each r_i is followed by e_i or N **surely offers** X **along** w where X consists of those e_j where some r_j in w is not followed by e_j.
3. Assume in w each r_j is followed by e_j and each e_j by l_j. Then either r_i occurs and each l_i is followed by another r_i in w or N **surely offers** $\{r_i\}$ **along** w.

An n-MUTEX net N is used in a complete system consisting of N and its environment comprising the users, and these two components synchronize over I_n. The first part of MUTEX-liveness says that, if user i enters (performs e_i together with the scheduler N), later tries to leave (enables an l_i-transition) and does not withdraw (does not disable the transition again), then he will indeed leave; otherwise l_i would be enabled continuously in the complete system violating fairness. (Technically, recall how the refusal sets of fair refusal pairs are composed according to Theorem 4: the complete system is fair, i.e. Σ is refused, only if one of the components refuses l_i.)

In other words, if user i does not leave again, then he is not willing to leave since l_i is offered to him. This is a user misbehaviour, but the behaviour of the scheduler N is correct. As a consequence, if N satisfies Part 1, we can assume that each e_j is followed by l_j. Under this assumption, the second part of MUTEX-liveness says that each request of i is satisfied, unless some requesting user is permanently offered to enter. In the latter case, that user is misbehaving by not accepting this offer, and again N is working correctly.

Now we can assume that each request is satisfied. Under this assumption, i requests infinitely often or N at least offers him to request. The latter is not a user misbehaviour because each user is free not to request again.

The following is obvious from the definitions.

Proposition 18. *If* $N_1 \leq_{\mathcal{FF}} N_2$ *and* N_2 *satisfies MUTEX-safety, MUTEX-liveness resp., then also* N_1 *satisfies MUTEX-safety, MUTEX-liveness resp.*

For token-passing solutions, MUTEX-safety is usually easy to prove with an S-invariant; for the two families of solutions we will treat, sufficient arguments for MUTEX-safety are given in [2]. Hence, we will concentrate on proving MUTEX-liveness, which is more difficult. As already mentioned, application of the fixed-point approach does not seem feasible, since the more users an n-MUTEX net has to serve, the more visible actions it has.

The essential result from [2] (Theorem 20 below) states that, for so-called *user-symmetric* nets (see [2] for the precise definition), it is enough to check a version where only the actions of one user are visible. For our two families of solutions, [2] also gives sufficient arguments that each of their nets are user-symmetric.

Our first family of solutions is attributed to Le Lann; each of its nets is a ring of components, one for each user. Figure 2(a) shows the component LLU of the first user (except that the actions r, e and l should be indexed with 1). This user owns the access-token, the token on the right, while the other users look the same, except that they do not have this token. The first user can *request* access with r (i.e. by firing the r-labelled transition t_r) and *enter* the critical section with e. When he *leaves* it with l, he passes the token to the next user, i.e. the l-transition must be merged with the p-transition of the next user; p stands for previous, the respective transition produces the access-token coming from the previous user.

If the first user is not interested in entering the critical section, the token is passed by the n-transition to the *next* user; i.e. the n-transition must also be merged with the p-transition of the next user and then hidden. It is important that the n-transition checks the token on $\bullet t_r$ with a read arc, since this way the user is not prevented from requesting in a firing sequence with infinitely many checks. Intuitively, Le Lann's solution is correct, since the access-token is always passed around, and if a user has requested and the token reaches his ring component, the user will enter and leave his critical section before passing the token on.

For a Le-Lann-ring built as just explained, it should be clear that after firing the n-transition we get a symmetric marking where the second user owns the access-token.

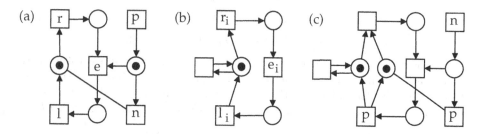

Fig. 2.

We next define the first-user view of a solution. For this we assume that each user except the first one has a standard behaviour modelled by the i-th *standard user* SU_i shown in Figure 2(b): when non-critical, such a user works internally or requests; after requesting, he enters and leaves as soon as possible. Then we abstract away all visible behaviour of these users with a suitable hiding.

Definition 19. The *first-user view* $FUV(N)$ of a net N is $(N\|(SU_2\|\ldots\|SU_n))/\{r_i,e_i,l_i\,|\,i=2,\ldots,n\}$.

The first-user view of Le Lann's ring is a ring where the first component looks like LLU in Figure 2(a), while all other components are the composition of LLU (with actions suitably indexed with i) and SU_i, i.e. they look essentially like LSU in (c). Actually, the labelling is different, but it has been chosen in Figure 2 such that we can directly construct the first-user view of each Le-Lann-ring from LLU and a suitable number of copies of LSU; also, LSU should have two additional unmarked places from SU_i, which are omitted since they are simply duplicates in LSU. We only have to study the first-user views due to the following theorem.

Theorem 20. *Assume that a safe n-MUTEX net N is user-symmetric and satisfies MUTEX-safety. Then, N is a correct n-MUTEX solution if $FUV(N)$ is a correct 1-MUTEX solution.*

6 Correctness Proofs

In this section, we will develop the fixed-point approach for two families of solutions to the MUTEX-problem already discussed in [2]. In both, an access-token is passed around which guarantees mutual exclusion.

6.1 Le Lann's Ring

We first describe formally how the first-user view of a Le-Lann-ring as explained in the previous section can be constructed from LLU and several copies of LSU, which are shown in Figure 2. We first define the 'Le-Lann-chain' LLC_n inductively, which is a chain of n copies of LSU.

$$LLC_1 = LSU$$

$$LLC_{n+1} = (LSU[n \to p']\|LLC_n[p \to p'])/p'$$

This chain has one n-labelled transition, which moves the access-token to the chain, and two p-labelled transitions 'at the other end', which remove the access-token from the chain. Clearly, this net is not safe, since the n-labelled transition can fire several times in a row. Now we close these chains to rings with LLU:

$$LL_n = (LLU\|LLC_{n-1}[n \to \{l,n\}])/\{p,n\}$$

We first observe:

Proposition 21.
1. LSU is covered by an $(\{n\},\{p\},0)$-invariant. LLU is covered by a $(\{p\},\{l,n\},1)$-invariant.
2. For all n, LLC_n is covered by an $(\{n\},\{p\},0)$-invariant and has alphabet $\{n,p\}$.
3. LL_n is safe for all n.

Proof. 1. There are two circuits in LSU, each containing one of the marked places; they are S-invariants. The places on the two paths from the n-labelled to the two p-labelled transitions form an $(\{n\}, \{p\}, 0)$-invariant. The case of LLU is very similar.

2. The claim about the alphabet follows by an easy induction. The first claim can also be shown by induction, where $i = 1$ follows from 1. By 1. and 14.1 and induction, $LSU[n \to p']$ is covered by a $(\{p'\}, \{p\}, 0)$-invariant and $LLC_n[p \to p']$ is covered by an $(\{n\}, \{p'\}, 0)$-invariant. Thus, by 12.1 and .2 and 14.2, LLC_{n+1} is covered by an $(\{n\}, \{p\}, 0)$-invariant.

3. By 1., LLU is covered by a $(\{p\}, \{l, n\}, 1)$-invariant. By 2. and 14.1 (observe that $l \notin \alpha(LLC_{n-1})$), $LLC_{n-1}[n \to \{l, n\}]$ is covered by an $(\{l, n\}, \{p\}, 0)$-invariant. Now by 12.1 and .2, $LLU \| LLC_{n-1}[n \to \{l, n\}]$ is covered by an $(\emptyset, \emptyset, 1)$, and by 14.2, 12.4 and 11, LL_n is safe. □

From the observations in [2] – underpinned by the safeness just shown – we can apply Theorem 20 to each Le-Lann-ring assuming that MUTEX-safety is satisfied; it is planned to refine the notion of partial S-invariant such that it supports the proof of MUTEX-safety. Hence, it remains to prove that LL_n satisfies MUTEX-liveness, where we identify r, e and l with r_1, e_1 and l_1; this has already been shown for $n = 2, 3, 4$ in [2]. We start with two lemmata.

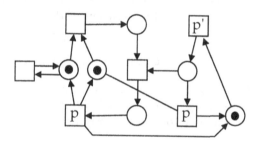

Fig. 3.

Lemma 22. $LSU[n \to p'] \| C(\{p\}, \{p'\}, 1) =_{\mathcal{FF}} C(\{p'\}, \{p\}, 0) \| C(\{p\}, \{p'\}, 1)$

The left-hand-side net is shown in Figure 3, the right-hand-side net is simply a circuit with a p'- followed by a p-transition; we omit the proof. The next lemma shows how to reduce one LSU-component. This only works due to the presence of C, which will arise from a partial S-invariant in the proof of Theorem 24; C also ensures safeness of the nets the lemma deals with.

Lemma 23. *Let* $C = C(\{l, n\}, \{p'\}, 0)$; *then* $(LLU \| LSU[n \to p'])/p \| C =_{\mathcal{FF}} LLU[p \to p'] \| C$.

Proof. By Law 7, $(LLU \| LSU[n \to p'])/p \| C =_{\mathcal{FF}} (LLU \| LSU[n \to p'] \| C)/p$. Now LLU has a $(\{p\}, \{l, n\}, 1)$-invariant by 21.1, C has a $(\{l, n\}, \{p'\}, 0)$-invariant by 12.5 and, hence, their parallel composition has a $(\{p\}, \{p'\}, 1)$-invariant by 12.1. Thus, applying Corollary 16, we can reduce $LSU[n \to p']$ to

$C(\{p'\}, \{p\}, 0)$ and arrive at $(LLU \| C(\{p'\}, \{p\}, 0) \| C)/p$, which by Law 7 again is fair-congruent to $(LLU \| C(\{p'\}, \{p\}, 0))/p \| C$. In the latter net, the unique p'-labelled transition has as postset the place of $C(\{p'\}, \{p\}, 0)$, which in turn has an internal transition as postset; now one sees that the net is simply an elongation of the right-hand-side net in the lemma, and we are done by Theorem 8. $\qquad \square$

Now we can apply the central Corollary 16 again to obtain the fixed-point result and the correctness we are aiming for in this subsection.

Theorem 24. *For $n > 2$, $LL_n =_{\mathcal{FF}} LL_{n-1}$. For $n > 1$, Le Lann's ring is a correct MUTEX-solution.*

Proof. Once we have shown the congruence, we have that each LL_n is fair-congruent to LL_2, which has been shown to satisfy MUTEX-liveness in [2]. Hence, each LL_n satisfies MUTEX-liveness by 18, and the second part follows with Theorem 20 as explained above.

Thus, we will transform

$$LL_n = (LLU \| (LSU[n \to p'] \| LLC_{n-2}[p \to p']) /p'[n \to \{l, n\}]) /\{p, n\}$$

preserving fair-congruence. First, we can commute the hiding of p' with the following renaming by Law 5a and move it out of the outer brackets by Law 7 (since $p' \notin \alpha(LLU)$) obtaining

$$(LLU \| (LSU[n \to p'] \| LLC_{n-2}[p \to p']) [n \to \{l, n\}]) /\{p', p, n\}.$$

Since $l, n \notin \alpha(LSU[n \to p'])$, we can move the right-most renaming to the left by Law 6; since the resulting component $L = LLC_{n-2}[p \to p'][n \to \{l, n\}]$ does not have p in its alphabet due to renaming, we can apply Law 7 to move $/p$ and get

$$((LLU \| LSU[n \to p']/p) \| LLC_{n-2}[p \to p'][n \to \{l, n\}]) /\{p', n\}.$$

The component L has an $(\{l, n\}, \{p'\}, 0)$-invariant by propositions 21.2 and 14.1. With Lemma 23 and Corollary 16, we obtain

$$(LLU[p \to p'] \| LLC_{n-2}[p \to p'][n \to \{l, n\}]) /\{p', n\}.$$

By Law 3, we can commute the second renaming with the third one and then suppress it by Law 9; with the definition we arrive at LL_{n-1}. $\qquad \square$

6.2 Dijkstra's Token-Ring

The second MUTEX-solution we want to consider is Dijkstra's token ring [4]; we will describe the first-user view of such a ring analogously to the above. Figure 4 shows the component DU of the first user who owns the access-token on place tok. In Dijkstra's token ring, the user keeps the token when leaving the critical section, so he can use it repeatedly until an order for the token is received (action ro) from the next component; observe the read arc from the now unmarked

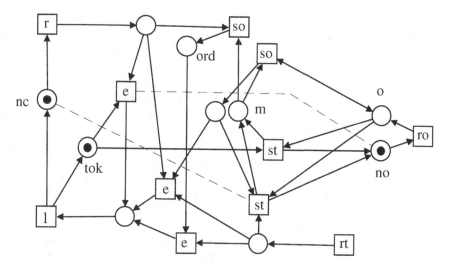

Fig. 4.

place *no* (no order). Then the token is sent (action *st*) to the next component, say clockwise. Now *m* is marked, indicating that this component misses the token. If the user requests again, an order is sent (upper *so*-transition) counterclockwise to the previous component. Alternatively, the token might have moved on clockwise from the next component such that another order is received from this component (*ro*) and forwarded counter-clockwise (lower *so*-transition). If consequently a token is received (*rt*) from the previous component, a request is served in case one is pending or otherwise the token is forwarded (*st*).

The other components *DSU* of the first-user view are obtained similarly as above, although this time we keep the actions for the ring communication as they are – in this case *so*, *ro*, *st* and *rt*. In detail: to get *DSU* from *DU* we hide *r*, *e* and *l* and move the token on *tok* to *m*; and we duplicate the place *nc*, but without the read arc and with an internal loop transition instead; compare the black part of Figure 5 below. Now we define a chain DC_n and a ring DTR_n as in the previous subsection, taking the first user component as chain of length 1 this time.

$$DC_1 = DU$$

$$DC_{n+1} = (DC_n[ro \to o, st \to t] \| DSU[so \to o, rt \to t])/\{o, t\}$$

Such a chain has actions *r*, *e*, *l*, *so* and *rt* 'at one end' and *ro* and *st* 'at the other end'. We close it to a ring with *DSU*:

$$DTR_n = (DC_{n-1}[ro \to so, so \to ro, rt \to st, st \to rt] \| DSU)/\{so, ro, st, rt\}$$

Again, we will reduce a sequence of components (this time all of type *DSU*) to a shorter one. Interestingly, we have to reduce a sequence of length 3, because reducing 2 components to 1 cannot work: in a sequence of 2 components, the first one can receive the token, send it to the second component internally and – since it is missing the token now – send an order as the next visible action; thus, it performs $rt\ so$ in sequence. This is not possible for one component, which after receiving the token must send it on (st) before performing so.

Since this more complicated application example is difficult to treat by hand, application of a tool is advisable. Since a tool for deciding fair (pre)congruence was not available, a tool for a related precongruence was used; this made some ad-hoc measures necessary as described below.

We make use of two partial S-invariants which we give now. DU is covered by the $(\{rt\}, \{st\}, 1)$-invariant $\{gt, c, tok\}$ making use e.g. of the S-invariant $P = \{c, tok, m, ord, ord'\}$, and it also has the $(\{rt\}, \{so\}, 0)$-invariant $\{gt, c, tok, m\}$. Use of this second partial S-invariant is not really necessary, but it slightly simplifies the behaviour of the nets below, and it makes a little smaller the reachability graphs the tool has to deal with.

Similarly to DU, DSU is covered by an $(\{rt\}, \{st\}, 0)$-invariant, and it also has an $(\{rt\}, \{so\}, 1)$-invariant. With induction, one shows from this that DC_n is covered by an $(\{rt\}, \{st\}, 1)$-invariant and has an $(\{rt\}, \{so\}, 0)$-invariant. A first consequence is then that each DTR_n is safe.

Formally, we will reduce D_3 to D_2 with the following definitions:

$$D_2 = (DSU[ro \to o, st \to t] \parallel DSU[so \to o, rt \to t])/\{o, t\}$$
$$D_3 = (D_2[ro \to o, st \to t] \parallel DSU[so \to o, rt \to t])/\{o, t\}$$

It is not surprising that D_2 (in a composition according to the above partial S-invariants, i.e. with $C = C(\{st\}, \{rt\}, 1) \parallel C(\{so\}, \{rt\}, 0)$) reacts faster than D_3, and we have verified this with FastAsy according to the notion of faster-than explained e.g. in [2]. (For this step, the help of Elmar Bihler is gratefully acknowledged.) As described in [2], this implies fair precongruence, where in the present setting we additionally check that the two nets have the same alphabet.

Also, we have essentially verified (again with FastAsy) that D_3 is faster than an elongation of D_2. This elongation concerns the transition that, in the second DSU-component, represents leaving the critical section and the ro-labelled transition; this slows down the reactions of st and so. Together, the verification results imply fair congruence.

More precisely, in order to perform the second verification step with FastAsy, the additional DSU-component of D_3 was replaced by the following net DSU' (just as the additional LSU-component was transformed in Lemma 22). Figure 5 shows in black $DSU \parallel C$; DSU' is DSU with the additional grey transition sc, which is some shortcut, hence the full figure shows $DSU' \parallel C$. The following lemma states the correctness of this replacement, whose proof we have to omit. The essential problem is that a firing sequence of $DSU \parallel C$ that is fair to all internal transitions can fire in $DSU' \parallel C$, but it might fail to be sc-fair.

Lemma 25. *Let* $C = C(\{st\}, \{rt\}, 1) \parallel C(\{so\}, \{rt\}, 0)$; *then we have* $DSU \parallel C =_{\mathcal{FF}} DSU' \parallel C$.

As described above, we have obtained the following lemma – using Lemma 25 and FastAsy:

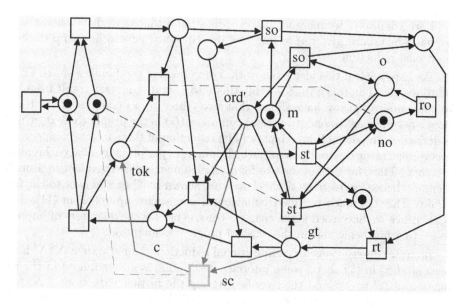

Fig. 5.

Lemma 26. *With C as above, $D_3 \parallel C =_{\mathcal{FF}} D_2 \parallel C$.*

We now proceed as in the proof of Theorem 24: We unfold DTR_{n+3} by definition such that we can isolate D_3 in it and then apply the reduction that is possible due to the above partial S-invariants and Lemma 26. Thus, we can show

$$
\begin{aligned}
DTR_{n+3} &= (DC_n[ro \to so, so \to ro, rt \to st, st \to rt] \parallel D_3)/\{so, ro, st, rt\} \\
&=_{\mathcal{FF}} (DC_n[ro \to so, so \to ro, rt \to st, st \to rt] \parallel D_2)/\{so, ro, st, rt\} \\
&= DTR_{n+2}.
\end{aligned}
$$

This way, we can reduce rings with at least 4 components to DTR_3; DTR_3 and DTR_2 were shown to satisfy MUTEX-liveness in [2]. This finishes our second application:

Theorem 27. *For $n > 3$, $DTR_n =_{\mathcal{FF}} DTR_{n-1}$. For $n > 1$, Dijkstra's token ring is a correct MUTEX-solution.*

7   Conclusion and Related Literature

In this paper, we have defined partial S-invariants for a setting where nets are composed by merging transitions – modelling a parallel composition with synchronous communication. We have shown how to derive from partial S-invariants of the components (partial) S-invariants and safety of composed nets. As already mentioned, this is analogous to the partial S-invariants of [8] for a setting where nets are composed by merging places. More importantly, we have shown how partial S-invariants give rise to interface descriptions that can be useful in compositional reduction.

We have applied this idea to show the correctness of two families of MUTEX-solutions based on token rings – exploiting also a vital symmetry result from [2]. In this application, we have shown that two (three resp.) components of these rings are equivalent to one (two resp.) component(s) in the proper context, which is derived from partial S-invariants . (The reduction of three to two components was verified using the tool FastAsy, which compares performance of asynchronous systems.) Thus, for each of the two families each member is equivalent to a small member; hence, correctness of this small net shown in [2] carries over to the full family. This approach is called behavioural fixed-point approach in [11]. The equivalence we have used is the coarsest congruence for the operators of interest respecting fair behaviour in the sense of progress assumption.

Interface descriptions for compositional reduction of some system $N\|N'$ have been studied in [5]: there, some interface description is guessed, used in the reduction of N' to N'' and then verified during the further reduction of $N\|N''$. The last step considers the behaviour of N, and this is not feasible in the fixed-point approach where N is the rest of the ring, i.e. not fixed. In contrast, partial S-invariants give verified interface descriptions; we have derived them with induction from a syntactic inspection of the ring components, i.e. without considering their behaviour by constructing a reachability graph and possibly encountering a state explosion. The latter points out the potential use of partial S-invariants in compositional reduction in general.

The fixed-point approach is very similar to the approach in [15]: there, a preorder is used instead of an equivalence; an invariant (or representative) process I has to be found manually, and then it is checked that P is less than I and that, whenever some Q is less than I, $P\|Q$ is less than I. This implies that the composition of any number of components P is less than I; with a suitable preorder and I, this implies that some relevant properties hold for all these compositions. As an example, MUTEX-safety of a version of Le Lann's ring is shown. More generally, [3] considers networks that might be built from various different types of components according to a network grammar, and a representative for each type is used. As an example, MUTEX-safety of Dijkstra's token ring is shown.

An important point is that the referenced papers use labelled transition systems while we use Petri nets that in themselves are usually smaller; this is in particular an advantage when determining partial S-invariants. Also, they allow to consider the subtleties of the progress assumption (i.e. the difference between loops and read arcs). Hence, we use a behavioural equivalence that takes the progress assumption into account, in contrast to the above papers that use failure-inclusion [15], a refinement thereof [11] or a simulation preorder [3]. Liveness properties often only hold under the progress assumption.

It is planned to verify also the third family of MUTEX-solutions considered in [2]. The problem is that in this family data from an infinite set are used (unique identifiers for the components), and it has to be checked whether these systems can be considered as data-independent in some sense. An interesting case study involving data-independence and a reduction as in [15] and [3] can be found in [7].

References

1. P. Baldan, N. Busi, A. Corradini, and M. Pinna. Functional concurrent semantics for Petri nets with read and inhibitor arcs. In C. Palamidessi, editor, *CONCUR 2000*, Lect. Notes Comp. Sci. 1877, 442–457. Springer, 2000.
2. E. Bihler and W. Vogler. Efficiency of token-passing MUTEX-solutions – some experiments. In J. Desel et al., editors, *Applications and Theory of Petri Nets 1998*, Lect. Notes Comp. Sci. 1420, 185–204. Springer, 1998.
3. E.M. Clarke, O. Grumberg, and S. Jha. Verifying parameterized networks using abstraction and regular languages. In I. Lee and S. Smolka, editors, *CONCUR 95*, Lect. Notes Comp. Sci. 962, 395–407. Springer, 1995.
4. E.W. Dijkstra. Invariance and non-determinacy. In C.A.R. Hoare and J.C. Sheperdson, editors, *Mathematical Logic and Programming Languages*, 157–165. Prentice-Hall, 1985.
5. S. Graf, B. Steffen, and G. Lüttgen. Compositional minimisation of finite state systems using interface specifications. *Formal Aspects of Computing*, 8:607–616, 1996.
6. R. Janicki and M. Koutny. Semantics of inhibitor nets. *Information and Computation*, 123:1–16, 1995.
7. R. Kaivola. Using compositional preorders in the verification of sliding window protocol. In *CAV 97*, Lect. Notes Comp. Sci. 1254, 48–59. Springer, 1997.
8. E. Kindler. *Modularer Entwurf verteilter Systeme mit Petrinetzen*. PhD thesis, Techn. Univ. München, Bertz-Verlag, 1995.
9. U. Montanari and F. Rossi. Contextual nets. *Acta Informatica*, 32:545–596, 1995.
10. W. Reisig. Partial order semantics versus interleaving semantics for CSP-like languages and its impact on fairness. In J. Paredaens, editor, *Automata, Languages and Programming*, Lect. Notes Comp. Sci. 172, 403–413. Springer, 1984.
11. A. Valmari and Kokkarinen. Unbounded verification results by finite-state compositional technique: 10^{any} states and beyond. In *Int. Conf. Application of Concurrency to System Design, 1998, Fukushima, Japan*, 75–87. IEEE Computer Society, 1998.
12. W. Vogler. Fairness and partial order semantics. *Information Processing Letters*, 55:33–39, 1995.
13. W. Vogler. Efficiency of asynchronous systems and read arcs in Petri nets. In P. Degano, R. Gorrieri, and A. Marchetti-Spaccamela, editors, *ICALP 97*, Lect. Notes Comp. Sci. 1256, 538–548. Springer, 1997. Full version at http://www.informatik.uni-augsburg.de/~vogler/ under the title 'Efficiency of Asynchronous Systems, Read Arcs, and the MUTEX-Problem'.
14. W. Vogler, A. Semenov, and A. Yakovlev. Unfolding and finite prefix for nets with read arcs. In D. Sangiorgi and R. de Simone, editors, *CONCUR 98*, Lect. Notes Comp. Sci. 1466, 501–516. Springer, 1998. Full version as Technical Report Series No. 634, Computing Science, University of Newcastle upon Tyne, February 1998; can be obtained from: ftp://sadko.ncl.ac.uk/pub/incoming/TRs/.
15. P. Wolper and V. Lovinfosse. Verifying properties of large sets of processes with network invariants. In *Automatic Verification Methods for Finite Systems*, Lect. Notes Comp. Sci. 407, 68–80. Springer, 1989.

Author Index

Lecture Notes in Computer Science

For information about Vols. 1–1998
please contact your bookseller or Springer-Verlag